MODEL SYSTEMS OF DEVELOPMENT AND AGING OF THE NERVOUS SYSTEM

Institute of
Developmental
Neuroscience &
Aging

MODEL SYSTEMS OF DEVELOPMENT AND AGING OF THE NERVOUS SYSTEM

EDITED BY:

ANTONIA VERNADAKIS
ALAIN PRIVAT
JEAN M. LAUDER
PAOLA S. TIMIRAS
EZIO GIACOBINI

MARTINUS NIJHOFF PUBLISHING
A MEMBER OF THE KLUWER ACADEMIC PUBLISHERS GROUP
BOSTON/DORDRECHT/LANCASTER

PRINTED IN THE UNITED STATES OF AMERICA

Distributors

for the United States and Canada: Kluwer Academic Publishers, 101 Philip Drive, Assinippi Park, Norwell, MA, 02061, USA

for the UK and Ireland: Kluwer Academic Publishers, MTP Press Limited, Falcon House, Queen Square, Lancaster LA1 1RN, UK

for all other countries: Kluwer Academic Publishers Group, Distribution Centre, P. O. Box 322, 3300 AH Dordrecht, The Netherlands

Library of Congress Cataloging-in-Publication Data

Model systems of development and aging of the nervous system.

1. Developmental neurology. 2. Nervous system—Aging. 3. Biological models. I. Vernadakis, Antonia, 1930- . [DNLM: 1. Aging 2. Models, Neurological. 3. Nervous System—growth & development. 4. Nervous System—physiology. WL 102 M6885]
QP361.M78 1986 599′.01′88 86-23549
ISBN 0-89838-838-4

CONTENTS

CONTRIBUTING AUTHORS

ADLER, Ruben
Wynn Center, Wilmer Institute, Johns Hopkins University School of Medicine, 600 North Wolfe Street, Baltimore, MD 21205, USA

AIZENMAN, Yair
Departments of Anatomy and Psychiatry, University of California School of Medicine, Los Angeles, CA 90024, USA

ALDINIO, Caterina
Fidia Neurobiological Research Laboratories, Via Ponte della Fabbrica 3/A, F-35031 Abano Terme, ITALY

ALLIOT, Françoise
INSERM U178, Bâtiment INSERM, 16 avenue Paul Vaillant Couturier, F-94807 Villejuif Cedex, FRANCE

ASSOULINE, José G.
Division of Neurochemistry and Neurobiology, Department of Neurology, University of Iowa College of Medicine, Iowa City, IA 52242, USA

BARON-VAN EVERCOOREN, Annick
INSERM U134, Hoptial de la Salpêtrière, F-75651 Paris Cedex 13, FRANCE

BAULAC, Michel
Laboratoire Charles Foix, Hôpital de la Salpêtrière, F-75651 Paris Cedex 13, FRANCE

BAUMANN, Nicole
INSERM U134, Hôpital de la Salpêtrière, F-75651 Paris Cedex 13, FRANCE

BAXTER, Claude F.
Department of Psychiatry and Biobehavioral Sciences, UCLA School of Medicine, Los Angeles, CA 90024, USA

BOSCH, E. Peter
Division of Neurochemistry and Neurobiology, Department of Neurology, University of Iowa College of Medicine, Iowa City, IA 52242, USA

BRENNEMAN, Douglas E.
Laboratory of Developmental Neurobiology, National Institute of Child Health and Human Development, National Institutes of Health, Bethesda, MD 20205, USA

CAVICCHIOLI, Lucio
Fidia Neurobiological Research Laboratories, Via Ponte della Fabbrica 3/A, I-35031 Abano Terme, ITALY

COLE, Gregory M.
Department of Physiology–Anatomy, University of California, Berkeley, CA 94720, USA

COLLINS, Janet M.
Department of Biochemistry/Biophysics/Genetics, University of Colorado Medical School, 4200 East 9th Avenue B-121, Denver, CO 80262, USA

CONSOLAZIONE, Adriana
Fidia Neurobiological Research Laboratories, Via Ponte della Fabbrica 3/A, I-35031 Abano Terme, ITALY

DANIELS, Mathew P.
Laboratory of Biochemical Genetics, National Heart, Lung and Blood Institute, National Institutes of Health, Bethesda, MD 20892, USA

DAVIES, David
Department of Anatomy no. 510, University of Arkansas for Medical Sciences, 4301 West Markham Street, Little Rock, AR 72205-7199, USA

DE VELLIS, Jean
68-177 Mental Retardation Research Center, Center for Health Sciences, University of California, 760 Westwood Plaza, Los Angeles, CA 90024, USA

FAFOUTIS, Dessie
Department of Biochemistry Biophysics Genetics, University of Colorado Medical School, 4200 East 9th Avenue B-121, Denver, CO 80262, USA

FILOGAMO, Guido
Department of Anatomy and Physiology, Corso Massimo D'Azeglio 52, I-10126 Torino, ITALY

GAGE, Fred H.
Department of Neurosciences (M-024), University of California at San Diego, La Jolla, CA 92093, USA

GANSMULLER, Annie
INSERM U134, Hôpital de la Salpêtrière, F-75651 Paris Cedex 13, FRANCE

GEFFARD, Michel
IBCN-CNRS, Rue Camille Saint-Saens, F-33077 Bordeaux Cedex, FRANCE

GENSBURGER, Claire
Centre de Neurochimie du CNRS and INSERM U44, 5 rue Blaise Pascal, F-67084 Strasbourg Cedex, FRANCE

GIACOBINI, Ezio
Department of Pharmacology, Southern Illinois University School of Medicine, PO Box 3926, Springfield, IL 62708, USA

GREMO, Fulvia
Institute of Human Anatomy, School of Medicine, Cagliari, ITALY

GUMPEL, Madeleine
INSERM U134, Hôpital de la Salpêtrière, F-75651 Paris Cedex 13, FRANCE

HAFFKE, Susan C.
Department of Biochemistry/Biophysics/Genetics, University of Colorado Medical School, 4200 East 9th Avenue B-121, Denver, CO 80262, USA

HAYDON, Philip G.
Department of Zoology, Iowa State University, Ames, IA 50010, USA

HERTZ, Leif
Department of Pharmacology, University of Saskatchewan, Saskatoon, Saskatchewan S7N 0W0, CANADA

JACQUE, Claude
INSERM U134, Hôpital de la Salpêtrière, F-75651 Paris Cedex 13, FRANCE

KATER, Stanley B.
Program in Neuronal Growth and Development, Colorado State University, Fort Collins, CO 80521, USA

KRIKORIAN, Jacqueline G.
Laboratory of Biochemical Genetics, National Heart, Lung and Blood Institute, National Institutes of Health, Bethesda, MD 20892, USA

KURIYAMA, Kinya
Department of Pharmacology, Kyoto Prefectural University of Medicine, Kawaramachi-Hirokoji, Kamikyo-ku, Kyoto 602, JAPAN

LABOURDETTE, Gérard
Centre de Neurochimie du CNRS and INSERM U44, 5 rue Blaise Pascal, F-67084 Strasbourg

LaCHAPELLE, François
INSERM U134, Hôpital de la Salpêtrière, F-75651 Paris Cedex 13, FRANCE

LAUDER, Jean M.
Department of Anatomy, University of North Carolina School of Medicine, Chapel Hill, NC 27514, USA

LEON, Alberta
Fidia Neurobiological Research Laboratories, Via Ponte della Fabbrica 3/A, I-35031 Abano Terme, ITALY

LIM, Raymon
Department of Neurology, University of Iowa Hospitals and Clinics, Iowa City, IA 52242, USA

LING, Alice
Laboratory of Biochemical Genetics, National Heart, Lung and Blood Institute, National Institutes of Health, Bethesda, MD 20892, USA

MANGOURA, Dimitra
Departments of Psychiatry and Pharmacology, University of Colorado School of Medicine, 4200 East 9th Avenue, Denver, CO 80262, USA

MANSOUR, Hussein
Neurobiologie du Développement, INSERM U249, F-34060 Montpellier Cedex, FRANCE

MANTHORPE, Marston
Department of Biology M-001, School of Medicine, University of California at San Diego, La Jolla, CA 92093, USA

MASSA, Paul T.
Laboratory of Neuromorphology, University of Connecticut, Box U-154, Storrs, CT 06268, USA

MUGNAINI, Enrico
Laboratory of Neuromorphology, University of Connecticut, Box U-154, Storrs, CT 06268, USA

MUSSINI, Isabella
C.S. Biologia e Fisiopatologia Muscolare, c/o Istituto di Patologia Generale, Via Loredan 16, I-35100 Padova, ITALY

MYTILINEOU, Catherine
Department of Neurology, Mount Sinai School of Medicine, New York, NY 10029, USA

NELSON, Phillip G.
Laboratory of Developmental Neurobiology, National Institute of Child Health and Human Development, National Institutes of Health, Bethesda, MD 20205, USA

OHKUMA, Seitaro
Department of Pharmacology, Kyoto Prefectural University of Medicine, Kawaramachi-Hirokoji, Kamikyo-ku, Kyoto 602, JAPAN

OLEK, Anthony J.
Department of Zoology, University of Maryland, College Park, MD 20742, USA

PEREZ-POLO, Regino
Department of Human Biological Chemistry and Genetics, Gail Borden Room 436 (DO3), University of Texas Medical Branch, Galveston, TX 77550, USA

PESSAC, Bernard
INSERM U178, Bâtiment INSERM, 16 avenue Paul Vaillant Couturier, F-94807 Villejuif Cedex, FRANCE

PRIVAT, Alain
Neurobiologie du Développement, INSERM U249, F-34060 Montpellier Cedex, FRANCE

PROCHIANTZ, Alain
Chaire de Neuropharmacologie, College de France, 11 pl. Marcelin Berthelot, F-75231 Paris Cedex 05, FRANCE

ROEDER, Lois M.
Department of Pediatrics, University of Maryland School of Medicine, 655 West Balimore Street, Baltimore, MD 21201, USA

SAKELLARIDIS, Nikos
Departments of Psychiatry and Pharmacology, University of Colorado School of Medicine, 4200 East 9th Avenue, Denver, CO 80262, USA

SCHOUSBOE, Arne
Department of Biochemistry A, Panum Institute, University of Copenhagen, DK-2200 Copenhagen, DENMARK

SEEDS, Nicholas W.
Department of Biochemistry/Biophysics/Genetics, University of Colorado Medical School, 4200 East 9th Avenue B-121, Denver, CO 80262, USA

SEGAL, Menahem
Center for Neuroscience, Weizmann Institute of Science, Rehovot, ISRAEL

SENSENBRENNER, Monique
Centre de Neurochimie du CRNS and INSERM U44, 5 rue Blaise Pascal, F-67084 Strasbourg Cedex, FRANCE

SKAPER, Stephen D.
Department of Biology M-001, School of Medicine, University of California at San Diego, La Jolla, CA 92093, USA

SPOERRI, Polyxenie E.
Department of Neurosurgery, University of Göttingen, Robert-Kochstr. 40, D-3400 Göttingen, FRG

SZUCHET, Sara
Department of Neurology, University of Chicago, Chicago, IL 60637, USA

TILDON, J. Tyson
Department of Pediatrics, University of Maryland School of Medicine, 655 West Baltimore Street, Baltimore, MD 21201, USA

TIMIRAS, Paola S.
Department of Physiology–Anatomy, University of California, Berkeley, CA 94720, USA

TOFFANO, Gino
Department of Biochemistry, Fidia Neurobiological Research Laboratories, Via Ponte della Fabbrica 3/A, I-35031 Abano Terme, ITALY

VARON, Silvio
Department of Biology M-001, School of Medicine, University of California at San Diego, La Jolla, CA 92093, USA

VERNADAKIS, Antonia
Departments of Psychiatry and Pharmacology, University of Colorado School of Medicine, 4200 East 9th Avenue, Denver, CO 80262, USA

VON HUNGEN, Kern
Brain Biochemistry and Neurochemistry Laboratories, Veterans Administration Medical Center, Sepulveda, CA 90272, USA

WERRBACH-PEREZ, Karin
Department of Human Biological Chemistry and Genetics, University of Texas Medical Branch, Galveston, TX 77550, USA

WHITE, H. Steve
Department of Pharmacology and Toxicology, College of Pharmacy, University of Utah, Salt Lake City, UT, 84112, USA

WOODBURY, Dixon M.
University of Utah School of Medicine, Department of Physiology, Division of Neuropharmacology/Epileptology, 410 Chipeta Way, Salt Lake City, UT 84108, USA

WU, Doris K.
Departments of Anatomy and Psychiatry, University of California School of Medicine, Los Angeles, CA 90024, USA

PREFACE

The last decade has generated a multitude of studies using in vitro model systems to explore growth and differentiation of the nervous system. Although the findings have been exciting and have revealed unique properties of neural cells, considerable concern continues to be expressed regarding the significance of in vitro findings in terms of their applicability to in vivo biological events. To examine this issue further, a group of scientists presented and discussed their findings at a conference sponsored by the Institute of Developmental Neuroscience and Aging held in Crete, Greece, 26–29 May 1985. The conference was cosponsored by the University of Crete and was generously supported by the Ministry of Research and Technology of Greece, Tourism Organization of Greece, and also Sandoz and FIDIA. The Directors of the Institute of Developmental Neuroscience and Aging are indebted to these Institutions for their support. For the success of this conference, the Directors owe much to Drs. Eleni Fleischer-Lambropoulos and Yiannis Tsouderos, who spent countless hours in making arrangements so that the participants would have not only a scientific, but also a unique cultural, experience.

Several chapters of this book focus on the complex phenomena of neurogenesis and gliogenesis, and the modulation of neuronal differentiation. The concept that neuronal differentiation has both genetic and epigenetic components is documented by elegant studies using both in vitro cultured cells and neurons transplanted in vivo. Epigenetic factors range from specific neurotrophic factors such as gangliosides, neurotransmitters, and neuropeptides to more general factors such as hormones, metabolic substrates, and neurotoxins. The complexity of cell–cell interactions is demonstrated by the exciting approaches of tissue transplantation. This new methodology more closely approximates the multidimensionality of cell–cell interactions that occur in vivo as opposed to the cell–cell interactions provided by a monolayer cell environment. For the first time, model systems have been considered for their potential usefulness in studying aging processes and nervous system degenerative disease of old age. One can expect that the entire spectrum of neuronal and glial life may eventually be explored in vitro. Changes associated with aging in vivo such as aging pigments and gliosis have been demonstrated in culture and in tissue transplantation, and thus provide evidence for the biological validity of model systems for research in aging.

At a time when animal research is challenged and, in several areas, severely curtailed, model systems such as those presented in this book expand the horizon for neurobiology research.

The Directors of the
Institute of Developmental
Neuroscience and Aging

I. NEURONS, NEUROGLIA AND NEURONAL–GLIAL INTERACTIONS

1. CELL CULTURE SYSTEMS FOR PURIFIED RETINAL NEURONS AND PHOTORECEPTORS

RUBEN ADLER

1. INTRODUCTION

Many of the cellular mechanisms that are particularly conspicuous during embryonic development are also prevalent in the adult, either under normal circumstances or in specific pathological conditions. For example, many of the severe degenerative disorders known to affect the adult nervous system involve selective neuronal death, thus resembling the phenomenon of developmental neuronal death that occurs in the embryo (Appel, 1981; Varon and Adler, 1980). Neurological deficits can also result from failures in the development or maintenance of the differentiated functions typical of nerve cells. Both types of degeneration are known to occur in the neural retina and often lead to blindness (c.f. Adler, 1986a). Thus, detailed knowledge of the molecular mechanisms that regulate neuronal survival and differentiation is important for understanding not only normal development but also pathological alterations of the nervous system in general and the retina in particular.

The intact organism is much too complex to allow the precise type of experimental analysis necessary to investigate these mechanisms. This is because one can seldom study the independent responses of the cells involved to regulatory molecules and pathological agents. These limitations of in vivo experimentation have stimulated interest in in vitro systems that would allow better control of experimental conditions, as well as more direct cell accessibility to analytical methods. Our laboratory has been dedicated for several

Vernadakis, A. (ed.), *Model Systems of Development and Aging of the Nervous System*.

years to the development of cell culture systems that would facilitate the investigation of cellular and molecular mechanisms involved in retinal development and in retinal degenerations. Before describing these culture systems in some detail, it is appropriate to review briefly some general considerations about the approaches used in our laboratory to study these questions (for more details, see Adler, 1986a–c). First, the *heterogenity* of central nervous system (CNS) tissues such as the retina is a complicating factor for in vitro experiments. Different cell types, coexisting within the same culture environment, can interact with each other in a manner that might resist detection by the observer. Hidden interaction can, in turn, blur the distinction between direct and indirect effects of a treatment. Overcoming this problem requires the availability of *cell separation* methods, in order to obtain separate cultures of the different cell types present in an organ. Moreover, *cell identification*, sometimes difficult even in the in vivo situation, can be even more problematic in cell culture, where topographical tissue landmarks are not available. This problem can only be overcome through the use of a battery of techniques, frequently including immunocytochemistry, autoradiography, and electron microscopy. With these considerations in mind, it is now appropriate to review the development of a number of different culture systems for neural retinal cells, in which cell separation and cell identification have been central considerations.

2. DEVELOPMENT OF PURIFIED POPULATIONS OF CHICK EMBRYO RETINAL CELLS IN MONOLAYER CULTURE

2.1 Blackground

The donor tissue for the experiments described here is the neural retina from the 8-day-old chick embryo. At this stage, most retinal cells still show an undifferentiated epithelial configuration, although they are already postmitotic (rev. Grun, 1982). A distinctive property of this tissue is the absence of blood vessels and connective tissue, so that chick retinal cultures are not contaminated with fibroblasts, endothelial cells, etc. The 8-day chick retina can be cleanly dissected from other ocular tissues, including the retinal pigment epithelium.

The 8-day chick embryo retina can be easily dissociated after brief incubation in trypsin, and is a generous source of cells (approximately 5×10^7 cells per retina, or 10^8 cells per embryo). These dissociated cells can be induced to reaggregate, by means of gyratory agitation, either with each other or with cells from other organs. These "reaggregation" cultures have generated useful information about the role of cell interactions in retinal development, which has already been reviewed (Moscona, 1965; Steinberg, 1963; Adler et al., 1976; Adler, 1982a). Therefore, the remainder of this chapter will be devoted to *stationary cultures* in which the dissociated cells grow as monolayers after attaching to a solid substratum. Particular attention will be paid to procedures that allow the generation of *enriched, separate cultures* of *identified* retinal cell types.

2.2 A starting point: complex, flat cell-, and clump-containing "monolayers"

The only type of chick retina monolayer cultures available until the end of the 1970s consisted of a confluent monolayer of "flat" cells, subjacent to multicellular clumps and process-bearing cells (figure 1-1). These complex cultures are obtained by seeding retinal cell suspensions at high density on untreated glass or plastic, in medium supplemented with fetal calf serum (Adler et al., 1982; Barr-Nea and Barishak, 1970; Crisanti-Combes et al., 1977; Dutt and Reif-Lehrer, 1981; Gremo et al., 1984; Itoh, 1976; Kaplowitz and Moscona, 1976; Okada, 1980; Pritchard et al., 1978; Pu[illegible] et al., 1977; Redfern et al., 1976; Thompson and Pelto, 1982; among others). The flat cells appear to be glial in nature, as suggested by their ultrastructure (Araki et al., 1982; Crisanti-Combes et al., 1977; Li and Sheffield, 1984), by some of their cytoskeletal features (Li and Sheffield, 1984), and by the presence of characteristic glial enzymes such as glutamine synthetase and carbonic anhydrase (Dutt and Reif-Lehrer, 1981; Linser and Moscona, 1983; Moscona and Linser, 1983). The multicellular clumps abundant in these cultures contain different cell types, including neurons and partially differentiated photoreceptor cells (Adler et al., 1982; Araki et al., 1982; Crisanti-Combes et al., 1977). They show morphological similarities with the "rosettes" that develop in reaggregation cultures (Sheffield and Moscona, 1970). These similarities also include the opportunity for cells to affect each other's development through contact-mediated interactions (c.f. Linser and Moscona, 1983). A limiting feature of these complex retinal cultures is the impossibility to study the cells present within the multicellular clumps without using histological techniques (see figure 1-1A).

2.3 Purified "glial" monolayers derived from complex cultures

The multicellular clumps and process-bearing neurons that develop on top of the flat cell layer progressively disappear from these "complex" retinal cultures after the first week in vitro. This process eventually leads to the development of a purified monolayer of flat cells, which could conceivably appear as a good model for the study of retinal glial cells. Unfortunately, this application is hampered by the tendency of flat cells to change their phenotype through a phenomenon known as "transdifferentiation" (rev. Moscona and Linser, 1983; Okada, 1980). In essence, this transformation implies a change in gene expression, leading to the loss of glial properties and to the appearance of gene products characteristic of other cell types (i.e., lens-specific crystalline proteins). The transdifferentiation phenomenon develops slowly and progressively, and therefore the number of nonglial cells present in the flat cell cultures increases with time in vitro.

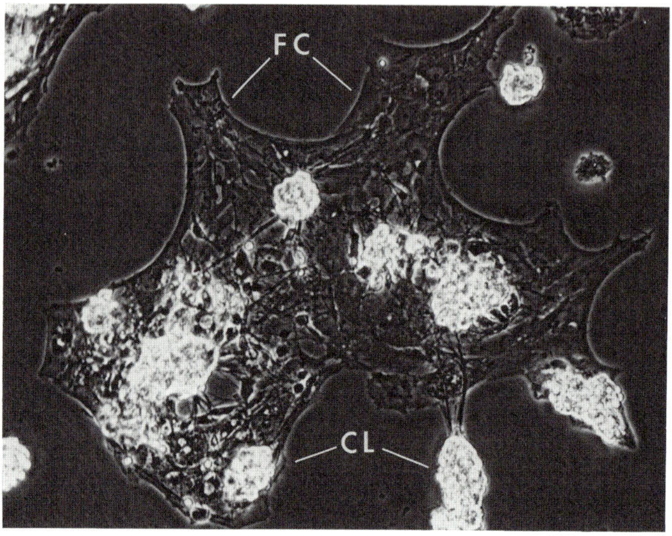

A

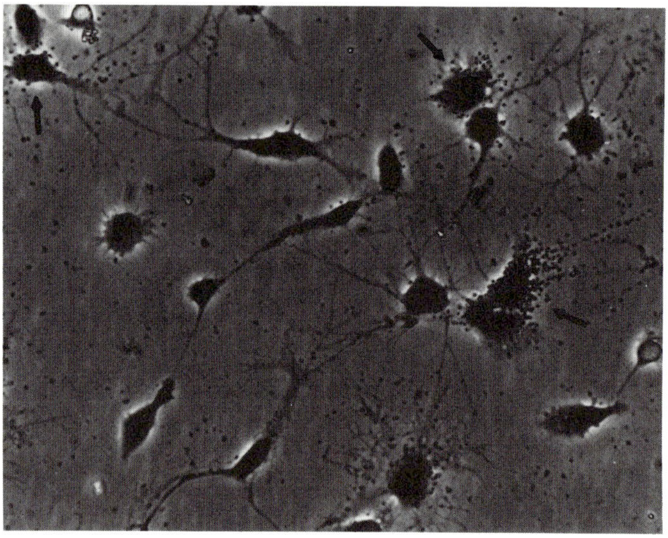

B

Figure 1-1. Eight-day chick embryo retinal cells in monolayer culture. (**A**) Cells seeded on uncoated tissue culture plastic and grown in fetal calf serum-containing medium for 5 days. Note the presence of large multicellular clumps (*CL*) and flat cells (*FC*). Flat cells migrate out of the clumps and upon the plastic substratum, giving rise to a confluent monolayer. (**B**) When identical suspension of retinal cells are seeded on polyornithine-coated plastic, they develop without multicellular clumps or flat cells. These cultures consist essentially of process-bearing neurons and some process-free round cells. This micrograph corresponds to an autoradiogram of a culture incubated with [^{3}H]-glutamate under conditions allowing the detection of high-affinity uptake mechanisms. Labeled cells are indicated with *arrows*. **B** is from Adler (1983), with permission.

There are several methods for the purification of glial cells from *early* retinal cultures (i.e., before the apparent onset of the transdifferentiation phenomenon). Mechanical removal of clumps and neurons can be accomplished by prolonged agitation of the cultures in a shaker, or by projecting a jet of sterile culture medium toward the surface of the cultures (Adler et al., 1982; De Pomerai and Gali, 1982; Koh et al., 1984; Li and Sheffield, 1984). Equivalent results can be obtained with selective cytolytic treatments, such as those described by Linser and Moscona (1983). The resulting flat cell populations are essentially free of contamination with clumps and process-bearing neuronal cells.

2.4 Clump-free, glia-free cultures of retinal neurons

Flat cells can not be seen in the above-mentioned complex cultures during the first few days in vitro. Rather, they start migrating out of multicellular clumps 3–4 days after culture onset (figure 1-1A). This observation suggested that the "induction" of flat cell development could require cell–cell interactions occurring within the clumps. If this were true, flat cell development could theoretically be avoided if dissociated retinal cells were prevented from reaggregating with each other before attaching to the culture substratum. Experimental verification of this hypothesis required the use of more adhesive substrata in order to accelerate attachment of the dissociated cells to the dish. This was accomplished by coating tissue culture plastic dishes with either polyornithine or with a highly adhesive collagen preparation described by Adler et al. (1979). Retinal cell suspensions grown on these substrata did indeed give rise to cultures devoid of both clumps and flat cells (Adler et al., 1982; see also Betz and Müller, 1982). As shown in figure 1-1B, dissociated retinal cells grown on these adhesive substrata developed as a flat cell-free, clump-free, sparse population of process-bearing neuronal cells accompanied by some process-free, round cells (Adler et al., 1982). These cultures were initially grown in Eagle's medium supplemented with either fetal calf or horse serum (Adler et al., 1982). However, the use of the chemically defined supplement "N1" of Bottenstein and Sato (1979) allowed growing retinal neurons under serum-free conditions (Skaper et al., 1979; Hyndman and Adler, 1982a).

The neuronal identity of the process-bearing cells, already suggested by their morphology, was supported by the cytochemical demonstration that these cells bind tetanus toxin (Adler et al., 1982). Quantitative morphological analysis of the cultures showed the presence of several subpopulations of neurons that could be identified on the basis of criteria such as the number and site of origin of the neurites (Hyndman and Adler, 1982a). However, these morphological criteria were not sufficient to establish correlations between the in vitro subpopulations and the different cell types present in the retina in vivo.

Some of the cytochemical "markers" available for neuronal identification are based on the autoradiographic or immunocytochemical investigation of cellular machineries involved in neurotransmitter-related activities (c.f. Adler, 1986b). Autoradiographic studies of the distribution of the high-affinity uptake mechanisms for γ-aminobutyric acid (GABA) (Hyndman and Adler, 1982b), glutamate (Hyndman and Adler, 1982c), and taurine and aspartate (Adler, 1983), however, did not yield conclusive evidence for the identification of cultured retinal neurons. Although apparent selectivity was seen when each of these precursors was presented individually to the neurons (figure 1-1B), a recent study using the same precursors in binary and tertiary combinations showed that a cultured neuron can actually have high-affinity uptake mechanisms for two or even three of these aminòacids (Pessin and Adler, 1985). Another complicating factor is that, in the in vivo retina, some neurotransmitter molecules are used by more than one neuronal type. In summary, then, neither morphology alone, studied by phase contrast microscopy, nor the investigation of neurotransmitter-related properties showed enough specificity to allow accurate and convincing identification of different subpopulations of cultured retinal neurons. As described in the next section, photoreceptor cells proved to be more amenable than other neurons to positive identification.

2.5 Development of photoreceptor cells in retinal neuronal cultures

Photoreceptor cells share many properties with other retinal neurons, but they also express a variety of unique morphological and chemical phenotypic features. In cultures such as those described in the preceding section, some 20%–30% of the neuronal cells showed a single neurite, as would be expected from a photoreceptor cell. This was certainly not sufficient evidence to establish conclusively their identity. However, when similar cultures were grown in medium 199 supplemented with fetal calf serum and the polyunsaturated fatty acid, linoleic acid, the vast majority of these monopolar cells developed a refractile vacuole, located in the cell region directly opposite to the point of origin of the neurite (Adler et al., 1984; see figure 1-2). This vacuole was reminiscent of the "lipid droplet" characteristically found in the inner segment of cone photoreceptor cells in birds. These elongated, monopolar cells showing a short neurite and a lipid droplet represented some 10%–20% of the process-bearing cells. Their identity as *cone photoreceptors* has now been demonstrated with a variety of techniques, including lectin cytochemistry, transmission and scanning electron microscopy, and immunocytochemistry (figure 1-2). The similarities between cultured photoreceptor cells and their in vivo counterparts of equivalent developmental age include the following properties (Adler et al., 1984; Adler, 1985 and 1986d): 1) an elongated, highly polarized cell body; 2) the presence of discrete cell compartments; 3) the absence of organelles in the nucleus-containing cell body, and the accumulation of mitochondria, golgi apparatus, and endoplasmic reticulum in the inner segment region; 4) the presence of a single, very

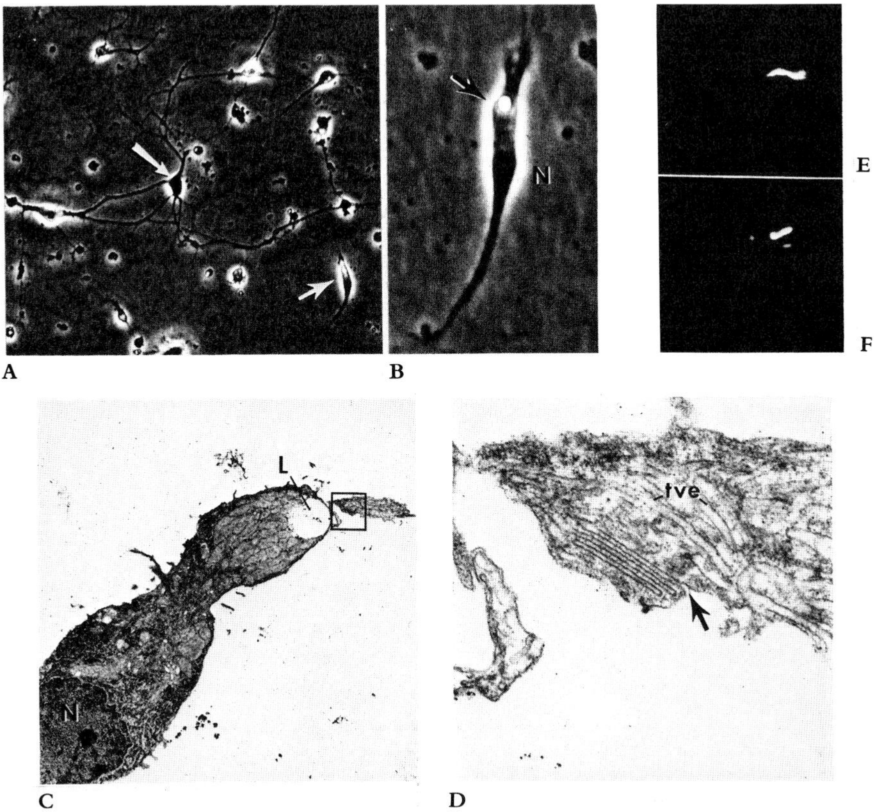

Figure 1-2. Eight-day chick embryo retinal cells grown on polyornithine-coated dishes under conditions that allow some of the cells to express a typical photoreceptor phenotype. (**A**) Phase contrast micrograph of a culture grown in vitro for 7 days. Some 80% of the process-bearing cells show a typical multipolar neuronal phenotype (*long arrow*). The remaining 20% of the process-bearing cells differentiate as photoreceptors (*short arrow*). (**B**) Photoreceptors are elongated, highly asymmetric, polarized cells. A single, characteristically short neurite emerges from one of the cell poles. The adjacent cell body is almost completely occupied by the nucleus (*N*). The organelle-rich inner segment region contains a conspicuous lipid droplet (*arrow*). (**C** and **D**) Transmission electron micrographs of a cultured photoreceptor cell. The photoreceptor appears highly polarized. Organelles are scarce in the vicinity of the nucleus (*N*), and are very abundant in the inner segment region that also shows a lipid droplet (*L*). A short, outer segment-like process emerges from the inner segment (enlarged in **C** from box in **B**). This outer segment process contains abundant smooth tubulovesicular elements (*tve*). Moreover, some compacted membranous profiles, reminiscent of the discs present in photoreceptor outer segments, are present near its base. One of these discs appears continuous with the plasma membrane (*arrow*). (**E** and **F**) Similar culture, stained by indirect immunofluorescence using antiopsin antiserum. Photoreceptor cells are the only positive cells in the cultures. Their neurite and cell body appear negative, whereas immunoreactive materials appear accumulated in the region occupied by the finger-shaped outer segment-like process. This pattern of immunostaining could be seen in unpermeated (**E**) as well as permeated (**F**) preparations. From Adler (1986d), with permission

short neurite; 5) the selective appearance and polarized accumulation of opsin, the visual pigment protein characteristic of photoreceptor cells; 6) the formation of a small outer segment process that accumulates opsin and shows compacted membranous elements similar to outer segment discs; and 7) the presence and polarized distribution of peanut lectin receptors, which, as shown by Blanks and Johnson (1983 and 1984) are also characteristic of cone photoreceptor cells. It is now possible to grow these cells in a serum-free, chemically defined medium (Lindsey and Adler, 1985 and 1986a).

When first isolated, all the cells from the 8-day chick embryo retina are spherical and lack the various morphological and cytochemical features described above. The fact that these features develop in only some of the cultured cells even while growing in the absence of intercellular contacts suggests that immature precursor cells present in the retina of the 8-day embryo are already *predetermined* to express many aspects of the photoreceptor phenotype. This phenotype develops even when photoreceptor precursors are dissociated and placed in a microenvironment in which other cells follow a completely different pattern of differentiation. On the other hand, these studies also suggest that some developmental signal(s), probably derived from the retinal pigment epithelium, might be necessary to stimulate further differentiation of the photoreceptor cells. Thus, cultured photoreceptors only develop an "abortive" outer segment, which remains very short and fails to reach the dimensions and organization observed in the adult. Given that pigment epithelial cells are not present in the cultures, this observation is consistent with in vivo studies in amphibians that showed similarly abortive outer segment development in pigment epithelium-free retinas (Hollyfield and Witkovsky, 1974).

2.6 Generation of enriched populations of cultured photoreceptor cells by selective neuronal destruction

The selective destruction of one or more of the cell types present in mixed cultures has been used to generate enriched cell subpopulations in vitro (rev. Schaffner and Schnaar, 1983; Varon and Manthorpe, 1980). The existence of toxins capable of destroying selectively some retinal cells suggested that this experimental paradigm could be applicable to cultures containing mixtures of retinal neurons and photoreceptors. For example, kainic acid (KA) is known to destroy amacrine, horizontal, and bipolar cells, without affecting retinal ganglion cells and photoreceptors (rev. Morgan, 1983). The snake venom product β-bungarotoxin (BT) causes selective destruction of retinal ganglion cells and some amacrine cells, while sparing photoreceptors and other retinal cells (Hirokawa, 1978; Rehm et al., 1982). We have recently used these two toxins to generate highly enriched populations of cultured photoreceptor cells (Politi and Adler, 1986a; see figure 1-3). The effects of KA and BT also appeared to be selective in vitro. For example, some 60% of the multipolar neurons could be destroyed by 2 m*M* KA and 70% of the neurons by 1 n*M*

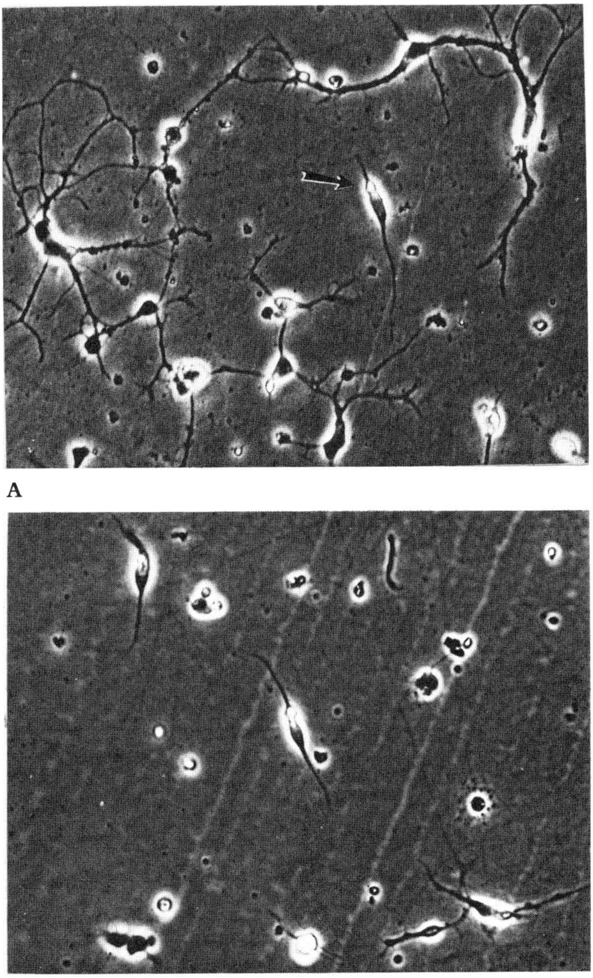

A

B

Figure 1-3. Chick embryo neural retina cells cultured for 7 days in glial-free monolayers. (**A**) Untreated cultures. Note the presence of multipolar neurons, a photoreceptor (*arrow*), and process-free cells. (**B**) Similar culture treated for 24 hours with 2 m*M* kainic acid. Note extensive neuronal degeneration. Photoreceptors appear unaffected.

BT, whereas the number of photoreceptor cells was not changed after either of these treatments. Neuronal disappearance from the cultures could also be detected by measuring some neuronal "markers." For example, both the cholinergic enzyme choline acetyltransferase and the high-affinity uptake mechanism for GABA were almost completely abolished by KA and BT. On the other hand, the lack of photoreceptor sensitivity to the toxins was corroborated by ultrastructural analysis using transmission electron microscopy. In summary, then, it is possible to generate highly enriched (though not completely pure) populations of cultured photoreceptor cells by selective neuronal destruction using either kainic acid or β-bungarotoxin.

2.7 Generation of enriched populations of cultured retinal neurons by complement-mediated photoreceptor lysis using antiopsin antibodies

Immunocytochemical studies already mentioned (figure 1-3) showed the presence of opsin in cultured photoreceptor cells, but not in other neurons (Adler, 1985 and 1986c and d). Opsin immunoreactive materials could be demonstrated without pretreatment of the photoreceptors with detergents, indicating that at least some of the opsin immunoreactive molecules were located at the cell surface. This observation suggested the possibility of applying to retinal cultures another established method for cell purification, namely, complement-mediated cell lysis using antibodies specific for cell surface antigens (rev. Schaffner and Schnaar, 1983; Varon and Manthorpe, 1980). When retinal cultures containing both photoreceptors and neurons were treated with an antiopsin antiserum in the presence of complement, approximately 85% of the photoreceptors were lysed, thus yielding a highly enriched neuronal population (Adler and Politi, 1986; Politi and Adler, 1987). Photoreceptor susceptibility to complement-mediated lysis by this antiserum developed in vitro in parallel with their expression of immunocytochemically recognizable opsin.

The effects of the antiopsin antiserum were concentration dependent, and lysis of 85% of the photoreceptors could be achieved with a 1:50 antiserum dilution. More than 90% of the neurons remained in the cultures after antibody treatment, and could be cultured for a further 24-hour period. Neuronal markers such as choline acetyltransferase or high-affinity GABA uptake remained essentially unchanged in antibody-treated cultures.

These observations demonstrated that it is possible to generate enriched populations of retinal neurons, almost completely free of contamination with cone photoreceptors. The demonstration that cultured photoreceptors are susceptible to complement-mediated lysis by antiopsin antibodies may also have some clinical relevance. Clinical observations and in vivo experiments in animals suggest that immune reactions against photoreceptor antigens could be involved in the genesis and/or spreading of degenerative disorders of the retina, such as retinitis pigmentosa. In vivo studies, however, have not elucidated whether photoreceptor cells are *directly* affected by abnormal im-

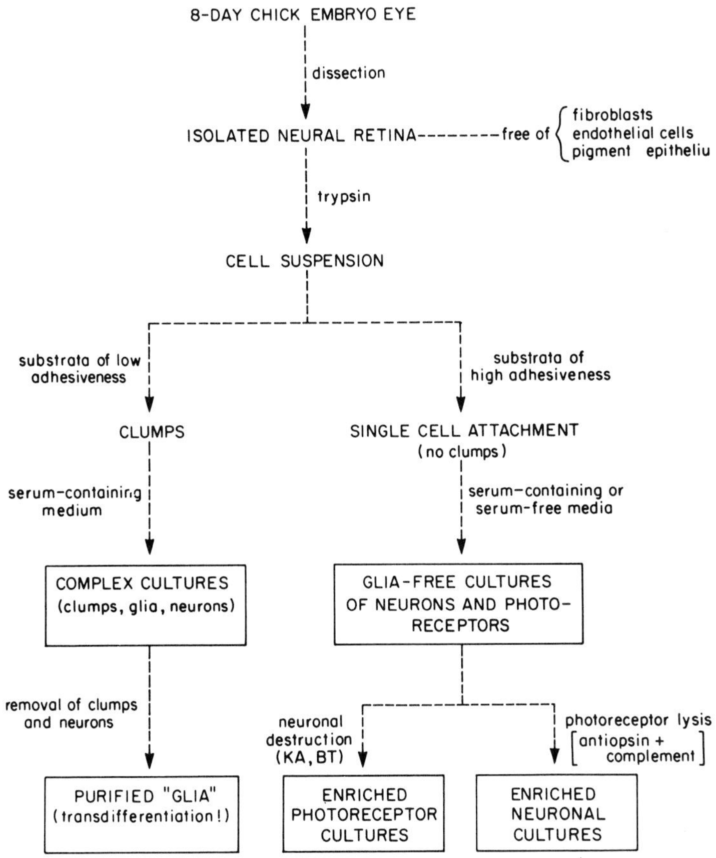

Figure 1-4. Summary of the different types of monolayer cultures that can be derived from the 8-day chick embryo neural retina. See text for details.

mune mechanisms. The in vitro studies reported here demonstrate that, even in the total absence of inflammatory cells, antibodies against an antigen abundant in photoreceptors can cause selective lysis of these cells.

3. CONCLUDING REMARKS

Figure 1-4 presents a summary of the different types of monolayer cultures that can be derived from a same cell suspension obtained by dissociating the neural retina of the 8-day chick embryo. These cultures range from complex populations containing a monolayer of flat "glial" cells and multicellular clumps, to simpler, sparser cultures of neurons and photoreceptors. It is now possible to generate enriched populations of *either* neurons *or* photoreceptors through the selective destruction of the other cell type. The studies described in the preceding sections illustrate how relatively minor changes in culture conditions can have profound effects upon the composition of the cultures. For example, the dramatic differences between complex, flat cell- and clump-containing cultures as compared with sparse, flat cell- and clump-free cultures of neurons and photoreceptors result from an increase in substratum adhesiveness and a decrease in seeding density.

We have made some progress toward developing *separate* populations of *identified* retinal cell subtypes, but there is still much ground for improvement in both aspects of the cultures. Enriched populations of photoreceptor cells can be generated by selective neuronal destruction, but they are certainly not pure as yet. Similar considerations apply to neuronal cultures in which the majority (but not all) of the photoreceptors are destroyed by complement-mediated lysis. Moreover, it would be important to develop methods for the identification and separation of different retinal neuronal subtypes, such as ganglionic, amacrine, horizontal, and bipolar cells. At present, retinal ganglion cells are the only ones for which there seems to be a clear identification approach, based on retrograde "backfilling" of the cells using fluorescent dyes. This method can be used not only for cell identification, but also for cell separation by fluorescence activated cell sorting (c.f. Armson and Bennett, 1983; Sarthy et al., 1983). Considerable effort is being invested in different laboratories to develop new tools, such as cell-specific monoclonal antibodies, which could allow further progress in this field (c.f. Barnstable et al., 1983; Lemmon and Gottlieb, 1982).

It seems appropriate to close this chapter with a very brief overview of the experimental possibilities open by the availability of purified cultures of identified retinal cells. For example, these culture systems allow the dynamic investigation of the differentiation of specific cell types, whose development can be monitored by time-lapse video recordings combined with cytochemical and ultrastructural techniques (i.e., Madreperla and Adler, 1986). The cells can be characterized in terms of metabolic requirements (Lindsey and Adler, 1986b), responsiveness to regulatory molecules (Adler et al., 1985), and sensitivity to neurotropic viruses (Politi et al., 1987) or lytic antibodies (Adler and Politi, 1986; Politi and Adler, 1987). The information derived from these and related studies will hopefully contribute to our understanding of normal and pathological mechanisms that can affect the survival and differentiated function of retinal neurons and photoreceptors.

ACKNOWLEDGMENTS

Research from the author's laboratory has been susported by USPHS grants EY02854 and 04859, Core grant EY01765, and a grant from Alcon Research Laboratories. R.A. is a William and Mary Greve Scholar from Research to Prevent Blindness. Many thanks are expressed to Dr. James D. Lindsey for editorial comments on the manuscript, to Mrs. Doris Golembieski for secretarial assistance, and to Mrs. Cindy Winter for artwork.

REFERENCES

Adler, R. (1982a) Trophic and neurite-promoting factors in eye development. In: Hollyfield, J. (ed.) The structure of the eye. Amsterdam: Elsevier/North Holland, pp. 215–228.

Adler, R. (1982b) Regulation of neurite growth in purified retina cultures: effects of PNPF, a substratum-bound neurite-promoting factor. J. Neurosci. Res. 8:165–177.

Adler, R. (1983) Taurine uptake by chick embryo retinal neurons and glial cells in purified culture. J. Neurosci. Res. 10:369–379.

Adler, R. (1985) Photoreceptor morphogensis and polarization in low density retinal cell cultures. J. Cell Biol. 101:222a.

Adler, R. (1986a) Trophic interactions in retinal development and in retinal degenerations: in vivo andin vitro studies. In: Adler, R., 2nd Farber, D. (eds.) The retina: a model for cell biology New York: Academic Press, pp. 112–150.

Adler, R. (1986b) In vitro techniques for the investigation of trophic factors active on CNS neurons. In: Perez-Polo, J.R. (ed.) Handbook of nervous system and muscle factors. Baca Rzton, FL:CRC Press (in press).

Adler, R. (1986c) The differentiation of retinal photoreceptors and neurons in virto. In: Chader, G., and Osborne, N. (eds.) Progress in retinal research. Vol. *6*, pp. 1–27.

Adler, R. (1986d) Developmental predetermination of the structural and molecular polarization of photoreceptor cells. Dev. Biol., *117*:520–527.

Adler, R., and Politi, L. (1986) Selective photoreceptor degeneration by antiopsin antisera: in vitro studies. Invest. Ophthalmol. Vis. Sci. [Suppl.] *27*:57.

Adler, R., Teitelman, G., and Suburo, A.M. (1976) Cell interactions and the regulation of cholinergic enzymes during neural differentiation "in vitro." Dev. Biol. 50:48–58.

Adler, R., Manthorpe, M., and Varon, S. (1979) Separation of neuronal and nonneuronal cells in monolayer cultures from chick embryo optic lobe. Dev. Biol. 69:424–435.

Adler, R., Magistretti, P.J., Hyndman, A.G., and Shoemaker, W.J. (1982) Purification and cytochemical identification of neuronal and nonneuronal cells in chick embryo retina cultures. Dev. Neurosci. 5:27–39.

Adler, R., Lindsey, J.D., and Elsner, C.L. (1984) Expression of cone-like properties by chick embryo neural retina cells in glial-free monolayer cultures. J. Cell Biol. 99:1173–1178.

Adler, R., Jerdan, J., and Hewitt, A.T. (1985) Responses of cultured neural retinal cells to substratum-bound laminin and other extracellular matrix molecules. Dev. Biol. 112:100–114.

Appel, S. (1981) A unifying hypothesis for the course of amyotrophic lateral sclerosis, Parkinsonism and Alzheimer's disease. Ann. Neurol. 10:499–505.

Araki, M., Ide, C., Saito, F., and Sato, F. (1982) Ultracytochemical study on in vitro differentiation of neural retina cells of chick embryos. Dev. Biol. 94:51–61.

Armson, P.F., and Bennett, M.R. (1983) Neonatal retinal ganglion cell cultures of high purity: effect of superior colliculus on their survival. Neurosci. Lett. 38:181–186.

Barnstable, C.J., Akagawa, K., Hofstein, R., and Horn, J.P. (1983) Monoclonal antibodies that label discrete cell types in the mammalian nervous system. Cold Spring Harbor Symp. Quant. Biol. 48:863–876.

Barr-Nea, L., and Barishak, R.Y. (1970) Tissue culture studies of the embryonal chicken retina. Invest. Ophthalmol. 9:447–457.

Betz, H., and Müller, U. (1982) Culture of chick embryo neural retina in serum-free medium: expression of cholinergic proteins. Exp. Cell Res. 138:297–302.

Blanks, J.C., and Johnson, L.V. (1983) Selective lectin binding of the developing mouse retina. J. Comp. Neurol. 221:31–41.

Blanks, J.C., and Johnson, L.V. (1984) Specific binding of peanut lectin to a class of retinal photoreceptor cells. Inv. Ophthalmol. Vis. Sci. 25:546–557.

Bottenstein, J.E., and Sato. G. (1979) Growth of a rat neuroblastoma cell line in serum-free supplemented medium. Proc. Natl. Acad. Sci. USA 76:514–517.

Crisanti-Combes, P., Privat, A., Pessac, B., and Calothy, G. (1977) Differentiation of chick embryo neuroretina cells in monolayer cultures: an ultrastructural study. I. Seven-day retina. Cell Tissue Res. 185:159–173.

De Pomerai, D.I., and Gali, M.A. (1982) Serum factors affecting transdifferentiation in chick embryo neuro-retinal cultures. Adv. Exp. Med. Biol. 158:199–207.

Dutt, K., and Reif-Lehrer, L. (1981) Glutamine synthetase induction in chick embryo retina monolayers. Cell Biophys. 3:1–17.

Gremo, F., Porru, S., and Vernadakis, A. (1984) Effects of corticosterone on chick embryonic retinal cells in culture. Brain Res. 317:45–52.

Grun, G. (1982) The development of the vertebrate retina: a comparative study. Adv. Anat. Embryol. Cell Biol. 78:7–85.

Hirokawa, N. (1978) Characterization of various nervous tissues of the chick embryo through responses to chronic application of immunocytochemistry of β-bungarotoxin. J. Comp. Neurol. 180:449–466.

Hollyfield, J.G., and Witkovsky, P. (1974) Pigmented retinal epithelium involvement in photoreceptor development and function. J. Exp. Zool. 189:357–378.

Hyndman, A.G., and Adler, R. (1982a) Neural retina development in vitro: effects of tissue extracts on survival and neuritic development in purified neuronal cultures. Dev. Neurosci. 5:40–53.

Hyndman, A.G., and Adler, R. (1982b) GABA uptake and release in purified neuronal and nonneuronal cultures from chick embryo retina. Dev. Brain Res. 3:167–180.

Hyndman, A.G., and Adler, R. (1982c) Analysis of glutamate uptake and monosodium glutamate toxicity in neural retina monolayer cultures. Dev. Brain Res. 2:303–314.

Itoh, Y. (1976) Enhancement of differentiation of lens and pigment cells by ascorbic acid in cultures of neural retinal cells of chick embryos. Dev. Biol. 54:157–162.

Kaplowitz, P.B., and Moscona, A.A. (1976) Lectin-mediated stimulation of DNA synthesis in cultures of embryonic neural retina cells. Exp. Cell Res. 100:177–189.

Koh, S.W.M., Kyritsis, A., and Chader, G. (1984) Interaction of neuropeptides and cultured glial (Müller) cells of the chick retina: elevation of intracellular cyclic AMP by vasoactive intestinal peptide and glucagon. J. Neurochem. 43:199–203.

Lemmon, V., and Gottlieb, D.I. (1982) Monoclonal antibodies selective for the inner portion of the chick retina. J. Neurosci. 2:531–535.

Li, H.P., and Sheffield, J.B. (1984) Isolation and characterization of flat cells, a subpopulation of the embryonic chick retina. Tissue Cell 16:843–857.

Lindsey, J.D., and Adler, R. (1985) A chemically defined medium that support photoreceptor differentiation in vitro. Invest. Ophthalmol. Vis. Sci. 26:335.

Lindsey, J.D., and Adler, R. (1986a) A chemically defined medium that supports photoreceptor differentiation in vitro. (In preparation.)

Lindsey, J.D., and Adler, R. (1986b) Comparative effects of serum-containing and serum-free media on the survival and maturation of retinal neurons and photoreceptors. Invest. Ophthalmol. Vis. Sci. [Suppl.] 27:329.

Linser, P., and Moscona, A.A. (1983) Hormonal induction of glutamine synthetase in cultures of embryonic retina cells: requirement for neuron–glia cell contacts. Dev. Biol. 96:529–534.

Madreperla, S., and Adler, R. (1986) Kinetic analysis of the differentiation and polarization of photoreceptor precursor cells. Invest. Ophthalmol. Vis. Sci. [Suppl.] (in press).

Morgan, I.G. (1983) Kainic acid as a tool in retinal research. Prog. Retinal Res. 2:249–266.

Moscona, A.A. (1965) Recombination of dissociated cells and the development of cell aggregates. In: Wilmer, E. (ed.) Cells and tissues in cultures. New York: Academic Press, pp. 489–529.

Moscona, A.A., and Linser, P.J. (1983) Developmental and experimental changes in retinal glia cells: cell interactions and control of phenotype expression and stability. Curr. Top. Dev. Biol. 18:155–188.

Okada, T.S. (1980) Cellular metaplasia or transdifferentiation as a model for retinal cell differentiation. Curr. Top. Dev. Biol. 16:349–380.

Pessin, M., and Adler, R. (1985) Coexistence of high affinity uptake mechanisms for putative neurotransmitter molecules in chick embryo retinal neurons in purified culture. J. Neurosci. Res. 14:317–328.

Politi, L.E., and Adler, R. (1986) Generation of enriched populations of cultured photoreceptor cells. Invest. Ophthalmol. Vis. Sci. 27:656–665.

Politi, L.E., and Adler, R. (1987) Selective destruction of photoreceptor cells by antiopsin antibodies. Invest. Ophthalmol. Vis. Sci. 28:118–125.

Politi, L., Adler, R., and Whittum-Hudson, J. (1987) Differential sensitivity of retinal neurons and photoreceptors to herpes simplex infection. Exp. Eye Res. (in press).

Pritchard, D.J., Clayton, R.M., and De Pomerai, D.I. (1978) Transdifferentiation of chicken neural retina into lens and pigment epithelium in culture: controlling influences. J. Embryol. Exp. Morphol. 48:1–21.

Puro, D.G., De Mello, F.G., and Nirenberg, M. (1977) Synapse turnover: the formation and termination of transient synapses. Proc. Natl. Acad. Sci. USA 74:4977–4981.

Redfern, N., Israel, P., Bergsma, D., Robinson, W.G.J., Jr., Whikehart, D., and Chader, G. (1976) Neural retinal and pigment epithelial cells in culture: patterns of differentiation and effects of prostaglandins and cyclic-AMP on pigmentation. Exp. Eye Res. 22:559–568.

Rehm, H., Schafer, T., and Betz, H. (1982) β-bungarotoxin-induced cell death of neurons in chick retina. Brain Res. 250:309–319.

Sarthy, P.V., Curtis, B.M., and Catterall, W.A. (1983) Retrograde labeling, enrichment, and characterization of retinal ganglion cells from the neonatal rat. J. Neurosci. 3:2532–2544.

Schaffner, A.E., and Schnaar, R.L. (1983) The isolation and purification of neurons from the vertebrate central nervous system. In: Barker, J., and McKeivy, J. (eds.) Current methods in cellular neurobiology. New York: Wiley and Sons.

Sheffield, J.B., and Moscona, A.A. (1970) Electron microscopic analysis of aggregation of embryonic cells: the structure and differentiation of aggregates of neural retinal cells. Dev. Biol. 23:36–61.

Skaper, S.D., Adler, R., and Varon, S. (1979) Generalized procedure for purifying neuron-like cells in cultures from central nervous tissues with a defined medium. Dev. Neurosci. 2:233–237.

Steinberg, M.S. (1963) Reconstitution of tissues by dissociated cells. Science 141:401–408.

Thompson, J.M., and Pelto, D.J. (1982) Attachment, survival and neurite extension of chick embryo retinal neurons on various culture substrates. Dev. Neurosci. 5:447–457.

Varon, S., and Mathorpe, M. (1980) Separation of neurons and glial cells by affinity methods. Adv. Cell Neurobiol. 1:405.

2. PRIMARY CULTURES OF GABAERGIC AND GLUTAMATERGIC NEURONS AS MODEL SYSTEMS TO STUDY NEUROTRANSMITTER FUNCTIONS I. DIFFERENTIATED CELLS

LEIF HERTZ AND ARNE SCHOUSBOE

1. INTRODUCTION

The complexity of the central nervous system (CNS) has necessitated the development of different methodologies, e.g., microdissection, separation by gradient centrifugation, and cell culturing, to study the characteristics of individual cell types. In the present review, we discuss results obtained using one of these preparations, i.e., neurons in primary cultures. By definition such cultures are obtained directly from the living animal, often at an immature stage, and maintained in vitro for at least 24 hours (Fedoroff, 1977). One advantage of the tissue culture system is that it is possible to obtain highly enriched populations of specific cell types. Since the cells are obtained from immature brain tissue, it is, however, essential to ensure that an appropriate development has occurred during the culturing period. Such alterations during development in vitro will be discussed in chapter 3. In this chapter, we discuss characteristics of neurons that have already differentiated during a prolonged culturing period ($\simeq$ 2 weeks). The selected characteristics are related to transmitter function, i.e., release, high-affinity uptake, biosynthesis, and receptor interactions of the neurotransmitters γ-aminobutyric acid (GABA) and glutamate. For this purpose, two different culture preparations, i.e., cerebral cortical interneurons and cerebellar granule cells, have been used. From in vivo studies it is known that the cortical interneurons use GABA (Ribak, 1978) and the cerebellar granule cells glutamate (Young et al., 1974; Hudson et al., 1976; Stone, 1979) as their transmitter.

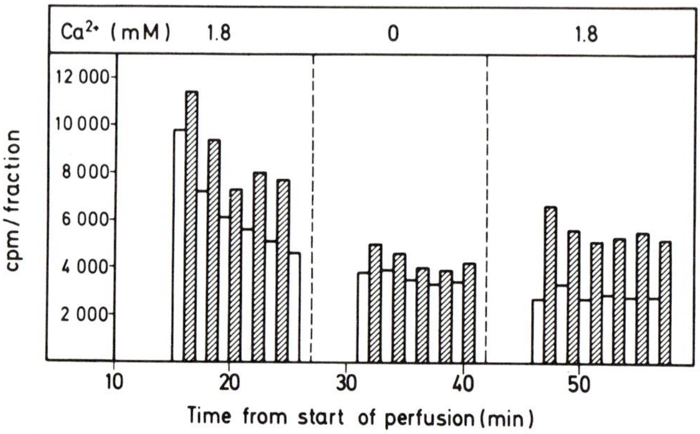

Figure 2-1. Release of radioactivity (cpm/fraction) from cultured cerebral cortical neurons superfusd with medium containing a low potassium [5 m*M* (□)] or a high potassium [55 m*M* (▨)] concentration. After preincubation of the cultures with 1 μM ^{3}H-GABA for 30 min at 37°C and three times wash with phosphate-buffered saline, the culture flask was cut open and placed in a superfusion system. The cell monolayer was covered with a piece of nylon cloth to facilitate dispersion of medium over the cells. The superfusion medium was continuously collected at intervals of 30 seconds. Throughout the experiment, superfusion medium was changed every 60 seconds between the medium with a low potassium concentration and that with a high potassium concentration. During the time interval, 27–41 minutes from the start of the superfusion, the superfusion medium contained no Ca^{2+}, and $MgCl_2$ was increased from the normal concentration of 0.6 m*M* to 15 m*M*. From Larsson et al. (1983).

2. TRANSMITTER RELEASE

2.1 GABA

A potassium-induced release of GABA from primary cultures of cerebral cortical interneurons has repeatedly been demonstrated (Snodgrass et al., 1980; Hauser et al., 1980; Hertz and Schousboe, 1980; Larsson et al., 1983; Yu et al., 1984a; Schousboe et al., 1986a). That this potassium-stimulated GABA release is dependent upon the presence of calcium in the extracellular fluid, as illustrated in figure 2-1, is an indication that GABA is, indeed, released in a transmitter-related fashion (Rubin, 1970; Carvalho, 1982). Nipecotic acid, a GABA analogue that is not metabolized, behaves in a similar manner (Larsson et al., 1983), showing that exogenously supplied GABA or GABA analogue has access to the transmitter pool that is released by elevated potassium. In contrast, there is little (Snodgrass et al., 1980) or no (Yu et al., 1984a) potassium-induced release of glutamate, and the release of glycine is also not affected by elevated potassium concentrations (Snodgrass et al., 1980). Thus it seems that cerebral cortical interneurons predominantly are GABAergic. This does not mean that this preparation does not contain other types of neurons, e.g., peptidergic (Staun-Olsen et al., 1985) or cholinergic (Thomas, 1985; see also chapter 4). Nor does it mean that no other primary cultures of neurons contain GABAergic cells. In most preparations of cerebellar granule cells, there seems to be a minor population of GABAergic neurons [presumably stellate and basket cells (Currie, 1980)] that shows a potassium-induced, calcium-dependent release of GABA (Pearce et al., 1981). This release occurs primarily from the distal part of the axon and its branches (Levi and Ciotti, 1983; Levi et al., 1984). In addition, evoked GABA release have been demonstrated in spinal cord neurons (Farb et al., 1979).

Astrocytes do not seem to contribute to the potassium-induced release of GABA, since it has been demonstrated in both cerebral cortical astrocytes (Hertz and Schousboe, 1980) and cerebellar astrocytes (Pearce et al., 1981) that elevated potassium concentrations do not enhance the release of GABA.

2.2 Glutamate

In analogy with the potassium-induced release of GABA from cerebral cortical neurons, elevated potassium concentrations cause a calcium-dependent release of glutamate from cerebellar granule cells (Drejer et al., 1982, 1983, and 1985b; Gallo et al., 1982; Levi and Ciotti, 1983; Levi et al., 1984; Meier et al., 1984; Schousboe et al., 1986a) as illustrated in figure 2-2. This release can be observed not only after labeling with glutamate itself but also after labeling with D-aspartate, a nonmetabolizable glutamate analogue, which, again, is an indication that exogenously supplied transmitter or transmitter analogue may be reutilized (Drejer et al., 1983; Levi and Ciotti, 1983; Meier et al., 1984; Levi et al., 1984). In contrast to what was observed in the case of GABA, the potassium-induced release of glutamate seems to occur both from perikarya and cell processes (Levi and Ciotti, 1983). It seems to be somewhat larger than the GABA release from cerebral cortical neurons as indicated by the wider arrow in figure 2-3.

Astrocytes also show a potassium-induced release of glutamate, but in contrast to the release from neurons, this release is extremely short lasting and is independent of the presence of calcium (Drejer et al., 1985b; Schousboe et al., 1986b).

3. TRANSMITTER UPTAKE

Inactivation of the transmitter amino acids GABA and glutamate is thought to be brought about by high-affinity uptake (Kuffler and Edwards, 1958; Curtis and Johnston, 1974) that may occur into either neurons or astroglial cells (Schousboe and Hertz, 1983). In the case of GABA, an efficient high-affinity uptake is found in mature cerebral cortical neurons (Larsson et al., 1983, 1985b; Yu et al., 1984a), whereas no high-affinity uptake is found in cerebellar granule cells (Messer, 1977; Currie, 1980; Yu and Hertz, 1982), and the high-affinity uptake into astrocytes is less intense than that into neurons (Schousboe and Hertz, 1983). This is illustrated in figure 2-3, which compares the release of GABA and glutamate from cerebral cortical neurons and cerebellar granule cells, respectively, with the uptake of these compounds

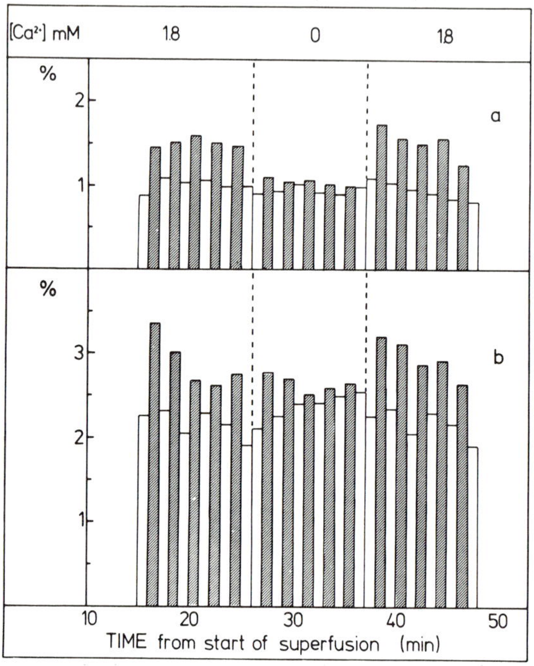

Figure 2-2. Release of radioactivity (percentage of remaining radioactivity/fraction) from cultured cerebellar granule cells superfused with medium containing a low potassium [5 m*M* (□)] or a high potassium (55 ml*M* (▨)] concentration. After preincubation of the cultures with 1 μ*M* D-[^{3}H] aspartate (*panel a*) or L-[^{3}H]glutamate (*panel b*) for 30 minutes at 37°C and three times wash with phosphate-buffered saline, the culture flask was cut open and placed in a superfusion system. The cell monolayer was covered with a piece of nylon cloth to facilitate dispersion of medium over the cells. The superfusion medium was continuously collected at intervals of 30 seconds. Throughout the experiment, superfusion medium was changed every 60 seconds between the medium with a low potassium concentration and that with a high potassium concentration. During the time interval, 26–37 minutes from the start of the superfusion, the superfusion medium contained no Ca^{2+}, and $MgCl_2$ was increased from the normal concentration of 0.6 m*M* to 15 m*M*. From Drejer et al. (1983).

into both astrocytes and neurons. The intensity of uptake and release processes is indicated by the width of the arrows so that the figure illustrates the fact that most of the GABA that has been released from cerebral cortical neurons is reaccumulated into these cells. Thus, there is a relatively modest drain of GABA from neurons to astrocytes. This is in contrast to the situation for glutamate release from glutamatergic neurons (figure 2-3). The absolute release rate is somewhat higher than that of GABA from GABA-ergic neurons (indicated by the wider arrow) and the most intense reuptake occurs into astrocytes (Hertz, 1979; Schousboe, 1982). There is a less intense uptake into glutamatergic neurons and a slight uptake into GABAergic neurons (Yu and Hertz, 1982; Drejer et al., 1982; Ramaharobandro et al., 1982; Schousboe et al., 1986b). Thus, the drain of glutamate from gluta-

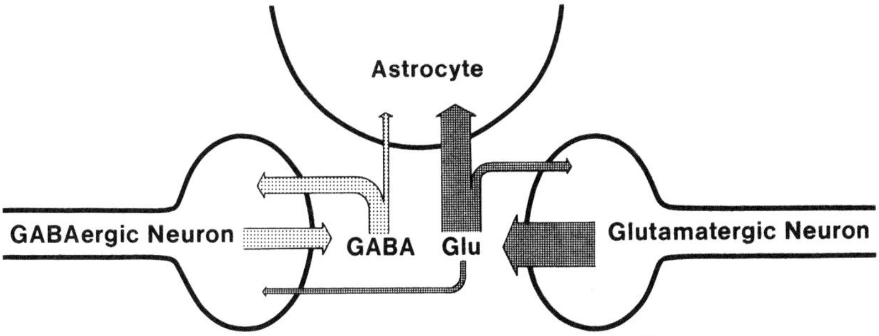

Figure 2-3. Schematic drawing of evoked release and uptake of glutamate and GABA in GABAergic or glutamatergic neurons and in astrocytes. The sizes of the *arrows* give an estimate of the relative magnitudes of the respective fluxes. It can be seen that neuronally released glutamate to a major extent is accumulated into astrocytes, whereas most of the released GABA is reaccumulated into neurons.

matergic neurons after evoked release is quantitatively much more important than the corresponding drain of GABA from GABAergic neurons.

4. BIOSYNTHETIC PATHWAYS

4.1 GABA

In the preceding sections it was concluded that, after accumulation, GABA can be released in a potassium-dependent manner. Although a large fraction of the released GABA is accumulated into neurons, some drain occurs by uptake into astrocytes. It is therefore essential that GABAergic neurons possess a synthetic machinery for the production of GABA. This occurs by conversion of glutamate to GABA catalyzed by glutamate decarboxylase (GAD). This enzyme is present at high activity in mature cerebral cortical neurons (table 2-1), whereas it is absent in cerebellar granule cells both in culture (table 2-1) and in vivo (Oertel et al., 1981), as well as in astrocytes (Schousboe et al., 1977; Hertz et al., 1983; Hertz and Schousboe, 1986a and b).

Another issue is the source of glutamate used for GABA synthesis. Several compounds seem to be able to function as glutamate precursors in cultured cerebral cortical neurons. These include exogenous glutamine (Borg et al., 1983; Yu et al., 1984b), but not exogenous glutamate (Borg et al., 1983; Yu et al., 1984a). However, glutamine is not the only GABA precursor and it may be of special interest that malate, a TCA cycle constituent, may also function as a glutamate precursor in cerebral cortical neurons (Hertz and Schousboe, 1986a). The formation of glutamate from malate is not abolished by AOAA, a transamination inhibitor, indicating that the process occurs at least partially as a reductive amination rather than as a transamination and that malate uptake does not solely depend on the operation of the malate–

Table 2-1. Activities (nmol × min^{-1} × mg^{-1} cell protein) of enzymes associated with the synthesis of GABA and glutamate in cultured neurons.

	Activity (nmol × min^{-1} × mg^{-1})			
Enzyme	Cerebral cortex neurons	Reference	Cerebellar granule cells	Reference
GAD	1.3	Hauser and Bernasconi (1980)	0[a]	Schousboe et al. (1985)
	10	Snodgrass et al. (1980)		
	14.2	Yu et al. (1984a)		
PAG	7.0 (18.8)[b]	Larsson et al. (1985a)	5.6 (19.0)[b]	Drejer et al. (1985a)
			21.8[c]	Patel et al. (1982)
γ-GT	1.3	Kvamme et al. (1985)	2.3	Kvamme et al. (1985)
GLDH	25.4	Larsson et al. (1985a)	60.8[d]	Patel et al. (1982)
			12.5	Drejer et al. (1985a)
AAT	280	Larsson et al. (1985a)	1657	Drejer et al. (1985a)
Orn-T	5.3	Drejer and Schousboe (1984)	3.1	Drejer and Schousboe (1984)

[a] The absence of GAD activity is based upon the finding that no AOAA-inhibitable CO_2 production occurs from [1-^{14}C]-glutamate.
[b] Values for PAG activity shown in parentheses indicate that the enzyme was assayed at 20 m*M* phosphate whereas 5 m*M* phosphate was used otherwise.
[c] The authors' value has been multiplied by 10 since they indicated that only 10% of the maximal enzyme activity was found at the glutamine concentration employed.
[d] GLDH was assayed in the presence of ADP, which stimulates the enzyme.

aspartate shuttle (see below). In addition, it has been shown that radioactivity from glucose can be incorporated into GABA in cultured cerebral neurons (Snodgrass et al., 1980), but, since pyruvate carboxylase is absent in neurons (Shank et al., 1981 and 1985; Yu et al., 1983), the neurons have no possibility to carry out a net synthesis of GABA from glucose.

Yoneda et al. (1982) have suggested that ornithine (and/or arginine) might serve as a precursor for glutamate used in GABA synthesis. This point of view is supported by the fact that ornithine-δ-aminotransferase (Orn-T) has a tenfold higher affinity for ornithine in cerebral cortical neurons than in cerebellar granule cells and astrocytes (Drejer and Schousboe, 1984).

The actual metabolic flux from glutamine toward GABA has been measured in cerebral cortical neurons by incubating the cells with labeled glutamine and measuring incorporation of radioactivity into glutamate and GABA. Such experiments have shown a flux from glutamine via glutamate to GABA of about $0.3\,\text{nmol} \times \text{min}^{-1} \times \text{mg}^{-1}$ protein (Yu et al., 1984b). This flux is well below the measured GAD activity and comparable with the rate of GABA release (Yu et al., 1984a).

4.2 Glutamate

It has not been established with certainty by which enzyme(s) the synthesis of transmitter glutamate is catalyzed. The possibilities are that glutamate either is synthesized from glutamine by phosphate-activated glutaminase (PAG) or that it is formed from α-ketoglutarate either by reductive amination, catalyzed by glutamate dehydrogenase (GLDH), or by transmination, catalyzed by aspartate aminotransferase (AAT). The activities of these enzymes are shown in table 2-1. It can be seen that PAG activity is similar in cerebral cortical interneurons and in cerebellar granule cells. Since the requirement for de novo synthesis of glutamate in cerebellar granule cells is much higher than that for de novo synthesis of GABA (and thus glutamate) in cerebral cortical neurons (figure 2-3), this does not suggest a key role for PAG in synthesis of transmitter glutamate. The GLDH activity in 2-week-old cultures is also similar in glutamatergic and GABAergic neurons (table 2-1). The AAT activity is, in contrast, one order of magnitude higher in the glutamatergic neurons, indicating that this enzyme may be of fundamental importance for the synthesis of transmitter glutamate as suggested by Wenthold and coworkers (Altschuler et al., 1981; Wenthold and Altschuler, 1983). It should be noted, however, that the activity of this enzyme is high in both cell types and it might seem peculiar that an enzyme that catalyzes a reaction in which glutamate is formed at the expense of aspartate can be involved in net synthesis of glutamate. An alternative possibility would be that malate uptake into the mitochondria of granule cells might be dependent upon the malate–aspartate shuttle (La Noue and Schoolwerth, 1979), i.e., the presence of AAT activity and that exogenous malate is an important precursor for transmitter glutamate. This concept is in agreement with the findings that

CO_2 production from [^{14}C]-malate as well as the incorporation of label from the compound into glutamate are partly inhibited by AOAA (J. Drejer, A. Schousboe, and L. Hertz, unpublished experiments).

No information is available about metabolic rates for production of transmitter glutamate from different potential precursors, but it is known that both glutamine and malate may serve as a precursor for glutamate released in a stimulus-related manner (Gallo et al., 1982; Levi and Ciotti, 1983; Meier et al., 1984; J. Drejer, A. Schousboe, and L. Hertz, unpublished experiments).

Cellular accumulation of glutamine might be rate limiting for glutamate synthesis from glutamine. Since it has been suggested that γ-glutamyltranspeptidase (γ-GT) may be of importance for the uptake of glutamine (Tate and Meister, 1981) the activity of this enzyme is also of interest. From table 2-1, it can be seen that the difference between cortical neurons and cerebellar granule cells is modest (twofold). Since the enzyme activity is of the same magnitude as the uptake rate for glutamine (Hertz et al., 1980; Yu and Hertz, 1982; Yu et al., 1984b), the twofold higher activity in cerebellar granule cells could, however, be of functional importance. This point of view is supported by the demonstration that the γ-GT activity is the only enzyme whose activity correlates well with glutamate content during the entire culturing period (Kvamme et al., 1985; see also chapter 3).

In the preceding section it was suggested that ornithine, by the action of Orn-T, might be a precursor for glutamate used for GABA synthesis. Conceivably, a similar mechanism might operate for glutamate synthesis in glutamatergic neurons as suggested by Wong et al. (1981). This hypothesis is contradicted by the lower Orn-T activity in cerebellar granule cells than in cerebral cortical neurons (table 2-1). In addition, the affinity for ornithine is tenfold lower in the former cells (Drejer and Schousboe, 1984).

5. RECEPTOR FUNCTION

5.1 GABA receptors

Both cerebral cortical neurons and cerebellar granule cells possess GABA receptors as demonstrated by GABA or muscimol binding (De Feudis et al., 1979; Ticku et al., 1980; Meier and Schousboe, 1982; Meier et al., 1984). In undifferentiated cells, there is only a high-affinity binding site, but differentiated cells have both a high-affinity and a low-affinity binding site (see table 3-2 in chapter 3) as does the mature brain in vivo (e.g., Wong and Horng, 1977; Meier and Schousboe, 1982).

The functional role of GABA receptors on cerebral cortical neurons is unknown, whereas the function of these receptors on cerebellar granule cells has been studied in some detail. The high-affinity receptors mediate the neurotrophic effect of GABA on these cells as well as the ability of GABA to induce low-affinity GABA receptors (see chapter 3), but they have no other

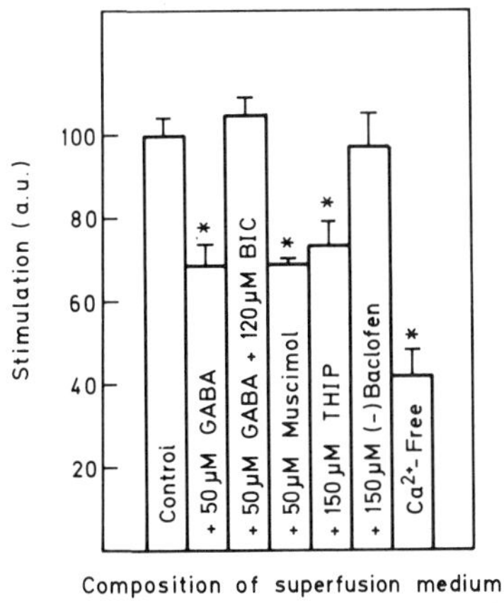

Figure 2-4. Pharmacological characterization of the GABA-mediated inhibition of potassium-evoked glutamate release from cerebellar granule cells cultured in the presence of 50 µ*M* GABA and preloaded with D-[^{3}H]aspartate. The composition of the superfusion media during different parts of the release experiment (cf. figure 2-2) is indicated in the individual columns [*BIC*, bicuculline; and *THIP*, 4,5,6,7,-tetrahydro-isoxalo (5,4-C)-pyridin-3-*ol*]. The *columns* represent the average evoked release (means + SEM) expressed in arbitrary units (a.u.), 100 a.u. being the release under control conditions (corresponding to an actual evoked release amounting to 80%–100% of the basal release; cf. figure 2-2). *Asterisks* indicate statistically significant differences from the release in the control experiment ($p < 0.01$). Results are from three successive stimulations of three individual cultures.

known function. The low-affinity receptors inhibit the release of glutamate evoked by 100 m*M* potassium (Meier et al., 1984). The pharmacological properties of these receptors have been investigated by the use of different GABA agonists and antagonists. It was found that the effect of GABA could be mimicked by THIP [4,5,6,7-tetrahydroisoxazolo(5,4-*c*)-pyridin-3-*ol*] and muscimol and antagonized by bicuculline whereas baclofen had no effect (figure 2-4). Accordingly, these receptors should be classified as $GABA_A$ receptors (Hill and Bowery, 1981).

5.2 Glutamate receptors

Glutamate receptors have also been demonstrated on both cerebral cortical neurons and cerebellar granule cells. They stimulate transmitter release, i.e., release of GABA from cerebral cortical neurons and release of glutamate from cerebellar granule cells (Drejer et al., 1986a and b). The cortical neurons possess all conventional glutamate receptor subtypes, i.e., NMDA (*N*-methyl-D-asparatate), quisqualate-, and kainate-preferring receptors, whereas the glutamate receptor on the cerebellar granule cells is not markedly affected by any of these substances. It can, however, be blocked by γ-D-glutamylglycine, indicating that it is not related to the high-affinity uptake site (Drejer et al., 1986b).

6. CONCLUDING REMARKS

This chapter has reviewed characteristics of differentiated cerebral cortical neurons and cerebellar granule cells in primary cultures. These characteristics suggest that the two preparations are valid models of GABAergic and glutamatergic neurons, respectively, and that there are considerable functional differences between these two nerve cell types. In chapter 3, the developmental aspects of the two culture systems will be discussed, in order to review information about the different ontogenetic development of the two neuronal classes, and to compare development of these two nerve cell types with developmental alterations occurring in the same regions of the CNS in vivo.

REFERENCES

Altschuler, R.A., Neises, G.R., Harmison, G.G., Wenthold, R.J., and Fex, J. (1981) Immunocytochemical localization of aspartate aminotransferase immunoreactivity in cochlear nucleus of the guinea pig. Proc. Natl. Acad. Sci. USA 78:6553–6557.

Borg, J., Hamel, G., Spitz, B., and Mark, J. (1983) Compartmentation of glutamate and glutamine metabolism in cultured nerve cells. In: Hertz, L., Kvamme, E., McGeer, E.G., and Schousboe, A. (eds.) Glutamine, glutamate and GABA in the central nervous system. New York: Alan R. Liss, pp. 317–325.

Carvalho, A.P. (1982) Calcium in the nerve cell. In: Lajtha, A. (ed.) Handbook of neurochemistry, vol. 1. New York: Plenum, pp. 69–116.

Currie, D.N. (1980) Identification of cell type by immunofluorescence in defined cell cultures of cerebellum. In: Giacobini, E., Vernadakis, A., and Shahar, A. (eds.) Tissue culture in neurobiology. New York: Raven, pp. 75–87.

Curtis, D.R., and Johnston, G.A.R. (1974) Amino acid transmitters in the mammalian central nervous system. Ergb. Physiol. 69:97–188.

De Feudis, F.V., Ossola, L., Schmitt, G., and Mandel, P. (1979) High-affinity binding of [^{3}H]-muscimol to subcellular particles of a neuron enriched culture of embryonic rat brain. Neurosci. Lett. 14:195–199.

Drejer, J., and Schousboe, A. (1984) Ornithine-δ-amino-transferase exhibits different kinetic properties in astrocytes, cerebral cortex interneurons, and cerebellar granule cells in primary culture. J. Neurochem. 42:1194–1197.

Drejer, J., Larsson, O.M., and Schousboe, A. (1982) Characterization of L-glutamate uptake into and release from astrocytes and neurons cultured from different brain regions. Exp. Brain Res. 47:259–269.

Drejer, J., Larsson, O.M., and Schousboe, A. (1983) Characterization of uptake and release processes for D- and L-aspartate in primary cultures of astrocytes and cerebellar granule cells. Neurochem. Res. 8:231–243.

Drejer, J., Larsson, O.M., Kvamme, E., Svenneby, G., Hertz, L., and Schousboe, A. (1985a) Ontogenetic development of glutamate metabolizing enzymes in cultured cerebellar granule cells and in cerebellum. Neurochem. Res. 10:49–62.

Drejer, J., Benveniste, H., Diemer, N.H., and Schousboe, A. (1985b) The cellular origin of ischemia-induced glutamate release from brain tissue in vivo and in vitro, J. Neurochem. 45:145–151.

Drejer, J., Honore, T., and Schousboe, A. (1986a) Excitatory amino acid receptors on neuronal cultures: pharmacological characterization of excitatory amino acid receptors on two neuronal primary cultures. In: Roberts, P.J. (ed.) Excitatory amino acids. Oxford: McMillan (in press).

Drejer, J., Honore, T., Meier, E., and Schousboe, A. (1986b) Pharmacologically distinct glutamate receptors on cerebellar granule cells. Life Sci. 38:2077–2085.

Farb, D.H., Berg, D.K., and Fischbach, G.D. (1979) Uptake and release of [^{3}H] γ-aminobutyric acid by embryonic spinal cord neurons in dissociated cell culture. J. Cell Biol. 80:651–661.

Fedoroff, S. (1977) Primary cultures, cell lines and cell strains: Terminology and characteristics.

In: Fedoroff, S., and Hertz, L. (eds.) Cell, tissue and organ cultures in neurobiology. New York: Academic Press, pp. 265–286.

Gallo, V., Ciotti, M.T., Coletti, A., Aloisi, F., and Levi, G. (1982) Selective release of glutamate from cerebellar granule cells differentiating in culture. Proc. Natl. Acad. Sci. USA 79:7919–7923.

Hauser, K., and Bernasconi, R. (1980) Rat cortical neurons in dissociated cell culture: changes in GABA and guanyl cyclase activity during development. In: Giacobini, E., Vernadakis, A., and Shahar, A. (eds.) Tissue culture in neurobiology. New York: Raven, pp. 205–219.

Hauser, K., Balcar, V.J., and Bernasconi, R. (1980) Development of GABA neurons in dissociated cell culture of rat cerebral cortex. Brain Res. Bull. [Suppl. 2] 5:37–41.

Hertz, L. (1979) Functional interactions between neurons and astrocytes. I. Turnover and metabolism of putative amino acid transmitters. Prog. Neurobiol. 13:277–323.

Hertz, L., and Schousboe, A. (1980). Interactions between neurons and astrocytes in the turnover of GABA and glutamate. Brain Res. Bull. [Suppl. 2] 5:389–395.

Hertz, L., and Schousboe, A. (1986a) Role of astrocytes in compartmentation of amino acid and energy metabolism. In: Astrocytes. Fedoroff, S., and Vernadakis, A. (eds.) New York: Academic Press, Vol. 2: 179–208.

Hertz, L., and Schousboe, A. (1986b) Metabolism of glutamate and glutamine in neurons and astrocytes in primary cultures. In: Kvamme, E. (ed.) Glutamine and glutamate in mammals. Boca Raton, FL: CRC Press (in press).

Hertz, L., Yu, A., Svenneby, G., Kvamme, E., Fosmark, H., and Schousboe, A. (1980) Absence of preferential glutamine uptake into neurons: an indication of a net transfer of TCA constituents from nerve endings to astrocytes? Neurosci. Lett. 16:103–109.

Hertz, L., Yu, A.C.H., Potter, R.L., Fisher, T.E., and Schousboe, A. (1983) Metabolic fluxes from glutamate and towards glutamate in neurons and astrocytes in primary cultures. In: Hertz, L., Kvamme, E., McGeer, E.G., and Schousboe, A. (eds.) Glutamine, glutamate and GABA in the central nervous system. New York: Alan R. Liss, pp. 327–342.

Hill, D.R., and Bowery, N.G. (1981) ^{3}H-Baclofen and ^{3}H-GABA bind to bicuculline-insensitive $GABA_B$ sites in rat brain. Nature (Lond.) 290:149–152.

Hudson, D.B., Valcana, T., Bean, G., and Timiras, P.A. (1976) Glutamic acid: a strong candidate as the neurotransmitter of the cerebellar granule cell. Neurochem. Res. 1:73–81.

Kuffler, S.V., and Edwards, C. (1958) Mechanism of γ-aminobutyric acid (GABA) action and its relation to synaptic inhibition. J. Neurophysiol. 21:589–610.

Kvamme, E., Schousboe, A., Hertz, L., Torgner, I.Aa., and Svenneby, G. (1985) Developmental change of endogenous glutamate and gamma-glutamyl transferase in cultured cerebral cortical interneurons and cerebellar granule cells and in mouse cerebral cortex and cerebellum in vivo. Neurochem. Res. 10:993–1008.

Là Noue, K.F., and Schoolwerth, A.C. (1979) Metabolite transport in mitochondria. Annu. Rev. Biochem. 48:871–922.

Larsson, O.M., Drejer, J., Hertz, L., and Schousboe, A. (1983) Ion dependency of uptake and release of GABA and (RS)-nipecotic acid studied in cultured mouse brain cortex neurons. J. Neurosci. Res. 9:291–302.

Larsson, O.M., Drejer, J., Kvamme, E., Svenneby, G., Hertz, L., and Schousboe, A. (1985a) Ontogenetic development of glutamate and GABA metabolizing enzymes in cultured cerebral cortex interneurons and in cerebral cortex in vivo. Int. J. Dev. Neurosci. 3:177–185.

Larsson, O.M., Krogsgaard-Larsen, P., and Schousboe, A. (1985b) Characterization of uptake of GABA and its analogues nipecotic acid and *cis*-4-OH-nipecotic acid in cultured neurons and astrocytes. Neurochem. Int. 7:853–860.

Levi, G., and Ciotti, M.T. (1983) Glutamate and GABA localization and evoked release in cerebellar cells differentiating in culture. In: Hertz, L., Kvamme, E., McGeer, E.G., and Schousboe, A. (eds.) Glutamine, glutamate and GABA in the central nervous system. New York: Alan R. Liss, 493–508.

Levi, G., Aloisi, F., Ciotti, M.T., and Gallo, V. (1984) Autoradiographic localization and depolarization-induced release of acidic amino acids in differentiating cerebellar granule cells. Brain Res. 290:77–86.

Meier, E., and Schousboe, A. (1982) Differences between GABA receptor binding to membranes from cerebellum during postnatal development and from cultured cerebellar granule cells. Dev. Neurosci. 5:546–553.

Meier, E., Drejer, J., and Schousboe, A. (1984) GABA induces functionally active low-affinitive GABA receptors on cultured cerebellar granule cells. J. Neurochem. 43:1737–1744.

Messer, A. (1977) The maintenance and identification of mouse cerebellar granule cells in monolayer culture. Brain Res. 130:1–12.

Oertel, W.H., Schmechel, D.E. Mugnaini, E., Tappaz, M.L., and Kopin, I.J. (1981) Immunocytochemical localization of glutamate decarboxylase in rat cerebellum with a new antiserum. Neuroscience 6:2715–2735.

Patel, A.J., Hunt, A., Gordon, R.D., and Balazs, R. (1982) The activities in different neural cell types of certain enzymes associated with the metabolic compartmentation of glutamate. Dev. Brain Res. 4:3–11.

Pearce, B.R., Currie, D.N., Beale, R., and Dutton, G.R. (1981) Potassium-stimulated, calcium-dependent release of [^{3}H] GABA from neuron- and glia-enriched cultures of cells dissociated from rat cerebellum. Brain Res. 206:485–489.

Ramaharobandro, N., Borg, J., Mandel, P., and Mark, J. (1982) Glutamine and glutamate transport in cultured neuronal and glial cells. Brain Res. 244:113–121.

Ribak, C.E. (1978) Aspinous and sparsely-spinous stellate neurons in the visual cortex of rats contain glutamic acid decarboxylase. J. Neurocytol. 7:461–478.

Rubin, R.P. (1970) The role of calcium in the release of neurotransmitter substances and hormones. Pharmacol. Rev. 22:389–428.

Schousboe, A. (1982) Neurotransmitter metabolism and function. In: Pfeiffer, S.E. (ed.) Neuroscience approached through cell culture, vol. 1. Boca Raton, FL: CRC Press, pp. 107–141.

Schousboe, A., and Hertz, L. (1983) Regulation of glutamatergic and GABAergic neuronal activity by astroglial cells. In: Osborne, N.N. (ed.) Dale's principle and communication between neurons. Oxford: Pergamon, pp. 113–141.

Schousboe, A., Hertz, L., and Svenneby, G. (1977) Uptake and metabolism of GABA in astrocytes cultured from dissociated mouse brain hemispheres. Neurochem. Res. 2:217–229.

Schousboe, A., Drejer, J., and Hertz, L. (1985) Comparison of the metabolism of glucose and glutamate in cultured cerebellar granule cells. J. Neurochem. [Suppl.] 44:S168.

Schousboe, A., Drejer, J., Hansen, G.H., and Meier, E. (1986a) Cultured neurons as model systems for biochemical and pharmacological studies on receptors for neurotransmitter amino acids. Dev. Neurosci. 7:252–262.

Schousboe, A., Drejer, J., and Hertz, L. (1986b) Uptake and release of glutamate and glutamine in neurons and astrocytes in primary cultures. In: Kvamme, E. (ed.) Glutamine and glutamate in mammals. Boca Raton, FL: CRC Press (in press).

Shank, R.P., Campbell, G.L., Freyteg, S.D., and Utter, M.F. (1981) Evidence that pyruvate carboxylase is an astrocyte specific enzyme in CNS tissues. Abstr. Soc. Neurosci. 7:936.

Shank, R.P., Bennet, G.S., Freyteg, S.O., Campbell, G.L., and Utter, M.F. (1985) Pyruvate carboxylase: an astrocyte-specific enzme implicated in the replenishment of amino acid neurotransmitter pools. Brain Res. 329:364–367.

Snodgrass, S.R., White, W.F., Biales, B., and Dichter, M. (1980) Biochemical correlates of GABA function in rat cortical neurons in culture. Brain Res. 190:123–138.

Staun-Olsen, P., Fahrenkrug, J., Gammeltoft, S., Ottesen, B., and Schousboe, A. (1985) Development of binding sites for vasoactive intestinal polypeptide in mouse cerebral cortex and cultured cortical neurons. Int. J. Dev. Neurosci. 3:609–616.

Stone, T.W. (1979) Glutamate as the neurotransmitter of cerebellar granule cells in the rat: electrophysiological evidence. Br. J. Pharmacol. 66:291–296.

Tate, S.S., and Meister, A. (1981) Gamma-glutamyl transpeptidase: catalytic, structural and functional aspects. Mol. Cell. Biochem. 39:357–368.

Thomas, W.E. (1985) Synthesis of acetylcholine and γ-aminobutyric acid by dissociated cerebral cortical cells in vitro. Brain Res. 332:79–89.

Ticku, M.K., Huang, A., and Barker, J.L. (1980) Characterization of γ-aminobutyric acid receptor binding in cultured brain cells. Mol. Pharmacol. 17:285–289.

Wenthold, R.J., and Altschuler, R.A. (1983) Immunocytochemistry of aspartate aminotransferase and glutaminase. In: Hertz, L., Kvamme, E., McGeer, E.G., and Schousboe, A. (eds.) Glutamine, glutamate, and GABA in the central nervous system. New York: Alan R. Liss, 33–50.

Wong, D.T., and Horng, J.S. (1977) Na^+-independent binding of GABA to the Triton X-100 treated synaptic membranes from cerebellum of rat brain. Life Sci. 20:445–452.

Wong, P.T.-H. McGeer, E., and McGeer, P.L. (1981) A sensitive radiometric assay for

ornithine aminotransferase: regional and subcellular distributions in rat brain. J. Neurochem. 36:501–505.
Yoneda, Y., Roberts, E., and Dietz, G.W., Jr. (1982) A new synaptosomal biosynthetic pathway of glutamate and GABA from ornithine and its negative feedback inhibition by GABA. J. Neurochem. 38:1686–1694.
Young, A.B., Oster-Granite, M.L., Herndon, R.M., and Snyder, S.H. (1974) Glutamic acid: selective depletion by viral induced granule cell loss in hamster cerebellum. Brain Res. 73:1–13.
Yu, A.C.H., and Hertz, L. (1982) Uptake of glutamate, GABA and glutamine into a predominantly GABAergic and a predominantly glutamatergic nerve cell population in culture. J. Neurosci. Res. 7:23–35.
Yu, A.C.H., Drejer, J., Hertz, L., and Schousboe, A. (1983) Pyruvate carboxylase activity in primary cultures of astrocytes and neurons. J. Neurochem. 41:1484–1487.
Yu, A.C.H., Hertz, E., and Hertz, L. (1984a) Alterations in uptake and release rates for GABA, glutamate and glutamine during biochemical maturation of highly purified cultures of cerebral cortical neurons, a GABAergic preparation. J. Neurochem. 42:951–960.
Yu, A.C.H., Fisher, T.E., Hertz, E., Tildon, J.T., and Schousboe, A. (1984b) Metabolic fate of [^{14}C]-glutamine in mouse cerebral neurons in primary cultures. J. Neurosci. Res. 11:351–357.

3. PRIMARY CULTURES OF GABAERGIC AND GLUTAMATERGIC NEURONS AS MODEL SYSTEMS TO STUDY NEUROTRANSMITTER FUNCTIONS II. DEVELOPMENTAL ASPECTS

ARNE SCHOUSBOE AND LEIF HERTZ

1. INTRODUCTION

In the preceding chapter, characteristics of purified populations of mature differentiated cerebral cortical neurons and cerebellar granule cells were described. These cultures are prepared from immature brain (e.g., Hertz et al., 1985a and b). When they are used as models for their differentiated counterparts in vivo, it must be ascertained that the cells undergo a differentiation during the culturing period. In addition, the immaturity of the tissues from which the cultures are prepared provides the opportunity to study aspects of neuronal development, e.g., the time course of ontogenetic alterations and the effect of trophic factors. In this chapter, we discuss development of transmitter-related functions (transmitter release and uptake, biosynthetic pathways, and receptor function) in each of these two preparations and compare these observations with results obtained in corresponding regions of the brain in vivo. This comparison is essential in order to evaluate the degree of functional differentiation that has occurred in vitro and it also provides information as to what extent in vivo alterations are due to the development of neurons or of other cells types.

2. TRANSMITTER RELEASE

The most obvious indication of functional differentiation in a neuron may be the presence of an evoked release of its neurotransmitter. Such a release of

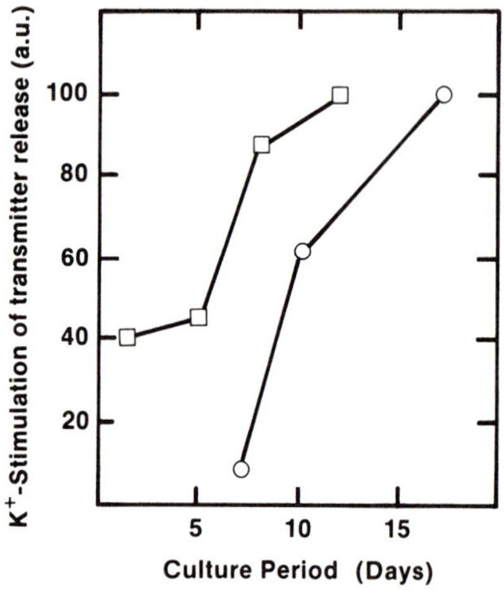

Figure 3-1. Potassium stimulation of GABA (○) or glutamate (□) release from, respectively, cerebral cortical neurons and cerebellar granule cells as a function of development of the neurons in culture. The values (expressed in arbitrary units, i.e., 100 indicating release in the most mature cultures) represent the release in the presence of calcium. Recalculated from original data of Larsson et al. (1983) and Yu et al. (1984) in the case of GABA release and of Levi and Ciotti (1983) in the case of glutamate release.

γ-aminobutyric acid (GABA) and of glutamate occurs in differentiated cerebral cortical neurons and cerebellar granule cells, respectively (see chapter 2). The development of this phenomenon is shown in figure 3-1. In mouse cerebral cortical cells, no potassium-induced stimulation is found in young cultures and it is only during the second week of culture that a potassium-induced release develops (figure 3-1). This developmental pattern is in agreement with results reported by Snodgrass et al. (1980) and Hauser et al. (1980) in rat cultures. (In general, the development of the brain is very similar in mouse and rat.) In rat cerebellar granule cells, there is a small potassium-induced release already after 2 days of culturing, but it is only marginally dependent upon the presence of calcium (Gallo et al., 1982; Levi and Ciotti, 1983). The development of transmitter release in these neurons may be slightly faster than in cerebral cortical cells since a pronounced, although not maximal, potassium-induced release of glutamate occurs already in 8-day-old cultures (figure 3-1).

This development in the cultured cells is in accordance with the pronounced potassium-induced release of GABA occuring in brain slices from 6- to 10-day-old rats (Davies et al., 1975; Schousboe et al., 1976) and in synaptosomes from 2- to 3-week-old rats (Redburn et al., 1978). In contrast, this response is absent or greatly reduced in slices from 1-day-old rats (Davies et al., 1975; Schousboe et al., 1976).

Table 3-1. Development of GABA uptake into cerebral cortical neurons.

Culture period (days)	K_m (μM)	V_{max} (nmol × min^{-1} × mg^{-1})	Sodium dependence (Hill coeff.)
1	11.7 ± 1.8	0.128 ± 0.007	1.05
3	6.8 ± 0.2	0.282 ± 0.002	1.02
10	21.0 ± 3.2	2.76 ± 0.24	1.77

O.M. Larsson, L. Hertz, and A. Schousboe, unpublished.

3. TRANSMITTER UPTAKE

Table 3-1 shows that, during the same period when the potassium-induced release of GABA from cerebral cortical cultures develops, there is an increase in uptake of GABA (Borg et al., 1980; Yu et al., 1984; Larsson et al., 1986). GABA uptake generally occurs as a cotransport with sodium (e.g., Martin, 1976) and the stoichiometry between GABA and sodium uptake in cerebral cortical neurons changes during development (Larsson et al., 1986). This is shown by the finding that the Hill coefficient of the sodium dependence changes from a value of 1.0 in 1- to 3-day-old cultures to a value of about 2.0 in 10-day-old cultures (table 3-1).

The increase in GABA uptake with age is similar to a corresponding increase in brain slices (Johnston and Davies, 1974; Schousboe et al., 1976; Wong and McGeer, 1981). However, in brain slices or synaptosomes, the GABA uptake is specially fast in 1- to 2-week-old animals and subsequently shows a decline (Schousboe et al., 1976; Redburn et al., 1978; Wong and McGeer, 1981).

4. BIOSYNTHETIC PATHWAYS

In agreement with the development of GABAergic properties in cerebral cortical neurons during the second week in culture, the activity of glutamate decarboxylase (GAD) more than quadruples during the same period (figure 3-2A). This development occurs at an earlier stage in the cultured cells than in the cerebral cortex in vivo (Yu et al., 1984). It also appears to be faster in our model than in cultures containing more glial cells (Snodgrass et al., 1980; Hauser et al., 1980), suggesting an enhanced rate of maturation in the absence of astrocytes.

The development of biosynthetic pathways for glutamate formation is of interest both in GABAergic and glutamatergic neurons since the GABAergic cells require glutamate as a GABA precursor. The only enzyme (figure 3-2B) that shows a perfect correlation with the glutamate content of both cerebral cortical neurons and cerebellar granule cells with time in culture is γ-glutamyltranspeptidase (γ-GT) (Kvamme et al., 1985), an enzyme that may regulate glutamine uptake and thus availability of glutamine as a glutamate

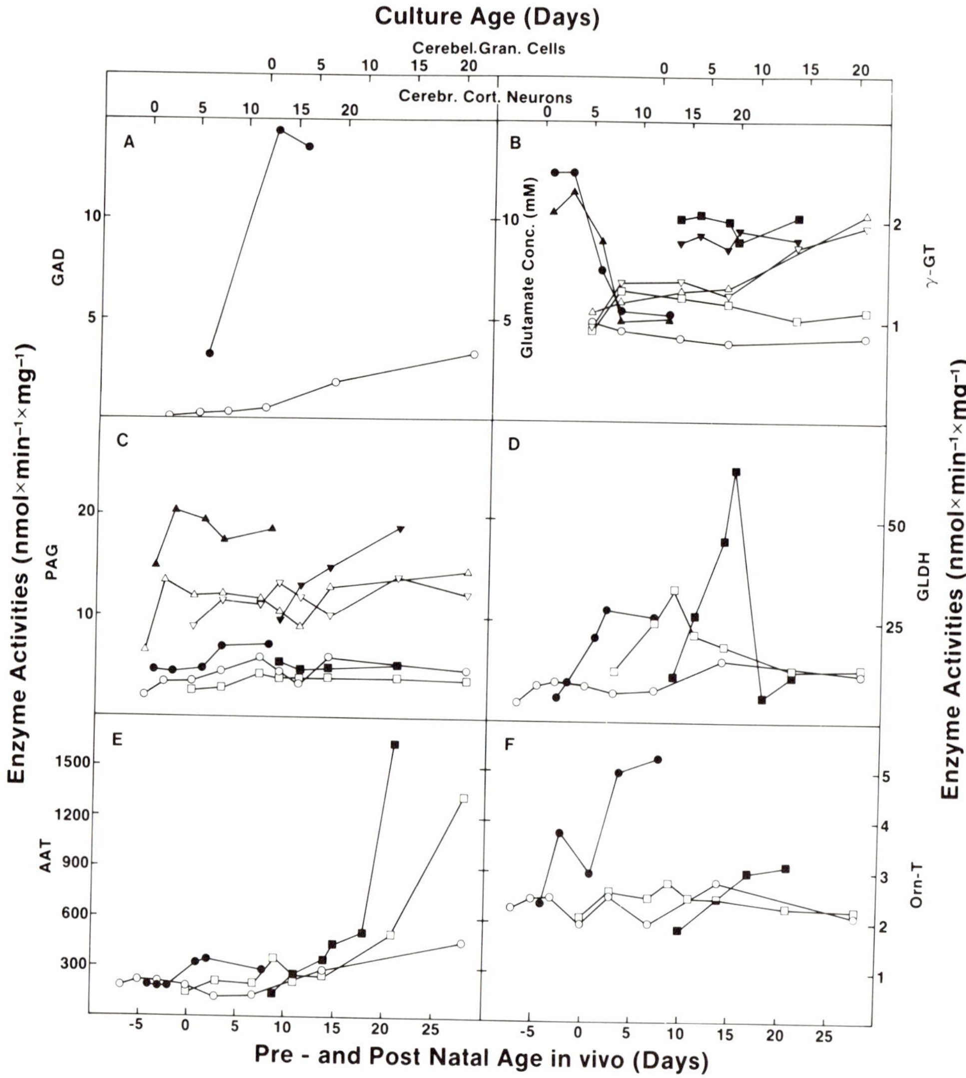

Figure 3-2. Development of GAD (**A**), γ-GT (**B**), PAG (**C**), GLDH (**D**), AAT (**E**), and Orn-T (**F**) activity in primary cultures of cerebral cortical neurons (●) or of cerebellar granule cells (■) and in the cerebral cortex (○) or cerebellum (□) in vivo as a function of age in culture or in vivo. Zero times for age in culture are shown corresponding to the in vivo age at which the cultures were prepared (15-day-old embryos for cerebral cortical cultures and 7-day-old animals for cerebellar granule cells) so that ages in vivo and ages in culture are directly comparable. In **B**, triangles indicate glutamate content in cerebral cortical neurons (▲) and cerebellar granule cells (▼) in primary cultures and in cerebral cortex (△) and cerebellum (▽) in vivo and in **C**, triangles indicate PAG activity in the presence of 20 m*M* phosphate in cerebral cortical neurons (▲) and cerebellar granule cells (▼) in primary cultures and in cerebral cortex (△) and cerebellum (▽) in vivo. Values are from Yu et al. (1984), Larsson et al. (1985), Drejer et al. (1985), and Kvamme et al. (1985). In vivo values in **A** are unpublished data of A. Schousboe.

precursor (chapter 2). Such a correlation between γ-GT activity and glutamate content is not observed in vivo (figure 3-2B). It should be noted that the γ-GT activity in cerebral cortical neurons decreases during development whereas all other enzyme activities measured show either an increase or no change (figure 3-2).

An enzyme that more directly regulates the formation of glutamate from glutamine is phosphate-activated glutaminase (PAG). In the presence of a low concentration of phosphate (5 m*M*), the development of this enzyme is the same in the cultured neurons as in vivo, both for cerebral cortical neurons and in the case of cerebellar granule cells (figure 3-2C). In the presence of a high concentration of phosphate (20 m*M*), the activity of the enzyme is enhanced. This phosphate stimulation is pronounced both in cerebral cortical neurons and in cerebellar granule cells. However, it reaches maximum activity already at an early stage in the cortical cells whereas the development occurs much more slowly in the glutamatergic cerebellar cells. Thus, in the granule cells there is a good correlation between the development of phosphate stimulation of PAG (figure 3-2C) and evoked release of glutamate, whereas no such correlation exists between PAG activity in cerebral cortical neurons and the release of GABA from these cells (figure 3-1). This does not indicate that glutamine may be a more important precursor for glutamate than for GABA since, in the GABAergic cells, glutamate is a prerequisite for GABA formation, whereas the regulation of the synthesis of the transmitter itself may depend on the development of GAD, which occurs at a later stage (figure 3-2A).

In both types of cultured neurons, glutamate dehydrogenase (GLDH) shows a maximum activity that is considerably higher than that in vivo (figure 3-2D). In cerebellar granule cells there is a peak of activity both in culture and in vivo (figure 3-2D), but in cerebral cortical neurons there is no obvious correlation between activity in culture and in vivo, and the in vivo development coincides much more closely with that found in cultured astrocytes (Hertz et al., 1978; Schousboe et al., 1980). There is a pronounced reduction of the activity of this enzyme in newly established cultures, which may be related to the fact that young cultures of neurons become depleted of most of their processes. There is no developmental correlation between the activity of GLDH (figure 3-2D) and the glutamate content (figure 3-2B). However, at all ages the enzyme activity is high enough to support the hypothesis that the GLDH reaction might be important for glutamate synthesis (see chapter 2).

Aspartate aminotransferase (AAT) activity is high under all conditions, both in cultures and in vivo (figure 3-2E). In cerebral cortical neurons, and especially in cerebellar granule cells, there is an increase in activity as maturation proceeds. This increase occurs earlier in culture than in vivo, and again the in vivo development in brain cortex coincides more closely with astrocytic development (Hertz et al., 1978; Schousboe et al., 1980). There is no correlation with the content of glutamate (figure 3-2B) in either cell type, and the increase in AAT activity in cerebellar granule cells occurs at a stage where evoked glutamate release has almost reached its maximum (figure 3-1). This does not support the concept of a correlation between AAT activity and production of transmitter glutamate (see chapter 2).

The activity of ornithine-δ-aminotransferase (Orn-T) is identical in cerebral cortex and cerebellum in vivo (figure 3-2F) and it is not modified during maturation. In cultures of cerebral cortical neurons and cerebellar granule cells there is a pronounced increase in enzyme activity (figure 3-2F). This may suggest a role for this enzyme in neuronal development (particularly during process formation). This is in line with the well-known fact that the related enzyme ornithine decarboxylase and the polyamines play a role in development of GABAergic neurons (Seiler et al., 1984). Again, there is no correlation between the activity of this enzyme and glutamate content.

5. RECEPTOR FUNCTION

Both cerebral cortical neurons and cerebellar granule cells express GABA receptors (see chapter 2). Table 3-2 shows the development of this receptor function in the two cell types. In both types of neurons there is a high-affinity binding site in immature cells. In untreated cerebellar granule cells this remains the only receptor site even after prolonged culturing. However, addition of GABA to the culture medium leads to the development of a GABA binding site with lower affinity (Meier and Schousboe, 1982; Meier et al., 1984). By the use of GABA-specific agonists and antagonists it has been established that GABA exerts its effect via the preexisting high-affinity GABA receptors. It was thus found that the effect of GABA could be mimicked by THIP [4,5,6,7-tetrahydro isoxazolo [5,4-*c*]-pyridin-3-*ol*] and muscimol, and the effect could be blocked by bicuculline methobromide (Belhage et al., 1986). This development of low-affinity receptors corresponds in retina to an induction of GABA receptors by GABA or muscimol (Madtes and Redburn, 1983a and b). In cortical neurons, prolonged culturing alone induces low-affinity GABA receptors (table 3-2), probably on account of endogenous GABA production and release.

In the brain in vivo, both the high- and low-affinity binding sites exist (see chapter 2). According to Meier et al. (1983) the situation in culture mimicks the in vivo counterpart since, in the latter, the binding site with a low affinity develops at a later stage than the high-affinity site. This difference was less pronounced in studies by Aldinio et al. (1980) and Skerritt and Johnston (1982).

In contrast to the GABA receptors with low affinity that play a functional role in the adult central nervous system by regulating release of transmitter glutamate (see chapter 2), the role of the high-affinity receptors may be associated with certain aspects of neuronal development. This has been discussed above in relation to development of the GABA binding site with

Table 3-2. K_D and B_{max} values for specific binding of GABA or muscimol to membranes from cerebral cortical neurons and cerebellar granule cells in primary cultures.

Cultured period (days)	Cortical neurons				Cerebellar granule cells			
	K_D (nM)		B_{max} (pmol × mg^{-1})		K_D (nM)		B_{max} (pmol × mg^{-1})	
	High	Low	High	Low	High	Low	High	Low
4	8.7[a]		0.27[a]	0[a]				
8					7.0[b]		0.47[b]	0[b]
8 (+GABA)[d]					7.0[b]	546[b]	0.62[b]	1.04[b]
21	9[c]	250[c]	0.38[c]	2.40[c]	5.6[b]		0.48[b]	0[b]

In studies (b) and (c), ^{3}H-GABA was used as the ligand and in (a) ^{3}H-muscimol.
[a] De Feudis et al. (1979).
[b] Meier et al. (1984).
[c] Ticku et al. (1980).
[d] Cells were grown in the presence of 50 μM GABA.

lower affinity. In addition, the high-affinity GABA receptor has been implicated in the ability of GABA to promote neuronal growth and development (Meier et al., 1985). It has thus been shown both in vivo and in culture that GABA acts as a neurotrophic factor (Wolff et al., 1978, 1979; Spoerri and Wolff, 1981; Madtes and Redburn, 1983b; Hansen et al., 1984; Meier et al., 1985).

6. CONCLUDING REMARKS

The present findings have methodological and theoretical implications. The methodological implication is that both cerebral cortical neurons and cerebellar granule cells are highly deficient in transmitter functions during the first week in culture. Thus, when these cells are used as models of their differentiated counterparts in vivo, the cultures should not be used until they have been kept in culture for a minimum of 9–10 days. Even then AAT activity in cerebellar granule cells has not reached its maximum.

The theoretical implication is that, when distinctly neuronal properties such as transmitter release or GAD activity were studied, there was a good correlation between development in vivo and in culture. This indicates that the cultured cells may be good models of their in vivo counterparts also for developmental studies. Profound differences were, however, found between development in many enzyme activities in cultured neurons and in corresponding regions of the brain. This might be expected since the in vivo results reflect the additional presence of other cells in the brain. This point of view is supported by the development of GLDH and AAT activity in vivo that coincides better with astrocytic (Hertz et al., 1978; Schousboe et al., 1980) than with neuronal development. Therefore, in order to obtain information regarding specifically neuronal development of parameters that are present both in astrocytes and in neurons, the culture system may be of great value. This is also true when neurotrophic factors are studied since the neuronal environment in the culture system can be manipulated with much greater accuracy and to a much larger extent than in vivo.

REFERENCES

Aldinio, C., Balzano, M.A., and Toffano, G. (1980) Ontogenetic development of GABA recognition sites in different brain areas. Pharmacol. Res. Commun. 12:495–500.

Belhage, B., Meier, E., and Schousboe, A. (1986) GABA-agonists induce the formation of low affinity GABA-receptors on cultured cerebellar granule cells via preexisting high affinity GABA-receptors. Neurochem. Res. 11:599–606.

Borg, J., Ramaharobandro, N., Mark, J., and Mandel, P. (1980) Changes in uptake of GABA and taurine during neuronal and glial maturation. J. Neurochem. 34:1113–1122.

Davies, L.P., Johnston, G.A.R., and Stephanson, A.L. (1975) Postnatal changes in the potassium-stimulated, calcium-dependent release of radioactive GABA and glycine from slices of rat central nervous tissue. J. Neurochem. 25:387–392.

De Feudis, F.V., Ossola, L., Schmitt, G., and Mandel, P. (1979) High-affinity binding of [^{3}H]muscimol to subcellular particles of a neuron enriched culture of embryonic rat brain. Neurosci. Lett. 14:195–199.

Drejer, J., Larsson, O.M., Kvamme, E., Svenneby, G., Hertz, L., and Schousboe, A. (1985)

Ontogenetic development of glutamate metabolizing enzymes in cultured cerebellar granule cells and in cerebellum. Neurochem. Res. 10:49–62.

Gallo, V., Ciotti, M.T., Coletti, A., Aloisi, F., and Levi, G. (1982) Selective release of glutamate from cerebellar granule cells differentiating in culture. Proc. Natl. Acad. Sci. USA 79:7919–7923.

Hansen, G.H., Meier, E., and Schousboe, A. (1984) GABA influences the ultra-structure composition of cerebellar granule cells during development in culture. Int. J. Dev. Neurosci. 2:247–257.

Hauser, K., Balcar, V.J., and Bernasconi, R. (1980) Development of GABA neurons in dissociated cell culture of rat cerebral cortex. Brain Res. Bull [Suppl. 2] 5:37–41.

Hertz, L., Bock, E., and Schousboe, A. (1978) GFA content, glutamate uptake and activity of glutamate metabolizing enzymes in differentiating mouse astrocytes in primary cultures. Dev. Neurosci. 1:226–238.

Hertz, L., Juurlink, B.H.J., and Szuchet, S. (1985a) Cell cultures. In: Laijtha, A. (ed.) Handbook of neurochemistry, 2nd edn., vol. 8. New York: Plenum, pp. 603–661.

Hertz, L., Juurlink, B.H.J., Szuchet, S., and Walz, W. (1985b) Cell and tissue cultures. In: Boulton, A.A., and Baker, G.B. (eds.) Neuromethods, Vol. 2 Clifton, NJ: Humana 117–167.

Johnston, G.A.R., and Davies, L.P. (1974) Postnatal changes in the high affinity uptake of glycine and GABA in the rat central nervous system. J. Neurochem. 22:101–105.

Kvamme, E., Schousboe, A., Hertz, L., Torgner, I.Aa., and Svenneby, G. (1985) Developmental change of endogeneous glutamate and gammaglutamyl transferase in cultured cerebral cortical interneurons and cerebellar granule cells and in mouse cerebral cortex and cerebellum in vivo. Neurochem. Res. 10:993–1008.

Larsson, O.M., Drejer, J., Hertz, L., and Schousboe, A. (1983) Ion dependency of uptake and release of GABA and (RS)-nipecotic acid studied in cultured mouse brain cortex neurons. J. Neurosci. Res. 9:291–302.

Larsson, O.M., Drejer, J., Kvamme, E., Svenneby, G., Hertz, L., and Schousboe, A. (1985) Ontogenetic development of glutamate and GABA metabolizing enzymes in cultured cerebral cortex interneurons and in cerebral cortex in vivo. Int. J. Dev. Neurosci. 3:177–185.

Larsson, O.M., Hertz, L., and Schousboe, A. (1986) Uptake of GABA and nipecotic acid in astrocytes and neurons in primary cultures: changes in the sodium coupling ratio during differentiation. J. Neurosci. Res. 16:699–709.

Levi, G., and Ciotti, M.T. (1983) Glutamate and GABA localization and evoked release in cerebellar cells differentiating in culture. In: Hertz, L., Kvamme, E., McGeer, E.G., and Schousboe, A. (eds.) Glutamine, glutamate and GABA in the central neurvous system. New York: Alan R. Liss, pp. 493–508.

Madtes, P.C., and Redburn, D.A. (1983a) GABA as a trophic factor during development. Life Sci. 33:979–984.

Madtes, P.C., and Redburn, D.A. (1983b) Synaptic interactions in the GABA system during postnatal development in retina. Brain Res. Bull. 10:741–745.

Martin, D.L. (1976) Carrier-mediated transport and removal of GABA from synaptic regions. In: Roberts, E., Chase, T.N., and Tower, D.B. (eds.) GABA in nervous system function. New York: Raven, pp. 347–386.

Meier, E., and Schousboe, A. (1982) GABA induces formation of low affinity GABA receptors. Trans. Am. Soc. Neurochem. 13:90.

Meier, E. Drejer, J., and Schousboe, A. (1983) Trophic actions of GABA on the development of physiologically active GABA receptors. In: Mandel P., and De Feudis, F.V. (eds.) CNS-receptors from molecular pharmacology to behavior. New York: Raven, pp. 47–58.

Meier, E., Drejer, J., and Schousboe, A. (1984) GABA induces functionally active low-affinitive GABA receptors on cultured cerebellar granule cells. J. Neurochem. 43:1737–1744.

Meier, E., Hansen, G.H., and Schousboe, A. (1985) The trophic effect of GABA on cerebellar granule cells is mediated by GABA-receptors. Int. J. Dev. Neurosci. 3:401–407.

Redburn, D.A., Broome, D., Ferkany, J., and Enna, S.J. (1978) Development of rat brain uptake and calcium-dependent release of GABA. Brain Res. 152:511–519.

Schousboe, A., Lisy, V., and Hertz, L. (1976) Postnatal alterations in effects of potassium on uptake and release of glutamate and GABA in rat brain cortex slices. J. Neurochem. 26:1023–1027.

Schousboe, A., Nissen, C., Book, E., Sapirstein, V.S., Juurlink, B.H.J., and Hertz, L. (1980)

Biochemical development of rodent astrocytes in primary cultures. In: Giacobini, E., Vernadakis, A., and Shahar, A. (eds.) Tissue culture in neurobiology. New York: Raven, pp. 397–409.

Seiler, N., Sarhan, S., and Roth-Schechter, B.F. (1984) Polyamines and the development of isolated neurons in cell culture. Neurochem. Res. 9:871–886.

Skerritt, J.H., and Johnston, G.A.R. (1982) Postnatal development of GABA binding sites and their endogenous inhibitors in rat brain. Dev. Neurosci. 5:189–197.

Snodgrass, S.R., White, W.F., Biales, B., and Dichter, M. (1980) Biochemical correlates of GABA function in rat cortical neurons in culture. Brain Res. 190:123–138.

Spoerri, P.E., and Wolff, J.R. (1981) Effect of GABA-administration on murine neuroblastoma cells in culture. Cell Tissue Res. 218:567–579.

Ticku, M.K., Huang, A., and Barker, J.L. (1980) Characterization of γ-aminobutyric acid receptor binding in cultured brain cells. Mol. Pharmacol. 17:285–289.

Wolff, J.R., Joo, F., and Dames, W. (1978) Plasticity in dendrites shown by continuous GABA administration in superior cervical ganglion of adult rat. Nature (Lond.) 274:72–74.

Wolff, J.R., Rickmann, M., and Chronwall, B.M. (1979) Axo-glial synapses and GABA-accumulating glial cells in the embryonic neocortex of the rat. Cell Tissue Res. 201:239–248.

Wong, P.T.-H., and McGeer, E.G. (1978) Postnatal changes of GABAergic and glutamatergic parameters. Dev. Brain Res. 1:519–529.

Yu, A.C.H., Hertz, E., and Hertz, L. (1984) Alterations in uptake and release rates for GABA, glutamate and glutamine during biochemical maturation of highly purified cultures of cerebral cortical neurons, a GABAergic preparation. J. Neurochem. 42:951–960.

4. DEVELOPMENT OF CEREBRAL CORTICAL GABAERGIC NEURONS IN VITRO

KINYA KURIYAMA AND SEITARO OHKUMA

1. INTRODUCTION

The central nervous system (CNS) is composed of several types of neural cells such as neurons, and glial and endothelial cells. This cellular heterogeneity is believed to be essential for the development of the brain and the maintenance of CNS functions. However, the use of cerebral mass, a mixture of these different types of cells, often causes difficulties for analyzing the direct effects of various centrally acting drugs and hormones on each type of cell. It is especially noteworthy that the proportion of neurons in the brain is only 40% (Pope, 1978), and more than 60% of protein found in brain homogenates is attributed to glial cells (Hauser et al., 1980). Therefore, it is reasonable to consider that the use of a pure population of neurons is more suitable than that of cerebral tissues for analyzing the effect of centrally acting drugs on neurons as well as on the developmental patterns of various neural cells. Recently, dissociation procedures to obtain homogeneous populations of dispersed neurons (Varon and Raiborn, 1969; Sensenbrenner et al., 1975) and glial cells (Booher and Sensenbrenner, 1972; Moonen, et al., 1975) have been developed, as well as procedures to maintain these cells for a long period of time in the presence of some growth factors and/or serum. In the present study, we have examined the developmental pattern of γ-aminobutyric acid (GABA)ergic neurons using a primary culture procedure, especially those of the GABA synthesizing system and GABA receptor.

2. PRIMARY CULTURE OF GABAERGIC NEURONS

It has been reported that the majority of neurons in the cerebral cortex are GABAergic (Ribak, 1978). Therefore, we have attempted to obtain dispersed brain cells from the neopallium of mouse fetus and to maintain these cells in primary culture.

Neurons subjected to primary culture were prepared from the neopallium of 15-day-old fetuses of STD:ddy strain mouse according to the methods of Varon and Raiborn (1969), Yavin and Menkes (1973), Dichter (1978), and Yu et al. (1984b) with a minor modification. After the removal of meninges, the neopallium was minced with scissors and digested by 0.1% trypsin (Sigma Chemical Company, USA) dissolved in Ca^{2+}-free Puck's solution at 37°C for 5 min. After the trypsin treatment, the reaction was terminated with 20% horse serum (Flow Laboratories, Australia) in modified Eagle's minimal essential medium [MEM: including twofold concentrations of amino acids, fourfold concentrations of vitamins, 30 m*M* glucose, 10 m*M* Hepes, 10 m*M* TES, and carbenicillin (0.1 mg/ml) and streptomycin (0.1 mg/ml)] and centrifuged at 900 *g* for 2 minutes following tritration by a Pasteur pipette. The pellet obtained was then resuspended with MEM and inoculated on a Corning plastic dish (60 mm in diameter) precoated with poly-L-lysine for 24 hours (Sensenbrenner et al., 1978). After incubation for 15 minutes, the culture medium was changed for MEM supplemented with 15% fetal calf serum (Flow Laboratories, Australia, and Granite Diagnostics, USA) and cells remaining on the culture dish were continuously cultured at 37°C in humidified 95% air–5% CO_2 for 3 days. The cells were then exposed to 20 μM cytosine arabinoside dissolved in MEM containing 15% horse serum. After the exposure of cells to cytosine arabinoside for 24 hours, the neurons were cultured in 15% horse serum in MEM under the same conditions as described above and the medium was changed every 4 days.

The morphological developmental changes of GABAergic neurons in primary culture were examined with phase contrast and scanning electron microscopy. Neurons grown on plastic dishes for 1, 3, 7, and 14 days were directly examined by a phase contrast microscope (Nikon Diaphoto-TMD). For scanning electron microscopy, the specimens of neurons cultured for 3 and 14 days were prepared as follows: the neuronal cells attached to culture dishes were fixed with 1% glutaraldehyde at room temperature for 30 minutes followed by three washings with Ca^{2+}- and Mg^{2+}-free Dulbecco's phosphate-buffered saline (PBS). After the rinse, the cells were again fixed with 1% OsO_4 at room temperature for 30 minutes and then washed with Ca^{2+}- and Mg^{2+}-free PBS. Following dehydration in ethanol and drying with a Hitachi HPC-2 critical-point dryer, the dried cells were coated with platinum–palladium and subjected to scanning electron microscopy.

By 24 hours after the seeding, the cells remaining in culture dishes were well dispersed. Many cells had newly formed neuronal processes (Figure

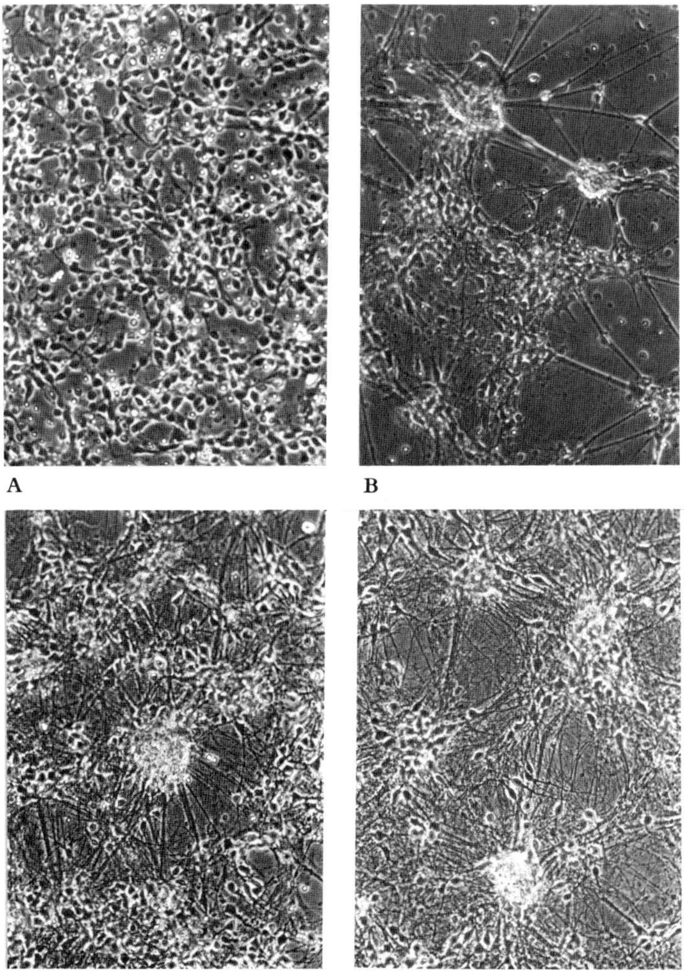

Figure 4-1. Phase contrast micrographs of primary cultured neurons. Neurons were dissociated by a trypsin treatment and were inoculated on poly-L-lysine-coated surface: (**A**) 1 day, (**B**) 3 days, (**C**) 7 days, and (**D**) 14 days after inoculation, respectively. Scale in **D** = 80 μm.

4-1A), and initiation of the aggregation of neurons was observed. On day 3 following the inoculation of cells, the cells were remarkably aggregated, the neuronal processes were multiplied and enlarged and ultimately formed fine networks that extended throughout the dish (figure 4-1B). On day 7 after seeding, the processes extruding from the aggregates formed, not only a more complex network, but also bundles connecting with adjacent aggregated neurons (figure 4-1C). Some bipolar or multipolar cells were also

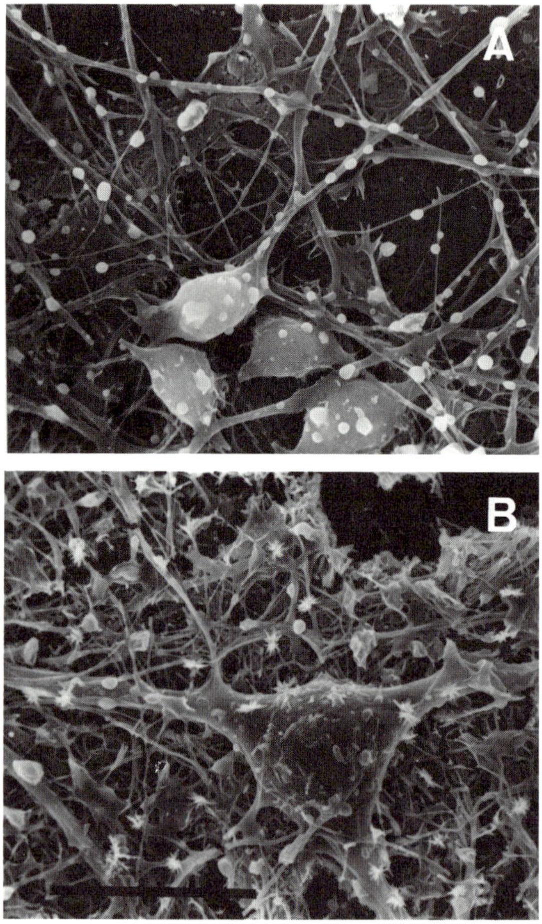

Figure 4-2. Scanning electron micrographs of primary cultured neurons: (**A**) 3-day-old culture, and (**B**) 14-day-old culture. Scale in **B** = 10 μm.

found to be dispersed around aggregated cells. Furthermore, several types of cells different from neurons, which consisted of polygonal and flattened epithelial and glial cells, were found in these cultures. These cells were few, however, and practically devoid of any reactivities to anti-GAD (glutamic acid decarboxylase) and anti-GFAP (glial fibrillary acidic protein) antibodies. Similar morphological features of cultured cells were found at days 14 and 21 after the initiation of the primary culture (figure 4-D).

In scanning electron microscopy, it was also found that the neurons cultured for 3 days had long neuronal processes (figure 4-2A). At day 14 after

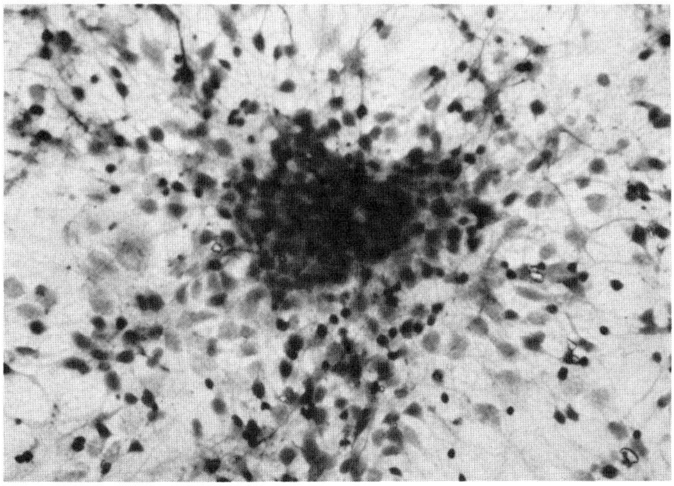

Figure 4-3. Immunohistochemical features of cells at 14 days after inoculation on poly-L-lysine-coated surface. Astroglias reactive to anti-GFAP antibody are *brown*. The cells were poststained by hemotoxylin.

the initation of primary culture, the size of the cell body was enlarged and numerous neuronal processes extruded from other neurons, forming a complex network around neurons (figure 4-2B). In addition, immunohistochemical investigations using an antibody to an astroglial marker, GFAP (Eng et al., 1971), revealed that the majority of cells cultured for 14 days showed no immunoreactivity to anti-GFAP antibody (figure 4-3). These results indicate that the morphological features of cultured cells used in this study were found to be similar to those reported previously (Yavin and Menkes, 1973; Sensenbrenner et al., 1978; Yu and Hertz, 1982). Immunohistochemical studies using anti-GFAP antibody also revealed that the contamination of astroglia in these cell populations was negligible.

In addition, it was also found in preliminary experiments that the release of [^{3}H]GABA from these primary cultured cells preloaded in the presence of 10^{-4} *M* aminooxyacetic acid was stimulated by the addition of 30 m*M* KCl; this concentration of KCl is known to induce the release of various neurotransmitters from neurons, but from glial cells. These morphological and neurochemical data combined suggest that the majority of cultured cells used in this study are indeed neuronal cells.

3. IMMUNOHISTOCHEMICAL OBSERVATIONS ON GABAERGIC NEURONS IN PRIMARY CULTURE

The neurons were immunohistochemically stained by anti-GAD and anti-GFAP antibodies, respectively. Neurons attached to culture dishes were rinsed with ice-cold Krebs-Ringer bicarbonate buffer (KRB) and fixed initially with a mixture of 4% paraformaldehyde, 0.3% glutaraldehyde, and 0.2% picric acid in 0.1 *M* PBS (pH 7.4) at 4°C for 10 minutes. After the fixation of neurons, the cells were washed three times with 0.1 *M* PBS containing 4% paraformaldehyde and 0.2% picric acid, placed on ice for 30 minutes, and washed once with 15% sucrose in 0.1 *M* PBS. After a few hours, the cells were washed with PBS containing 0.3% Triton X-100 and stored in the same medium at 4°C for 4 days. The treated neurons were finally subjected to immunohistochemical studies using the peroxide–antiperoxide method (Kimura et al., 1981). Namely, they were incubated with the rabbit IgG against human GFAP (Dako Corporation, Denmark; 1/5000) or with the rabbit IgG against GAD (a gift from Dr. J.-Y. Wu, Department of Physiology, Pennsylvania State University; 1/400) at 4°C for 4 days, and incubated successively with the goat antirabbit IgG (Miles-Yeda, Israel; 1/200) for 2 hours, followed by the incubation with PAP (rabbit, Dako, Denmark; 1/200) for 1 hour, finally it was reacted with 0.02% 3,3′-diaminobenzidine and 0.005% H_2O_2 dissolved in 50 m*M* Tris–HCl buffer (ph 7.4) for 10 min. After these immunohistochemical treatments, the cells were stained with hematoxylin, dehydrated in ethanol and finally examined microscopically.

The immunoreactivity to anti-GAD antibody was localized clearly in cell bodies and neuronal processes of neurons cultured for 3–21 days (figure 4-4A–C). The intensity of immunoreactivity on the neuronal processes in 3-day-old cultured cells was, however, weaker compared to that in neurons 21 days after seeding. In adition, the immunohistochemical investigations revealed that the majority of cells used were reactive to anti-GAD antibody, whereas less than 30% of cells were immunoreactive to anti-GFAP antibody. On the other hand, several cells, which exhibited no reactivity to either anti-GFAP or anti-GAD antibodies, had enlarged polygonal and flattened cell bodies with large nuclei as compared with other cells, suggesting that these cells may be endothelial cells, although they were not identified by morphological and biochemical methods.

4. DEVELOPMENTAL PATTERNS OF GABA CONTENT AND GAD ACTIVITY IN PRIMARY CULTURED NEURONS

The content of several neuroactive amino acids such as γ-aminobutyric acid (GABA) and glutamic acid in neurons in primary culture as well as in the cerebral cortices obtained from age-matched fetal and neonatal mice was determined by the method of Ida and Kuriyama (1983). Following the washing of neurons three times with ice-cold KRB (pH 7.4), amino acids were extracted by 5% trichloracetic acid and the extract was directly applied to a HPLC (high-performance liquid chromatography) system equipped with a cation-exchange column. The effluents used were of three types: 0.067 *N* sodium citrate (pH 3.25 and 4.25) and 0.2 *N* sodium citrate (pH 9.00); the flow rate was 0.5 ml/minute. The effluents from the column equipped with

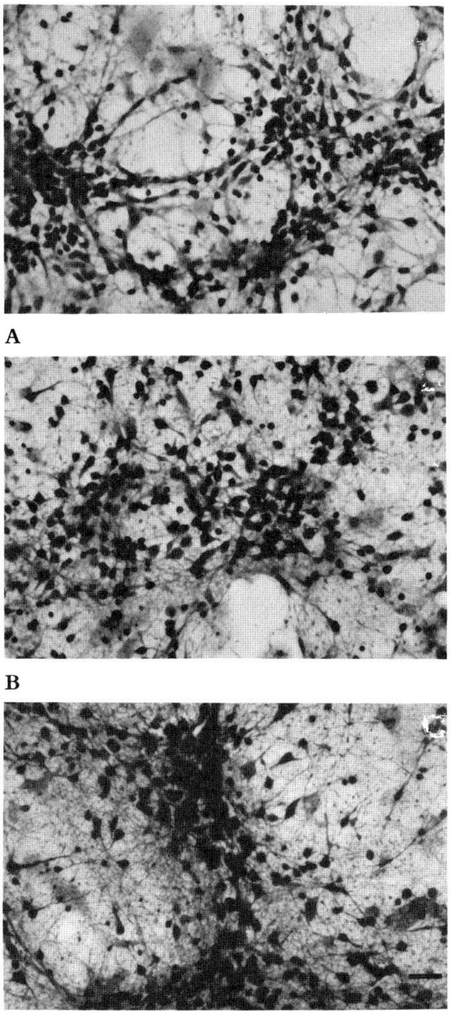

A B C

Figure 4-4. Immunohistochemical stain using anti-GAD antibody. The cells were stained by hematoxylin following the immunohistochemical staining with anti-GAD antibody. Neurons used are: (**A**) 3 days, (**B**) 7 days, and (**C**) 21 days after the inoculation, respectively. Scale in **C** = 50 μm.

HPLC were mixed with *o*-phthalaldehyde, and the fluorescence of each effluent was measured spectrofluorometrically at excitation and emission wavelengths of 348 and 450 nm, respectively.

The activities of GAD (EC 4.1.1.15) in primary cultured neurons and in the cerebral cortices in vivo were determined according to a method de-

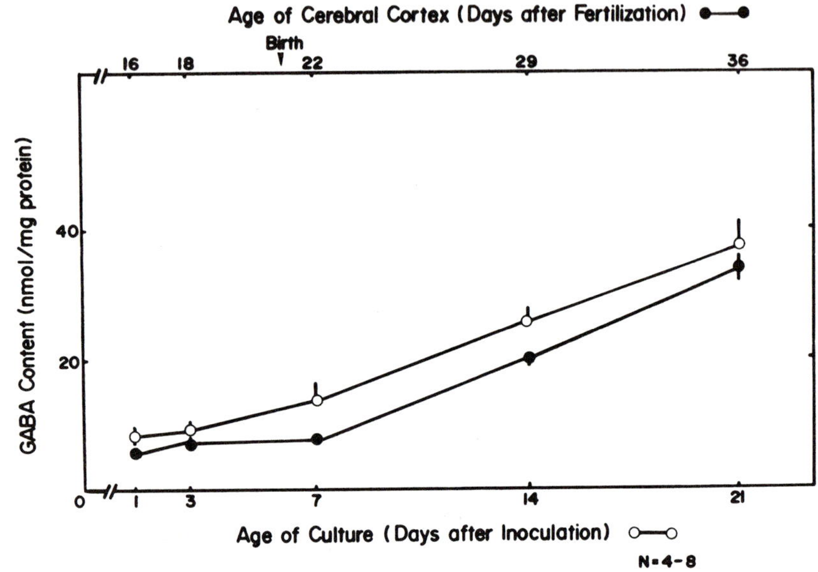

Figure 4-5. Developmental pattern of GABA content in primary cultured neurons and in cerebral cortices.

scribed previously (Kimura and Kuriyama, 1975). The cells, which were washed three times with ice-cold KRB and scraped from cultured dishes, were homogenized in nitrogen-gassed distilled water containing 0.25% (vol/vol) Triton X-100 at 2°C, and an aliquot of the homogenate was examined for GAD activity by measuring the [^{14}C]CO_2 evolved from L-[1-^{14}C]glutamic acid (New England Nuclear, USA; 47.3 mCi/mmol).

GABA content in neurons in vitro gradually increased with the cell growth, although the level of GABA in these neurons during the first week after inoculation remained very low (figure 4-5). The developmental pattern of GABA content in primary cultured neurons was similar to that found in the cerebral cortices obtained from age-matched fetuses or neonates. Similarly, GAD activity in primary cultured neurons exhibited a progressive increase similar to that found in the brains of the fetus and the neonate, although the former increase was found to be retarded at days 29 and 36 after fertilization as compared with that in the latter case (figure 4-6). In contrast to the increase in GABA content with the development of neurons in vitro, the content of glutamic acid, a precursor of GABA, showed a remarkable decrease in these cells. Glutamic acid content in the cerebral cortices obtained from fetuses or neonates, however, was elevated during brain development (figure 4-7).

These developmental patterns of GABA content and GAD activity in neurons in primary culture agreed with the data previously reported (Hauser

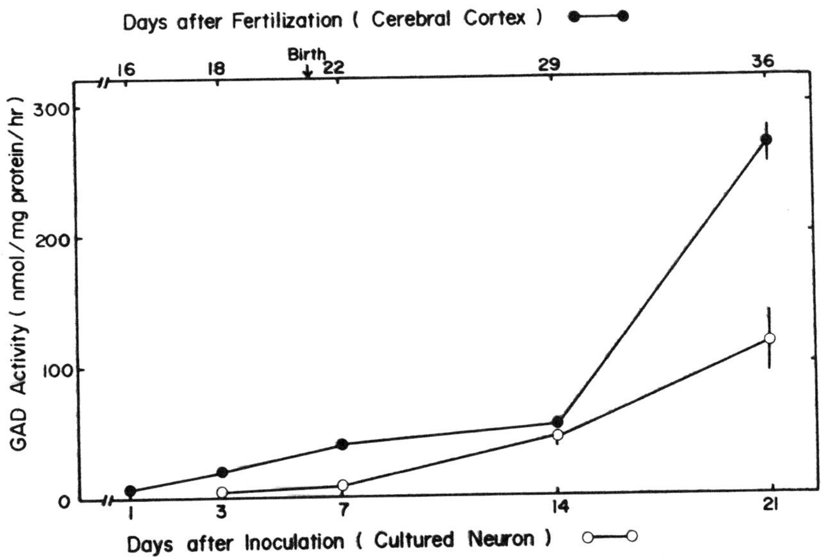

Figure 4-6. Developmental pattern of GAD activity in primary cultured neurons and in cerebral cortices.

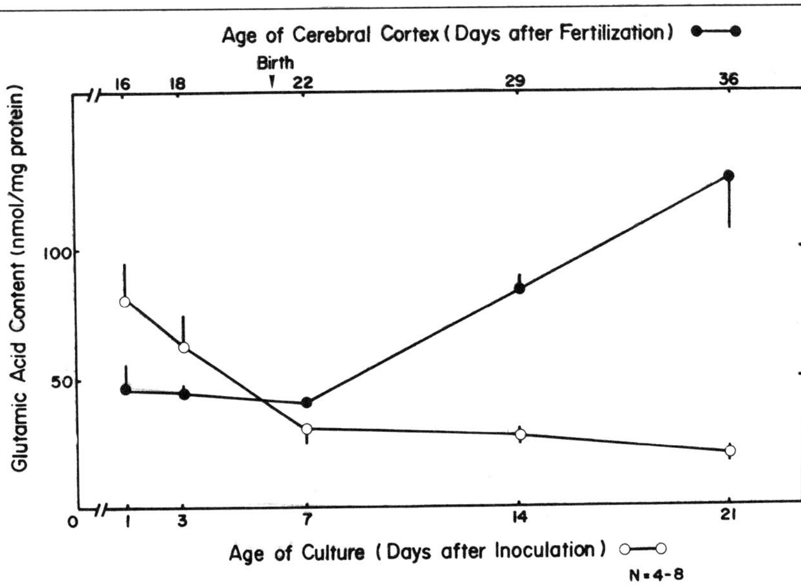

Figure 4-7. Developmental pattern of glutamic acid content in primary cultured neurons and in cerebral cortices.

et al., 1980; Snodgrass et al., 1980; Yu et al., 1984a) and also resembled those found in the cerebral cortices obtained from age-matched fetuses and neonates. The rate of GAD increase in the cerebral cortex of neonatal brains, however, was faster than that found in primary cultured neurons, especially at the later stage of development. This discrepancy between neurons in vitro and the cerebral cortices may be due, at least in part, to the involvement of growth of glial cells in the cerebral cortex in vivo, since rapid glial maturation occurs by the end of third postnatal week in rodent brain (Ling and Leblond, 1973).

In contrast to gradual increases in GABA content and GAD activity during neuronal development in vitro, the content of glutamic acid, a major precursor of GABA, showed a gradual reduction in primary cultured neurons during the development of neurons. A similar pattern with glutamic acid level in neurons in vitro was also reproted by Hertz and his colleagues (Yu et al., 1984a). On the other hand, Snodgrass et al. (1980) reported that glutamic acid content in primary cultured neurons increased with the time in culture, resembling the developmental changes of glutamic acid content in the cerebral cortex in vivo found in this study (figure 4-7). The reason for such a discrepancy is unclear at present. However, highly purified neuronal populations without astroglial contamination may metabolize glutamic acid to GABA to a greater extent than primary cultured cells containing large amounts of astroglia, due to the lack of recycling system for glutamic acid such as the glutamine–glutamate shunt found in astroglia.

5. DEVELOPMENTAL PATTERNS OF SPECIFIC BINDINGS OF [^{3}H]MUSCIMOL AND [^{3}H]DIAZEPAM TO PARTICULATE FRACTION FROM PRIMARY CULTURED NEURONS

To investigate the developmental changes of GABA and benzodiazepine (BZP) receptors in neurons in primary culture, [^{3}H]muscimol (New England Nuclear, USA; 11.5 Ci/mmol), a $GABA_A$ agonist, and [^{3}H]diazepam ([^{3}H]DZP; New England Nuclear, USA; 76.4 Ci/mmol) were used, respectively. The specific bindings of [^{3}H]muscimol and [^{3}H]DZP to particulate fractions obtained from cultured neurons were calculated by subtracting nonspecific bindings of 40 n*M* [^{3}H]muscimol or 1 n*M* [^{3}H]DZP found in the presence of 1 *M* GABA or 10 μ*M* DZP from the total bindings found in the absence of 1 *M* GABA or 10 μ*M* DZP, respectively.

The specific binding of [^{3}H]muscimol, a $GABA_A$ receptor agonist, to the particulate fraction obtained from primary cultured neurons increased in conjunction with the development of these cells. A similar developmental pattern was also observed in the particulate fraction prepared from the cerebral cortices of age-matched fetuses and neonates (figure 4-8). Furthermore, the increase in [^{3}H]DZP binding to the particulate fraction from primary cultured neurons was also observed during days 3–10 after inoculation, but it reached a plateau at 10 days in culture. In contrast, [^{3}H]DZP binding to

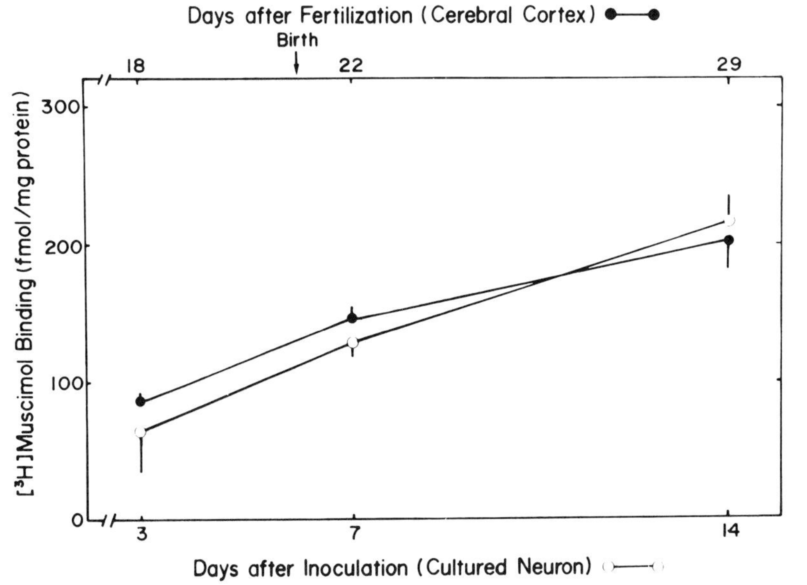

Figure 4-8. Developmental pattern of [^{3}H]muscimol binding to particulate fractions obtained from primary cultured neurons and cerebral cortices.

the particulate fraction obtained from the cerebral cortices from neonate exhibited further development until 21 days after fertilization (figure 4-9). Since it has been well established that GABA and BZP receptors are functionally coupled to each other (Olsen, 1981), we examined the effect of exogenously added GABA on [^{3}H]DZP binding. As shown in figure 4-10, the activation of [^{3}H]DZP binding by 10^{-4}–10^{-5} M GABA was indeed detected in the particulate fraction obtained from primary cultured neurons for 14 days.

Several groups of investigators have reproted that the iontophoretic application of GABA as well as GABA agonist to the dissociated neurons from fetal brains of rodents evokes IPSP (Godfrey et al., 1975; White et al., 1981), suggesting that neurons in primary culture possess postsynaptic GABA receptor. In order to investigate the developmental pattern of GABA receptor–benzodiazepine receptor complex in primary cultured neurons, the binding capacities for [^{3}H]muscimol and [^{3}H]DZP were examined using particulate fractions prepared from cultured neurons at various developmental stages. Both [^{3}H]muscimol and [^{3}H]DZP bindings in neurons in vitro showed progressive increases as found in developing brain in vivo. The [^{3}H]muscimol binding to the particulate fraction obtained from 3-day-old neurons in culture was similar to that reported previously by De Feudis et al. (1980). It has been well established that binding of BZPs to cerebral BZP receptor is enhanced by GABA as well as $GABA_A$ agonists. It is important to note that the enhancement of [^{3}H]DZP binding by GABA and $GABA_A$ agonists indeed observed in the particulate fractions was obtained from primary cul-

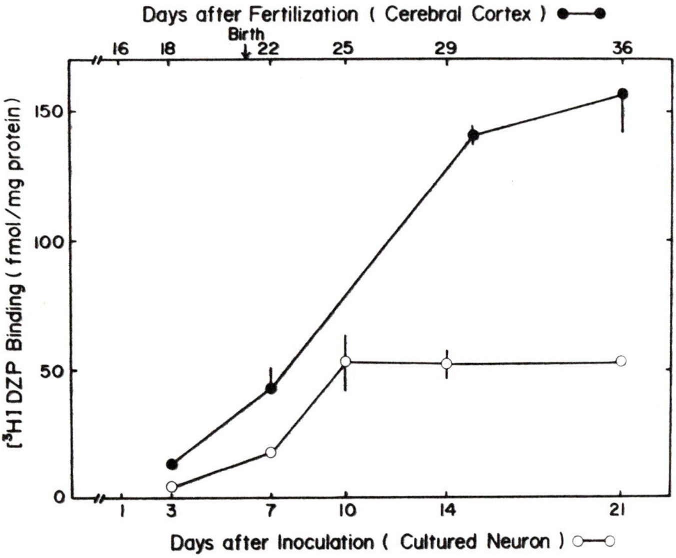

Figure 4-9. Developmental pattern of [^{3}H]diazepam binding to particulate fractions obtained from primary cultured neurons and cerebral cortices.

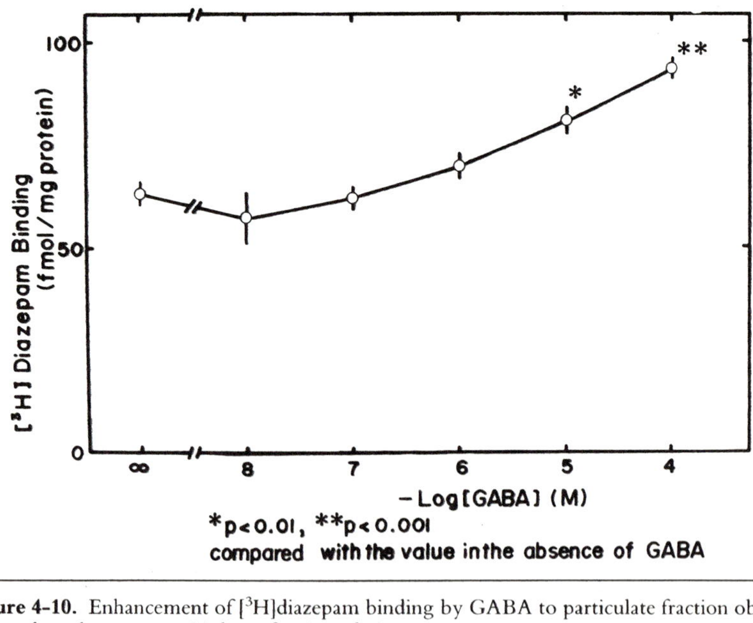

Figure 4-10. Enhancement of [^{3}H]diazepam binding by GABA to particulate fraction obtained from cultured neurons at 14 days after inoculation.

tured neurons used in this study. These results strongly suggest that the functional coupling between $GABA_A$ receptor and BZP receptor may operate in neurons in primary culture from a relatively early stage of development. Similar results have been reported using the assay BZP receptor binding as well as the electrophysiological recording of potential changes in these cultured cells (White et al., 1981).

6. CONCLUSION

Developmental patterns of γ-aminobutyric acid (GABA) synthesizing system and of GABA receptor were investigated using primary cultured neurons obtained from the neopallium of 15-day-old fetuses of mice by trypsin treatment in comparison with those in cerebral cortices prepared from age-matched fetuses and neonates. The content of GABA in these neurons in vitro showed a gradual increase with the time of culture, and the pattern of these increases was essentially similar to that found in vivo. Similarly, L-glutamic acid decarboxylase (GAD) activities in both neurons in primary culture and cerebral cortices in vivo showed a gradual elevation in conjunction with neuronal maturation, and it was observed in primary cultured neurons that the immunoreactivity to anti-GAD antibody was mainly localized in cell bodies and cellular processes.

The specific bindings of $[^3H]$muscimol and $[^3H]$diazepam to the particulate fraction obtained from primary cultured neurons also showed a progressive increase during the development of neurons in vitro. In addition, exogenously added GABA induced a significant enhancement of $[^3H]$diazepam binding to the particulate fraction obtained from primary cultured neurons. The present results indicate that primary cultured neurons used in this study possess not ony a GABA-synthesizing system but also GABA receptor coupled with benzodiazepine receptor. The results also suggest that GABAergic neurons in primary culture may be a useful experimental tool for elucidating the effects of various centrally acting drugs on the development of cerebral GABAergic neurons.

ACKNOWLEDGMENT

This work was supported in part by grants-in-aid (nos. 59480120, 60126005, and 60770175, 1985) from the Ministry of Education and Culture, Japan. The authors wish to express our thanks to Dr. J.-Y., Wu (Pennsylvania State University, Hershey Medical Center, USA) for kindly providing the anti-GAD antibody used in this study. The collaboration of S. Tomono, M. Hirouchi, T. Oishi, and T. Mukainaka in this department is also gratefully acknowledged.

REFERENCES

Booher, J., and Sensenbrenner, M. (1972) Growth and cultivation of dissociated neurons and glial cells from embryonic chick, rat and human brain in fransk cultures. Neurobiology 2:97–105.

De Feudis, F.V., Ossola, L., Schmitt, G., Wolff, P., and Mandel, P. (1980) Na^+-independent binding of [^{3}H]GABA and [^{3}H]muscimol to subcellular particles of neural primary cultures and whole brain. J. Neurochem. 34:216–218.

Dichter, M.A. (1978) Rat cortical neurons in cell culture: culture methods, cell morphology, electrophysiology, and synapse formation. Brain Res. 149:279–293.

Eng, L.F., Vanderhaeghen, J.J., Bignami, A., and Gerstl, B. (1971) An acidic protein isolated from fibrous astrocytes. Brain Res. 28:351–354.

Godfrey, E.W., Nelson, P.G., Schrier, B.K., Breuer, A.C., and Ransom, B.R. (1975) Neurons from fetal rat brain in a new cell culture system: a multidisciplinary analysis. Brain Res. 90:1–21.

Hauser, K., Balcar, V.J., and Bernasconi, R. (1980) Development of GABA neurons in dissociated cell culture of rat cerebral cortex. Brain Res. Bull. [Suppl. 2] 5:37–41.

Ida, S., and Kuriyama, K. (1983) Simultaneous determination of cysteine sulfinic acid and cysteic acid in rat brain by high-performance liquid chromatography. Anal. Biochem. 130:95–101.

Kimura, H., and Kuriyama, K. (1975) A new microassay method for L-glutamic acid decarboxylase (GAD) activity, Jpn. J. Pharmacol. 25:189–195.

Kimura, H., McGeer, P.L., Peng, J.H., and McGeer, E.G. (1981) The central cholinergic system studied by choline acetyltransferase immunohistochemistry in the cat. J. Comp. Neurol. 200:151–201.

Ling, E.A., and Leblond, C.P. (1973) Investigation of glial cells in semithin sections. II. Variation with age in the numbers of the various glial cell types in rat cortex and corpus callosum. J. Comp. Neurol. 149:73–82.

Moonen, G., Cam, Y., Sensenbrenner, M., and Mandel, P. (1975) Variability of the effects of serum-free medium, dibutyryl cyclic AMP or theophylline on the morphology of cultured new-born rat astroblasts. Cell Tissue Res. 163:365–372.

Olsen, R.W. (1981) GABA–benzodiazepine–barbiturate receptor interactions. J. Neurochem. 37:1–13.

Pope, A. (1978) Neuroglia: quantitative aspects. In: Schoffeniels, E., Frank, G., Hertz, L., and Tower, D.B. (eds.) Dynmic properties of glial cells. Oxford. Pergamon, pp. 13–20.

Ribak, C. (1978) Aspinous and sparasely-spinous stellate neurons in the visual cortex of rats contain glutamic acid decarboxylase. J. Neurocytol. 7:461–478.

Sensenbrenner, M. (1977) Dissociated brain cells in primary cultures. In: Fedoroff, S., and Hertz, L. (eds.) Cell, tissue and organ cultures in neurobiology. New York: Academic Press, pp. 191–213.

Sensenbrenner, M., Jaros, G.G., Moonen, G., and Mandel, P. (1975) Effects of synthetic tripeptide on the differentiation of dissociated cerebral hemisphere nerve cells in culture. Neurobiology 5:207–213.

Sensenbrenner, M., Maderspach, K., Latzkovits, L., and Jaros, G.G. (1978) Neuronal cells from chick embryo cerebral hemispheres cultivated on polylysine-coated surfaces. Dev. Neurosci. 1:90–101.

Snodgrass, S.R., White, W.F., Biales, B., and Dichter, M. (1980) Biochemical correlates of GABA function in rat cortical neurons in culture. Brain Res. 190:123–138.

Varon, S., and Raiborn, C.W. Jr. (1969) Dissociation, fractionation, and culture of embryonic brain cells. Brain Res. 12:180–199.

White, W.F., Dichter, M.A., and Snodgrass, S.R. (1981) Benzodiazepine binding and interactions with the GABA receptor complex in living cultures of rat cerebral cortex. Brain Res. 215:162–176.

Yavin, E., and Menkes, J.H. (1973) The culture of dissociated cells from rat cerebral cortex. J. Cell Biol. 57:232–237.

Yu, A.C.H., and Hertz, L. (1982) Uptake of glutamate, GABA and glutamine into a predominantly GABAergic and a predominantly glutaminergic nerve cell population in culture. J. Neurosci. Res. 7:23–35.

Yu, A.C.H., Hertz, E., and Hertz, L. (1984a) Alterations in uptake and release rates for GABA, glutamate, and glutamine during biochemical maturation of highly purified cultures of cerebral cortical neurons, a GABAergic preparation. J. Neurochem. 42:951–960.

Yu, A.C.H., Fisher, T.E., Hertz, E., Tildon, J.T., Schousboe, A., and Hertz, L. (1984b) Metabolic fate of [^{14}C]glutamine in mouse cerebral neurons in primary cultures. J. Neurosci. Res. 11:351–357.

5. MORPHOLOGICAL ANALYSIS OF 5-HT NEURONS TRANSPLANTED INTO THE OLFACTORY BULB OF THE ADULT RAT

ALAIN PRIVAT, HUSSEIN MANSOUR AND MICHEL GEFFARD

1. INTRODUCTION

Transplantation of monoaminergic neurons has been extensively developed during the last decade, owing to the availability of specific techniques for the study of grafted tissues (see review in Gash, 1983; Björklund and Stenevi, 1985). Many of these studies concentrated on the dopaminergic system, due to the underlying pathology of Parkinson's disease (Stenevi et al., 1976; Björklund et al., 1976; Björklund and Stenevi, 1979; Dunnett et al., 1985) and the possible application to human subjects (Backlund et al., 1985).

Less attention has been paid to the serotonergic systems, probably because of the lack of reliable techniques for the detection of 5-HT. Recently, the availability of specific anti-5-HT (serotonin) antibodies (Steinbusch, 1978; Lauder et al., 1982) has filled the gap. In their pilot study, Azmitia et al. (1981) demonstrated the ability of fetal 5-HT neurons to survive and develop after transplantation into the hippocampus of adult recipients, and we (McRae-Degueurce et al., 1983 and 1984) have shown that fragments of raphe nuclei injected into the IVth and lateral ventricles integrated into the host's parenchyma and released 5-HT. Both of these studies used solid transplants, the principal drawbacks of which are either extensive lesions when placed directly into the parenchyma (Stenevi et al., 1976) or limited outgrowth into the host tissue when placed into the ventricles (McRae-Degueurce et al., 1983).

Recently, cell suspensions of dopaminergic and noradrenergic neurons have been successfully transplanted intracerebrally (Björklund et al., 1983), and we have shown (Privat, 1982) that serotonergic neurons from fetal rats can grow and differentiate in vitro after appropriate dissociation.

2. MODEL SYSTEM

2.1 Choice of the model

The aim of this chapter is to present the cytological integration of dissociated fetal serotonergic neurons in the adult host. For that purpose, the olfactory bulb was chosen as the site of transplantation, for two reasons: first, the 5-HT innervation of the olfactory bulb is precisely restricted to the glomerular layer, thus allowing a good evaluation of the specificity of the projections; second, one aspect of the influence of the serotonergic innervation upon the excitability of bulbar neurons has recently been investigated (Lerner-Natoli et al., 1986), thus potentially allowing control of a functional parameter of the grafted neurons.

2.2 Experimental conditions

Antibodies against 5-HT are prepared according to the method described by Geffard et al. (1985). They can be used at a dilution of 20,000–30,000. Transplants are prepared as follows: 14 to 15-day-old Sprague-Dawley rat fetuses, obtained by laparotomy, are transferred into Hank's buffered solution, enriched with 0.5% glucose, and the brain is dissected in order to isolate the region of the brain stem between the pontine and mesencephalic flexures. The raphe region is then isolated with parasagittal sections (Privat, 1982). Tissue blocks including the raphe nuclei are then mechanically dissociated by gentle pipetting in Ca^{2+}- and Mg^{2+}-free Puck saline, centrifuged at 70 *g* and resuspended in the same vehicle in order to reach a concentration of 25,000–30,000 cells/μl. Then, adult Sprague-Dawley rats are anesthetized with Equithesin, and stereotaxically receive an injection into the medullary layer of the right olfactory bulb of 1 μl of the above suspension, with a 10 μl Hamilton syringe. They are then left to survive up to six months. At 1½ hours prior to sacrifice, the rats are pretreated with pargylin (100 mg/kg) and tryptophan (100 mg/kg). They are perfused with 5% glutaradehyde and 1% sodium metabisulfite in 0.05 *M* Cacodylate buffer. The brain is then dissected and immersed for 1 hour in the same perfusate. Vibratome sections (50 μm thick) are then collected, and reacted using PAP method (Privat et al., 1986). After reaction with diaminobenzidine, some of the sections can be mounted on slides for light microscopy; others are treated with osmium tetroxyde and flat embedded in araldite for electron microscopy. After reembedding in gelatin capsules, 1-μm sections are stained with toluidine blue for light microscopy, and ultrathin sections contrasted with uranyl and lead for electron microscopy.

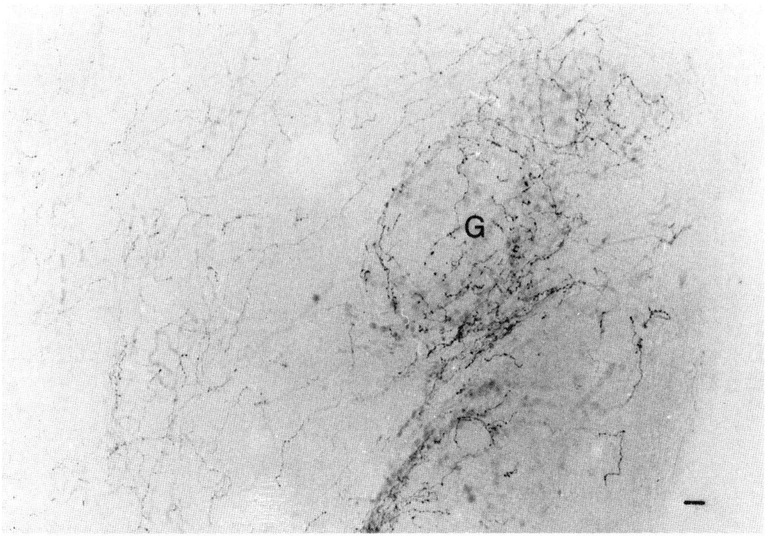

Figure 5-1. Control rat. Immunocytochemical detection of 5-HT on a 50-μm vibratome section. In the olfactory bulb, glomeruli (*G*) are the specific target of 5-HT axons. Bar = 20 μm.

3. DISTRIBUTION OF SEROTONERGIC INNERVATION IN THE OLFACTORY BULB

The distribution of serotonergic axons in the olfactory bulb follows a distinct pattern. The fibers run in the medullary layer, ascend through the granular, mitral, and external plexiform layers, and they end up in a bushy fashion in the glomerular layer (figure 5-1). On a 1-μm section, the fibers appear to end up at the periphery of the glomeruli, without invading their center (figure 5-2). A preliminary ultrastructural study shows that immunoreactive boutons contact dendrites, either large or small (figure 5-3 and 5-4), and also, in some cases, axons at the periphery of the glomeruli. They most often form typical synapses but, in some cases, varicosities filled with vesicles are seen without corresponding postsynaptic densities in the neighboring profiles (figure 5-5). These fibers contact neither the perikaraya of mitral, tufted and horizontal cells, nor on the periglomerular cells (figure 5-6).

4. ORGANIZATION OF THE TRANSPLANTS

In all of the grafted olfactory bulbs, 5-HT immunoreactive perikarya are found after various survival intervals. The quantity of immunoreactive perikarya detected on 50-μm sections ranges from a dozen to more than a hundred per bulb. Taking into account the fact that only the superficial 10μm on each surface of the section is penetrated by the antibodies, the actual

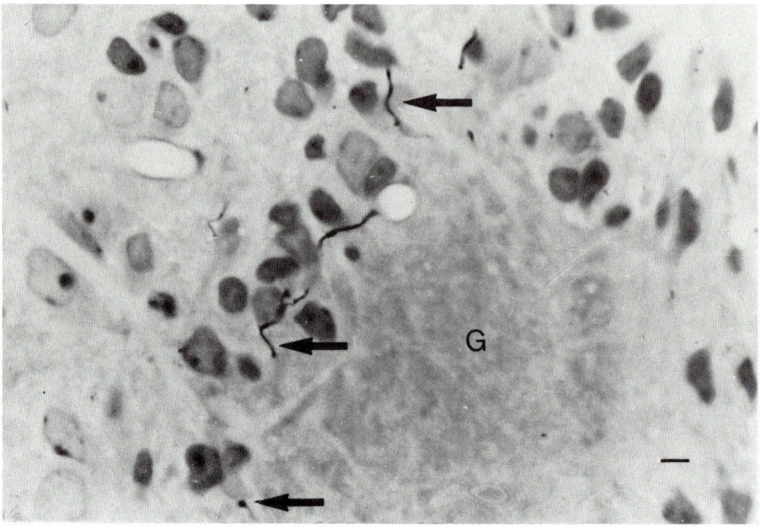

Figure 5-2. Control rat. Immunochemical detection of 5-HT on a 1-μm section. The immunoreactive fibers (*arrows*) are in the periphery of the glomerulus (*G*). Bar = 10 μm.

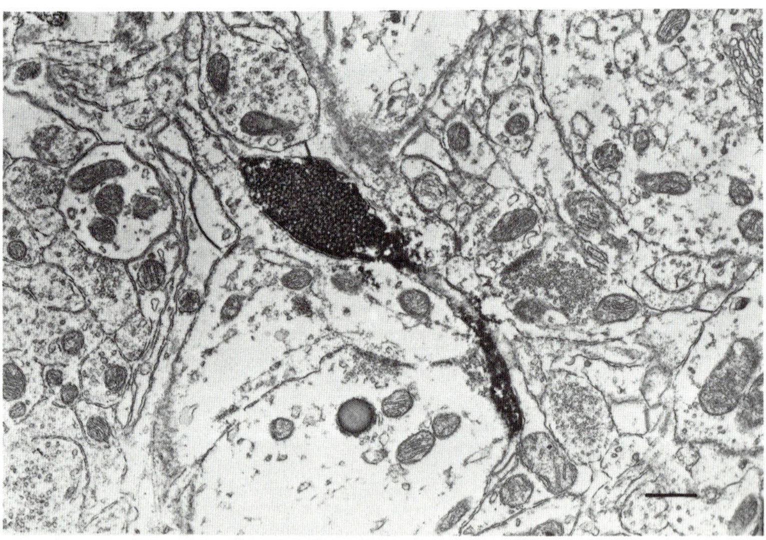

Figure 5-3. Control rat. Immunoelectron- microscopic detection of 5-HT axons. A varicosity filled with 40-nm vesicles makes a type-II synapse with a small dendrite. Bar = 500 nm.

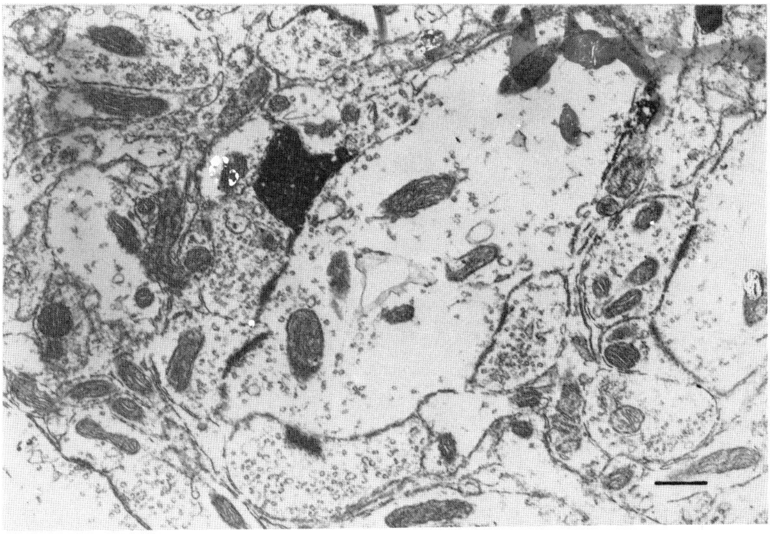

Figure 5-4. Control rat. An immunoreactive 5-HT bouton makes a type-II synapse with a large dendrite, possibly belonging to a mitral cell. Bar = 500 nm.

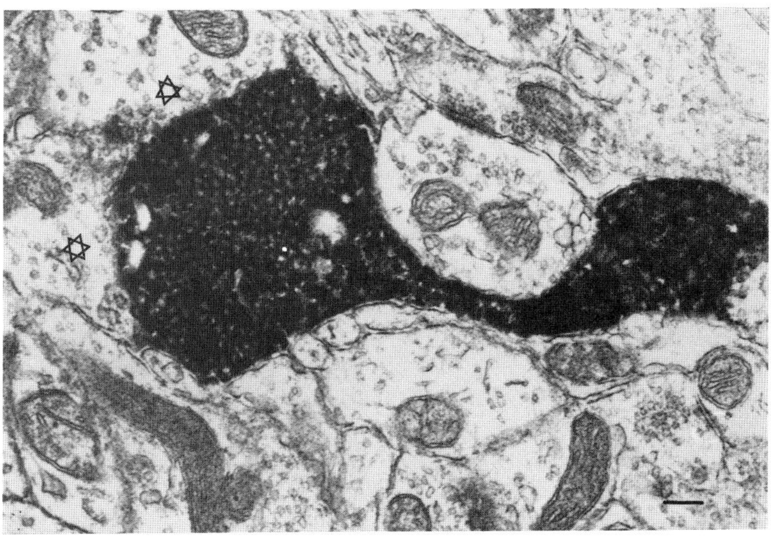

Figure 5-5. Control rat. In some instances, 5-HT immunoreactive varicosities are not facing postsynaptic differentiations. The only neighboring structures are axonal profiles (⋆). Bar = 200 nm.

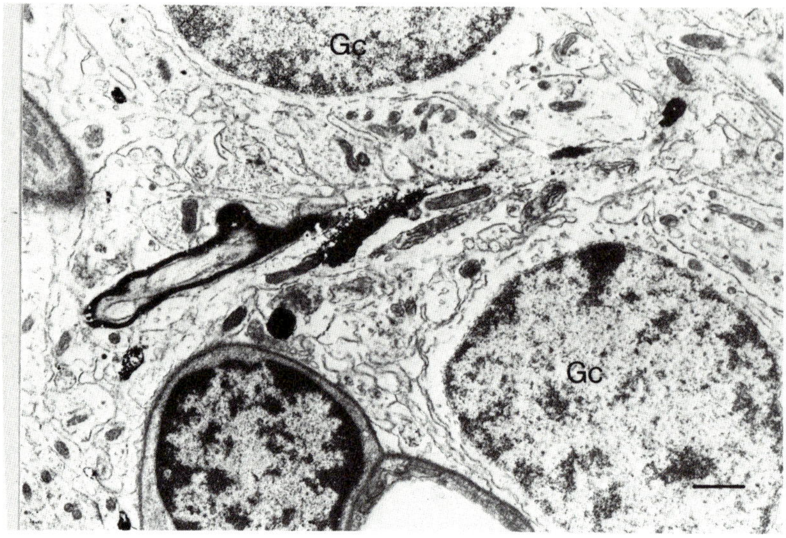

Figure 5-6. Control rat. 5-HT immunoreactive fibers meander around glomerular cells (*Gc*), but do not contact their perikarya. Bar = 1 μm.

number of 5-HT neurons must be at least twice the number detected. Cells appear most often clustered at the site of injection (figure 5-7) but, in some cases, immunoreactive cells are more dispersed (figure 5-8).

Most of the *perikarya* are strongly immunoreactive and display a wide range of shapes and sizes, like those of the raphe nuclei in the intact animal (Steinbusch, 1984). Cells are small and round (figure 5-10), small and medium fusiform (figure 5-9), or large and polygonal (figure 5-10). Dendrites do not usually extend over great distances but, in some instances (figure 5-7), they can be followed over several hundred micrometers. Axons arise either from the perikarya or from the initial segment of a dendrite (figure 5-12). They immediately branch profusely (figure 5-11), but most often do not invade the host tissue to a great extent. In several instances, however, large bundles of axons are directed toward the surface of the bulb and invade the glomerular layer (figure 5-13); some of them are seen to encircle the perikarya of the host's neurons (figure 5-15). On 1-μm sections, counterstained with toludine blue, one can see that the profuse innervation provided by grafted neurons to the glomerular layer does not invade the center of the glomeruli (figure 5-14), but remains at the periphery, like that of the control animals (figure 5-2).

5. ULTRASTRUCTURE OF THE TRANSPLANT

Immunoreactive perikarya are readily identified as medium- to large-sized neurons with a notched nucleus, an abundance of Golgi and endoplasmic

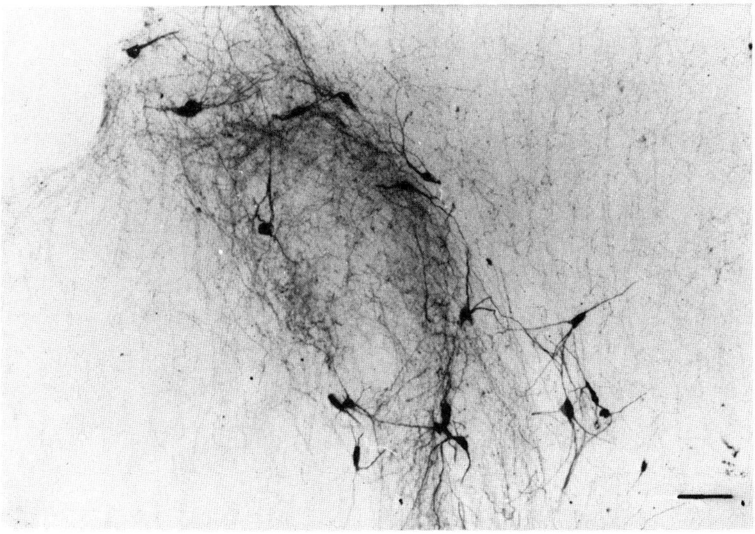

Figure 5-7. Transplanted rat: 6-weeks of survival. 5-HT immunoreactive neurons 6 weeks after transplantation. Most often perikarya remain clumped around the site of the injection, and their axons and dendrites do not extend more than a few hundred micrometers. Bar = 100 μm.

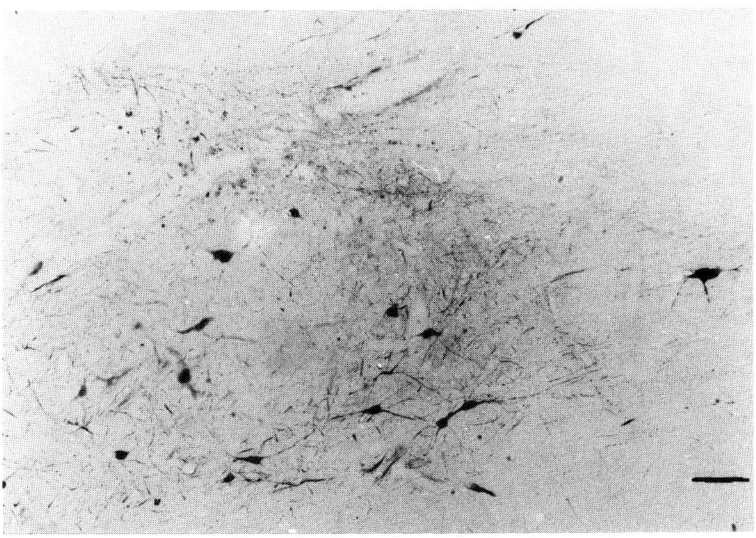

Figure 5-8. Transplanted rat: 6-weeks of survival. In some instances, 5-HT immunoreactive neurons are more dispersed in the olfactory bulb and the innervation of the host's tissue is more diffuse. Bar = 10 μm.

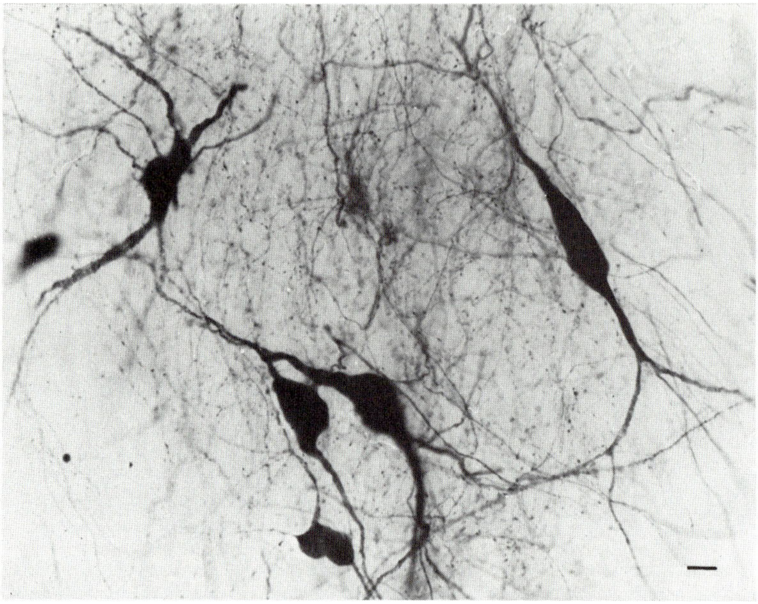

Figure 5-9. Transplanted rat: 3-months of survival. Grafted neurons show a wide variety of shapes and sizes, as those of the raphe nuclei in adult animals. However, large bipolar cells are the most common type. Bar = 10 μm.

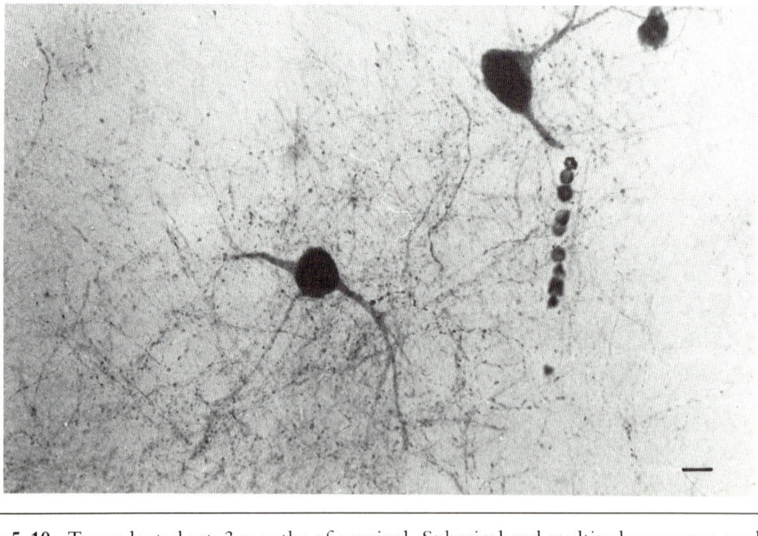

Figure 5-10. Transplanted rat: 3 months of survival. Spherical and multipolar neurons are less common than bipolar cells. Notice, as in figure 5-9, the dense immunoreactive neuropile. (Bar = 10 μm)

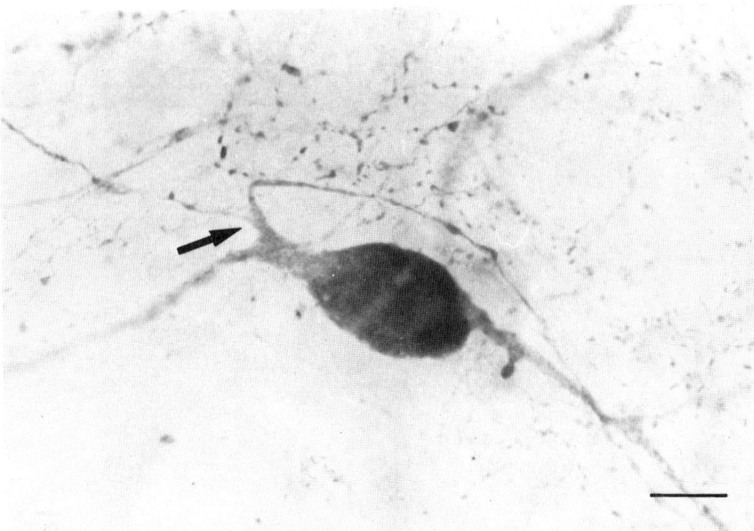

Figure 5-11. Transplanted rat: 3 months of survival. Axons of transplanted 5-HT neurons arise most often from the perikarya and immediately branch profusely (*arrow*). Bar = 10 μm.

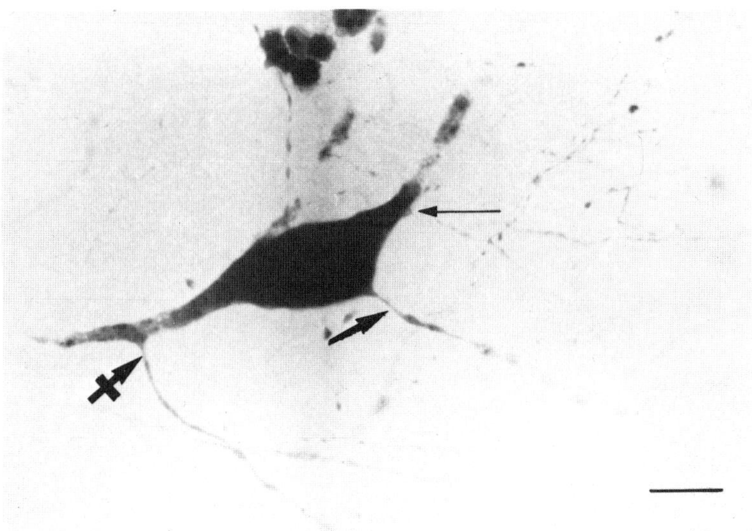

Figure 5-12. Transplanted rat: 3 months of survival. In rare instances, axons arise from a dendritic trunk (*crossed arrow*). In this exceptional case, another axon is seen arising directly from the pericaryon (*arrow*). The *small arrow* indicates an afferent contact. Bar = 10 μm.

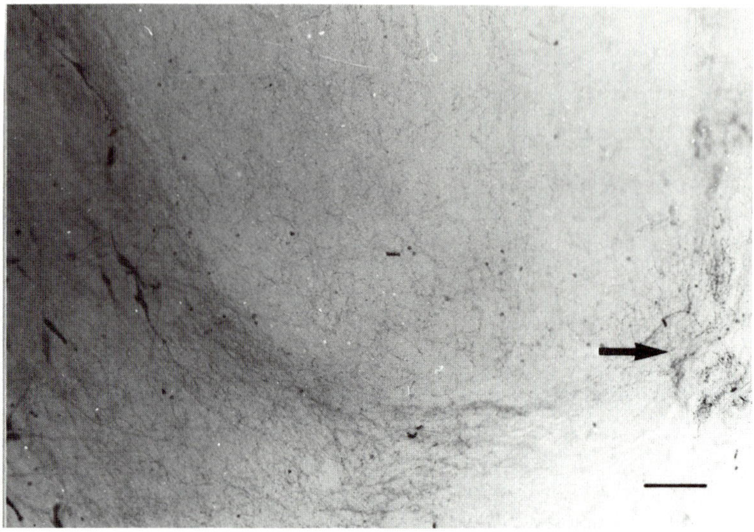

Figure 5-13. Transplanted rat: 3 months of survival. From the injection site, on the *left* of the photograph, a large bundle of axons extends toward the glomerular layer (*arrow*). Bar = 100 μm.

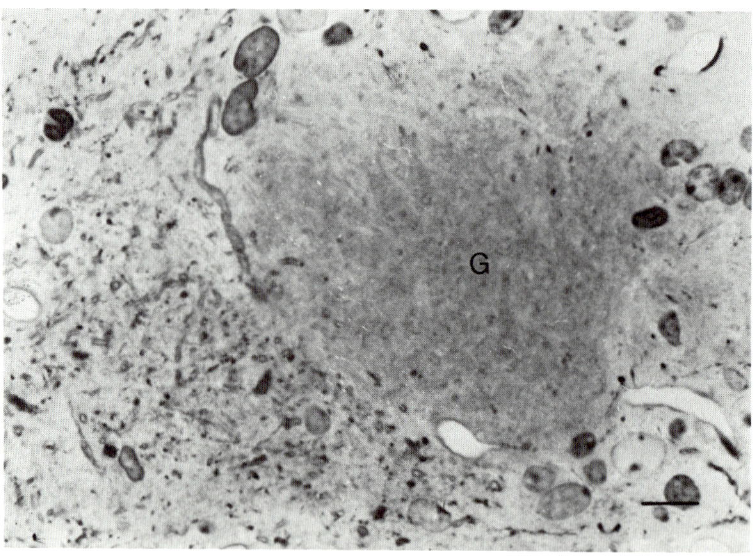

Figure 5-14. Transplanted rat: 3 months of survival. In the same olfactory bulb, a 1-U section shows that immunoreactive axons remain at the periphery of the glomerulus (*G*), as in the control animal (figure 15-2). Notice, however, the higher density of 5-HT terminals around the glomerulus. Bar 10 = μm.

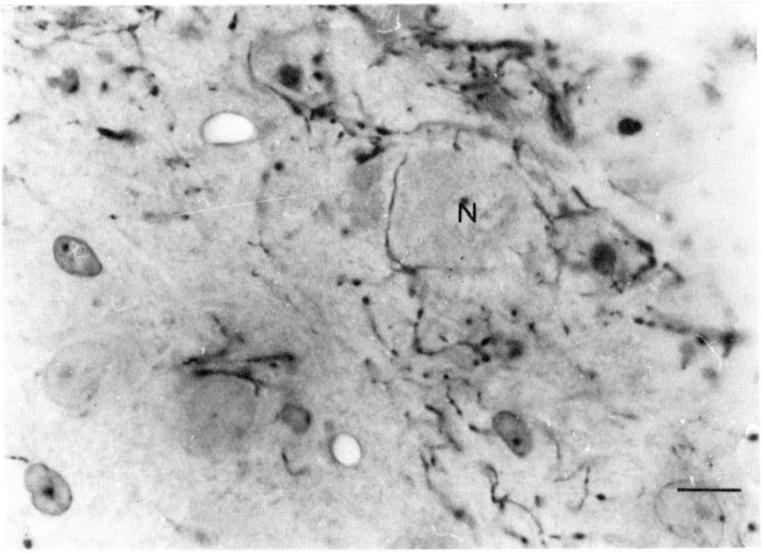

Figure 5-15. Transplanted rat: 3 months of survival. In the external plexiform layer, large neuronal perikarya (*N*) are contacted by immunoreactive varicose fibers. Bar = 10 μm.

reticulum, and a few afferent synaptic boutons (figure 5-16). In contrast, immunoreactive dendrites are contacted by many boutons (figure 5-17), which as a rule make Gray's type-I synapses, with spherical vesicles and conspicuous postsynaptic densities (figure 5-18).

Immunoreactive axons exhibit medium-sized boutons that contact unlabeled perikarya, dendrites (figure 5-19), and, in several instances, contribute typical Gray's type-II synapses. In several instances, however, immunoreactive varicosities, filled with 40-nm vesicles, are seen devoid of facing postsynaptic densities (figure 5-21). Finally, some immunoreactive boutons face immunoreactive dendrites (figure 5-20). The frequent clumping of several immunoreactive 5-HT neurons precludes the identification of those contacts as autapses. It is more likely that the pre- and postsynaptic elements belong to different cells.

6. CONCLUSIONS

We have seen that 5-HT fetal neurons transplanted into the olfactory bulb of adult rats can survive, grow axons and dendrites, and establish synaptic connections with the host neurons. Since the intrinsic 5-HT innervation was left intact in our model, this condition is similar to that already described by Azmitia et al. (1981) in the hippocampus. These authors concluded that innervation provided by the transplant followed a pattern similar to that of the intrinsic innervation.

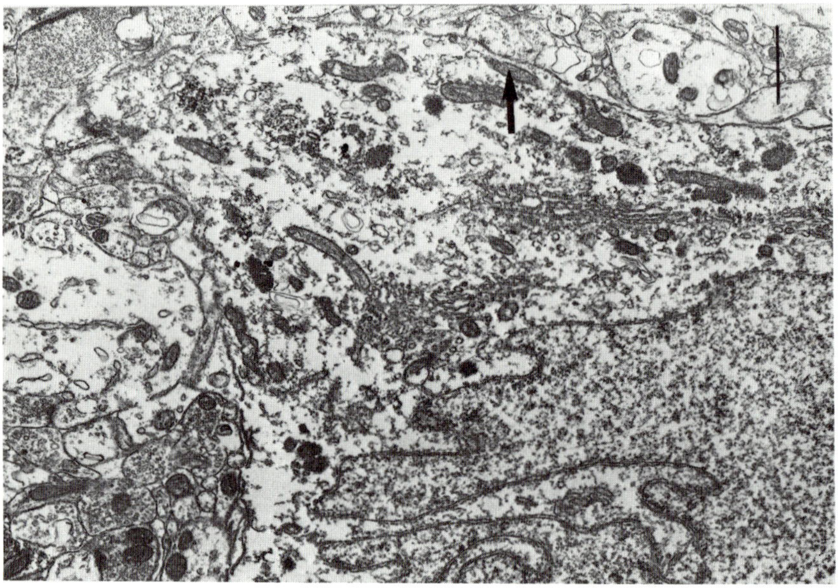

Figure 5-16. Transplanted rat: 3 months of survival. A transplanted immunoreactive neuron is surrounded by many axonal profiles, one of which (*arrow*) makes a synapse on its surface. Bar = 1 μm.

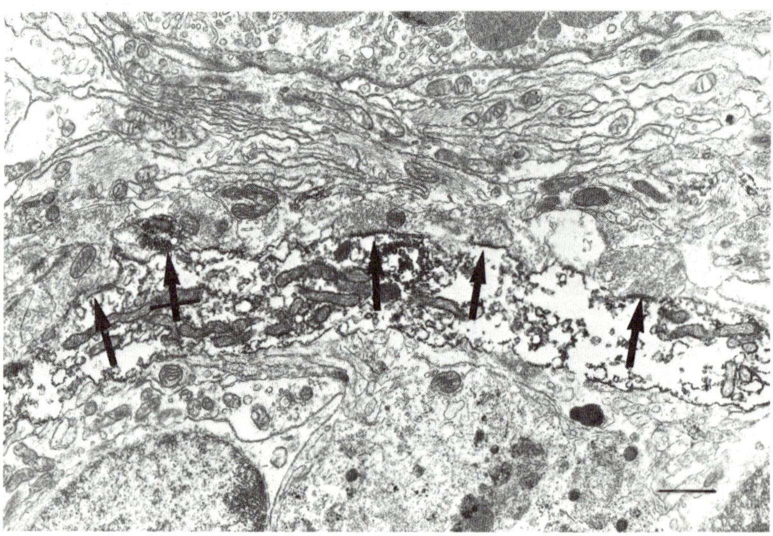

Figure 5-17. Transplanted rat: 3 months of survival. An immunoreactive dendrite is profusely innervated by type-I boutons (*arrows*); the *crossed arrow* points to an immunoreactive axon. Bar = 1 μm.

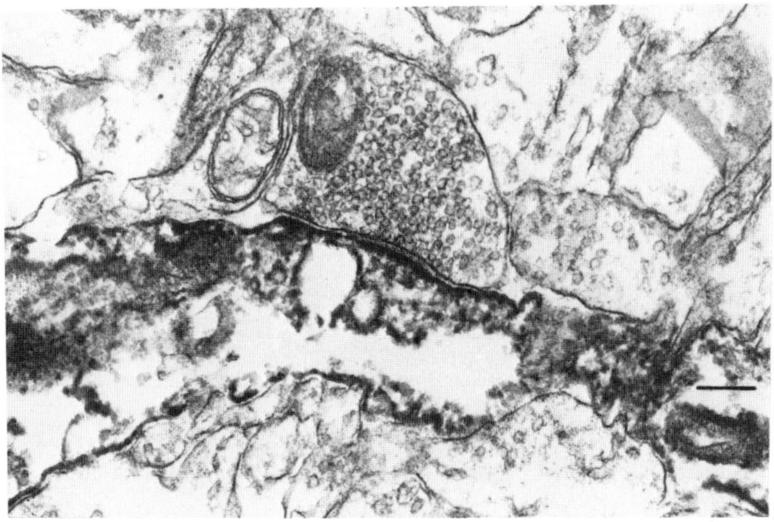

Figure 5-18. Transplanted rat: 3 months of survival. Higher magnification of an afferent synapse similar to those in figure 5-17. The typical Gray's type-I synapse exhibits spherical vesicles, widening of the extracellular cleft, and conspicuous postsynaptic density. Bar = 200 nm.

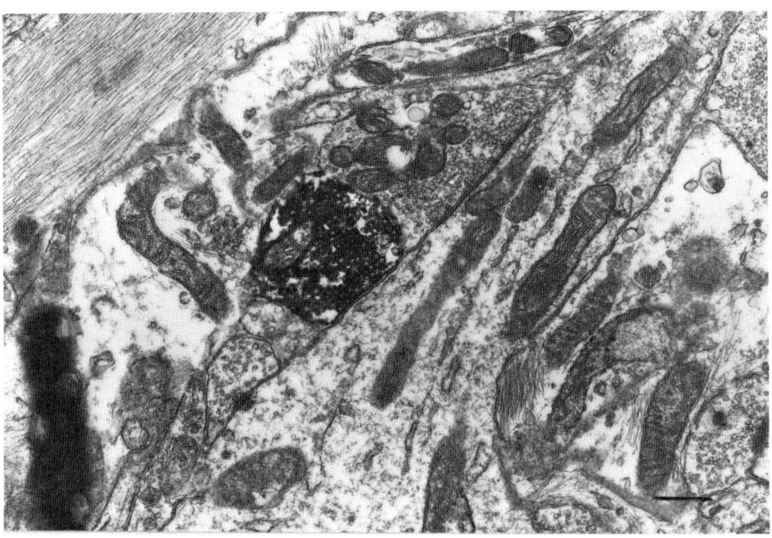

Figure 5-19. Transplanted rat: 3 months of survival. In the external plexiform layer, an immunoreactive bouton makes a type-II synapse with the initial part of the dendrite of a large host's neuron. Bar = 500 nm.

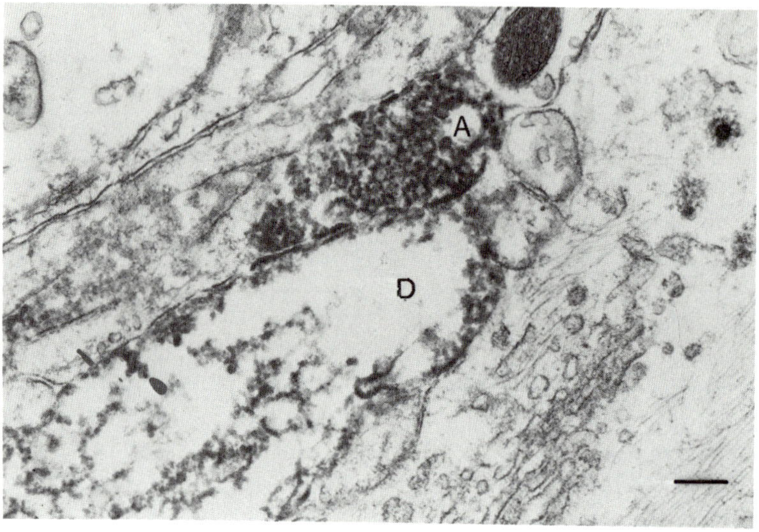

Figure 5-20. Transplanted rat: 3 months of survival. In the area of the transplant, an immunoreactive axon (*A*) makes a type-II synapse with an immunoreactive dendrite (*D*). Bar = 200 nm.

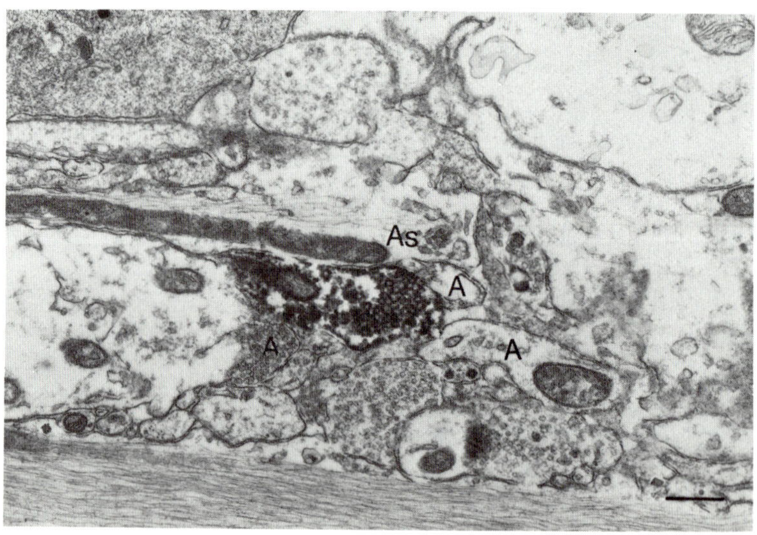

Figure 5-21. Transplanted rat: 3 months of survival. In the glomerular layer, an immunoreactive varicosity filled with 40-nm vesicles is surrounded by astrocytes (*As*) and axonal (*A*) profiles, a configuration also seen in control animals (figure 5-5). Bar = 500 nm.

In our model, the results are slightly different, as we could detect terminals in regions of the olfactory bulb that are not normally innervated by 5-HT axons. However, interestingly enough, those axons that reach the normal target of 5-HT terminals, i.e., the glomerular layer, show a pattern of innervation similar to that of the control, large varicose fibers being confined at the periphery of the glomeruli. Thus, some specifity is preserved, at least at the level of the glomerulus. This relative difference with the results of Azmitia et al. (1981) may be due to the fact that the hippocampus is profusely innervated by 5-HT fibers, whereas the innervation of the olfactory bulb is precisely limited to the glomerular layer. Thus, 5-HT neurons transplanted into the hippocampus may not have a chance to show an heterotopic innervation, at least at the light-microscopic level.

Few ultrastructural studies have been carried out on transplanted monoaminergic neurons. Freund et al. (1985), Mahalik et al. (1985), and Jaeger et al. (1984) have studied dopaminergic neurons transplanted into the striatum of adult rats. These authors identified synaptic arrangements that are never seen in the striatum, despite the fact that the rats had been previously deprived of intrinsic dopaminergic innervation by 6-OHDA lesions. In our studies, the ultrastructural characteristics of grafted 5-HT neurons are similar to those of the raphe nuclei (Beaudet and Descarries, 1979; Privat, unpublished), and these neurons bear many afferent synapses and make efferent contacts. Whereas some of the afferent synapses may be contributed by other neurons from the transplant, efferent projections either to the external plexiform layer (figure 5-15) or to the glomerular layer (figures 5-13 and 5-14) clearly involve host neurons. A detailed study of these projections, in comparison with the intact animal, is beyond the scope of this chapter, but suffice it to say that, in both cases, 5-HT axons form typical synaptic differentiations, as well as varicosities lacking synaptic relationships (Descarries et al., 1975; Beaudet and Sotelo, 1981), as in cerebral and cerebellar cortices.

Regarding the functional influence of transplanted 5-HT neurons, ongoing studies (Lerner-Natoli et al., in preparation) show that they can modify the electrical excitability of the intrinsic neurons of the bulb. We have recently shown (Lerner-Natoli et al., 1986) that chemical lesioning of 5-HT innervation of the bulb with the neurotoxin 5-7 DHT induces a facilitation and acceleration of kindling (Racine and Coscina, 1979), which suggests that 5-HT exerts an inhibitory tone upon bulbar neurons. Some of the transplanted rats have been subsequently kindled from the bulb, and we recorded an increase of the threshold of the after-discharge, versus nongrafted controls and controls grafted with cortical neurons. Work is now in progress to compare the parameters of kindling in between control animals, unlesioned grafted, 5-7 DHT-lesioned ungrafted, and 5-7-DHT-lesioned grafted.

In conclusion, fetal 5-HT neurons transplanted into the olfactory bulb of adult rats can survive, grow, and acquire the 5-HT transmitter phenotype. Moreover, these neurons establish synaptic connections with the host's neurons, both afferent and efferent. The efferent connections are more widespread than those of the intrinsic 5-HT neurons, indicating a relative absence of specificity. Preliminary evidence from electrophysiological studies indicates some influence of grafted neurons upon the host's neurons, and the exact mechanism of this process awaits further clarification.

ACKNOWLEDGMENTS

Part of the work reported in this chapter has been supported by DRET grant 84126. The authors acknowledge the expert help of Mrs. F. Sandillon, M.J. Drian, V. Kimbor, and M. Bonnefoy.

REFERENCES

Azmitia, E.C., Perlow, M.J., Brennan, M.J., and Lauder, J.M. (1981) Fetal raphe and hippocampal transplants into adult and aged C57BL/6N mice: a preliminary immunocytochemical study. Brain Res. Bull. 7:703–710.

Backlund, E.O., Granberg, P-O., Hamberger, B., Sedvall, G., Seiger, A., and Olson L. (1985) Transplantation of adrenal medullary tissue to striatum in Parkinsonism. In: Björklund, A., and Stenevi, U. (eds.) Neural grafting in the mammalian CNS. Amsterdam: Elsevier, pp. 551–556.

Beaudet, A., and Descarries, L. (1979) Radioautographic characterization of a serotonin accumulating nerve cell group in adult rat hypothalamus. Brain Res. 160:231–243.

Beaudet, A., and Sotelo, C. (1981) Synaptic remodeling of serotonin axon terminals in rat agranular cerebellum. Brain Res. 206:305–329.

Björklund, A., and Stenevi, U. (1979) Reconstruction of the nigrostriatal dopamine pathway by intracerebral nigral transplants. Brain Res. 177:555–560.

Björklund, A., and Stenevi U. (1985) Intracerebral neural grafting: a historical perspective. In: Björklund, A., and Stenevi, U. (eds.) Neural grafting in the mammalian CNS. Amsterdam: Elsevier, pp. 3–14.

Bjöklund, A., Stenevi, U., and Svendgaard, N. -Aa. (1976) Growth of transplanted monoamiergic neurones into the adult hippocampus along the perforant path. Nature (Lond.) 262: 787–790.

Björklund, A., Stenevi, U., Schmidt, R.H., Dunnett, S.B., and Gage, F.H. (1983) Intracerebral grafting of neuronal cell suspensions. I., Introduction and general methods of preparation. Acta Physiol. Scand. Suppl. 522:1–7.

Descarries, L., Beaudet, A., and Watkins, K.C. (1975) Serotonin nerve terminals in adult rat neocortex. Brain Res. 100:563–588.

Dunnett, S.B., Björklund, A., Gage, F.H., and Stenevi, ULF (1985) Transplantation of mesencephalic dopamine neurons to the striatum of adult rats. In: Björklund, A., and Stenevi, U. (eds.) Neural grafting in the mammalian CNS. Amsterdam: Elsevier, 451–469.

Freund, T.F., Bolam, J.P., Björklund, A., Stenevi, U., and Dunnett, S.B. (1985) Efferent synaptic connections of grafted dopaminergic neurons reinnervating the host neostriatum: a tyrosine hydroxylase immunocytochemical study. J. Neurosci. 5:603–616.

Gash, D.M. (1983) Neural transplants in mammals: a historical overview. In: Sladek, J.R., and Gash, D.M. (eds.) Neural transplants: development and function. New York: Plenum pp. 1–11.

Geffard, M., Henrich-Roch, A-M., Dulluc, J., and Seguela, P.H. (1985) Antisera against small neurotransmitter-like molecules. Neurochem. Int. 7:403–413.

Jaeger, C.B. (1985) Cytoarchitectonics of substantia nigra grafts: a light and electron microscopic study of immunocytochemically identified dopaminergic neurons and fibrous astrocytes. J. Comp. Neurol. 231:121–135.

Lauder, J.M., Wallace, J.A., Krebs, H., Peterusz, P., and McCarthy, K. (1982) In vivo and in vitro development of serotonergic neurons. Brain Res. Bull. 9:605–625.

Lerner-Natoli, M., Rondouin, G., Malafosse, A., Sandillon, F., Privat, A., and Baldy-

Moulinier M. (1986) Facilitation of olfactory bulb kindling after specific destruction of serotoninergic terminals in the olfactory bulb of rat. Neurosc. Lett (in press).

Mahalik, T.J., Finger, T.E., Stromberg, I., and Olson, L. (1985) Substantia nigra transplants into denervated striatum of the rat: ultrastructure of graft and host interconnections. J. Comp. Neurol. 240:60–70.

McRae-Degueurce, A., Lauder, J.M., and Privat, A. (1983) Adult age destruction versus neonatal age destruction of the serotonin system: two models to investigate the survival of serotonin neurons transplanted in adult rats. Neurosci. Lett. 40:27–32.

McRae-Degueurce, A., Serrano, A., Sandillon, F., Privat, A., and Scatton, B. (1984) In vivo voltametric measurement of extra-cellular 5-HIAA in the denervated striatum after transplantation of mesencephalic raphe neurons. Neurosci. Lett. 48:97–102.

Privat, A. (1982) In vitro cultures of serotonergic neurons from fetal rat brain. J. Histochem. Cytochem. 30:185–187.

Privat, A., Mansour, H., Pavy, A., Geffard, M., and Sandillon, F. (1986) Transplantation of foetal 5-HT neurons into the transected spinal cord of adult rats. Neurosci. Lett. (in press).

Racine, R.J., and Coscina, D.V. (1979) Effects of midbrain raphe lesions or systemic *p*-chlorophenylalaline on the development of kindles seizures in rats. Brain Res. Bull. 4:1–7.

Steinbusch, H.W.M. (1984) Serotonin-immunoreactive neurons and their projections in the CNS. In: Bjorklund, A., Hökfelt, T., and Kuhar, M.J. (eds.) Handbook of chemical neuroanatomy. Amsterdam: Elsevier, pp. 68–125.

Steinbusch, H.W.M., Verhofstad, A.J., and Joosten, H.W.J. (1978) Localization of serotonin in the central nervous system by immunohistochemistry: description of a specific and sensitive technique and some applications. Neuroscience 3:811–819.

Stenevi, U., Björklund, A., and Svengaard, N.-Aa. (1976) Transplantation of central and peripheral monoamine neurons to the adult rat brain: technique and conditions for survival. Brain Res. 114:1–20.

6. AN ANALYSIS OF INTERACTIONS BETWEEN GRAFTED SEROTONIN NEURONS AND AN ADULT HOST HIPPOCAMPUS

MENAHEM SEGAL

1. INTRODUCTION

The past few years have witnessed a rapid expansion of research of plastic properties of transplanted central nervous system (CNS) neurons. Two major developments contributed to this expansion. First was the demonstrated ability to identify chemically specific transplanted neurons in a host brain and follow their growth using standard histochemical methodologies (Björklund et al., 1979 and 1983a). Second was the suggestion that grafted neurons may substitute for damaged ones in providing the brain with specific chemicals needed for its normal operation (Gash et al., 1980). The prospects of using brain grafts for the treatment of neurodegenerative diseases of the CNS have ignited the imagination of physicians and scientists alike. Indeed, preliminary experiments with Parkinsonian patients have already been conducted, albeit with little proven success (Backlund et al., 1985). Our current usage of brain grafting methods is motivated by a different research goal: 1) An analysis of the factors underlying developmental specificity of neural connections, e.g., will the grafted neurons form connections in a new brain akin to those formed by the native neurons and under what conditions are such connections made (Azmitia et al., 1981; Björklund et al., 1983a)? 2) Analysis of behavioral functions of a neuron group; certain behavioral deficits can result from specific neurochemical damage. A most direct way relating structure to function is the illustration that the grafting of a neuron structure into a deficient brain can restore the damaged function (e.g., Dunnett et al., 1982;

Vernadakis, A. (ed.), *Model Systems of Development and Aging of the Nervous System*.

Drucker-Colin et al., 1984). 3) Reconstruction of neuronal connections for the simplification of their physiological analysis. Such is the case with long axon pathways that cannot be studied in an in vitro slice preparation (Segal et al., 1985). 4) Clinical application of the graft methodology for treatment of degenerative diseases, e.g., Parkinsonism. These studies concentrate on the optimalization of the graft conditions, sources of grafted tissue, and the form of its application to yield best possible clinical results (see Björklund et al., 1983a).

Regardless of the specific objectives, it is quite evident that grafted tissue of various origins can grow in a host brain and send axons that will innervate the host tissue. The results obtained thus far allow several tentative generalizations: 1) The origin of the grafted tissue is a major determinant of its successful growth in the host. Brain nuclei that normally have long diffuse efferent projections, e.g., the monoamines, will send such long axons into the host tissue whereas brain regions have more precise, specific laminated efferents will exhibit only a limited growth. 2) The availability of blood supply to the grafted tissue will determine its survival. Thus a graft placed in a region where blood vessels can reach it easily (e.g., near the choroid plexus) will have a better chance of survival. 3) The age of the grafted tissue as well as the age of the recipient contribute to the success of grafting. The grafted tissue is usually taken from embryonic brain at about the time when cell division ceases. Adult tissue does not survive as well in a host. 4) The form of tissue transplanted is different in various experiments. A small piece of intact tissue can be placed in a cavity, can be minced and injected into the recipient tissue using a wide-bored needle, or can be dissociated enzymatically into a cell suspension and injected into the host tissue using fine-tip pipettes. 5) The deprivation of the host tissue of its normal innervation will determine to a large extent the spread and specificity of the graft innervation of the host. It is believed that the deprived tissue might produce growth-promoting factors that would facilitate proliferation of the grafted neurons. The specificity of such action has been claimed recently for some neurotransmitter systems (Zhou et al., 1985).

While the morphological analysis of graft–host interaction has been conducted extensively for several neuronal pathways and a functional recovery has been demonstrated elegantly for some neuroendocrine motor functions and perhaps some aspects of cognitive functions (Drucker-Colin et al., 1984; Gash et al., 1980; Dunnett et al., 1984), the physiological analysis of graft–host interaction lags behind. Three classes of questions need to be examined before we can assume that the grafted neurons integrate into the host brain. First, the graft has to receive afferents comparable to its normal counterpart such that stimulation of these afferents produces the same postsynaptic responses of the grafted neurons as seen normally. Second, the grafted neurons have to develop (or retain) physiological properties akin to those of their normal counterparts. Morphologically distinct neuronal types are known to possess characteristic ionic conductances, properties that determine their reactivity to incoming afferents and to pharmocological and pathological conditions. These might or might not develop under new and distinctly different conditions inherent in a graft–host relationship. Finally, we need to know whether the grafted neurons produce the same postsynaptic effects on their efferent targets as normal cells do. Once an equivalence is established at the three levels between the grafted and the homotypic neurons, one can then assume that the grafted neurons indeed replaced the lost ones.

2. PROPERTIES OF RAPHE NEURONS

We have begun a physiological analysis of graft–host interaction using serotonergic neurons grafted into the rat hippocampus. The latter is a major target area of serotonin-containing fibers of midbrain raphe origin (Azmitia and Segal, 1978). It has already been shown that raphe neurons that contain serotonin at a relatively young embryonic age (E13; Lidov and Molliver, 1982; Wallace and Lauder, 1983) can be transplanted into the hippocampus at E14, where they will innervate the host in a pattern similar to that of the normal innervation pattern of raphe-originated serotonin neurons. In the initial series of experiments, we investigated the properties of normal adult serotonin neurons recorded in a slice taken from the midbrain raphe. These neurons possess unique physiological properties that distinguish them from adjacent midbrain neurons and from hippocampal neurons recorded in a slice. Some of these properties have been described elsewhere (Aghajanian and Vandermaelen, 1982; Crunelli et al., 1983). These cells have a relatively high input resistance (100–300 mohm), higher than that of most central neurons. The high input resistance cannot be explained as being simply a function of the size of these neurons. It probably represents a genuine difference in membrane composition between raphe and other neurons. Passage of depolarizing or hyperpolarizing current pulses across the recorded membrane can uncover voltage-dependent membrane currents that govern the behavior of the cell at this membrane potential range. Most neurons exhibit inward rectification in response to large (> 25 mV) hyperpolarizing current pulses (Purpura et al., 1968). This is assumed to result from the hyperpolarization-induced opening of cationic channels that causes current to flow into the cell. This property is conspicuously absent in serotonin neurons. The significance of this absence will be discussed below. Action potentials discharge spontaneously or in response to brief depolarizing current pulses. They are broad and contain a distinct Ca component. The spikes in serotonergic neurons are followed by a large after-hyperpolarization that is associated with an increase in conductance and appears to consist of fast and slow components. This large after-hyperpolarization is caused by activation of a Ca-dependent K current (Brown and Griffith, 1983; Hoston and Prince, 1980) that brings the membrane close to K equilibrium potential. The lack of inward rectification allows the membrane to reach such a polarized potential.

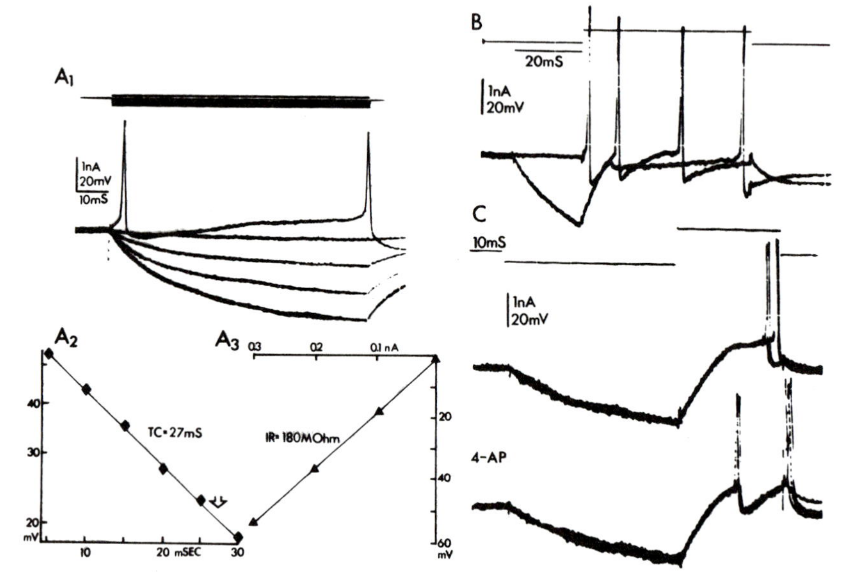

Figure 6-1. Electrical properties of a dorsal raphe (DR) neuron. Intracellular recording was made from a putative 5-HT containing DR neuron in the brain stem. Hyperpolarizing and depolarizing current pulses were applied to the neuron using a bridge circuit. (**A2**) An analysis of the hyperpolarizing charging curve indicated the presence of a single exponential curve with a time constant of 27 milliseconds. (**A3**) The current–voltage curve is linear down to 60 mV below resting potential (−65 mV) and indicates an input resistance of 180 mohm. No sag in the voltage response can be seen at any of the potentials. (**B**) A 50-millisecond depolarizing current pulse produces a train of action potentials. A brief hyperpolarization prevents the cell from discharging action potentials to the same current pulse for about 45 milliseconds. (**C**) A conditioning hyperpolarization delays the discharge of action potentials in response to a depolarization pulse (*top*). This delay is blocked after topical application of a microdrop of 5 m*M* 4-AP (*bottom*); 5–6 traces overlapped. Reproduced from Segal and Azmitia (1986).

During the hyperolarization, a new species of K current becomes activable; this is a transient outward current (TOC; Gustafsson et al., 1982), that is activated by depolarization and prevents the membrane from reaching firing threshold (figure 6-1).

In fact, activation of this current clamps the membrane at a potential slightly below firing threshold for up to several hundreds of milliseconds (Segal, 1985). The presence of a potent TOC is unique to serotonin neurons and is not present as such in most central neurons, e.g., the hippocampus. The TOC restricts the number of spikes discharged by a serotonergic neuron under normal conditions. The coexistence of fast Na and Ca currents, followed by fast outward K currents and the lack of inward rectification, produces a unique pattern of electrical activity that characterizes serotonergic neurons. Two questions related to grafting of central neurons are pertinent: Do these cells possess the unique physiological characteristics already de-

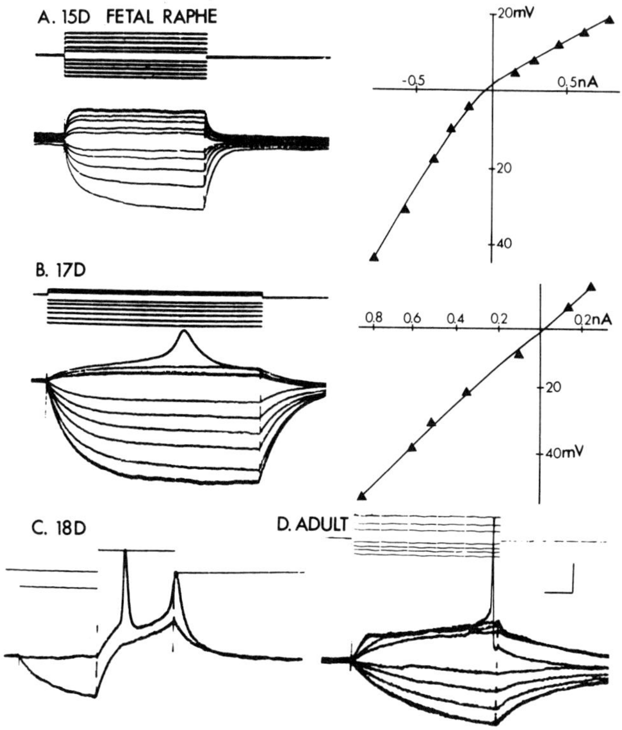

Figure 6-2. Properties of immature raphe neurons. Slices from embryonic mesencephalon were placed in the recording chamber and intracellular recording was made from putative raphe neurons. Series of hyperpolarizing and depolarizing current pulses were applied through the recording electrode. In a neuron recorded at E15 (**A**), there was no evidence for regenerative Na or Ca spikes, but there was a clear presence of outward rectification when the cell was depolarized. This is evident in the current–voltage plot (*right*) of the same cell. (**B**) Recroding from a cell at E17. A small regenerative spike is evoked by depolarization. The lack of inward rectification is clear. (**C**) Recording from a E18 raphe cell exhibiting adult neuronal properties. These include high input resistance, broad spike, and a strong transient outward rectification. (**D**) Adult neuron. The typical raphe serotonergic properties are evident. Depolarizing current pulses evoke a transient outward rectification which delays the spike firing. Calibration 10 milliseconds, 20 mV, and 0.5 nA for **A–D**.

scribed above when taken out of an embryo and grafted into the host brain, and do they develop (or maintain) these properties when they grow in a host brain?

3. PROPERTIES OF IMMATURE AND GRAFTED RAPHE NEURONS

We have recorded activity of neurons from the raphe in slices taken from brains of day-15 to day-19 day embryos. At the day of grafting (E15), the recorded cells did not exhibit any of the characteristics of the adult serotonergic neurons, with the exception, perhaps, of the lack of inward rectification (figure 6-2).

Within only two additional days (E17), cells in this region begin to express some of the identifying properties of serotonin neurons. Some cells begin to have adult physiological properties already by day E18. It thus appears that, on the day of grafting (E15), the serotonin neurons do not have, as yet, clear physiological properties. At this age they do produce serotonin and appear to project serotonin fibers into terminal regions in the cortex. (Wallace and Lauder, 1983; Lidov and Molliver, 1982). It is evident that, if they possess adult physiological properties in a host brain, they must have developed them in the host brain after grafting. Recording was made from regions in the hippocampal slice taken from a brain that was previously treated with a drug that destroys serotonin fibers (5,7-dihydroxytryptamine, 5 μg, was injected into the formix-fimbria and cingulum bundle 5–7 days before grafting) and subsequently implanted with serotonin-rich E15 mesencephalic neurons (Segal and Azmitia, 1986). In some slices, where careful inspection indicated the presence of grafted tissue, we could impale neurons that possess identical properties to those of midbrain serotonin-containing neurons. These properties include high input resistance, slow membrane time constant, lack of inward rectification, broad spikes, large after-hyperpolarization, presence of a large transient outward rectification, and lack of accomodation. Furthermore, preliminary experiments suggest that these neurons have the same pharmacolgical properties as those of normal cells. These experiments indicate that neurons of the raphe nuclei that contain serotonin can be transplanted into the adult hippocampus where they will develop physiological properties similar to those seen in their natural place in the brain. These observations have important implications for the putative roles of growth conditions, afterent innervation, juvenile hormones, and environmental factors in the development of properties of serotonin neurons. The present studies do not rule out the significance of such factors in supporting the growth and proliferation of these neurons in the host tissue.

4. GRAFT–HOST INTERACTIONS

The grafted neurons project serotonin-containing fibers into the host hippocampus. These fibers reach the same zones originally innervated by serotonin fibers of raphe origin. When these areas are stimulated electrically, an antidromic spike can be recorded from grafted neurons present in the same slice at some distance away (figure 6-3). The properties of grafted central serotonin-containing fibers can thus be studied in vitro. These axons have slow conduction velocity and refractoriness. The antidromicity can be confirmd using a collision test: the spike traveling from the axon to the soma can be shown to collide with a spike evoked in the soma by a depolarizing current pulse. The conduction velocity (< 1 m/second) indicates that the activated axons are unmyelinated. Stimulation at high frequencies can cause a dissociation between a soma-dendritic spike and an axon spike that can follow higher frequencies of stimulation. This observation indicates that the

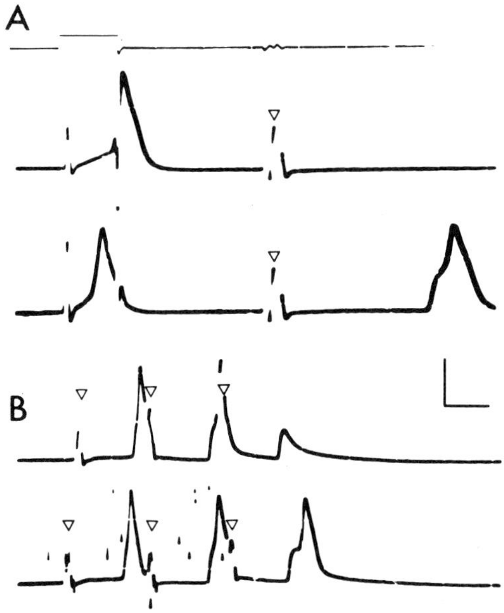

Figure 6-3. Antidromic invasion of grafted raphe neuron by stimulation of the host hippocampus. (**A**) A depolarizing current pulse (*top*) evokes an action potential that collides with a spike evoked by stimulation of the host tissue (*open triangle*) and blocks response to the latter. If, however, (*bottom*) the depolarizing 2-millisecond pulse evokes a spike with a 1-millisecond shorter delay, the collision fails and the antidromic response is seen; 5–10 repetitions for each trace. (**B**) Estimates of refractoriness of a transplanted raphe neuron. A train of three antidromic stimuli applied at a rate of 166 Hz (*top*) or 142 Hz (*bottom*). The cell can follow the slower rate, but only the axon spike follows reliably the faster rate. Calibration 50 mV and 5 milliseconds for **B**, and 2 milliseconds for **A**.

soma is under a strong inhibitory control that can block invasion of axon spikes.

The presence of functional connections between the raphe graft and the host hippocampus was assessed by electrical stimulation of the graft and recording intracellular activity in the host hippocampus. It has been previously shown that topically applied serotonin hyperpolarizes hippocampal neurons (Segal, 1980). This response is due to an increase in postsynaptic K conductance. The possibility that graft stimulation will hyperpolarize hippocampal neurons was therefore tested. A distinct long-lasting hyperpolarization was detected in hippocampal neurons located some 2–3 mm away from the graft. This hyperpolarization could be enhanced by loading the slice with a serotonin precursor, 5-hydroxytryptophan (5-HTP) (figure 6-4). The effect is distinctly different from that seen in response to stimulation of a graft taken from the septal region where a pronounced and long-lasting depolarization is evident.

The present results suggest that a raphe graft can make specific functional

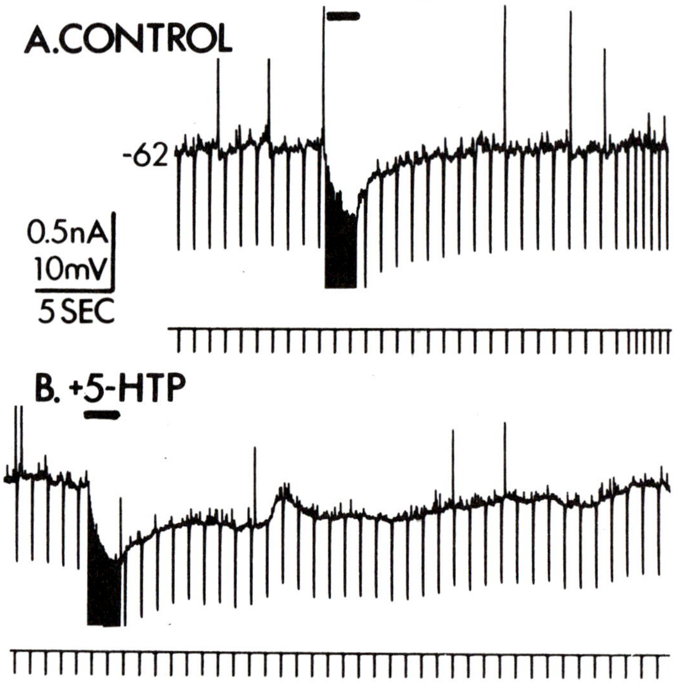

Figure 6-4. Postsynaptic effect of stimulation of a raphe graft on activity of a hippocampal neuron. (**A**) in normal medium, stimulation of the graft (10 Hz for 2 seconds) produces a slow 1- to 3-mV hyperpolarization. (**B**) After addition of 5-HTP to the slice (0.1 m*M* drop applied 2 minutes prior to stimulation), the same stimulation produces a much larger hyperpolarizing response.

connections with the host hippocampus. One major question related to the functional restoration of raphe–hippocampal connections is what controls raphe firing. Raphe neurons in situ fire spontaneously at different rates in relation to sleep–wakefulness cycles (Trulson and Jacobs, 1979). It is unlikely that grafted raphe neurons receive the same afferents as their normal counterparts do. One of the afferents that is likely to still be present in the new location is a noradrenergic input arising from the nucleus locus coeruleus, which innervates the host hippocampus. In any case, the serotonin neurons discharge spontaneously and do possess the same intrinsic mechanisms required to generate rhythmic activity when grafted into the hippocampus. Whether these mechanisms are sufficient to modulate hippocampal activity in the normal sleep–wake cycles remains to be determined. Preliminary experiments indicate that grafted neurons do receive afferent input and some of them exhibit spontaneous postsynaptic potentials. The origin of these afferents remains unclear.

While the present and related studies indicate that grafting can, under certain conditions, restore function connections with the host, it is quite clear that this property is shared with only a small number of brain nuclei. Most neuronal systems, where precision of afferent location and timing is crucial for information transfer, are not likely to be restored by grafting. This method can still be very useful for analysis of the rules governing the developmental specificity of these structures. The common denominator for the successful grafting of some nuclei is the diffuse and relatively slow efferent systems they exhibit. This is the case for some monoamines and acetylcholine nuclei in the brain. This is perhaps also the case for a number of peptide-containing nuclei in the hypothalamus. It is interesting to note that neurological and psychiatric disorders are associated with these nuclei and thus the pursuit of grafting methods for these can be of eventual considerable help in understanding their mechanisms of action.

5. CONCLUSIONS

Embryonic pieces of brain tissue can be grafted into an adult host brain where they can grow and form connections with the host. We have followed the development of physiological properties of serotonin-containing raphe neurons grafted into an adult host hippocampus. These cells have no clear identifying properties on day of grafting and develop them while growing in the host. Raphe neurons, recorded one month after grafting, possess adult normal physiological properties. These include high input resistance, slow membrane time-constant, lack of inward rectification, a transient outward rectification, broad spikes having a Ca component, lack of accomodation, and a large after-hyperpolarization. The graft innervates the host and, when stimulated, hyperpolarizes postsynaptic hippocampal neurons. These studies illustate that grafted serotonin neurons devleop functional relations with a host brain.

ACKNOWLEDGMENTS

This work was supported by a US–Israel Binational Science Foundation. I would like to thank Dr. E. Azmitia for this collaboration.

REFERENCES

Aghajanian, G.K., and Vandermaelen, C.P. (1982) Intracellular recording from serotonergic dorsal raphe neurons: pacemaker potentials and the effects of LSD. Brain Res. 238:463–469.

Azmitia, E.C., and Segal, M. (1978) The efferent connections of the dorsal and median raphe nuclei in the rat brain. J. Comp. Neurol. 179:641–668.

Azmitia, E.C., Perlow, M.J., Brennan, M.J., and Lauder, J.M. (1981) Fetal raphe and hippocampal transplants into adult and aged C57BL/GN mice: a preliminary immunocytochemical study. Brain Res. Bull. 7:703–710.

Backlund, E.O., Granberg, P.O., Hamberger, B., Sedvall, G., Seiger, A., and Olson, L. (1985) Transplantation of adrenal medullary tissue to striatum in parkinsonism. In: Björklund, A., and Stenevi, U. (eds.) Neural grafting in the mammalian CNS. Amsterdam: Elsevier, pp. 551–556.

Björklund, A., Segal, M., and Stenevi, U. (1979) Functional reinnervation of rat hippocampus by locus coeruleus implants. Brain Res. 170:409–426.

Björklund, A., Stenevi, U., Schmidt, R.H., Dunnett, S.B., and Gage, F.H. (1983b) Introduction and general methods of preparation. Acta Physiol. Scand. Suppl. 522:1–7.

Björklund, A., Gage, F.H., Stenevi, U., and Dunnett, S.B. (1983a) Survival and growth of intrahippocampal implants of septal cell suspensions. Acta Physiol. Scand, Suppl. 522:49–58.

Brown, D.A., and Griffith, W.H. (1983) Calcium activated outward current in voltage clamped hippocampal neurons of the guinea pig. J. Physiol. 337:287–301.

Crunelli, V., Forda, S., Brooks, P.A., Wilson, K.C.P., Wise, J.C.M., and Kelly, J.S. (1983) Passive membrane properties of neurons in the dorsal raphe and periacquaductal grey recorded in vitro. Neurosci. Lett. 40:263–268.

Drucker-Colin, R., Aguilar-Roblero. R., Garcia Hernandez, F., Fernandez-Cancino, F., and Bermudez Rattioni, F. (1984) Fetal suprachiasmatic nucleus transplants: diurnal rhythm recovery of lesioned rats. Brain Res. 311:353–357.

Dunnett, S.B., Low, W.C., Iversen, S.D., Stenevi, U., and Björklund, A. (1982) Septal transplants restore maze learning in rats with fornix-fimbria lesions. Brain Res. 251:335–348.

Dunnett, S.B., Bunch, S.T., Gage, F.H., and Björklund, A. (1984) Dopamine-rich transplants in rats with 6-OHDA lesions of the ventral tegmental area. 1. Effects on spontaneous and drug-induced locomotor activity. Behav. Brain Res. 13:71–82.

Gash, D., Sladek, J.R., and Sladek, S.D. (1980) Functional development of grafted vasopressin neurons. Science 210:1267–1369.

Gustafsson, B., Galvan, M., Grafe, P., and Wigstrom, H. (1982) A transient outward current in a mammalian central neurone blocked by 4-aminopyridine. Nature 299:252–254.

Hoston, J.R., and Prince, D.A. (1980) A calcium activated hyperpolarization follows repetitive firing in hippocampal neurons. J. Neurophysiol. 43:409–419.

Lidov, H.G.W., and Molliver, M.E. (1982) Immunohistochemical study of the development of serotogenergic neurons in the rat CNS. Brain Res. Bull. 9:559–604.

Purpura, D.P., Prelevic, S., and Santini, M. (1968) Hyperpolarizing increase in membrane conductance in hippocampal neurons. Brain Res. 7:310–312.

Segal, M. (1980) The action of serotonin in the rat hippocampal slice preparation. J. Physiol. (Lond.) 303:423–439.

Segal, M. (1985) A potent transient outward current regulates excitability of dorsal raphe neurons. Brain Res. 359:347–350.

Segal, M., and Azmitia, E.C. (1986) Fetal raphe neurons grafted into the hippocampus develop normal adult physiological properties. Brain Res. (in press).

Segal, M., Rogawski, M.A., and Barker, J.L. (1984) A transient potassium conductance depresses the excitability of cultured hippocampal and spinal neurons. Neuroscience 4:604–609.

Segal, M., Björklund, A., and Gage, F.H. (1985) Transplanted septal neurons make viable cholinergic synapses with a host hippocampus. Brain Res. 336:308–312.

Trulson, M.E., and Jacobs, B.L. (1979) Raphe unit activity in freely moving cats: correlation with level of behavioral arousal. Brain Res. 163:135–150.

Wallace, J.A., and Lauder, J.M. (1983) Development of the serotonergic system in the rat embryo: an immunocytochemical study. Brain Res. Bull. 10:459–479.

Zhou, F.C., Azmitia, E.C., Auerbach, S., and Jacobs, B.L. (1985) A specific serotonergic growth factor from 5, 7 DHT lesioned hippocampus: in vivo evidence from fetal transplantation of raphe and locus coeruleus neurons. Soc. Neurosci. Abstr. 11:1084.

7. NEURONAL MODULATION BY NUMBER AND BY VOLUME: A REVIEW AND CRITICAL ANALYSIS

GUIDO FILOGAMO

1. INTRODUCTION

Developmental neurology has been a subject of interest for over a hundred years and is associated with the names of His, Kölliker, Van Geuchten, Golgi, Cajal, and many others. Much information has been gathered toward a fuller understanding of the prenatal development mechanisms that become active at an early stage to produce functionally appropriate neuronal connection patterns. Focuses of current research are morphological and functional events such as induction, proliferation, migration, cell differentiation, cell death, connectivity, and synaptogenesis. Closely related to developmental biology is the degree of plasticity and repair of the nervous system in postnatal life, especially in the adult and the aging. Neuroplasticity, namely the ability of the nervous system to adapt to physiological and other events arising when development is completed, has been one of the most important topics in current biomedical research (see reviews by Bloom, 1985; Macchi, 1985). Cajal was the first to describe a regenerative process for the cerebral axons (Cajal, 1907) and adopted the view that "nerve paths are fixed, final and immutable. . .nothing may be regenerated" (Cajal, 1928). More recently, Björklund and associates (Björklund and Stenevi, 1984; Björklund et al., 1985) have expressed the opposite view that the adult mammalian brain has "the capacity for sprouting, synaptogenesis and of severed connections." The difference between these two opinions lies in the fact that Cajal could only refer to about 60 studies; today there are thousands from which to choose.

Vernadakis, A. (ed.), *Model Systems of Development and Aging of the Nervous System*.

In order to provide an overview of the current state of the subject, it is important to establish which of the many mechanisms operative during prenatal development continue after birth and how far and at what intensity they can alter existing connections or repair what has been damaged or lost. In this brief review, we consider only changes occurring in the volume and number of the neurons in the adult organism and their implication to plasticity and repair of the nervous system.

2. NEURONAL VOLUME

Variation of the volume of neurons is an active or potentially active process even in the adult and until very late in life. According to Pilati (1938), the volume of the human sensory nerve cells is greater at age 81 than at age 60. The pyramidal cells of the cerebral cortex also become appreciably larger after birth and continue to grow, like spinal ganglia cells, until the age of 50–60 years in man (Olivo and Gagliano, 1926). The active increase in volume when corporeal growth is complete and the peripheral field is not involved is undoubtedly surprising, and its significance is uncertain (Levi, 1954). Perhaps it correlates with the loss of neurons possibly occurring during aging.

Experiments have shown on several occasions that body growth and neural volume are proportional. Reference can be made to Terni's brilliant demonstration that, when a lizard's tail is regenerated after amputation, the tegument and muscles, but not the spinal cord or ganglia, are regenerated. Innervation of the tegument and muscles is provided by the cord and ganglion neurons immediately above the mutilation. Under these conditions, uniform hypertrophy of the neurons takes place and they increase by more than a third in volume (Terni, 1919 and 1922). Luria (1935), on the other hand, has shown that the spinal ganglion cells of hypoplasic rats are consistenty smaller than those of animals with normal growth. Benninghoff (1951) and Filogamo and Vigliani (1953 and 1954) have provided ample evidence of intestinal neuron hypertrophy due to hyperfunction in both adult and aged subjects, and these findings have been corroborated by Gabella (1984).

Many other studies (Greenough, 1975: Greenough et al., 1979; Chang and Greenough, 1982) have reported changes in the volume of the perikarion and dendrites of cortical neurons in animals subjected to intensive training immediately after birth. In adult animals, such training and electrical stimulation of the brain had similar, though less striking, effects on all central nervous system areas studied. Breedlove and Arnold (1983) have observed that androgen manipulation in the adult alters the volume of the spinal motor neurons directed to the musculus bulbocavernosus, whereas their number is only changed if manipulation is carried out immediately after birth. According to Vernadakis et al. (1982), neuronal plasticity in the adult may be associated with glial cell function. Enhancement of neural volume has also been noted in vitro. When chick embryo spinal ganglia were kept in culture for 5 months, neural volume increased to a greater size than had been observed in the adult bird (Meyer, 1938; Meyer and Jablonsky, 1938). Pedicled lobes also formed in some cultures and the axons always put out large numbers of collaterals. The in vitro and in vivo models for regeneration studies have provided a basis for the study of neurotrophic factors like NGF that may sustain anabolic activites in the neuronal soma (Varon and Manthorpe, 1985). Enhancement of the volume of the perikarion in the adult is frequently accompanied by the emission of dendrites, of collaterals to the neurite, and of new branches of the axon terminals. Fenestrate apparatuses and cell prolongations with clubbed terminals [paraphytes, according to Nageotte (1907)] appear in the sensory neurons of vertebrates.

The plasticity of neuronal prolongations, and of the axon in particular, in relation to age was apparent in the observations of Perroncito (1905), Lugaro (1906), Cajal (1907 and 1910), Marinesco (1907), Nageotte (1907), and Tello (1923). According to Terni (1922), De Castro (1922), Slavich (1932), and Amprino (1938), intracapsular dendrites equivalent to paraphytes form in the human superior cervical ganglion: dendrites ramify in a complicated fashion and wrap themselves around the perikarion. The pyramidal cells of the cortex also emit a large number of collaterals (Olivo and Gangliano, 1926), while short, thin prolongations form in the polymorphous cells of the gyrus dentatus (Rio-Hortega, 1918). According to Levi (1946), since the neuron is a morphological unit, progressive extension of arborizations and the formations of new collaterals can be regarded as expressions of an increase in individual mass that begins during, and continues after, somatic growth is completed (figure 7-1). Levi (1954) also suggests that the ends of thin prologations are endowed with amoeboid activity in the mature nervous system. This activity may be supposed to initiate, limit, or interrupt the synaptic relationship.

Much research has recently been conducted on changes in the prolongations in both the peripheral and central nervous systems as part of the investigation of synaptic plasticity. According to Smith (1982), in the adult rat, complexity of the terminal arborization of the motor neurons of the phrenic nerve results in motor end plates of the diaphragm richer terminals and more ultrastructurally complex motor end plates of the diaphragm in the old as compared to the young animal. Smith relates this process to an alteration of the equilibrium between sprouting and degeneration of terminals. Guth (1974) has expressed the view that "the preceding studies have revealed the central nervous system's rather remarkable capacity for morphological and physiological plasticity."

Marked variations have been noted by Sladek (1983) in the innervation pattern of catecholaminergic neurons on oxytocin and dopamine neurons in the rat starting from the 12th month of life. The neurons of the supraoptic and paraventricular nuclei are richer in synaptic contacts in aged as opposed

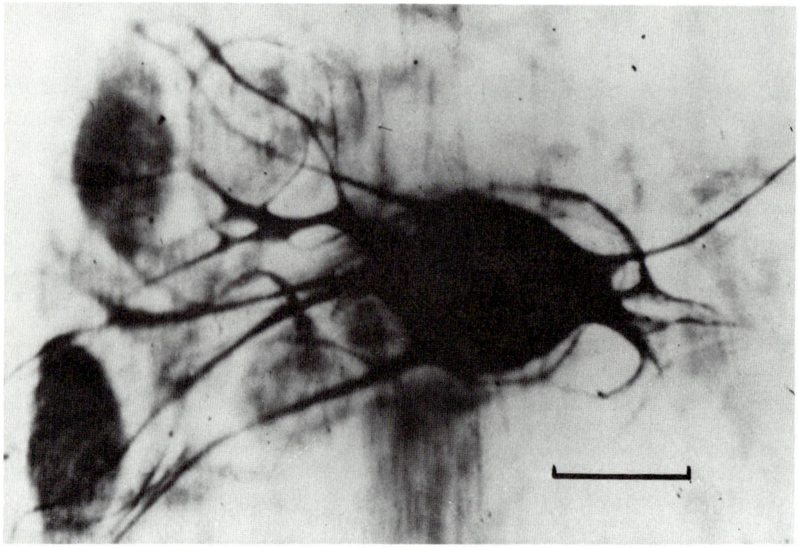

Figure 7-1. Large neuron of goat Auerbach's plexus. Short and long arborizations are visible. Bar = 20 μm.

to young animals. These data can be considered as further evidence of the notable plasticity of the central nervous system, even during senescence.

Enrichment of the terminal arborization and neosynaptogenesis have also been described in the hippocampus (Lynch and Cotman, 1975; Lynch et al., 1975) and other structures, such as the red nucleus after deafferentation (Murakami et al., 1982). According to McWilliams and Lynch (1978 and 1979), reinnervation in the molecular layer of the gyrus dentatus in 3- to 6-month old rats is complete 8 weeks after lesion of the commissural afferents. It is accompanied by the formation of new terminal boutons, expansion of those already present, and genesis of new synaptic sites. Purves and Hadley (1985) have recently employed a brilliant original technique for visualizing individual superior cervical ganglion neurons in the young rat. The dendrites changed their morphology every 2 weeks or at longer intervals: some shrank, others lengthened, while others seemed to form for the first time. These authors also maintain that the dendrite changes are probably accompanied by changes in the synaptic connections of the neurons. On the whole, synapses are continuously reworked (Soetelo and Palay, 1971) and may also be formed ex novo in the adult as an aspect of an age-dependent synaptic reorganization process.

The plasticity of the neuronal arborizations is also well illustrated by the regeneration processes: injured peripheral nerves can regenerate and reestablish anatomical and functional connections with distant targets (Gorio et al.,

1981). In the central nervous system, the ingrowth of nerve fibers was long regarded as limited to a few millimeters. However, recent studies have partially modified this view. Many types of neurons have a high regeneration capacity, and the formation of new synaptic contacts is possible following damage to the central nervous system in adult laboratory animals. The central monaminergic and cholinergic systems in particular have a high regenerative capacity as well as specific guidance mechanisms (Weiss, 1968; Björklund and Stenevi, 1984; Björklund et al., 1985).

3. NEURONAL NUMBER

In addition to changes in the volume and plasticity of the arborizations, a decrease or an increase in neuron number could be a modulatory factor in organization of the nervous system, even in the adult. As far as decrease in neuron number is concerned, cell counts carried out in different areas and at different ages produced contrasting results (see reviews by Levi, 1946; and Oppenheim, 1981). Morpurgo and Tirelli (1892) were of the opinion that modification of the specific cells of the spinal ganglia in the rabbit ceases when the fetus reaches 4 mm; its growing functional requirements are met by subsequent differentiation of immature cells acting as a reserve. Olivo et al. (1932) and Pilati (1938), however, reject the existence of these reserve cells. In their view, the functional requirements imposed by somatic growth are supplied by uniform hypertrophy of all the ganglionic cells. According to Milone (1923), in the spinal cord of chick embryos, mitotic activity is exhausted by the 12th incubation day, after which the number of neurons remains the same. This and other works initially established the opinion that the number of nerve cells is determined at an early stage of prenatal development or immediately after birth, and is no longer open to further modification. No undifferentiated cells capable of multiplying remained, nor were there any reserve cells capable of differentiating later. Around 1940, however, this view began to be questioned. Evidence accumulated to show that both the central and peripheral nervous systems contained areas in which a certain degree of plasticity appeared to be associated not only with an increase in neuron volume, but also with an increase in their number. Celestino da Costa (1939) and Picard and Chambost (1953) demonstrated the existence of nerve cells in a rudimentary stage of morphological and functional differentiation in the adrenal paraganglia, even below the capsule cells surrounding the bodies of the ganglionic neurons.

3.1 Peripheral Nervous System

Miura (1913) and Benninghoff (1951) were the first to provide experimental evidence of an increase in the number of neurons in the peripheral nervous system. Benninghoff induced hypertrophy and hyperplasia in the adult rat small intestine smooth muscle cells upstream from a stenosis imposed by ligation. The thickness of the layer of circular musculature was increased threefold at the boundary with the longitudinal layer. An increase in the size and number of the ganglionic cell bodies of Auerbach's plexus was also evident, although the underlying process involved in this increase could not be determined. The segment of intestine 4 cm distal to the stenosis displayed signs of atrophy, and the volume of the nuclei of its ganglion cells was reduced by half. In the light of these data, Benninghoff drew the conclusion that a morphologically demonstrable capacity for adaptation of the autonomic system exists and can be attributed to both Auerbach's and Meissner's plexuses.

Filogamo and Vigliani (1953 and 1954) have repeated Benninghoff's experiments in the dog and have also carried out an extensive statistical assessment of the correlation between the size of the innervated area and the number of ganglion cells in Auerbach's plexus. Operated animals were killed at intervals up to day 40 after surgery. In each case, graphic representations were made of the neurons of Auerbach's plexus in the dilated, hypertrophic loop upstream from stenosis, and the downstream loop, starting serial 1-cm^2 sections of the intestinal wall. It was found that increased hypertrophy of the neurons of Auerbach's plexus was pronounced in the upstream loop. This was not, however, the only expression of adaptation of the ganglionic mass to the increased peripheral field. It was clear, in fact, that as early as 4–5 days after the operation a marked increase in the number of ganglion neurons could be seen at high magnification under the light microscope. Some groups of neurons were newcomers in the web of the plexus and constituted new, small ganglia. In some cases, there was a fourfold increase in nerve cells compared with the downstream loop. It was later shown under the same experimental conditions in the rabbit (Filogamo, 1960) that, in addition to the large plexus neurons evidently already present, small cells become intensely cholinesterase positive using Koelle reaction for acetylcholinesterase and are readily distinguishable from cholinesterase-negative cells of other kinds. Moreover, when examined on preparations of delaminated intestinal wall silver impregnated according to Bielchowsky-Gross, close to large neurons of plexus those same cells display very short, needle-like prolongations and are the exact image of the neuroblasts noted at an early stage in the plexus ganglia during prenatal development (figure 7-2). At first these prolongations do not exceed the confines of the ganglion. In animals observed later, on the other hand, they are longer and ramify into the meshes of the plexus. The network thus becomes thicker as a result of both the prolongations and the emission of collaterals by neurons already present.

The nerve cells thus exploited are probably the result of poorly differentiated or undifferentiated cells capable of eventually turning into neurons. They are cells that differentiate, not during normal ontogenetic development, but only under the influence of exceptional formative stimuli. This interpretation stems from the observation that the increase in number is not

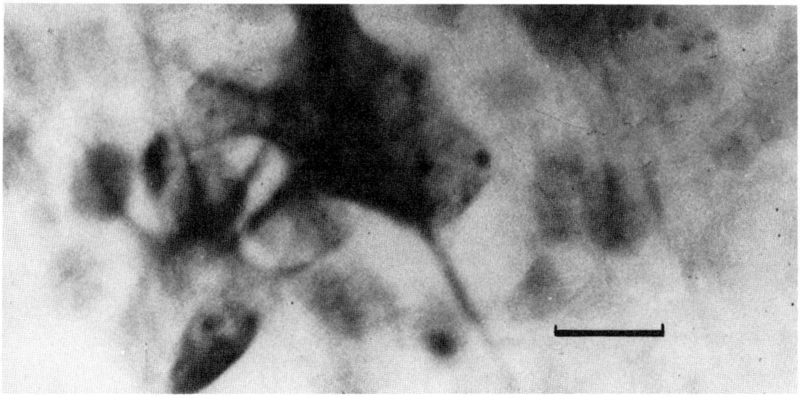

Figure 7-2. Two small neurons with short needle-like prolongations in goat plexus ganglia. Bar = 10 μm.

accompanied by evidence of nerve cell division, nor cells of any other kind present in the ganglion. Using colchicine, Filogamo and Vigliani (1953 and 1954) and Filogamo and Lievre (1955a) never observed mitotic figures in the ganglia. Gabella and Gaia (1967) also report an absence of blocked mitoses in nearly all the ganglia of animals stenosed and treated with colchicine. No uptake of thymidine by the ganglia was noted following treatment with ^{3}H-thymidine (Gabella and Gaia, 1967). Yet the total number of nerve cells is four times greater in the dilated, hypertrophic loop compared with the loop downstream. These findings led us to suggest that undifferentiated cells capable of becoming neurons persist in the adult plexus. Moreover it has been shown (Filogamo and Lievre, 1955a) that, when a normal segment is excluded from the transit of intestinal contents for approximately 40 days by connecting it to the abdominal wall, the nerve cells of intraneural ganglia decrease considerably in size, but not in number. If this experiment is carried out on a segment with signs of hypertrophy following previous stenosis (nervous cells being more numerous than in a normal segment), all cells decrease in volume, but not in number.

Similar conclusions have been drawn by Dal Forna and Rua (1961), who observed an increase in the number of ganglion cells when the muscle tunica hypertrophized in an adult human appendix. This finding supports the view that neurogenesis may be the outcome of functional stress. Gentilli (1955) has noted an increase in both the volume and the number of Auerbach's plexus neurons in two of four cases of stenosing tumor of the human small intestine. Levi (1954), Bairati (1954), Amprino (1955), Sthör (1957), Lorenz (1962), and Schofield (1962) are all of the opinion that "reserve" nerve cells, already decribed by Coujard (1949), may well remain in the peripheral nervous system of fully grown higher vertebrates. However, Corelli (1958) and Gabella (1984) maintain the view that there is no "firm evidence of increase in the number of nerve cells (neuronal hyperplasia). Increase in the number of neurons counted is attributable to the fact that large neurons were more easily detected." In the absence of mitosis and of an exact knowledge of the degree of initial differentiation of the cells, a morphologist cannot speak of a numerical increase. However, it should be noted that, whatever the degree of initial differentiation, synthesis of acetylcholinesterase only takes place in response to an exceptional stimulus, as it occurs in all neuroblasts in the early stages of their development (Filogamo and Marchisio, 1971). More importantly, these new cells put out prolongations both inside and outside the ganglion and probably establish synapses, so that they become part of the functional structure of the plexus. In this way, a new organization is established.

Considering another area, the basal cells of the olfactory neuropithelium have been identified as stem elements of sensory neurons. Moulton et al. (1970), Graziadei and Monti Graziadei (1978 and 1979), and Samanen and Forbes (1984) have reported that olfactory receptor cells retain a capacity for neurogenesis in adulthood. Studies of ^{3}H-thymidine uptake have revealed spontaneous stem cell division and neural differentiation in otherwise untreated adult rodents.

3.2 Central Nervous System

Postnatal neurogenesis involving microneurons and mitoses has been described since the 1960s. Altman and associates (Altman, 1966; Altman and Das, 1965) are of the opinion that "the prenatally formed afferent relay and efferent macroneurons represent the unvaried or fixed components of the central nervous system whereas the interposed microneurons, which develop postnatally as the animal comes to face and respond to the varied external environment, could be its modulatory and plastic elements." The number of cells displaying thymidine uptake in the gyrus dentatus is very high in newborn rats and falls to a level that can still be detected in adults; the same phenomenon appears in the cerebellar cortex and the olfactory bulb. Altman's work on the central nervous system is new and in opposition to Cajal's (1928) view that "everything may die, nothing may be regenerated," which has been upheld by several workers in view of the lack of thymidine uptake. Penfield (1932), Smart (1961), Noetzel and Rox (1964), Fujita (1874), Warwick and Williams (1973), and Jacobson (1978) deny the existence of neurogenesis in the mammalian neocortex in the adult.

Bayer et al. (1982) maintain that the number of granule cells in the rat hippocampus increases up to the 120th day of life. Kaplan and Hinds (1977 and 1981), while stressing the impact of the technical problems associated with counting mitoses, conclude that cells displaying ^{3}H-thymidine labeling in the olfactory bulb and gyrus dentatus in the adult rat are, in effect, new neurons. In their veiw "growth and plasticity including neurogenesis and synaptogenesis can also occur in mature mammalian brain." Bayer et al. (1982) have estimated that the number of granule cells in the rat increases by 770,000 between the 30th day and the first year of life. Further observations on the gyrus dentatus were made by Schlessinger et al. (1975) and Cowan et al. (1981).

In 1983, Goldman and Nottebohm observed labeled neurons in the hyperstriated ventral nucleus of the adult female canary after parenteral administration of ^{3}H-thymidine, and in particular labeled cells in the paraventricular area near the nucleus; these cells are regarded as precursors. They proliferate, differentiate, and migrate into the nucleus. The neurogenesis mechanism has a rate of neuron production of 0.9%–2.0%. Paton and Nottebohm (1984) observed that they are recruiting into functional brain circuits.

Rakic (1985) has concluded that in Rhesus monkey the presence of slight neuronal labeling of the numerous central nervous system centers examined after thymidine administration "may reflect a technical aspect of DNA repair particularly in the cases subjected to multiple injections of the isotope. Furthermore, immunohistochemistry at the light and electron microscopic level with antibody to glial fibrillary acidic protein showed that none of the radiolabelled cells in the mature dentate gyrus had neuronal properties." His conclusion is that in primates the increase in neuron number is completed during a brief developmental period that ends during the first half of gestation, except for some granule cells in the cerebellum and hippocampus.

It is in the light of these results in the central nervous system that we have further investigated stenosis-induced intestinal wall hypertrophy model to see whether thymidine is taken up by newly generated cells (Giacobini, Robecchi et al., 1985). Under physiological conditions, the cells of Auerbach's plexus have been observed to take up thymidine only occasionally (Messier and Leblond, 1960), and no uptake was noted even in stenosed rats by Gabella and Gaia (1967), although the administration modality of the thymidine was not such as to ensure sufficient cumulative labeling.

In our studies, 5 days after stenosis, some rats received four injections containing 1 μCi/g body weight ^{3}H-thymidine (New England Nuclear, specific activity 15–25 Ci/mmol) at 6-hour intervals and were killed 6 hours after the last injection. This 24-hour cumulative labeling resulted in autoradiographic evidence of significant uptake by the Auerbach's plexus in an area of the dilated and hypertrophic intestinal wall 5–6 cm upstream from the stenosis. This labeling was confined to some very large cells (figure7-3). As can be seen in figure 7-4, Nissl staining on a tight series of slices also reveals well-differentiated nerve cells. There was no evidence of uptake on the part of small and undifferentiated cells and mitotic figures.

It is universally agreed that thymidine uptake is an index of DNA synthesis. Data in the literature indicate that in the adult nervous system it can be accomplished by cells in the mitotic cycle, whether they are glial cells (Smart

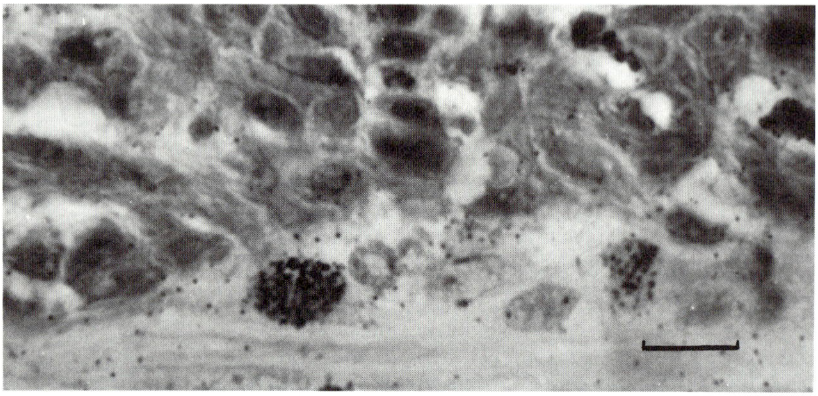

Figure 7-3. Autoradiography after injection of 3H-thymidine. A heavily labeled large nucleus of a ganglionic cell of rat Auerbach's plexus upstream from stenosis. Bar = 20μm.

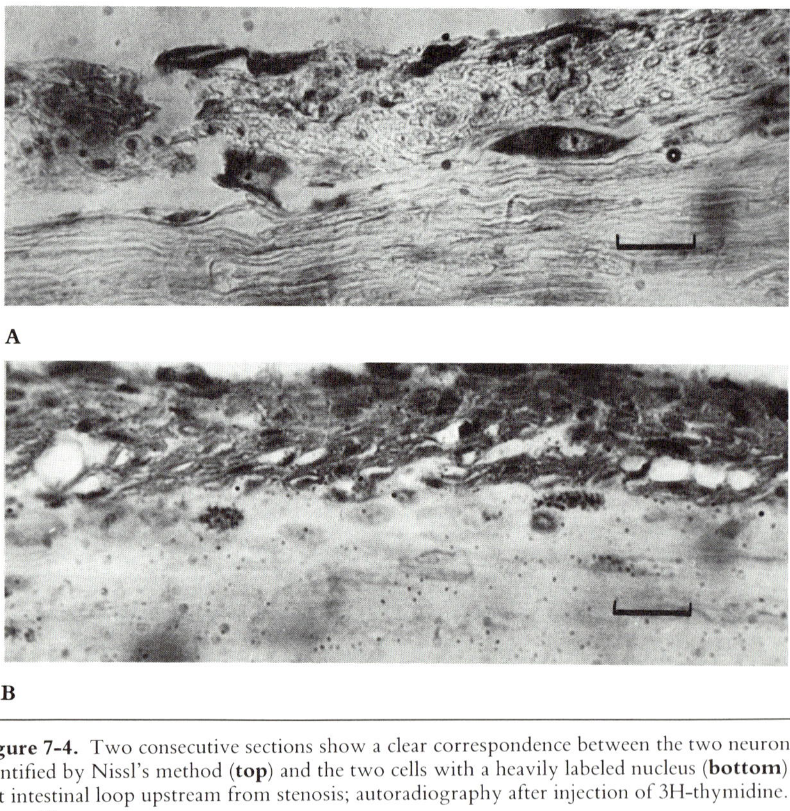

A

B

Figure 7-4. Two consecutive sections show a clear correspondence between the two neurons identified by Nissl's method (**top**) and the two cells with a heavily labeled nucleus (**bottom**). Rat intestinal loop upstream from stenosis; autoradiography after injection of 3H-thymidine. Bars = 50 μm.

and Leblond, 1961) or precursors of nerve cells (Altman and Das, 1965a and b; Bayer and Altman, 1974; Kaplan, 1981; Kaplan and Hinds, 1977; Bayer et al., 1982; Goldman and Nottebohm, 1983), or as a result of DNA repair (Rakic, 1985) or polyploidization (for references, see Swartz and Bathnagar, 1981). In our experimental system, nonuptake of thymidine by small cells and the absence of mitotic figures imply that cell multiplication can be ruled out. Furthermore, the extent of the labeling shows that DNA repair was not involved. It may thus be suggested that thymidine uptake by differentiated nerve cells is more probably linked to polyploidization. Polyploidy in nerve cells is both affirmed (Lapham, 1968; Herman and Lapham, 1969; Novàkovà et al., 1970; Bregnard et al., 1975; Bohn and Mitchell, 1976; Swartz and Bathnagar, 1981) and denied (Mann and Yates, 1973; Mares et al., 1973; Cohen et al., 1973; Fujita, 1974; McIlwain and Capbs-Covey, 1976; Fukada et al., 1978). The controversial results may be attributed to technical difficulties as suggested by Swartz and Bathnagar (1982). Our experimental model is particularly suitable in this respect: by causing a massive and very rapid increase in the peripheral field, it may trigger in already differentiated neurons DNA synthesis that is so rapid and pronounced that it can be evidenced through thymidine uptake. The above results support the studies by Filogamo and Vigliani (1953 and 1954) and reveal that newly generated cells are not present, suggesting that the increase in neuron volume is accompanied by DNA neosynthesis.

4. GENERAL CONCLUSIONS

Intense research over the last ten years, especially at the molecular level, and continuous integrated evaluation of results have permitted an in-depth investigation of many questions relating to the fine structures of the nervous system. In contrast, the major dynamic problems, like quantitative adjustment of neuron size and number in the adult, have been questioned for a century and are still not fully understood.

An analysis—inevitably incomplete—of the literature shows that, with the exception of the olfactory neuroepithelium and the ventral hyperstriated nucleus of the canary, the ability of the nerve cells to reproduce is bound within very narrow spatial and temporal limits. In primates, this ability is lacking in the adult. In some areas, all of them peripheral, it seems to have been demonstrated that some poorly differentiated or undifferentiated cells are ready to become neurons and enter the nerve circuits due to the effect of various stimuli. No one has found such cells in the central nervous system, and few have looked for them. All other cells that differentiate into neurons before and immediately after birth selectively and specifically modulate their volume, arborizations, and terminals and are influenced by extension of the field of innervation, by specific training, by the action of hormones and growth factors, by microenvironment, and possibly by injury to neighboring areas.

The richness and ubiquity of changes in volume, when compared with the paucity of changes in number (at least as far as the central nervous system!) implies that upgrading of "old" structures is preferred to the insertion of new ones in the dynamics of postnatal neural organization. The question of why and how still appears to be open.

ACKNOWLEDGMENT

This investigation was supported by grants from Italian MPI and CNR.

REFERENCES

Altman, J. (1966) Autoradiographic and histological studies of postnatal neurogenesis. J. Comp. Neurol. 128:431–474.

Altman, J., and Das, G.D. (1965a) Post-natal origin of microneurones in the rat brain. Nature 207:953–956.

Altman, J., and Das, G.D. (1965b) Autoradiographic and histological evidence of post-natal hippocampal neurogenesis in rats. J. Comp. Neurol,. 124:319–336.

Amprino, R. (1938) Modifications de la structure des neurones sympathiques pendant l'accroissement et la sénescence: recherches sur le ganglion cervical supérieur. C.R. Assoc. Anat. 33:1–16.

Amprino, R. (1955) Morphological and functional aspects of the innervation of blood vessels. Minerua Cardioangiol. Int. 2:5–19.

Bairati, A. (1954) In: Cacucci-Bari (ed.) Trattato di Istologia.

Bayer, S.A., and Altman, J. (1974) Hippocampal development in the rat: cytogenesis and morphogenesis examined with autoradiography and low-level x-irradiation. J. Comp. Neurol. 158:55–80.

Bayer, S.A., Yackel, J.W., and Puri, P.S. (1982) Neurons in the rat dentate gyrus granular layer substantially increase during juvenile and adult life. Science 216:890–892.

Benninghoff, A. (1951) Vermehrung und Vergrösserung von Nervezellen bei Hypertrophie des Innervationsgebietes. Z. Naturforsch. 6:38–44.

Björklund, A., and Stenevi, U. (1984) Intracerebral neural implants: neuronal replacement and reconstruction of damaged circuitries. Annu Rev. Neurosci. 7:279–308.

Björklund, A., Gage, F.N., Dunnett, S.B., and Stenevi, U. (1985) Regenerative capacity of central neurons as revealed by intracerebral grafting experiments. In: Bignami, A., Bloom, F.E., Bolis, C.L., and Adeloye, A. (eds.) Nervous system plasticity and repair. New York: Raven, pp. 57–62.

Bloom, F.E. (1985) Central nervous system plasticity: a survey of opportunities. In: Bignami, A., Bloom, F.E., Bolis, C.L., and Adeloye, A. (eds.) Central nervous system plasticity and repair. New York: Raven.

Bohn, R.C., and Mitchell, R.B. (1976) Cytophotometric identification of tetraploid Purkinje cells in young and aged rats. J. Neurobiol. 7:255–258.

Breedlove, S.M., and Arnold, A.P. (1983) Hormonal control of a developing neuromuscular system. J. Neurosci. 3:424–432.

Bregnard, A., Knüsel, A., and Kuenzle, C.C. (1975) All are the neuronal neclei polyploid? Histochemistry 43:59–61.

Cajal, S.R. (1907) Note sur la dégénérescence traumatique des fibres nerveuses du cervelet et du cerveau. Trab. Lab. Invest. Biol. 5:105–115.

Cajal, S.R. (1910) Algunas observaciones favorables a la hipotesis neurotropica. Trab. Lab. Invest. Biol. 8:63–134.

Cajal, S.R. (1928) In: Degeneration and regeneration of the nervous system. London: Oxford University Press.

Celestino da Costa, A. (1939) Paraganglions et sympathique. Ann. Endocrinol 1:449–464.

Chang, F.F., and Greenough, W.T. (1982) Lateralized effects of monocular training on dendritic branching in adult split-brain rats Brain Res. 232:283–292.

Cohen, J., Mares, V., and Lodin, Z. (1973) DNA content of purified preparations of mouse

Purkinje neurons isolated by a velocity sedimentation technique. J. Neurochem. 20:651–657.
Corelli, D. (1958) Studio sulla rigenerazione delle cellule gangliari dell'intestino con il metodo di Cannon. Biol. Latina 11:395–405.
Coujard, R. (1949) Degrès de différenciation varisbles des cellules des plexus intestinaux. C.R. Assoc. Anat. 36:151–160.
Cowan, W.M., Stanfield B.B., and Amaral, D.G. (1981) Further observations on the development of the dentate gyrus. In: Cowan, W.M. (ed.) Studies in developmental neurobiology. New York: Oxford University Press, pp. 395–435.
Dal Forno, and Ruà, S. (1961) Effetti plastici del rimaneggiamento del tessuto linforeticolare nell'appendice. II. Modificazioni dei plessi nervosi intramurali. Biol. Latina 14:383–416.
De Castro, F. (1922) Evoluciòn de los ganglios simpàticos vertebrales y prevertebrales Connexiones y citoarquitectonia de algunos grupos de ganglios en el nino y hombre adulto. Trab. Lab. Invest. Biol. 20:113–208.
Filogamo, G. (1960) Il plesso di Auerbach nella stenosi sperimentale dell'intestino studiato con il metodo di Koelle per l'AChE nel coniglio. Giorn. Acc. Med. Torino 123:75–81.
Filogamo, G., and Lievre, C. (1955a) Aumento di numero delle cellule nervose e moltiplicazione cellulere nei gangli del plesso di Auerbach, in condizioni di esaltata attività funzionale. Boll. Soc. Ital. Biol. Sper. 31:717–719.
Filogamo, G., and Lievre, C. (1955b) Comportamento delle cellule nervose del plesso mienterico, dell'intestino normale ed ipertrofico, nell'ansa alla Thity-Vella. Boll. Soc. Piem. Chir. 25:1–3.
Filogamo, G., and Marchisio, P.C. (1971) Acetylcholine system and neural development. Neurosci. Res. 4:29–64.
Filogamo, G., and Vigliani, F. (1953) Les cellules du plexus myentérique (d'Auerbach) dans la sténose expérimentale de l'intestin. C.R. Assoc. Anat. 78:683–692.
Filogamo, G., and Vigliani, F. (1954) Ricerche sperimentali sulla correlazione tra estensione del territorio di innervazione e grandezza e numero delle cellule gangliari del plesso mienterico (di Auerbach) nel cane. Riv. Patol. Nerv. Ment. 75:1–32.
Fujita, S. (1974) DNA constancy in neurons of the human cerebellum and spinal cord as revealed by Feulgen cytophotometry and cytofluorometry. J. Comp. Neurol. 155:195–202.
Fukuda, M., Böhm N., and Fujita, S. (1978) Cytophotometry and its biological application. In: Progress in histochemistry and cytochemistry, vol. II. Stuttgart: Gustav Fischer, pp. 1–119.
Gabella, G. (1984) Size of neurons and glial cells in the intramural ganglia of the hypertrophic intestine of the guinea pig. J. Neurocytol. 13:73–84.
Gabella, G., and Gaia, E. (1967) La proliferazione cellulare nel plesso di Auerbach di ratto in accrescimento e in condizioni sperimental. Boll. Soc. Ital. Biol. Sper. 43:1584–1586.
Gentilli, R. (1955) Ipertrofia e iperplasia delle cellule nervose del plesso mienterico (di Auerbach) nell'intestino ipertrofico in seguito a stenosi, nell'uomo. Boll. Soc. Piem. Chir. 25:13–25.
Giacobini Robecchi, M.G., Cannas, M., and Filogamo, G. (1985) Increase in the number and volume of mysenteric neurons in the adult rat. Int. J. Dev. Neurosci. 6:673–676.
Goldman, S.A., and Nottebohm, F. (1983) Neuronal production, migration and differentiation in a vocal control nucleus of the adult female canary brain. Proc. Natl. Acad. Sci. USA 80:2390–2394.
Gorio, A., Carmignoto, G., and Ferrari, G. (1981) Axon sprouting stimulated by gangliosides: a new model for elongation and sprouting. In: Rapport, M.M., and Gorio, A. (eds.) Gangliosides in neurological and neuromuscular function, development and repair. New York: Raven, pp. 177–195.
Graziadei, P.P.C., and Monti Graziadei, G.A. (1978) Continuous nerve cell renewal in the olfactory system. In: Jacobson, M. (ed.) Handbook of sensory physiology, vol. 9. New York: Springer, pp. 55–82.
Graziadei, P.P.C., and Monti Graziadei, G.A. (1979) Neurogenesis and neuron regeneration in the olfactory system of mammals. I. Morphological aspects of differentiation and structural organization of the olfactory sensory neurons. J. Neurocytol. 8:1–18.
Greenough, W.T. (1975) Experiential modification of the developing brain. Am. Sci. 63:37–46.
Greenough, W.T., Juraska, J.M., and Volkmar, F.R. (1979) Maze training effects in dendritic branching in occipital cortex of adult rats. Behav. Neural Biol. 26:287–297.
Guth, L. (1974) Axonal regeneration and functional plasticity in the central nervous system. Exp. Neurol. 45:606–654.

Herman, C.J., and Lapham, L.W. (1969) Neuronal poliploidy and nuclear volumes in the cat central nervous system. Brain Res. 15:35–48.

Jacobson, M. (1978) Developmental neurobiology, 2nd edn. New York: Plenum.

Kaplan, M.S. (1981) Neurogenesis in the 3-month-old rat visual cortex, J. Comp. Neurol. 195:323–338.

Kaplan, M.S., and Hinds, J.W. (1977) Neurogenesis in the adult rat: electron microscopic analysis of light radioautographs. Science 197:1092–1094.

Lapham, L.W. (1968) Tetrapolid DNA content of Purkinje neurons of human cerebellar cortex. Science 159:310–312.

Levi, G. (1946) Accrescimento e senescenza. Firenze: La Nuova Italia.

Levi, G. (1954) Trattato di Istologia. Torino: Unione Tipografica Editrice.

Lorenz, J. (1962) Observations comparatives sur l'innervation intramurale du cardia, du pylor et de la valvule iléo-coecale chez l'homme normal au cours de l'âge. Z. Mikrosk. Anat. Forsch. 68:540–563.

Lugaro, E. (1906) Sulla presunta rigenerazione autogena delle radici posteriori. Riv. Patol. Nerv. Ment. 11:8–19.

Luria, S. (1935) La grandezza delle cellule nervose in animali ad accrescimento artificialmente arrestato. Riv. Patol. Nerv. Ment. 11:8–19.

Lynch, G., and Cotman, C.W. (1975) The hippocampus as a model for studying anatomical plasticity in the adult brain. In: Issacson, R.L., and Pribram, K.H. (eds.) The hippocampus. New York: Plenum, pp. 123–154.

Lynch, G., Rose, G., Gall, C., and Cotman, C.W. (1975) The response of dentate gyrus to partial deafferentation. In: Santini, M. (ed.) Golgi cenntennial symposium proceedings. New York: Raven, pp. 305–317.

Macchi, G. (1985) Regeneration and plasticity in human central nervous system: anatomical clinical approaches. In: Bignami, A., Bloom, E.E., Bolis, C.L., and Adeloye A. (eds.) Central nervous system plasticity and repair. New York: Raven, pp. 107–114.

Mann, D.M.A., and Yates, P.O. (1973) Poliploidy in the human nervous system. I. The DNA content of neurones and glia of the cerebellum. J. Neurol. Sci. 18:183–196.

Mares, V.L., Lodin, Z., and Sacha, J. (1973) A cytochemical and autoradiographic study of nuclear DNA in mouse Purkinje cells. Brain Res. 53:273–289.

Marinesco, N. (1907) Quelques recherches sur la transplantation des ganglions nerveux. Rev. Neurol. 6:17–36.

McIlwain, D.L., and Capps-Covey, P. (1976) The nuclear DNA content of large ventral spinal neurons. J. Neurochem. 27:109–112.

McWilliams, R., and Lynch, G. (1978) Terminal proliferation and synaptogenesis following partial deafferentiation: the reinnervation of the inner molecular layer of the dentate gyrus following removal of its commissural afferents. J. Comp. Neurol. 180:581–616.

McWilliams, R., and Lynch, G. (1979) Terminal proliferation in the partially deafferented dentate gyrus: time course for the appearance and removal of degeneration and the replacement of lost terminals. J. Comp. Neurol. 187:191–198.

Messier, B., and Leblond. C.P. (1960) Cell proliferation and migration as revealed by radioautography after injection of thymidine-H3 into male rats and mice. Am. J. Anat. 106:247–285.

Meyer, H. (1938) Nenere Beobachtungen bei der Züchtung von Rautenhirn-Explantaten aus hünerembryonen vorgerückter Stadien. Riv. Biol. 26:1–11.

Meyer, H., and Jablonsky, W. (1938) Coltivazione protratta di cellule nervose in vitro. Monit. Zool. Ital 48:70–73.

Milone, S. (1923) La frequenza delle divisioni mitotiche negli elementi del midollo spinale embrionale di animali di differente mole corporea. Arch. Ital. Anat. Embriol. 20:417–432.

Miura, M. (1913) Aus notizien meiner phisiologischen und patologischen forschung: XXVI. muscularis mucosae et muscularis propria gastrointestinalis. Toko: Nakodo.

Morpurgo, B. and Tirelli, V. (1892) Sullo sviluppo dei gangli intervetebrali del coniglio. Ann. Fren. Sci. Affini 3:1–32.

Moulton. D.G., Celebi, G., and Fink, R.P. (1970) Olfaction in mammals: proliferation of cells in the olfactory epithelium and sensitivity to odours in taste and smell in vertebrates. In: Wolstenholme, G.E.W., and Knight, J. (eds.) Ciba foundation simposium. London: Churchill.

Murakami, F., Katsumaru, H., Saito, K., and Tsukahara, N. (1982) A quantitative study of synaptic reorganization in red nucleus neurons after lesion of the nucleus interpositus of the cat: an electron microscopic study involving intracellular injection of horseradish peroxidase. Brain Res. 242:41–53.

Nageotte, J. (1907) Etude sur la greffe des ganglions rachidiens. Anat. Anz. 31:225–245.

Noetzel, J., and Rox, J. (1964) Autoradiographische Untersuchungen uber Zellteilung und Zellentwicklung im Gemhirn der erwachsenen Maus und der erwachsenen Rhesus-affen nach Injection von radioakitivim Thymidin. Acta Neuropathol. 3:326–342.

Novàkovà, V., Sandritter, W., and Schlueter, G. (1970) DNA content of neurons in rat central nervous system. Exp. Cell Res. 60:454–456.

Olivo, O., and Gangliano, R. (1926) Modificazioni di forma e di grandezza delle cellule piramidali della circonvoluzione centrale anteriore umana durante l'accrescimento somatico. Boll. Soc. Ital. Biol. Sper. 1:111–114.

Olivo, O., Porta, E., and Barberis, L. (1932) Ricerche sulla velocità di accrescimento delle cellule e degli organi. IV. Modalità di accrescimento delle cellule dei gangli spinali nel pollo durante la vita embrionale e postnatale. Arch. Ital. Anat. 30:34–71.

Oppenheim, R.W. (1981) Neuronal cell death and some related regressive phenomena during neurogenesis: a selective historical review and progress report. In: Cowan, W.M. (ed.) Studies in developmental neurobiology. New York: Oxford Unviersity Press. pp. 74–133.

Paton, J.A., and Nottebohm, F.N. (1984) Neurons generated in the adult brain are recruited into functional circuits. Science 225:1046–1048.

Penfield, W. (1932) Neuroglia: normal and pathological. In: Penfield, W. (ed.) Cytology and cellular pathology of the nervous system. New York: Hoeber, pp. 423–479.

Perroncito, A. (1905) La rigenerazione delle fibre nervose. Boll. Soc. Med. Chir. Pavia 3:15–27.

Picard, D. and Chambost, M. (1953) Les cellules de remplacement dans le système nerveux vegetatif et les paraganglions: observations sur les formations nerveuses intrasurrénaliennes. Arch. Anat. Microsc. Morphol. Exp. 42:86–101.

Pilati, L. (1938) Statistische Untersuchungen über das Wachstum der Nervenzellen der menschlichen Spinal Ganglien. Mikrosk. Anat. Forsh. 44:1–44.

Purvs, D, and Hadley, R.D. (1985) Changes in the dendritic branching of adult mammalian neurons revealed by repeated imaging in situ. Nature 315:404–406.

Rakic, P. (1985) Limits of neurogenesis in primates. Science 227:1054–1056.

Rio-Hortega, P. (1918) Particularidades histologicas de la fascia dentate en algunos mamiferos. Trab. Lab. Invest. Biol. 21:291–308.

Samanen, D.W., and Forbes, W.B. (1984) Replication and differentiation of olfactory receptor neurons following axotomy in the adult hamster: a morphometric analysis of postnatal neurogenesis. J. Comp. Neurol. 225:201–211.

Schlessinger, A.R., Cowan, W.M., and Gottlieb, D.I. (1975) An autoradiographic study of the time of origin and the pattern of granule cell migration in the dentate girus of the rat. J. Comp. Neurol. 159:149–175.

Schofield, G.C. (1962) Experimental studies on the myenteric plexus in mannals. J. Comp. Neurol. 119:159–185.

Sladek, J.R. (1983) Aging of central neuropeptide systems. Psychopharmacol. Bull. 19:317–320.

Slavich, E. (1932) Confronti fra la morfologia di gangli del parasimpatico encefalico e del simpatico cervicale con speciale riguardo alla struttura del ganglio ciliare. Z. Zellforsch. 15:687–730.

Smart, I. (1961) The subsependimal layer of the mouse brain and its cell production as shown by radioautography after thymidine-H3 injection. J. Comp. Neurol. 116:325–348.

Smart, I., and Leblond, C.P. (1961) Evidence for division and transformations of neuroglia cells in the mouse brain, as derived for radioautography after injection of thymidine-H3. J. Comp. Neurol. 116:349–367.

Smith, D.O. (1982) Physiological and structural changes of the neuromuscular junctions during aging. In: Giacobini, E., Filogamo, G., Giacobini, G., and Vernadakis, A. (eds.) The aging brain: cellular and molecular mechanisms of aging in the central nervous system. New York: Raven, pp. 123–137.

Sotelo, C., and Palay, S.L. (1971) Altered axons and axon terminals in the lateral vestibular nucleus of the rat. Lab. Invest. 25:635–671.

Sthör, P. (1957) Die Innervation des Dünndorrms. In: Mikroskopische Anatomie des vegetativen Nervensystems. Berlin: Springer, p. 373.

Swartz, F.J., and Bathnagar, K.P. (1981) Are CNS neurons polyploid? A critical analysis based upon cytophotometric study of the DNA content of cerebellar and olfactory neurons of the bat. Brain Res. 208:267–281.

Tello, F. (1923) Les différenciations neuronales dans l'embryon de poulet, pendant les premiers jours d'incubation. Tab. Lab. Invest. Biol. 21:1–93.

Terni, T. (1919) Sulla correlazione fra ampiezza del territorio di innervazione e grandezza delle cellule gangliari. 2. Ricerche sui gangli spinali che innervano la coda rigenerata nei sauri (Gongylus ocellatus). Arch. Ital. Anat. Embriol 17:507–544.

Terni, T. (1922) Ricerche istologiche sul midollo spinale dei rettili, con particolare riguardo ai componenti spinali del fascicolo longitudinale mediale. Arch. Ital. Anat. Embriol 18:183–243.

Varon, S., and Manthorpe, M. (1985) In vitro models for neuroplasticity and repair. In: Bignami, A., Bloom, E.E., Bolis, C.L., and Adeloye, A. (eds.) Central nervous system plasticity and repair. New York: Raven. pp. 13–23.

Weiss, P.A. (1968) Dynamics of development: experiments and inferences. New York: Academic Press.

Vernadakis, A., Parker, K., Arnold, E.B., and Noremberg, M. (1982) Role of glial cells in central nervous system again. In: Giacobini, E., Filogamo, G., Giacobini, G., and Vernadakis, A. (eds.) The aging brain: cellular and molecular mechanisms of aging in the nervous system.

Warwick, R., and Williams, P. (1973) Gray's anatomy, 35th edn. Philadelphia: W.B. Saunders.

8. OLIGODENDROCYTE DEVELOPMENT: IMMUNOLOGICAL STUDIES

NICHOLAS W. SEEDS, JANET M. COLLINS,
SUSAN C. HAFFKE, AND DESSIE FAFOUTIS

1. INTRODUCTION

Oligodendrocytes are responsible for myelination of axons in the vertebrate central nervous system (CNS). These cells are thought to originate from a precursor cell, glioblast, that gives rise to both oligodendrocytes and astrocytes (Schaper, 1897). More recent studies by Raff and coworkers (Raff et al., 1983a and b, 1984) with dissociated cultures of rat optic nerve have shown that both oligodendrocytes and type-2 astrocytes are derived from a common precursor cell, distinct from the more common type-1 astrocyte. Furthermore the differentiation of this precursor cell into oligodendrocytes or type-2 astrocytes can be influenced by components in the cell culture environment (Raff et al., 1983b).

Although neuronal membrane components have been implicated in peripheral nervous system myelination and Schwann cell proliferation (Salzer et al., 1980; Ratner et al., 1985), the role of neuron and astrocyte interactions with oligodendrocytes for oligodendrocyte proliferation and myelination is not well characterized. Hopefully the use of neural cell cultures and the development of antibodies that recognize early stages of oligodendrocyte development will help provide answers to these questions. In this chapter, we describe the use of a monoclonal antibody that reacts will a cell surface determinant on oligodendrocytes to monitor oligodendrocyte development in cultures of dissociated cerebellar cells.

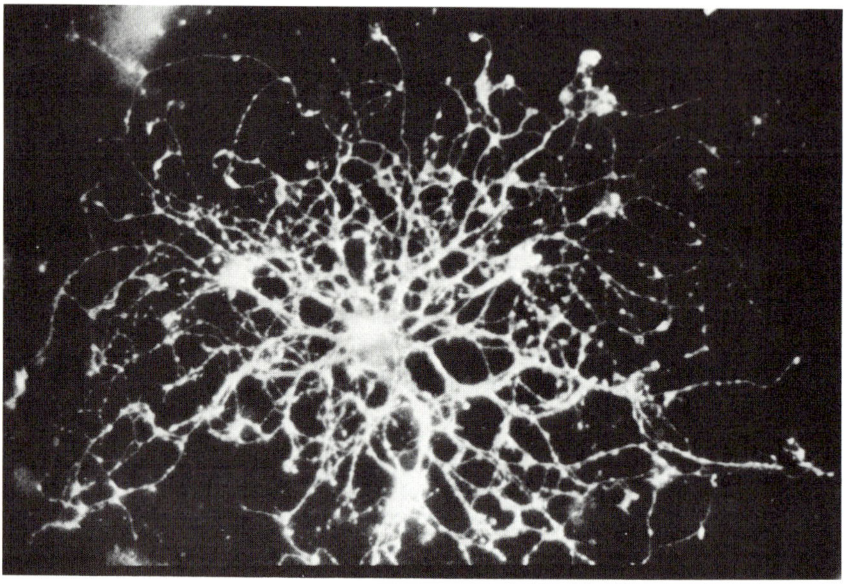

Figure 8-1. Immunofluorescence of monoclonal antibody mOg-1 binding to cerebella cell culture.

2. MONOCLONAL ANTIBODY mOg-1

As part of a study to prepare antibodies specific for neural cell surface determinants of dissociated cerebellar cells, we have previously reported on several neuronal specific antisera (Seeds, 1975; Stuhlfauth and Seeds, 1983). However, the polyclonal nature of these sera necessitated extensive adsorptions with nonneuronal tissues. Therefore, we initiated the preparation of monoclonal antibodies. Since C57B1/6 mouse brain membranes were not a good antigen when injected into Balb/c mice, we used purified plasma membranes (Stuhlfauth and Seeds, 1983) from sheep cerebella. The spleens of Balb/c mice showing circulating antibodies to mouse cerebellar cells were fused with the mouse NS-1 myeloma cell, and hybridomas selected in HAT medium (Collins, 1983). Live cell cultures prepared from 4- to 6-day-old C57B1/6 mouse cerebella were used to screen for cell-surface-specific antibodies. Several antibodies with different cell specificities were found and the hybridomas cloned twice by limiting dilution.

One of the most interesting antibodies (mOg-1) is an IgM that reacts with the surface of a minor population of cells present in cultures of 6-day-old cerebella (figure 8-1). The identity of these cells was pursued by double immunofluorescence labeling of dissociated cell populations. The binding of mOg-1 did not coincide with binding of a neuronal marker tetanus toxin (Collins and Seeds, 1986) or fibronectin, a fibroblast marker, nor with those

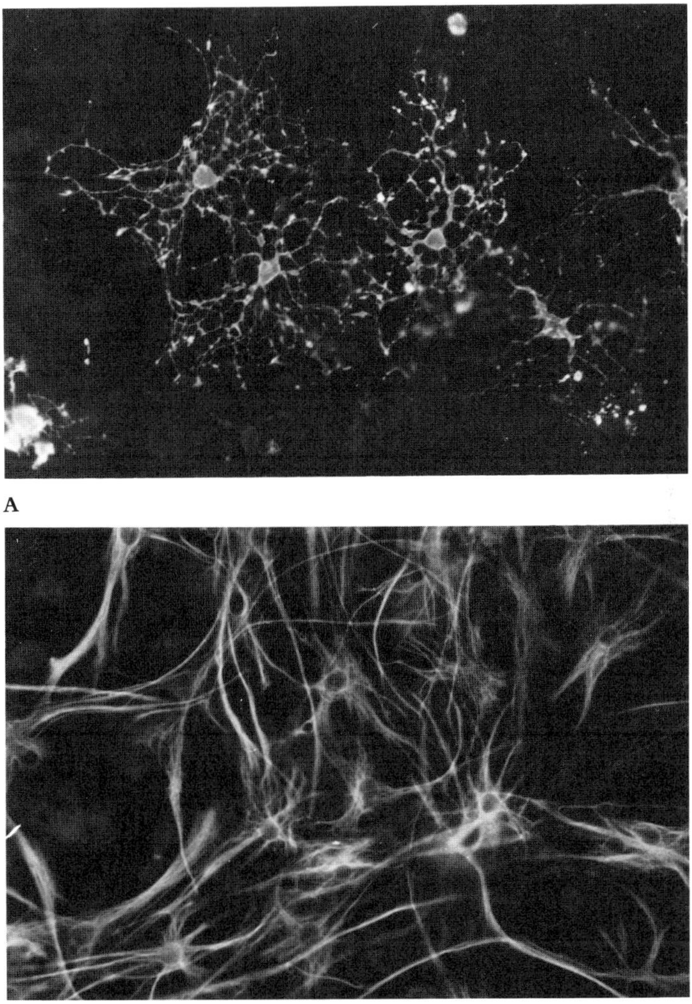

A

B

Figure 8-2. Double label immunofluorescence of oligodendrocytes and fibrous astrocytes in a cerebellar cell culture, showing the nonidentity of these cell types. (**A**) mOg-1 binding and fluorescein goat antimouse Ig. (**B**) GFAP binding visualized by rhodamine goat antirabbit Ig.

cells containing vimentin or glia fibrillary acidic protein (figure 8-2), intermediate filament proteins found in astrocytes. However, many of the cells that bound mOg-1 also bound the oligodendrocyte marker, antigalactocerebroside (figure 8-3C). Many of these same cells also reacted with antibody to myelin basic protein (Collins and Seeds, 1986). These findings suggested that mOg-1 was interacting with the surface of oligodendrocytes. This suggestion

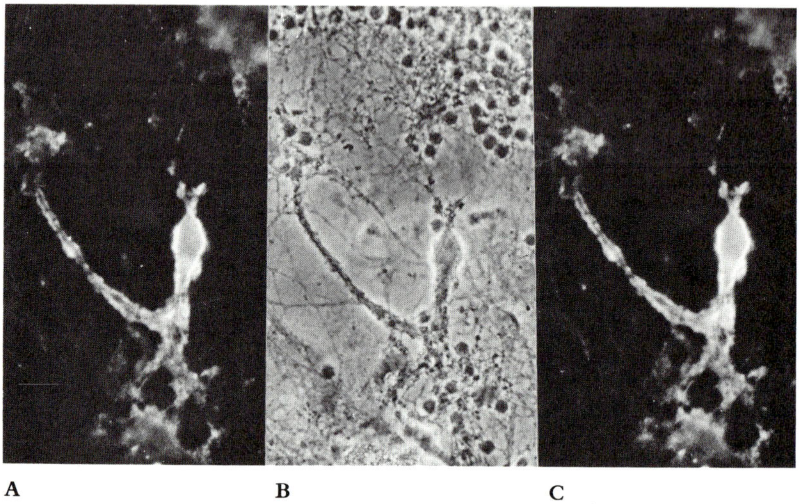

Figure 8-3. Identity of mOg-1 and antigalactocerebroside binding to an oligodendrocyte in a cerebellar cell culture. (**A**) mOg-1 binding. (**B**) Phase contrast of the same field. (**C**) Rabbit antigalactocerebroside binding.

is supported by studies with vibratome sections of neonatal (8 day) mouse cerebellum, where mOg-1 binding is concentrated in future white matter tracks (figure 8-4). Analysis of horseradish peroxidase product in the electron microscope showed it was localized to the surface of presumptive oligodendrocytes (Collins and Seeds, 1986). Analysis of adult cerebellum indicated that mOg-1 did not bind to compact myelin, indicating that the antigen had been excluded from the oligodendrocyte membrane during compaction or was not accessible to the antibody. In addition, mOg-1 binds to the surface of Schwann cells prepared from neonatal dorsal root ganglia, but only during the first 4 days of culture, after which it is apparently lost (Collins and Seeds, 1986).

Other studies showed that the mOg-1 did not bind to cells in the thymus, kidney, or spleen. Furthermore, several cloned cell lines of neural origin, including an oligodendroglioma, G26, the astrocytoma C6, and a Schwannoma TRM6B, did not bind mOg-1.

3. DEVELOPMENT IN CULTURE

Cells binding mOg-1 were first seen in suspensions prepared from 19-day fetal mouse cerebellum. Similarly, cultures prepared from 15-day fetal mice did not show mOg-1 binding until 3–5 days of culture, suggesting that cells expressing the Og-1 antigen first appear at fetal day 19.

These mOg-1 binding cells represent a minor population in dissociated 6-day cerebellar cultures, approximately 1% after 4 days of culture. After

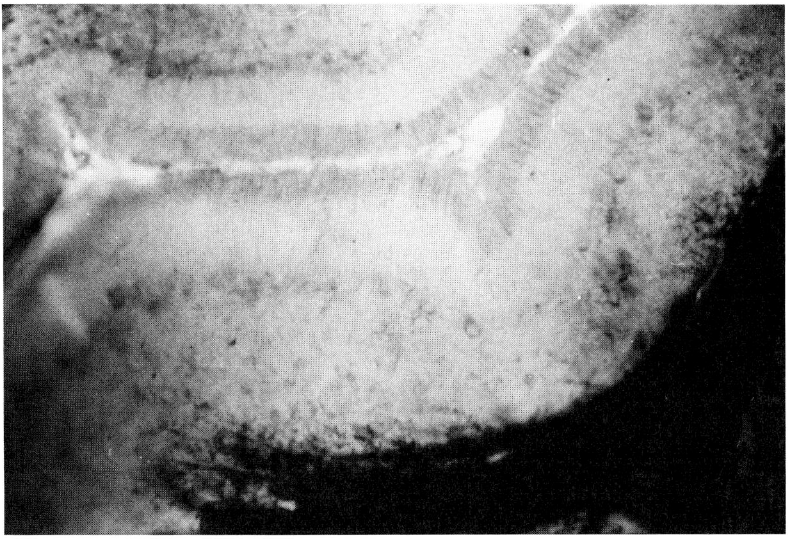

Figure 8-4. Binding of mOg-1 to future white matter tracks of 8-day rat cerebellum. A vibratome section was incubated with mOg-1 and followed by horseradish peroxidase-conjugated goat antimouse IgM.

several weeks in culture this population increased to 5%. This increase reflects, in part, the loss of other cell types, namely, granule neurons during this period. Whether this increase in precentage of cells binding mOg-1 reflects proliferation of mOg-1-binding cells or the acquisition of mOg-1 binding by previously Og-1-negative precursor cells was examined: 3-day cultures of dissociated cerebellum (5 day) cells were incubated with mOg-1 to saturate all the available Og-1 molecules on the cell surface; the cultures were rinsed to remove excess antibody and incubated for 4 days in ^{3}H-thymidine (1 μCi/ml). On day 4, some cultures were incubated for a second time (postlabel) with mOg-1, and all the cultures were labeled with rhodamine goat antimouse IgM and processed for autoradiography. Cultures prelabeled with mOg-1 showed only 0.5% labeled cells of which less than 1% had incorporated ^{3}H-thymidine during culture (figure 8-5A and B). In contrast, mOg-1-positive cells represented 1.3% of the population in postlabeled cells and about 5% of these Og-1-positive cells had incorporated ^{3}H-thymidine (figure 8-5C and D). Although some Og-1-positive cells may have lost the antigen during the 4 days of culture, this study suggests that very few cells divide once they acquire the Og-1 antigen; furthermore, a precursor cell capable of cell division probably exists in these cultures that acquires the Og-1 antigen during the 4-day culture period. This possibility is supported by studies with complement-mediated cell lysis of Og-1-positive cells in culture on day 1, which showed that all Og-1-positive cells were

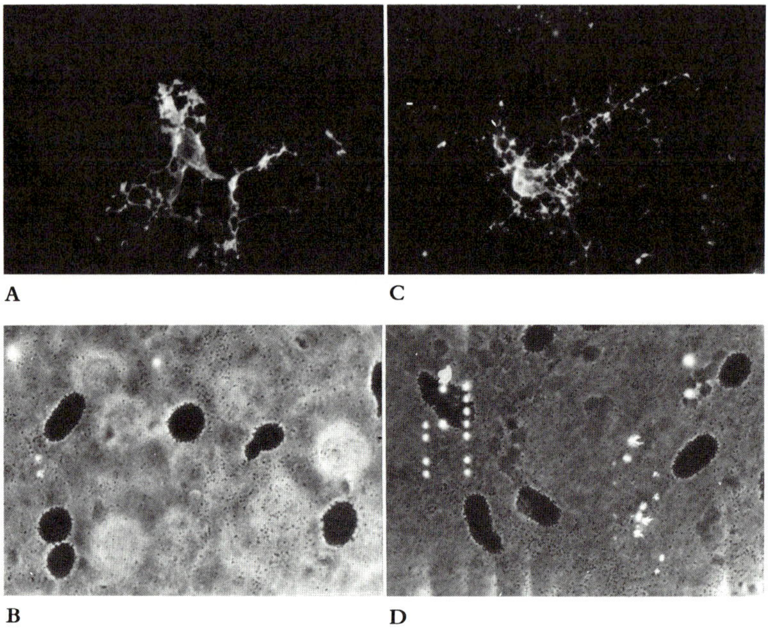

Figure 8-5. ^{3}H-thymidine labeling and mOg-1 binding to cerebellar cell cultures. (**A** and **B**). Typical Og-1 cell with bound mOg-1 (day 3) and no ^{3}H-thymidine incorporation by day 7. (**C** and **D**) A cell binding mOg-1 at day 7 and having incorporated ^{3}H-thymidine between days 3 and 7 of culture.

lysed at 3 hours after treatment; however, there was a steady increase in the number of Og-1-positive cells over the next 4 days in culture at which time Og-1-positive cells represented 1.2% of the population (Collins and Seeds, 1986). Thus a precursor cell is present in these cultures. Furthermore, complement-lysis studies with $10^{-5} M$ cytosine arabinoside also led to an increase, albeit reduced number, in Og-1-positive cells, indicating that additional cell division is not required for many of these precursor cells to acquire Og-1 antigen.

As indicated above and in figure 8-3, many of the Og-1-positive cells also bound antibody to galactocerebroside and myelin basic protein. However, the percentage of Og-1-positive cells that also bound these oligodendroglia markers increased with time in culture. At 4 days of culture, 54% of the Og-1 cells also bound galactocerebroside antibody and only 27% bound a monoclonal antibody to myelin basic protein. By 14 days of culture, 69% of the Og-1 cells bound galactocerebroside antibody and 65% bound myelin basic protein antibody (Collins and Seeds, 1986). All cells binding galactocerebroside or myelin basic protein antibodies also bound antibody mOg-1. Thus Og-1 cells appear to mature during culture and acquire additional

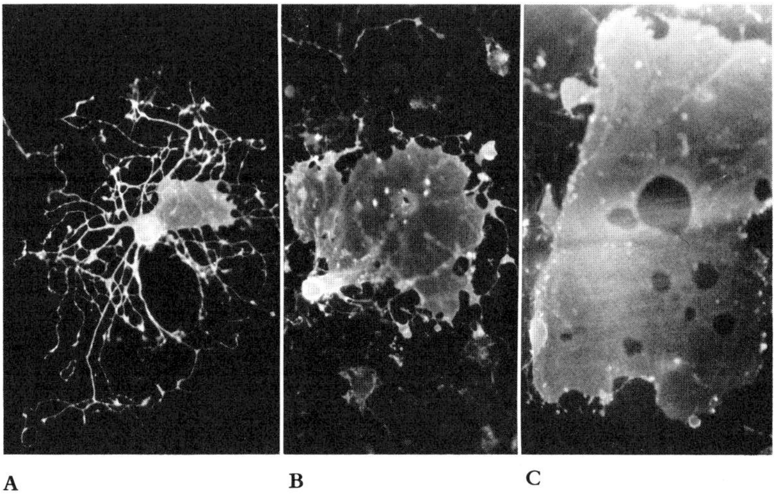

Figure 8-6. Sheath formation by oligodendrocytes as visualized by mOg-1 binding to cerebellar cells after 11 days in culture.

oligodendrocyte characteristics. The inability to reach 100% coincidence with mOg-1 and antigalactocerebroside binding may reflect the continual formation of immature Og-1 cells in the population, or indicate another subpopulation of cells that differentiate into some other cell type. Future studies with fluorescent-activated cell sorting of Og-1-positive, galactocerebroside-negative cells and their subsequent culture should distinguish these possibilities.

Since the mutant mouse Jimpy is deficient in CNS myelin, it was of interest to examine whether Og-1 was expressed on dissociated cerebellar cells from this mutant. When compared to cells from normal littermates, the Jimpy had consistently fewer Og-1-positive cells, 3.5% versus 5% of the population. In both the normal and mutant, approximately 60% of the Og-1 cells also bound galactocerebroside antibody; however, none of the Og-1 cells in Jimpy cultures expressed myelin basic protein, while 57% of the Og-1 cells from normal animals displayed myelin basic protein antibody binding (Collins and Seeds, 1986). Thus the block in oligodendroglial development in the Jimpy is at a stage after Og-1 and galactocerebroside acquisition and before myelin basic protein.

One of the most striking features of these Og-1-positive cells is their formation of large lamellar sheaths after 1–2 weeks in culture (figure 8-6). These sheaths appear to form by lamellipodial filling-in between the thin cellular processes (or extensions). These structures suggest the formation of aberrant myelin membrane; compact myelin surrounding axons is a rare event in dissociated cell cultures of the CNS (Norton, 1983). Although Og-1 and galactocerebroside are found on the sheaths, myelin basic protein is

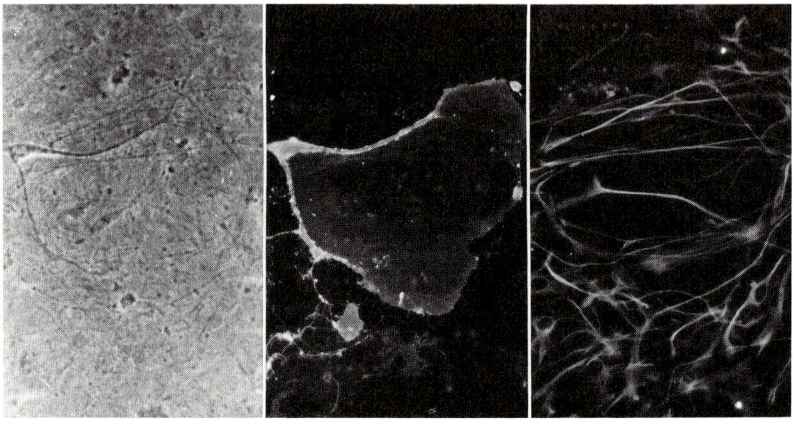

A B C

Figure 8-7. Oligodendrocyte sheath formation and its relationship to adjacent fibrous astrocytes. (**A**) Phase contrast. (**B**) mOg-I binding. (**C**) Anti-GFAP binding.

restricted to the main cellular processes. However, some sheath formation is apparent in cells from Jimpy mice. The factors regulating sheath formation are not well understood, although culture medium components may influence sheath formation. It does not require the immediate presence of a neuronal cell process, but sheaths tend to occur over areas of minimal astrocyte density (figure 8-7).

The elegant studies of Raff and coworkers described above showed that serum-free conditions favored oligodendrocyte differentiation in optic nerve culture. Therefore, we examined the effect of culture in serum-free N2 medium (Bottenstein and Sato, 1979) and the recently described ODM (oligodendrocyte-defined medium) (Saneto and De Vellis, 1985) on the percentage of Og-1-positive cells and glial fibrillary, acidic protein (GFAP)-positive cells. Cultures of 4-day cerebellum were established in 10% FCS–BEM (fetal calf serum and basal Eagle's medium) and after 1 day rinsed three times in BEM and transferred to either N2 or ODM. Table 8-1 shows that, after 4 days of culture, there was a higher percentage of Og-1-positive cells in the N2 cultures compared to the 10% FCS–BEM control, although the total cell number was reduced. After 11 days of culture, Og-1-positive cells had increased to 4.2% in the presence of serum while the percentage in N2 culture did not increase over the 4-day value. Cell survival in ODM was poor although Og-1-positive cells were a major component of the surviving population. As expected, the percentage of Og-1-positive cells was higher in serum-free N2 medium at day 4 and astrocytes represented a smaller proportion of the cells. The failure to see an increase in the percentage of Og-1 cells between days 4 and 11 may reflect an absence of Og-1 precursor cell pro-

Table 8-1. Influence of serum-free medium on oligodendrocytes.

Culture conditions	Days	Relative cell no.	% Og-1$^+$	% w̄ Sheaths	% GFAP$^+$
10% FCS–BEM	4	1	0.8	0	4
10% FCS–BEM	11	1	4.2	14	14
N2	4	0.7	1.8	0	2.6
N2	11	1.2	1.9	2	10
ODM	4	0.2	0.7	0	3
ODM	11	0.03	21	0	4

Dissociated cells from 8-day mouse cerebella were plated on polylysine-coated glass coverslips in basal Eagle's medium with 10% fetal calf serum. After 24 hours, some cultures were rinsed three times in BEM and switched to N2 or oligodendrocyte-defined medium. After 4 or 11 days, the coverslips were removed and incubated with mOg-1 antibody, fixed in 4% paraformaldehyde for 5 minutes and 4 minutes in −10°C methanol, and then incubated with rabbit anti-GFAP followed by fluorescein-conjugated goat antimouse Ig and rhodamine goat antirabbit Ig.

liferation in N2 medium. Although ODM has been shown to be a good medium for differentiated oligodendrocytes (Saneto and De Vellis, 1985), it may lack components necessary for the survival and maturation of immature oligodendrocytes in these cultures of neonatal cerebellar cells. Interestingly, sheath formation was minimal in serum-free cultures, suggesting that certain serum components may influence sheath formation.

4. INFLUENCE OF mOg-1 ON OLIGODENDROCYTE DEVELOPMENT

The presence of the Og-1 antigen on the surface of developing oligodendrocytes and its apparent absence from compact myelin suggest that Og-1 could play a role in some form of cell recognition during oligodendrocyte maturation or possibly during axon–oligodendrocyte interactions. Preliminary studies have been initiated to examine the effect of mOg-1 on oligodendrocyte development in dissociated cell cultures and in brain cell reaggregates (Seeds, 1973, 1983) where myelination occurs naturally (Seeds and Vatter, 1971; Seeds and Haffke, 1978; Trapp et al., 1979).

The monoclonal antibody mOg-1 was added to dissociated cerebellar cultures at the time of plating the cells in complement-inactivated 10% FCS–BEM onto polylysine-coated coverslips and at 2 days after plating. When the cultures were examined at day 4, a similar percentage of Og-1-positive cells (0.8%) was seen compared to the control but, at day 11, only 1.5% Og-1-positive cells were present, less than the control's 4.2%. However, sheath formation was normal in the presence of mOg-1.

Reaggregate cell cultures of 18-day fetal mouse brain were incubated with mOg-1 from day 1 of culture to day 17, prior to myelination, or from day 17 to day 35 of culture, a time of active myelination. The mOg-1 readily diffuses into and binds to oligodendrocytes in 17-day reaggregate cultures as confirmed by immunofluorescence of frozen sections (figure 8-8). Electron microscopy of thin sections from 35-day aggregate cultures (in 10% FCS–

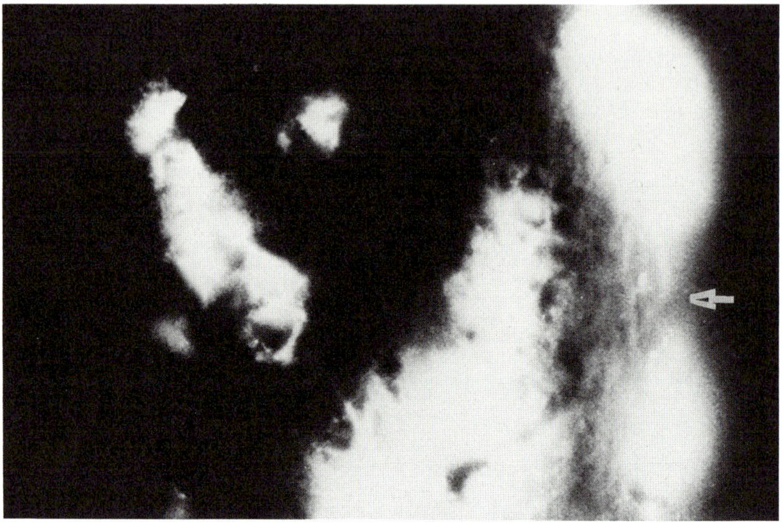

Figure 8-8. Diffusability of mOg-1 into brain cell reaggregate culture. A center 10-μm frozen section of a 17-day cultured reaggregate incubated in mOg-1 prior to sectioning and visualized by incubation with rhodamine goat antimouse IgM. The *arrow* indicates the surface of the reaggregate.

BEM) showed good myelination and synaptogenesis (figure 8-9A). Reaggregates incubated with the antibody during oligodendrocyte maturation days 1–17 showed normal myelin except for a rare loose wrapping. However, loose or noncompact myelin (figure 8-9B) was common in reaggregates cultured with mOg-1 during active myelination, although synaptogenesis in these aggregates appears normal. Thus these studies suggest that Og-1 may play a role in the process of myelin compaction; however, these are preliminary findings and more extensive studies will have to be carried out to elucidate this process.

5. COMPARISON OF Og-1 TO OTHER OLIGODENDROCYTE ANTIGENS

The Og-1 antigen first appears in mouse brain at fetal day 19, several days before galactocerebroside (Schachner and Willinger, 1979) and more than a week before myelin basic protein (Sternberger et al., 1978). Double labeling studies (Collins and Seeds, 1986) with a monoclonal antibody to galactocerebroside (Ranscht et al., 1982) or a monoclonal antibody to myelin basic protein (Hybritech, Inc.) show that the mOg-1 antibody is distinct from these two oligodendrocyte antigens, since Og-1 consistently binds to a larger cell population and appears earlier in developing brain.

The Og-1 antigen most closely resembles an oligodendroglial antigen 04 described by Schachner and coworkers (Sommer and Schachner, 1981),

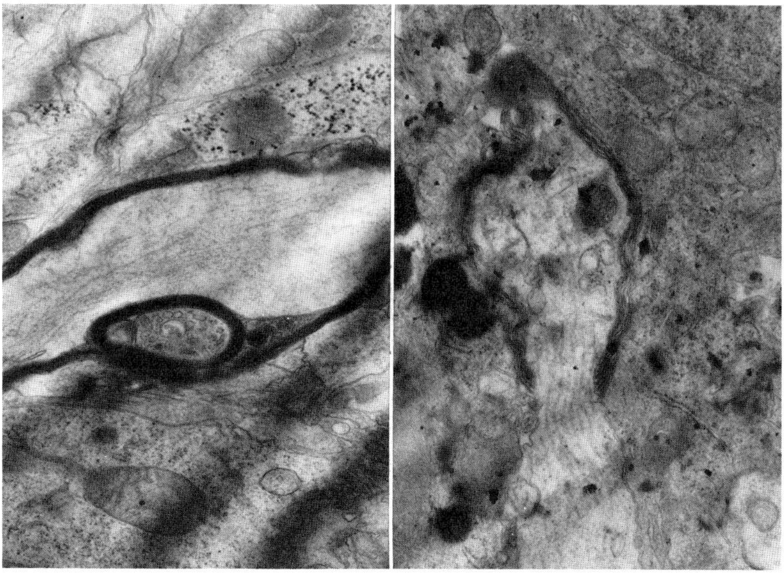

A B

Figure 8-9. Electron micrograph of myelinated neurites in 35-day brain cell reaggregates cultured in the absence (**A**) and the presence (**B**) of mOg-1 from days 17 to 35. Note the lack of compact myelin in **B**, typical of reaggregates cultured in the presence of mOg-1; however, synaptic profiles are normal.

although the 04 antigen appears in mouse brain at fetal day 16. Since both the antibody to 04 and Og-1 are IgM subclass, it has not been possible to perform double labeling studies with these two monoclonal antibodies. The 04 antigen has not been identified, but has been suggested to be a glycolipid (Schachner, 1982).

The Og-1 antigen has been difficult to identify. In part, IgMs show extensive nonspecific interaction with both lipids and proteins. Furthermore, the Og-1 antigen is irreversibly denatured by SDS. However, a protein of 90,000 Mr that binds mOg-1 has been extracted from cultured cerebellar cell membranes and purified on a goat antimouse IgM affinity column, as seen in figure 8-10. This 90,000-Mr protein appears to be the Og-1 antigen. The molecular weight of Og-1 is similar to the myelin-associated glycoprotein (MAG) (Quarles et al., 1983); however, mOg-1 does not appear to react with rat or human MAG (Noronha, Quarles, and Seeds, unpublished results).

6. CONCLUSIONS

A monoclonal antibody mOg-1 has been isolated that reacts with an antigen, Og-1, on the surface of immature oligodendrocytes. This antigen appears to be a 90,000-Mr protein that is distinct from previously described

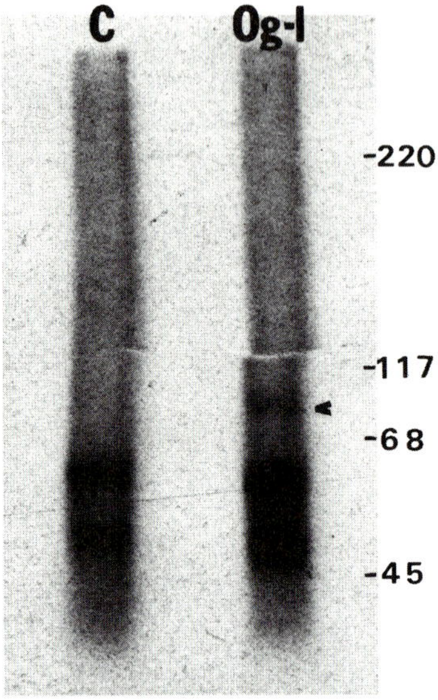

Figure 8-10. Metabolically labeled cerebellar cell proteins that interact with mOg-1. Cerebellar cell cultures were labeled with ^{35}S-methionine and mOg-1 was bound to the cells. Complexes of mOg-1 and cellular proteins were solubilized with NP-40 and isolated on goat antimouse IgM affinity columns. The eluted radioactive materials is compared to a control (no mOg-1 binding) sample. Although some radioactivity is seen in both samples, only the mOg-1 sample contains a specific band of approximately 90,000 Mr (presumably the Og-1 antigen).

oligodendroglial-specific antigens. The Og-1 antigen is present in neonatal brain regions destined to become white matter tracks; however, it appears to be excluded from compact myelin. Its cell surface localization suggests a possible role for Og-1 in cell recognition events in oligodendrocyte differentiation. Preliminary studies indicate that mOg-1 can interfere with normal compaction of myelin in brain cell reaggregate cultures. Future studies will be directed at elucidating the role of Og-1 antigen in oligodendrocyte development and function.

ACKNOWLEDGMENTS

The authors would like to thank Drs. D. Dahl, B. Ranscht, and M. Rapport for their generous gift of antisera and antibodies used in these studies. This work was supported in part by USPHS grant NS-09818.

REFERENCES

Bottenstein, J.E., and Sato, G. (1979) Growth of a rat neuroblastoma cell line in serum-free supplemented medium. Proc. Natl. Acad. Sci. USA 76:514–517.

Collins, J.M. (1983) Monoclonal antibodies to cerebellar antigens. PhD thesis, University of Colorado Medical School, Denver.

Collins, J.M., and Seeds, N.W. (1986) Oligodendroglia development in cell culture as monitored with a monoclonal antibody. J. Neurosci. 6:2635–2643.

Norton, W.T. (1983) Recent advances in the neurobiology of oligodendroglia. In: Fedoroff, S., and Hertz, L. (eds.) Advances in cellular neurobiology IV. New York: Academic Press, pp. 3–56.

Quarles, R.H., Barbarash, G., Figlewicz, D., and McIntyre, L. (1983) Purification and partial characterization of the myelin associate glycoprotein from adult rat brain. Biochem. Biophys. Acta. 75:140–143.

Raff, M.C., and Miller, R.H. (1984) Glial cell development in rat optic nerve. TINS 469–472.

Raff, M.C., Abney, E., Cohen, J., Lindsay, R., and Noble, M. (1983a) Two types of astrocytes in cultures of developing rat white matter: differences in morphology, surface gangliosides, and growth characteristics. J. Neurosci 3:1289–1300.

Raff, M.C., Miller, R., and Noble, M. (1983b) A glial progenitor cell that develops in vitro into an astrocyte or an oligodendrocyte depending on culture medium. Nature 303:390–396.

Raff, M.C., Abney, E., and Miller, R. (1984) Two glial cell lineages diverge prenatally in rat optic nerve. Dev. Biol. 106:53–60.

Ranscht, B., Clapshaw, P., Price, J., Noble, M., and Seifert, W. (1982) Development of oligodendrocytes and Schwann cells studied with a monoclonal antibody against galactocerebroside. Proc. Natl. Acad. Sci. USA 79:2709–2713.

Ratner, N. Bunge, R., and Glaser, L. (1985) A neuronal cell surface heparan sulfate proteoglycan is required for dorsal root ganglion neuron stimulation of Schwann cell proliferation. J. Cell Biol. 101:744–754.

Salzer, J., Bunge, R., and Glaser, L. (1980) Studies of Schwann cell proliferation. III. Evidence for the surface localization of the neurite mitogen. J. Cell Biol. 84:767–778.

Saneto, R., and De Vellis, J. (1985) Characterization of cultured rat oligodendrocytes proliferating in a serum-free chemically defined medium. Proc. Natl. Acad. Sci. USA 82:3509–3513.

Schachner, M. (1982) Cell type-specific surface antigens in the mammalian nervous system. J. Neurochem. 39:1–8.

Schachner, M., and Willinger, M. (1979) Developmental expression of oligodendrocyte-specific cell surface markers: nervous system antigen NS-1, cerebroside and myelin basic protein. In: The Menarini series of immunopathology, vol. 2. Basel: Schwabe, pp. 37–60.

Schaper, A. (1897) The earliest differentiation in the central nervous system of vertebrates. Science 5:430–431.

Seeds, N.W. (1973) Differentiation of aggregating brain cell cultures. In: Sato, G. (ed.) Tissue culture of the nervous system. New York: Plenum, pp. 35–53.

Seeds, N.W. (1975) Cerebellar cell surface antigens of mouse brain. Proc. Natl. Acad. Sci. USA 72:4110–4114.

Seeds, N.W. (1983) Neuronal differentiation in reaggregate cell cultures. In: Fedoroff, S., and Hertz, L. (eds.) Advance in cellular neurobiology 4. New York: Academic Press, pp. 57–79.

Seeds, N.W., and Haffke, S. (1978) Cell junction and ultra-structural development in reaggregated mouse brain cultures. Dev. Neurosci. 1:69–79.

Seeds, N.W., and Vatter, A. (1971) Synaptogenesis in reaggregating brain cell culture. Proc. Natl. Acad. Sci. USA 68:3219–3222.

Sommer, I., and Schachner, M. (1981) Monoclonal antibodies (01–04) to oligodendrocyte cell surfaces: an immunocytological study in the CNS. Dev. Biol. 83:311–327.

Sternberger, N.H., Itoyama, Y., Kies, M., and De F. Webster, H. (1978) Myelin basic protein demonstrated immunocytochemically in oligodendroglia prior to myelin sheath formation. Proc. Natl. Acad. Sci. USA 75:2521–2524.

Stuhlfauth, I., and Seeds, N.W. (1983) Cerebellar plasma membrane proteins and their antisera. Dev. Biol. 98:80–100.

Trapp. B.D., Honegger, P., Richelson, E., and De F. Webster, H. (1979) Morphological differentiation of mechanically dissociated fetal rat brain in aggregating cell cultures. Brain Res. 150:117–130.

9. INTRACEREBRAL TRANSPLANTATION AS A MODEL FOR STUDYING THE MYELINATION PROCESS

MADELEINE GUMPEL, FRANÇOIS LACHAPELLE, MICHEL BAULAC, ANNICK BARON-VAN EVERCOOREN, ANNIE GANSMULLER, CLAUDE JACQUE, AND NICOLE BAUMANN

1. INTRODUCTION

Early studies on myelination performed most often in the human brain clearly established that myelination does not proceed in all parts of the nervous system at the same time. It seems to follow the order of phylogenic development: portions of peripheral nervous system (PNS) myelinate first, then the spinal cord, and then the brain (reviewed by Norton, 1982). In the brain, myelination proceeds caudorostrally. In the human central nervous system (CNS), the process starts around the fifth month of intrauterine development in the motor roots (Yakovlev and Lecours, 1967) and continues for a long time since it is assumed that the intracortical associative areas are fully myelinated only 20 years later. Because of this lengthy developmental period in humans the myelination process has been extensively studied in small animal models like mouse or rat. In these models, myelin deposition occurs postnatally in the brain so that it is possible to follow the whole process for a short period of the animal's life. The maturation of glial cell lines and the different steps of myelin formation in these animal models have been studied simultaneously by morphologists and biochemists (reviewed by Webster et al., 1981). The development of microchemical methods allowed two distinct and interesting approaches. First it has been possible to analyze the different constituents of myelin and to study their sequential deposition in the myelin sheath. Mutations affecting myelination in the CNS have proved to be an excellent tool for understanding some mechanisms of myelin synthesis (Baumann, 1980a and b). Second, the purification of constituents of the myelin made it possible to obtain polyclonal antibodies, allowing the precise detection of the antigens on sections of the brain or on cell suspensions. The development of immunostaining techniques, using polyclonal and then monoclonal antibodies, has been essential in identifying the sequential synthesis of the myelin constituents in the myelin-forming cells; moreover, it made possible labeling myelin-forming cells very early in the development (M. Monge et al., 1986) and even following up the different glial cell lines during the development (Raff and Miller, 1984).

In vitro culture techniques have provided new approaches for elucidating developmental interactions between oligodendrocytes and neurons. Organotypic cultures or dissociated brain cell cultures have been used as models (see review in Vernadakis and Culver, 1980). Myelin develops in culture of brain explants and, if fragments of cerebellum in which oligodendrocytic differentiation has been suppressed are associated with fragments of cerebellum devoid of myelin receptive axons, myelination is observed. In this case, diffusible myelin-stimulating factor is excluded by the absence of myelination if the two fragments of CNS are not in direct contact (Seil and Blank, 1981). In the PNS, Schwann cells myelinate dorsal root ganglia (DRG) in culture (reviewed by Varon and Manthorpe, 1982). Oligodendrocytes from the rat optic nerve myelinate neurites from DRG, the myelin formed being of CNS type (Wood et al., 1980). Similarly, adult isolated oligodendrocytes have recently been demonstrated to myelinate DRG neurites (Bunge and Waksman, 1985). Although CNS and PNS myelin differ in a number of biochemical and morphological characteristics (reviewed by Morell, 1984), peripheral neurites can be myelinated by PNS or CNS myelin-forming cells. Oligodendrocytes and Schwann cells respond to PNS or CNS axonal stimulus, but the type of myelin formed depends upon the glial cell.

Schwann cells seem to be more dependent upon the presence of axons to myelinate and to synthesize myelin constituents. Isolated oligodendrocytes are able to synthesize in culture some of the most important constituents of the myelin (basic protein, cerebrosides, sulfatides, CNPase) and even to produce myelin-like structures (Poduslo et al., 1982; reviewed by Sarlieve et al., 1983, Szuchet et al., 1986). However, the problematic mechanisms of myelination cannot be approached through culture techniques. The anatomy of CNS transplants is not well preserved for extended periods in culture, and the best culture conditions for survival and differentiation are different for each type of neural cells.

In this chapter, we present three examples in which the in vivo intracerebral transplantation technique was useful in approaching some of the unsolved problems of myelination mechanisms. In the three examples to be discussed, the transplant containing oligodendrocytes is introduced into the newborn brain. Since myelination starts postnatally in the mouse brain (Folch-Pi, 1955), transplanted myelin-forming cells are present in the host during the myelination process. The transplanation technique and immunohistochemical procedure have been described in detail elsewhere (Gumpel et al., 1983 and 1985; Lachapelle et al., 1983–84). We used a classic electron-microscopic technique (Gumpel et al., 1986; Gansmüller et al., 1986).

2. THE SHIVERER MODEL

The shiverer (shi/shi) mouse is a mutant first described by Bird et al. (1977). The mutation, autosomal recessive, is characterized by tremor appearing around 13 days postnatally and convulsions appearing from 30 to 35 days. Most of the animals die around 120 days, but in some cases they survive up to 150 days. The shiverer myelin is biochemically deprived of myelin basic protein (MBP) (Jacque et al., 1978; Dupouey et al., 1979). Thus myelin formed by MBP-positive oligodendrocytes transplanted into the shiverer brain can be distinguished from the host shiverer myelin by immunohistochemistry, using an anti-MBP antiserum. Moreover, the shiverer myelin presents a defect of compaction with an absence of the major dense line (Privat et al., 1979). MBP-positive myelin can thus be detected in the shiverer host brain by electron microscopy (Gumpel et al., 1986; Gansmüller et al., 1986).

2.1 Transplantation of newborn normal mouse olfactory bulb into newborn shiverer brain

In this series of experiments, the graft was a fragment of olfactory bulb from newborn normal B6CBA, C57-BL6, and C3H/SW/V mice. At the time of implantation, oligodendrocytes are immature in the olfactory bulb (Jacque et al., 1985). Galactosylceramide (GC) appears only 3 days after birth, and MBP is expressed 4 days later. First, the graft was systematically introduced in the rostral thalamus (Lachapelle et al., 1983–84). Under these conditions, we detected MBP-positive myelin in the shiverer host brain from 20 days after grafting onward. This myelin formed by transplanted oligodendrocytes was found near the graft, but also far from the site of implantation, forming patches along what seemed to be the preferential routes to the pons and the cerebellum (Lachapelle et al., 1983–84; Gumpel et al., 1984). In "sham" experiments in which shiverer newborn olfactory bulb was implanted into shiverer newborn host brain, we never detected any MBP-positive myelin (Lachapelle et al., 1983–84). These results were fully confirmed by electron microscopy. Normal myelin appears to be scattered among that of shiverer, and the same axon can be successively ensheathed by both types of myelin. In this case, the node of Ranvier seems to form normally (Gumpel et al., 1986; Gansmüller et al., 1986).

Some considerations arising from these experiments are: 1) Whether the "routes" are determined by the point of implantation in the host environment or whether they depend upon the population of implanted oligodendrocytes. 2) What the true extension and aspect of myelination due to implanted oligodendrocytes are in the whole host brain. In the series described above, we sectioned the grafted brain only partially, at the level where the graft marked by charcoal was visible. 3) Migration appeared to be essentially rostrocaudal. Is there a possible migration toward the anterior part of the brain? To answer these questions, we designed two series of experiments in which 1) fragments of normal newborn CNS from another region were grafted in the rostral thalamus and 2) fragments of newborn olfactory bulb were grafted in various sites in the newborn shiverer brain.

The first series of experiments demonstrated that the oligodendrocytes contained in a fragment of newborn cerebellum or spinal cord were able to survive, migrate, and myelinate in the shiverer brain (unpublished results). However, it is difficult to compare accurately the behavior of these different oligodendrocyte populations due to the variations of several parameters. For instance, the maturation of oligodendrocytes in newborn spinal cord is equivalent to the oligodendrocyte maturation around 5 days in the olfactory bulb, and we have demonstrated that the migrative properties of the oligodendrocytes contained in a fragment of olfactory bulb decrease very rapidly after birth (Gumpel et al., 1984). This decrease may be due to progressive attachment of the graft cells to each other. Nevertheless, other experiments such as grafting of adult oligodendrocytes isolated from rat forebrain (Gumpel et al., 1986; Lubetzki et al., unpublished) or grafting of the subventricular region of human embryo hemispheres (see below) demonstrate that migrative and myelinating properties of the oligodendrocytes are not restricted to the olfactory bulb population.

In the second series of experiments, fragments of normal mouse olfactory bulb were grafted in different regions of the newborn shiverer brain (Gumpel et al., 1986; Baulac et al., in press). In this series, the implanted brains were cryostat sectioned serially, and one out of three sections were treated for immunohistological detection of MBP. Thus the extension of MBP-positive myelination could be studied in relation with the graft position. The results obtained in this series are as follows: 1) Transplanted oligodendrocytes are able to migrate and myelinate several centimeters from the graft. However, the MBP-positive myelination can be restricted to the region around the graft. The spinal cord can be myelinated by oligodendrocytes grafted in the posterior brain. 2) MBP-positive myelination appears as patches distributed sometimes at distances from each other, 3) The graft being in one hemisphere, patches of MBP-positive myelin are most often present in the two hemispheres and in the commissures. In the two hemispheres, patches of myelin are symmetrical or asymmetrical. 4) The migration toward the caudal part of the brain seems predominant. However, caudorostral migration has also been observed. 5) The transplanted oligodendrocytes seem to migrate or to be transported along the axonal pathway. Myelination by transplanted oligodendrocytes is more impressive and more dispersed when the graft is placed in a crossroad structure. This allows demonstration of the long-distance migration of grafted oligodendrocytes and examination of this migration in relation to the structures of the brain in situ.

The results suggest the existence during normal brain development of large migration of oligodendrocytes during the postnatal period preceding myelination. By labeling the graft it is possible to study the relation of implanted oligodendrocytes with all the other cells of the brain in normal conditions. It also allows investigation of the relationship between migration, maturation (presence of maturation markers), and division (^{3}H-thymidine incorporation) as well as the role of extracellular matrix (ECM) in their movements. By implanting Schwann cells and oligodendrocytes, the technique could allow comparison of the behavior (migration and myelination) of both types of myelin-forming cells during the process of myelination in the CNS. This should promote the understanding of the mechanisms that control the limit of the PNS–CNS fields, myelinated by Schwann cells and oligodendrocytes, respectively.

2.2 Transplantation of jimpy mouse olfactory bulb into newborn shiverer brain

In the isolated Jimpy mutant mouse, Philipps (1954) and Sidman et al. (1964) have shown that myelin is almost completely absent in the CNS, whereas the PNS is unaffected. Farkas-Bargeton et al. (1972) have reported the failure of oligodendroglia to differentiate in the white matter. Skoff (1976 and 1982) has shown that the proliferation of oligodendroblasts in Jimpy was increased, but that their total number in the white matter remained the same as that in control animals during development and up to the death of the animal, between 20 and 35 days. The number of dying cells suggests that the oligodendrocytes proliferate, but die sometimes afterward. Matthieu and Omlin (1984) insist on the fact that jimpy mice present severe astrocytic alterations prior to myelin formation and therefore that abnormalities of oligodendrocytes or even axons could be a secondary phenomenon. Omlin et al. (1979) and Matthieu and Omlin (1984) have noted that the fine structure of most of jimpy oligodendrocytes was not altered. Cultures of the total neural population of jimpy cerebellum showed that hypomyelination is unchanged (Billings-Gagliardi et al., 1980). Regarding the cell interactions involved in myelination, Billings-Gagliardi et al. (1983) have demonstrated that, in culture, normal glia myelinate jimpy axons. Jimpy abnormalities are thus inherent in their glial cells "probably although not definitely to the oligodendrocytes."

The lengthy migration that we obtained outside of the graft when newborn normal mouse brain fragments were grafted in the newborn shiverer brain might suggest the very interesting possibility of placing Jimpy oligodendrocytes in a shiverer environment in situ. Shiverer myelin is not normal, but the myelination process takes place. Moreover, the myelin defect in the shiverer mutant being directly linked to the oligodendrocyte itself, the environmental factors can be considered as normal. The jimpy myelin is MBP positive, so that the "shiverer model" can be used: myelin eventually formed by jimpy oligodendrocytes in shiverer environment can be detected in the shiverer brain by immunohistochemistry using anti-MBP antiserum. Although the phenotype of the mutation (severe tremor) appears only at days 11–12, the use of the Tabby (Ta) mutation as a marker allows the detection of Ta-jp/y hemizygotes at birth (absence of post-oral and post-orbital vibrissae). We thus grafted in newborn shiverer brain fragments of olfactory bulb 1) from newborn presumably Jimpy animals (the presence of very few recombinants makes their recognition slightly uncertain), and 2) from 11- to 35-day Jimpy mice recognized through a clear phenotypical expression (Gumpel et al., 1984).

Unexpectedly, a very important MBP-positive myelination was observed in the host shiverer brain especially when the graft consisted of newborn or 11- to 20-day CNS tissues. Under immunohistological observation, the myelin is comparable to normal myelin and is much more heavily labeled by the antiserum than jimpy myelin in situ (Gumpel et al., 1984). The MBP-positive myelin appears in the host brain 20–30 days after grafting newborn or older mouse Jimpy CNS. The ultrastructural study of the myelin formed by transplanted jimpy oligodendrocytes is presently under investigation. Thus, jimpy oligodendrocytes migrate out of the graft and, as expected, the transplantation technique allowed studying the differentiation of the Jimpy myelin-forming cells in a non-Jimpy environment. We observed that Jimpy oligodendrocyte populations differentiate better in shiverer environment than in situ. This confirms the opinion of other authors that the defect of the Jimpy mutant may not be inherent to oligodendrocytes only.

Skoff's (1982) hypothesis that the shortened life span of the mutant and the failure of differentiation of oligodendrocytes are possibly related is not confirmed by our experiments. Within 20–30 days, newborn jimpy oligodendrocytes are able to differentiate in the shiverer brain, and they never reach the same level of differentiation in the mutant, which survives frequently up to 35 days. On the other hand, if normal oligodendrocytes are able to myelinate in vitro, Jimpy axons (Billings-Gagliardi et al., 1983), the expression of the axonal myelination signal does not seem to be affected in this mutant. A possible hypothesis is that jimpy oligodendrocytes do not differentiate because of a defect in interactions between nonneuronal cells and oligodendrocytes. Astrocytic differentiation is known to be very disturbed in Jimpy mutant before myelination starts (Matthieu and Omlin, 1984). In Raff's hypothesis (Raff and Miller, 1984), oligodendrocytes and so-called type-2 astrocytes originate from the same precursor cell O2A. Raff et al. (1985) have demonstrated in in vitro cultures that the differentiation of this precursor cell depended upon growth factor(s) produced by a so-called type-1 astrocytes. If type-1 astrocytes were abnormal in Jimpy, which has not been demonstrated, this could account for the failure of the majority of oligodendrocytes to differentiate in this mutant. To clarify this specific point, transplantation technique should be relayed by in vitro culture experiments.

2.3 Transplantation of human embryonic CNS in the newborn shiverer brain

One of the purposes of this experiment was to define conditions in which human embryonic oligodendrocytes were able to survive and differentiate when grafted into a host brain. Grafting of human cells into mouse brain could also provide information about the possibilities of developmental interactions between very phylogenetically divergent species. Moreover, as no signs of rejection were noted in the two previously described experiments, it was interesting to look at the rejection problem in such heterografts. We thus transplanted, into the brain of newborn shiverer mouse, fragments of CNS containing the subventricular layer in which oligodendroblasts would be localized before maturation (reviewed by Privat, 1975). By 17–20 days after grafting human embryonic CNS tissue, extended normal myelin patches can be detected by immunocytochemistry (Gumpel et al., 1986 and unpublished), and the results are comparable to those of mouse–mouse graft. Moreover, 150 days after grafting, normal myelin is still present in the host brain without signs of degeneration or rejection. These results have been confirmed by electron microscopy.

Thus, human grafted oligodendrocytes behave like mouse oligodendrocytes in their long-distance migration and myelination of host axons. This suggests that molecules implied in axon–oligodendrocyte recognition, which allows preferential migration of the transplanted oligodendrocytes along the axonal pathways and exchange of myelination signal, are preserved during phylogenetic evolution. The timing of myelination is so different in man and mouse that this type of transplantation could provide information about the role of the maturation of the environment in the timing of transplanted oligodendrocyte differentiation. This question is presently under investigation.

We have demonstrated in this experiment that, under precise conditions, human oligodendrocytes can be transplanted at least in newborn mice. Under these conditions, they do differentiate. Schwann cells have already been transplanted in adult CNS to repair demyelinated lesions (Duncan et al., 1983; Blakemore, 1977, 1981, 1983, 1984 and 1985). However, transplantation of oligodendrocytes has never been attempted. Because of the importance of the demyelinating pathology in man, the ability of oligodendrocytes to be transplanted in adult to repair lesions is of particular interest and will be pursued in future studies.

3. CONCLUSIONS

The intracerebral transplantation technique should be considered as a new tool for investigating the myelination process during development as well as the remyelination process in adult. Like other techniques, intracerebral transplantation alone cannot reconcile the problems encountered in studying myelination. Moreover, the conclusions obtained particularly during the development should be confirmed by in vitro experiments which allows simplification of the models since interaction between only two types of cells is theoretically possible. The transplantation technique, however, can be combined with all types of markers (species-specific markers, markers of maturation, markers of mitotic behavior, and markers of ECM) in conditions as normal as possible; it thereby opens a large field of investigation.

The possibilities that transplantation techniques offer for studying the ability of myelin-forming cells to repair demyelinating lesions in adult are unique. The evolution of a demyelinated lesion, the effect of the presence of this lesion on transplanted myelin-forming cells, and their capabilities to migrate toward the lesion and to myelinate the axons in the lesions and the verification of the restoration of secure conditions by remyelination must be studied in vivo, using a marker that allows recognition of the origin of the cells responsible for remyelination.

ACKNOWLEDGMENTS

This work was supported by grant RG 1773-AI from the National Multiple Sclerosis Society and by INSERM. We are grateful to P. Coulongeon and C. Burté for typing the manuscript.

REFERENCES

Baulac, M., Lachapelle, F., Gout, O., Berger, B., Baumann, N., and Gumpel, M. (1987). Transplantations of oligodendrocytes in the new-born mouse brain. Extention of myelination by transplanted cells. Anatomical study. Brain Res. (in press).

Baumann, N. (ed.) (1980a) Neurological mutations affecting myelination. In INSERM symposium 14. Amsterdam: North Holland, p. 565.

Baumann, N. (1980b) Mutations affecting myelination in the central nervous system: research tool in neurobiology. Trends Neurosci. :82–85.

Billings-Gagliardi, S., Adcock, L.H., Schwing, G.B., and Wolf, N.K. (1980) Hypomeylinated mutant mice. II. Myelination in vitro. Brain Res. 200:135–150.

Billings-Gagliardi, S., Adcock, L.H., Lamperti, E.D., Schwing-Standhope, G., and Wolf, M.K. (1983) Myelination of jp, jp/mld and qk axons by normal glia in vitro: ultrastructural and autoradiographic evidence. Brain Res. 268:255–266.

Bird, T., Farell, D.F., and Sumi, S.M. (1977) Genetic developmental myelin defect in shiverer mouse. Trans. Am. Soc. Neurochem. 8:153.

Blakemore, W.F. (1977) Remyelination of CNS axons by Schwann cells transplanted from the sciatic nerve. Nature 266:68–69.

Blakemore, W.F. (1981) Observations on myelination and remyelination in the central nervous system. In: Garrod, D.R., and Feldman, J.D. (eds.) Development in the nervous system. British Society for Developmental Biology symposium 5. London: Cambridge University Press, pp. 289–308.

Blakemore, W.F. (1984) Limited remyelination of CNS axons by Schwann cells transplanted into the sub-arachnoid space. Neurol. Sci. 64:265–276.

Blakemore, W.F., and Cran, G. (1985) The use of cultured autologous Schwann cells to myelinate areas of persistent demyelination in the central nervous system. J. Neurol. Sci. 70:207–223.

Bunge, P.R., and Waksman, B.H. (1985) Glial development and interactions. TINS :424–427.

Duncan, J.P., Aguayo, A.J., Bunge, R.P., Bray, G.M., and Wood, P.M. (1983) Transplantation of cultured xenogenic Schwann cells into peripheral nerve and spinal cord of immunosuppressed mice. In: Kao, C.C., Bunge, R.P., and Reier, P.J. (eds.) In: Kao, C.C., Bunge, R.P., Reier, P.J. (eds.) New York: Raven.

Dupouey, J., Jacque, C., Bourre, J.M., Cesselin, E., Privat, A., and Baumann, N. (1979) Immunochemical studies of myelin basic protein in Shiverer mouse devoid of majoř dense line of myelin. Neurosci. Lett. 12:113–118.

Farkas-Bargeton, E., Robain, O., and Mandel, P. (1972) Abnormal glial maturation in the white mattr in Jimpy mice. Acta Neuropathol 21:272–281.

Folch-Pi, J. (1955) Composition of the brain in relation to maturation. In: Waelch (ed.) Biochemistry of the developing nervous system. New York: Academic Press, pp. 121–136.

Gansmüller, A., Lachapelle, F., Baron van Evercooven, A., Hauw, J.J., Baumann, N., and Gumpel, M. (1986) Transplantations of new-born CNS Fragments into the brain of shiverer mutant mice-extensive myelination by transplanted oligodendrocytes II. Electron microscopic study. Dev. Neurosci. 8:197–207.

Gumpel, M., Baumann, N., Raoul, M., and Jacque, C. (1983) Survival and differentiation of oligodendrocytes from neural tissue transplanted into newborn mouse brain. Neurosci. Lett. 37:307–312.

Gumpel, M., Lachapelle, F., Jacque, C., and Baumann, N. (1985) Central nervous tissue transplantation into mouse brain: differentiation of myelin from transplanted oligodendrocytes. In: Bjorklund, A. and Stenevi, U. (eds.) Neural grafting in the mammalian CNS. Fernström symposium 5. Amsterdam: Elsevier, pp. 151–158.

Gumpel, M., Lachapelle, F., Baulac, M., Baron-van Evercooren, A., Lubetzki, C., Cänsmuller, A., Lombrail, P., Jacque, C., and Baumann, N. (1986) Myelination in the mouse by transplanted oligoden-drocytes. In: Althaus, H.H., and Seifert, W. (eds.) Neuronal communication in development and regeneration. (In press.)

Jacque, C., Privat, A., Dupoucy, P., Bouvre, J.M., Bird, T., Baumann, N. (1978) Shiverer mouse: a dysmyclinating mutant with absence of major dense line and basic protein in myelin. Proc. Eur. Soc. Neurochem. 1:136.

Jacque, C., Collet, A., Raoul, M., Monge, M., and Gumpel, M. (1985) Functional maturation of the oligodendrocyte and myelin basic protein expression in the olfactory bulb of the mouse. Dev. Brain Res. 21:277:282.

Lachapelle, F., Gumpel, M., Baulac, M., Jacque, C., Duc, Ph., and Baumann, N. (1983–1984) Transplantation of fragments of CNS into the brain of shiverer mutant mice: extensive myelination by implanted oligodendrocytes. Dev. Neurosci. 6:326–334.

Matthieu, J.M., and Omlin, F.S. (1984) Murine leukodystrophies as tool to study myelinogenesis in normal and pathological conditions. Neuropediatrics. 15:37–52.

Monge, M., Ladiisky, D., Jacque, C.M., and Zale, B. (1986). Oligodendroglial expression and deposition of four major myelin constituents in the myelin sheath during development — An in vivo study. Dev. Neurosci. 8:222–235.

Morell, P. (ed.) (1984) Myelin, 2nd edn. New York: Plenum.

Norton, W.T. (1982) Formation, structure and biochemistry of myelin. In: Silfel, G.L., Albers, R.W., Agranoff, B.W., and Katzman, R. Basic neurochemistry, 3rd eds., pp. 63–82.

Omlin, F.X., Bischoff, A., Spycher, A.A., Wismann, V.N. (1979) Ultrastructure of lipids in the optic nerve of the mouse mutant "Jimpy": a study of histochemistry freeze etching and thin section technique, Acta Neuropathol. 45:221–229.

Philipps, R.J.S. (1954) Jimpy a new totally sex-linked gene in the mouse. Z. Induckt. Abstamm. U. Vererb. L. 86:322–326.

Poduslo, S.E., Miller, K., and Molinsky, J.S. (1982) The production of a membrane by purified oligodendroglia in culture. Exp. Cell Res. 137:203–215.

Privat, A. (1975) Postnatal gliogenesis in the mammalian brain. Int. Rev. Cytol. 40:281–323.

Privat, A., Jacque, C., Bourre, J.M., Dupouey, P., and Baumann, N. (1979) Absence of the major dense line in the mutant mouse shiverer. Neurosci. Lett. 12:107–112.

Raff, M.C., and Miller, R.L.T. (1984) Glial cell development in the rat optic nerve. TINS: 649–472.

Raff, M.C., Abney, E.R., and Fok Seang, M. (1985) Reconstitution of a developmental clock in vitro: a critical role for astrocytes in the timing of oligodendrocyte differenciation. Cell 42:61–69.

Sarlieve, L., Fabre, M., Susz, J., and Matthieu, J.J.M. (1983) Investigations on myelination in vitro. IV. "Myelin-like" or premyelin structures in cultures of dissociated brain cells from 14–15 day old embryonic mice. J. Neurosci. Res. 10:191–210.

Seil, F.S., and Blank, N.K. (1981) Myelination of central nervous system axons in tissue culture by transplanted oligodendrocytes. Science 212:1407–1408.

Skoff, R.P. (1976) Myelin deficit in the Jimpy mouse may be due to cellular abnormalities in astroglia. Nature 264:560–562.

Skoff, R.P. (1982) Increased proliferation of oligodendrocytes in the hypomyelinated mouse mutant Jimpy. Brain Res. 248:19–23.

Szuchet, S., Polak, P.E., and Yim, S.H. (1986) Mature oligodendrocytes cultured in the absence of neurons recapitulate the ontogenic development of myelin membranes. Dev. Neurosci. 8:208–221.

Varon, S., and Manthorpe, M. (1982) Schwann cell: an in vitro perspective. Adv. Cell. Neurobiol. 3:35–95.

Vernadakis, A., and Culver, B. (1980) Neural tissue culture: a biochemical tool. In: Kuman, S. (ed) Biochemistry of the brain. Oxford: Pergamon, pp. 407–452.

Webster, H., de F., Trapp, B., Sternberger, N.H., and Quarles, R.H. (1981) Myelin forming glial cells: morphological and immunochemical observations. In: Garrod, D.R., and Feldman, J.D., (eds.) Development of the nervous system. British Society for Developmental Biology Symposium 5. London: Cambridge University Press, pp. 265–288.

Wood, P.L., Okada, E., and Bunge, R. (1980) The use of networks of dissociated rat dorsal root ganglion neurons to induce myelination by oligodendrocytes in culture. Brain Res. 196:297–252.

Yakovlev, P.I., and Lecours, A.R. (1976) In: Minokowski, A. (ed.) Regional development of the brain in early life. Oxford: Blackwell.

10. IN VITRO EFFECTS OF GLIA MATURATION FACTOR ON BIPOTENTIAL GLIAL PROGENITOR CELLS

E. PETER BOSCH, JOSÉ G. ASSOULINE AND RAMON LIM

1. INTRODUCTION

The application of conventional and monoclonal antibodies as cell type specific markers has facilitated the identification of the different types of neural cells in culture (Raff et al., 1979). Astrocytes are identified with antibody against glial fibrillary acidic protein (GFAP) and oligodendrocytes with antibody against galactocerebroside (GC), the major glycolipid of myelin. Recently Raff et al. (1983a) distinguished two kinds of GFAP-positive astrocytes (type 1 and type 2) in cultures of developing rat optic nerve on the basis of morphology, growth characteristics, and binding to gangliosides. Type-1 astrocytes are flat, fibroblast-like cells that proliferate in response to growth factors, but do not bind tetanus toxin or the monoclonal antibody A2B5. In contrast, type-2 astrocytes have round cell bodies with multipolar branched processes resembling neurons or oligodendrocytes that divide infrequently in response to growth factors and bind both tetanus toxin and A2B5 antibody. Similar to tetanus toxin, the monoclonal antibody A2B5 recognizes specific gangliosides and was originally used as a neuron-specific marker in central nervous system cultures (Eisenbarth et al., 1979). In vitro studies have further indicated that the two types of astrocytes develop from two distinct precursor cells (Raff et al., 1984b). Oligodendrocytes and type-2 astrocytes, but not type-1 astrocytes, develop from a common A2B5-positive progenitor cell that has been identified in neonatal rat optic nerve cultures (Raff et al., 1983b). This bipotential glial progenitor cell, which is A2B5+, GFAP−,

Vernadakis, A. (ed.), *Model Systems of Development and Aging of the Nervous System*.

GC−, rapidly differentiates into an oligodendrocyte if cultured in serum-free medium, or into a type-2 astrocyte if cultured in the presence of calf serum (Raff et al., 1983b and 1984a; Temple and Raff, 1985).

Glia maturation factor (GMF), an acidic protein purified from adult bovine brains, stimulates the proliferation, phenotypic expression, and chemical differentiation of cultured brain astroblasts (Lim et al., 1972 and 1973; Lim and Mitsunobu, 1974; Lim et al., 1976 and 1977a and b; Lim, 1985) and triggers the release of interleukin-1 and prostaglandins from astrocytes (Fontana et al., 1983; Miller and Lim, 1984). In the experiments reported here, we examined the influence of GMF on the oligodendroglia/type-2 astrocyte (O-2A) progenitor cell in rat neonatal optic nerve and fetal brain-derived glioblast cultures.

2. EFFECTS OF GMF ON OPTIC NERVE-DERIVED GLIAL CELLS

Optic nerves were removed from 7-day-old Sprague-Dawley (SD) rats. The intracranial portion of the optic nerve and chiasm were minced and dissociated with trypsin and collagenase. Approximately 45,000 cells were plated in 80 μl base medium (a 1:1 mixture of F12 and Dulbecco's modified Eagle's medium) containing 10% fetal calf serum (FCS) in the centre of poly-L-lysine-coated 12-mm glass coverslips and incubated at 37°C in 5% CO_2 in air and saturated humidity. After 2 hours, 2 ml of the appropriate medium were added. Cultures were grown in the serum-free medium N2 (Bottenstein and Sato, 1979), in N2 supplemented with GMF, or in N2 containing 10% FCS. After 4 days in culture, the cells were removed with trypsin–EDTA and counted in a hemocytometer. Sister cultures on coverslips were studied by indirect immunofluorescence using two fluorochromes. The coverslips were incubated with either monoclonal antigalactocerebroside (GC) antibody (mouse ascites fluid diluted 1:100) or monoclonal A2B5 antibody (mouse ascites fluid diluted 1:50), followed by affinity-purified goat antimouse immunoglobulin conjugated to rhodamine (G anti-MIg-Rd, diluted 1:40; Cappel). After washing, the cells were fixed in 5% acetic acid in ethanol and incubated with rabbit anti-GFAP serum (diluted 1:50), followed by affinity-purified goat antirabbit immunoglobulin conjugated to fluorescein isothiocyanate (G anti-RIg-Fl, diluted 1:40; Cappel).

When 7-day-old rat optic nerve cells were studied by immunofluorescence after 4 days in culture, the cell type of the developing glial cells was noted to be influenced by different culture conditions (table 10-1). In serum-free N2 medium, the percentage of oligodendrocytes (GC+) was 28, while 40% of the cells were astrocytes (GFAP+). The addition of GMF to N2 medium resulted in an increase in the number of astrocytes and a decrease in oligodendrocytes (table 10-1 and figure 10-1). Serum had a similar effect (table 10-1). The difference between serum and GMF lies in the number of type-2 astrocytes (table 10-1). When the cells were exposed to GMF, the majority of GFAP-positive cells formed a cell net consisting of multipolar, interconnect-

Table 10-1. Effect of GMF on development of astrocytes and oligodendrocytes in optic nerve-derived cultures.

		Astrocytes (GFAP+)			Oligodendrocytes (GC+)	
Culture medium	Total cell no./coverslip	Cell no./coverslip	% of total cells	% of GFAP+ cells labeled with A2B5 (type 2)	cell no./coverslip	% of total cells
N2	13,400 ± 1,800	5,414 ± 509	40.4 ± 4	20.8 ± 6	3,765 ± 992	28.1 ± 7
N2 + GMF[a]	21,300 ± 4,550	8,286 ± 746	38.9 ± 4	5.7 ± 3	1,981 ± 554	9.3 ± 3
N2 + 10% FCS	31,700 ± 3,400	13,631 ± 2,124	43.0 ± 7	50.0 ± 8	2,885 ± 1,141	9.1 ± 3

Results are mean ± SD of four coverslips.

[a]For GMF, a sample from bovine brain purified 10,000-fold (Lim and Miller, 1984) was initially used at a concentration of 200 ng protein/ml. The results were confirmed with a 100,000-fold purified GMF at 20 ng/ml (Lim et al., 1985). Human serum fibronectin (2 μg/ml) was included in all media lacking FCS. The media were renewed daily. After immunostaining, the coverslips were mounted in glycerol/PBS (1/1, vol/vol) and examined with an Olympus fluorescence microscope equipped with Nomarski and fluorescence optics and selective filters for fluorescein and rhodamine, using a 40× objective. The proportions of type-1 astrocytes (GFAP+, A2B5−), type-2 astrocytes (GFAP+, A2B5+), or oligodendrocytes (GC+) in each coverslip were determined by counting more than 200 cells.

Under the experimental conditions, the progenitor cells (GFAP−, A2B5+) were estimated to be <1% of total cells.

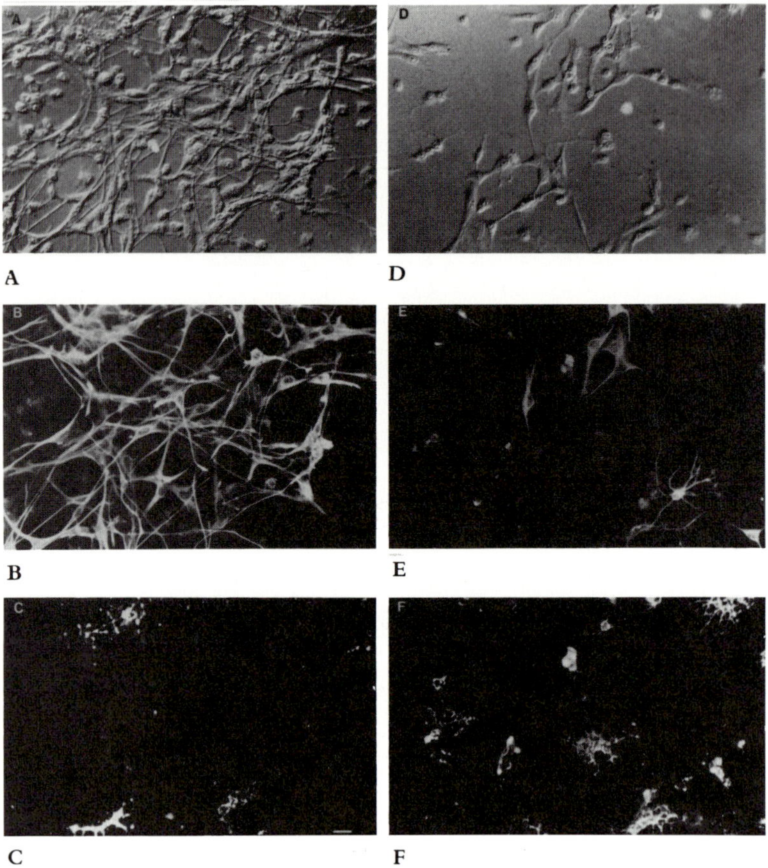

Figure 10-1. Effect of GMF on optic nerve-derived glial cells. Cells were grown in medium N2 supplemented with GMF (**A–C**) or in N2 alone (**D–F**). After 4 days in culture, cells were labeled with monoclonal anti-GC followed by G anti-MIg-Rd. After fixation they were labeled with rabbit anti-GFAP serum followed by G anti-RIg-Fl and viewed with Normarski (**A** and **D**), fluorescein (**B** and **E**), and rhodamine (**C** and **F**) optics. All antibodies were diluted in F10 with 3% goat serum and all incubations were for 30 minutes at room temperature. Control studies consisted of cells incubated with normal mouse ascites fluid in place of the specific monoclonal antibodies or of preparations incubated only with rhodamine and FITC conjugates. Note increased cell density of process-bearing astrocytes and paucity of oligodendrocytes in GMF-stimulated cultures. Bar = 20 μm.

ing astrocytes (figure 10-1). Although the morphological characteristics of these cells resembled process-bearing type-2 astrocytes, only a few of the GFAP+ cells were concomitantly labeled with A2B5 antibody (figure 10-2). GMF actually decreased the number and proportion of A2B5-positive type-2 astrocytes in optic nerve cultures (table 10-1). These cultures contained sparse type-1 astrocytes (< 10,000 cells/coverslip), the significance of which will be

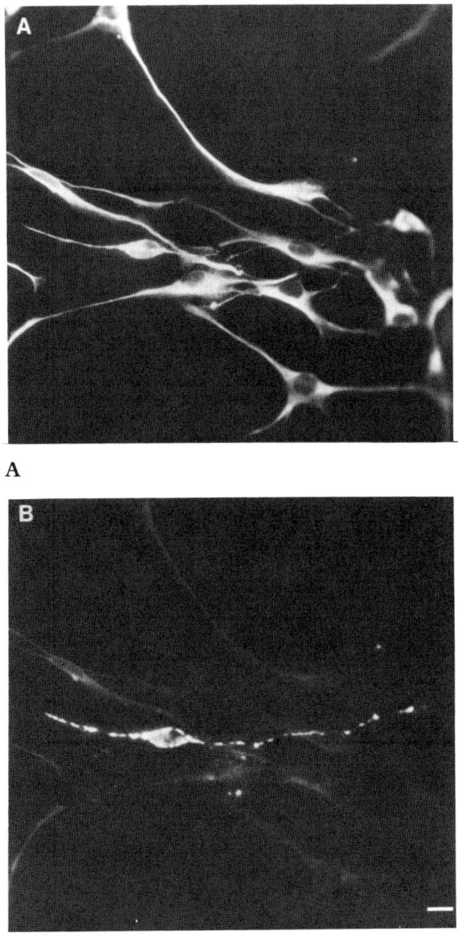

A

B

Figure 10-2. A single type-2 astrocyte (GFAP+, A2B5+) among several morphologically identical process-bearing type-1 astrocytes (GFAP+, A2B5−) in GMF-stimulated optic nerve culture. Cells were labeled with A2B5 antibody followed by G anti-MIg-Rd. After fixation they were labeled with anti-GFAP serum, followed by G anti-RIg-Fl and viewed with fluorescein (**A**) and rhodamine (**B**) optics. Bar = 20 μm.

discussed below. In contrast, the addition of serum increased the number and percentage of type-2 astroyctes (table 10-1).

3. EFFECTS OF GMF ON BRAIN-DERIVED GLIAL CELLS

To further study the effect of GMF on astrocyte development, we used purified secondary astrocyte cultures derived from embyronal rat brains (table 10-2). At 8 days in vitro and before stimulation, 97% of the total cells

Table 10-2. Effect of GMF on glial progenitor cells in brain-derived astrocyte cultures.

		Percentage of total cells		
Culture medium	Total cell no./coverslip	Type-1 astrocytes (GFAP+, A2B5−)	Type-2 astrocytes (GFAP+, A2B5+)	Progenitor cells (GFAP−, A2B5+)
Before stimulation	—	97 ± 4	< 0.1	<0.1
N2	48,167 ± 4,163	94 ± 4	2 ± 0.6	4.6 ± 2 (2,216)[a]
N2 + GMF	176,833 ± 21,572	93 ± 5	2 ± 0.9	12.0 ± 0.8 (21,220)
N2 + 10% FCS	378,500 ± 27,260	93 ± 3	9 ± 2.0	2.0 ± 0.6 (7,570)

Results are mean ± SD of six coverslips.
Brain glioblasts were obtained from 18-day-old SD rat fetuses by modifying a method previously described (Lim and Miller, 1984). At subculture after 1 week, approximately 10^5 cells were seeded on each poly-L-lysine-coated 18-mm glass coverslip in the base medium containing 10% FCS and incubated at 37°C in 5% CO_2 in air; 12 hours later the medium was changed to the appropriate test medium. The early medium change following the subculture eliminated neuroblasts and oligodendrocytes, leaving behind mainly type-1 astrocytes (97% of total cells). After 4 days in culture the cells were removed with trypsin–EDTA and counted on a hemocytometer. Sister cultures on coverslips were immunostained as described in the text.
[a] Numbers in parentheses are actual cell number per coverslip.

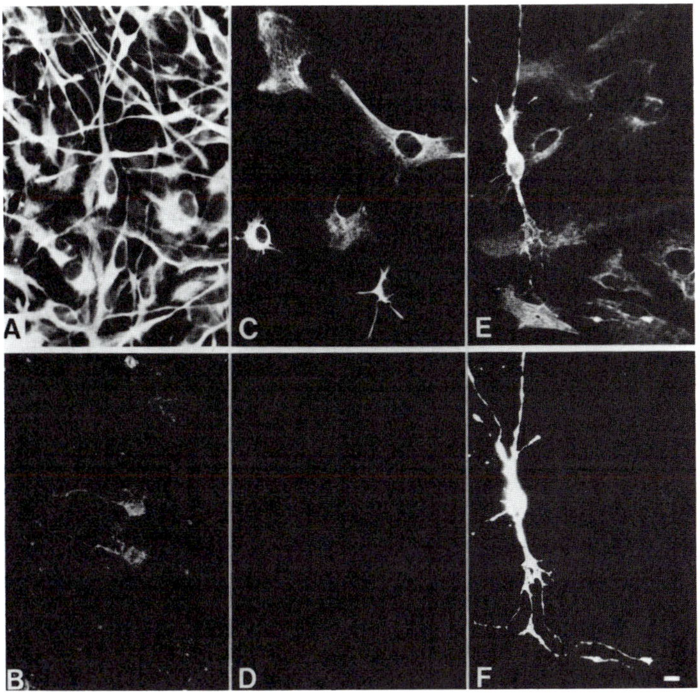

Figure 10-3. Effect of GMF on brain-derived glial cells. Glioblasts in secondary cultures were grown in GMF-supplemented N2 medium (**A** and **B**), N2 alone (**C** and **D**), and N2 with 10% FCS (**E** and **F**). After 4 days in culture, the cells were labeled with A2B5 and anti-GFAP antibodies and viewed as described in figure 10-2. (**A**, **C**, and **E**) Fluorescein. (**B**, **D**, and **F**) Rhodamine. Note two progenitor cells (A2B5+, GFAP−) in **B**, but none in **D** and **F**. A single type-2 astrocyte (GFAP+, A2B5+) is seen in **E** and **F**. Bar = 10 μm.

were type-1 astrocytes (GFAP+, A2B5−) and less than 1% oligodendrocytes (GC+) or A2B5+ progenitor cells. GMF stimulated the proliferation of type-1 astrocytes and transformed the majority of flat, fibroblast-like type-1 astrocytes to process-bearing astrocytes (table 10-2 and figure 10-3). The morphologically altered cells, which were invariably GFAP+, did not bind A2B5 antibody. This confirms the observation in optic nerve-derived astrocytes that GMF did not convert type-1 to type-2 astrocytes. We were surprised, however, to find an increase in O-2A progenitor cells (GFAP−, A2B5+) in our brain cell cultures exposed to GMF for 4 days (table 10-2 and figure 10-3). Since the astrocyte cultures before GMF stimulation consisted mainly of type-1 astrocytes and less than 1% oligodendrocytes or A2B5+ cells, the increase in A2B5+ progenitor cells can only be explained by proliferation and not by preferential attachment or survival of these cells.

Table 10-3. Stimulation of DNA synthesis in Brain-derived glial cultures by GMF

	Labeling index (percentage)		
Culture medium	Type-1 astrocytes (GFAP+, A2B5−)	Type-2 astrocytes (GFAP+, A2B5+)	Progenitor cells (GFAP−, A2B5+)
N2	19 ± 3	<0.1	25 (1/4)
N2 + GMF	34 ± 13	<0.1	53 (59/102)
N2 + 10% FCS	66 ± 1.5	18 (2/11)	<0.1

Results are mean ± SD of triplicate coverslips.
During the last 24 hours of culture on glass coverslips, the cultured cells were exposed to tritiated thymidine (specific activity 80 Ci/mmol; New England Nuclear) at 2 μCi/ml medium. After washing, the coverslips were processed for immunofluorescence as described before. Thereafter, the coverslips were dehydrated in ethanol, air dried, and mounted face up on microscope slides. The slides were coated with Ilford K5D emulsion, exposed for 24 hours at 4°C, developed with Dektol 1:1, and covered with another coverslip. At least 200 GFAP+ astrocytes and all A2B5+ progenitor cells were scored on each coverslip. The labeling index in percentage was calculated as the ratio of labeled to total GFAP+ or A2B5+ cells × 100. If less than 200 cells could be scored, the actual number of labeled and total cells are indicated in parentheses.

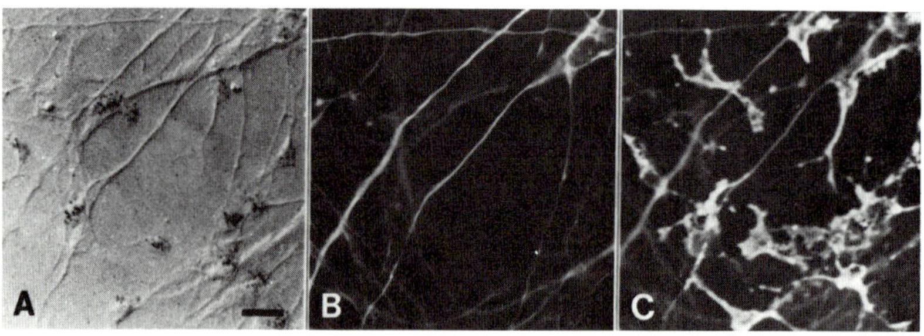

Figure 10-4. Thymidine uptake by progenitor cells after GMF stimulation. Brain-derived glioblasts were grown in GMF-supplemented N2 medium for 4 days and pulsed for the final 24 hours. The cells were immunostained and viewed as in figure 10-2. (**A**) Nomarski. (**B**) Fluorescein. (**C**) Rhodamine. Most progenitor cells (GFAP−, A2B5+) have labeled nuclei. Bar = 20 μm.

4. PROLIFERATION OF O-2A PROGENITOR CELLS BY INTERACTION OF GMF WITH TYPE-1 ASTROCYTES

In order to confirm this, tritiated thymidine was added to the brain cell cultures for the final 24 hours after 3 days on coverslips. As shown in table 10-3, 34% of the type-1 and more than 50% of the O-2A progenitor cells were radiolabeled after stimulation with GMF, indicating proliferation of these cells (figure 10-4 and 10-5). The presence of FCS promoted mainly the division of type-1 astrocytes with less dramatic effect on type-2 astrocytes and progenitor cells. It should be noted that GMF promoted proliferation of progenitor cells only in the presence of large numbers of type-1 astrocytes ($>1 \times 10^5$ cells/coverslip) in the brain-derived cultures, but not in optic nerve cultures where less than 1×10^4 cells/coverslip were stimulated with GMF.

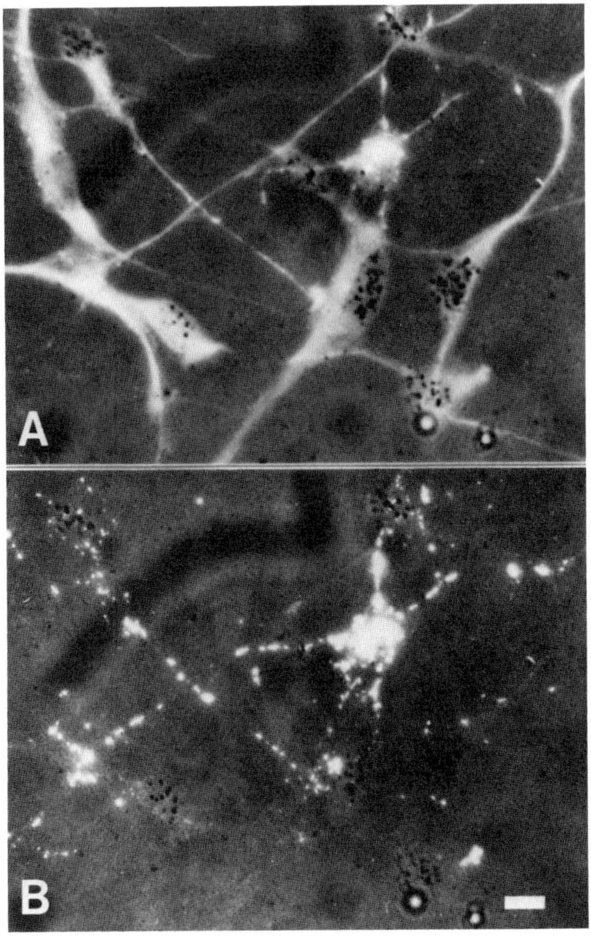

Figure 10-5. Thymidine uptake by type-1 astrocytes after GMF stimulation. A different area of the same culture as in figure 10-4 was chosen to demonstrate several type-1 astrocytes with labeled nuclei. A single type-2 astrocyte, unlabeled with isotope, is seen in the field. (**A**) Fluorescein. (**B**) Rhodamine. Bar = 10 μm.

The high density of type-1 astrocytes in the former instance may have contributed to the effect of GMF on the progenitor cells.

5. IMPLICATIONS IN GLIAL DEVELOPMENT

Glial cell development has previously been studied in the perinatal rat optic nerve (Skoff, 1980; Valat et al., 1983). In vivo morphological studies showed that glioblasts have the potential to differentiate into either astrocytes or oligodendrocytes (Valat et al., 1983). Autordiographic experiments (Skoff et al., 1976; Skoff, 1980; Valat et al., 1983) suggested that astrocyte proliferation begins during late embryonic development and continues after birth, while oligodendrocytes appear after the fifth postnatal day and proliferate prior to the onset of myelination. Studies of neonatally transected optic nerve further indicated the dependence of oligodendrocyte development on axonal signals (Privat et al., 1981). Although these studies contributed to our understanding of gliogenesis, they did not distinguish the different types of glial precursors and inadequately addressed the question of cell lineage.

Thus, the study of dissociated cell cultures from developing rat optic nerve with cell type specific immunologic markers has provided new insights into the development of central nervous system glia. The rat optic nerve contains three types of macroglial cells that develop in strict sequence from two immunologically distinct precursor cells. Type-1 (GFAP+, A2B5−) astrocytes first appear before birth at embryonal day 16, oligodendrocytes at birth, and type-2 (GFAP+, A2B5+) astrocytes in the second postnatal week (Miller et al., 1985). Oligodendrocytes and type-2 astrocytes develop from a common bipotential, A2B5-positive progenitor cell that can differentiate into either an oligodendrocyte or type-2 astrocyte, depending on the absence or presence of serum in the culture medium (Raff et al., 1983b). The GFAP-negative precursor cells of type-1 astrocytes can be identified with Ran-2, a monoclonal antibody that reacts with a cell surface antigen called rat neural antigen-2 (Raff et al., 1984b). Thus, the Ran-2-positive and A2B5-positive stem cells represent two different glial lineage. Type-1 astrocytes are found in the entire CNS, while type-2 astrocytes are restricted to white matter, suggesting that type-1 and type-2 astrocytes correspond in vivo to protoplasmic and fibrous astrocytes, respectively (Miller and Raff, 1984).

Although we demonstrated in previous studies, most recently reviewed by Lim (1985), the maturation and mitogenic effects of GMF on astroblast cultures derived from fetal rat brain, these new developments have raised additional questions concerning the role of GMF in glial development. Our present studies indicate that GMF stimulates mainly the proliferation of type-1 astrocytes both in optic nerve and brain-derived glial cell cultures. When optic nerve cultures were stimulated by GMF, they contained fewer oligodendrocytes and type-2 astrocytes than those cultures in serum-supplemented or serum-free medium. The influence of the culture medium on the developmental pathway taken by the progenitor cell was confirmed in our experiments, demonstrating an increase in oligodendrocytes in serum-free defined medium and an increase in type-2 astrocytes in serum-containing medium. GMF stimulation does not convert type-1 to type-2 astrocytes as determined by the immunological marker, A2B5, despite the consistent morphological transformation from falt, fibroblast-like type-1 astrocytes to process-bearing astrocytes with type-2 morphology. This confirms the previous findings by Raff et al. (1983a) that astrocytes do not convert from type

1 to type 2, despite similar morphological alterations induced by cyclic AMP, brain or pituitary extracts, and defined medium.

Perhaps the most interesting finding reported here is that GMF stimulation of confluent type-1 astrocytes derived from secondary fetal brain cell cultures resulted in an increase in A2B5-positive progenitor cells. These progenitor cells incorporated tritiated thymidine during a 24-hour pulse, providing evidence for cell proliferation during GMF stimulation.

The question arises as to why the proliferative effect of GMF was observed in the brain-derived glioblast cultures, but not in the optic nerve cultures. One major difference between the two is the cell density of type-1 astrocytes, being confluent in the former but sparse in the latter. Although alternative hypotheses may be possible, it is likely that GMF exerted its effect on the progenitor cells through the mediation of type-1 astrocytes and that, the more type-1 astrocytes there are, the stronger will be this mediation. One should be cautioned that the cell number of type-1 astrocytes may not be the sufficient cause for progenitor cell proliferation, as the phenomenon is not observed in serum-stimulated cultures where the density of type-1 astrocytes is the highest (table 10-2). It appears that the interaction of GMF with type-1 astrocytes is essential in bringing out this effect.

Recently, Noble and Murray (1984) showed that type-1 astrocytes in vitro produce a factor (or factors) that stimulates the proliferation of O-2A progenitor cells. In the absence of this factor, the progenitor cells stop dividing and differentiate prematurely into oligodendrocytes (Raff et al., 1985). In light of this finding and the fact that GMF can stimulate the secretion of some biologically active substances (e.g., interleukin-1 and prostaglandins) from astrocytes in culture (Fontana et al., 1983; Miller and Lim, 1984), we suspect that GMF may trigger the release of a growth factor from type-1 astrocytes that in turn stimulates the division of O-2A progenitor cells. As type-1 astrocytes are one of the major cell types in the CNS, and GMF is ubiquitously found in brain tissue, it seems likely that interactions similar to those described here play an important role in glial development.

6. CONCLUSIONS

Using immunostaining for the glial markers GFAP, A2B5, and galactocerebroside, the effects of glia maturation factor (GMF) on neonatal optic nerve and fetal brain-derived glioblast cultures were studied in the serum-free defined medium N2. GMF stimulated the proliferation of type-1 astrocytes in both low- and high-density cultures. Despite the consistently observed morphological transformation of flat, polygonal astroblasts to astrocytes with type-2 morphology, GMF did not actually convert type-1 to type-2 astrocytes. In the presence of confluent type-1 astrocytes, GMF promoted the proliferation of the A2B5-positive bipotential glial progenitor cells. As the proliferative effect of GMF was only apparent in the presence of large numbers of type-1 astrocytes, it seems likely that GMF exerted its stimulating action on the glial progenitor cells through the mediation of type-1 astrocytes.

ACKNOWLEDGMENTS

This work was supported by the following research grants to R.L.: VA Merit Review and Career Investigatorship Awards; NSF grant BNS-8308341; NCI grant CA-31796; Diabetes–Endocrinology Research Center grant (AM-25295); and University of Iowa Cancer Center grant.

We thank Drs. B. Ranscht (monoclonal anti-galactocerebroside antibody), F.S. Walsh, G.S. Eisenbarth (monoclonal A2B5 antibody), A. Bignami and D. Dahl (rabbit anti-GFAP serum) for kindly supplying antibodies, Dr. Xue-Mei Han for assistance in cell culture, and Mr. Shelly Plattner for autoradiography.

REFERENCES

Bottenstein, J.E., and Sato, G.H. (1979) Growth of a rat neuroblastoma cell line in serum-free supplemented medium. Proc. Natl. Acad. Sci. USA 76:514–517.

Eisenbarth, G.S., Walsh, F.S., and Nirenberg, M. (1979) Moncolonal antibody to a plasma membrane antigen of neurons. Proc. Natl. Acad. Sci. USA 76:4913–4917.

Fontana, A., Weber, E., Grob, P.J., Lim, R., and Miller, J.F. (1983) Dual effect of glia maturation factor on astrocytes: differentiation and release of interleukin-1 like factors. J. Neuroimmunol. 5:261–269.

Lim, R. (1985) Glia maturation factor and other factors acting on glia. In: Guroff, G. (ed.) Growth and maturation factors, vol. 3. New York: John Wiley and Sons, pp. 119–147.

Lim, R., and Miller, J.F. (1984) An improved procedure for the isolation of glia maturation factor. J. Cell Physiol. 119:255–259.

Lim, R., and Mitsunobu, K. (1974) Brain cells in culture: morphological transformation by a protein. Science 185:63–66.

Lim, R., Li, W.K.P., and Mitsunobu, K. (1972) Morphologic transformation of dissociated embryonic brain cells in the presence of brain extracts. Soc. Neurosci. Abstr. 2:181.

Lim, R., Mitsunobu, K., and Li, W.K.P. (1973) Maturation stimulating effect of brain extract and dibutyryl cyclic AMP on dissociated embryonic brain cells in culture. Exp. Cell Res. 79:243–246.

Lim, R., Turriff, D.E., and Troy, S.S. (1976) Response of glioblasts to a morphological transforming factor: cinematographic and chemical correlations. Brain Res. 113:165–170.

Lim, R., Troy, S.S., and Turiff, D.E. (1977a) Fine structure of glioblasts before and after stimulation by a glia maturation factor. Exp. Cell Res. 106:357–372.

Lim, R., Turiff, D.E., Troy, S.S., More B.W., and Eng. L.F. (1977b) Glia maturation factor: effect on chemical differentiation of glioblasts in culture. Science 195:195–196.

Lim, R., Miller, J.F., Hicklin, D.J., and Andresen, A.A. (1985) Purification of bovine glia maturation factor and characterization with monoclonal antibody. Biochemistry 24:8070–8074.

Miller, J.F., and Lim, R. (1984) Effect of glia maturation factor on the production of prostaglandins by rat astrocytes. Fed. Proc. 43:2004.

Miller, R.H., and Raff, M.C. (1984) Fibrous and protoplasmic astrocytes are biochemically and developmentally distinct. J. Neurosci. 4:585–592.

Miller, R.H. David, S., Patel, R., Abney, E.R., and Raff, M.C. (1985) A quantitative immunohistochemical study of macroglial cell development in the rat optic nerve: in vivo evidence for two distinct astrocyte lineages. Dev. Biol. 111:35–41.

Noble, M., and Murray, K. (1984) Purified astrocytes promote the in vitro division of a bipotential glial progenitor cell. EMBO J. 3:2243–2247.

Privat, A., Valat, J., and Fulcrand, J. (1981) Proliferation of neuroglial cell lines in degenerating optic nerve of young rats: a radioautographic study. J. Neuropathol. Exp. Neurol. 40:46–60.
Raff, M.C., Field, K.L., Hakomori, S.J., Mirsky, R., Pruss, R.M., and Winter, J. (1979) Cell type specific markers for distinguishing and studying neurons and the major classes of glial cells in culture. Brain Res. 174:283–308.
Raff, M.C., Abney, E.R., Cohen, J., Lindsay, R., and Noble, M. (1983a) Two types of astrocytes in cultures of developing rat white matter: differences in morphology, surface gangliosides, and growth characteristics. J. Neurosci. 3:1289–1300.
Raff, M.C., Miller, R.H., and Noble, M. (1983b) A glial progenitor cell that develops in vitro into an astrocyte or an oligodendrocyte depending on culture medium. Nature 303:390–396.
Raff, M.C., Williams, B.P., and Miller, R.H. (1984a) The in vitro differentiation of a bipotential glial progenitor cell. EMBO J. 3:1857–1864.
Raff, M.C., Abney, E.R., and Miller, R.H. (1984b) Two glial cell lineages diverge prenatally in rat optic nerve. Dev. Biol. 106:53–60.
Raff, M.C., Abney, E.R., and Fok-Seang, J. (1985) Reconstitution of a developmental clock in vitro: a critical role for astrocytes in the timing of oligodendrocyte differentiation. Cell 42:61–69.
Skoff, R.P. (1980) Neuroglia: a re-evaluation of their origin and development. Pathol. Res. Pract. 168:279–300.
Skoff, R.P., Price, D.L., and Stocks, A. (1976) Electron microscopic autoradiographic studies of gliogenesis in rat optic nerve. I. Cell proliferation. J. Comp. Neurol. 1969:291–311.
Temple, S., and Raff, M.C. (1985) Differentiation of a biopotential glial progenitor cell in a single cell microculture. Nature 313:223–225.
Valat, J., Privat, A., and Fulcrand, J. (1983) Multiplication and differentiation of glial cells in the optic nerve of the postnatal rat: a reassessment. Anat. Embryol. 167:335–346.

11. SPECIALIZED JUNCTIONS OF GLIAL CELLS IN SITU AND IN VITRO

ENRICO MUGNAINI, PAUL T. MASSA AND SARA SZUCHET

1. INTRODUCTION

Glial cells, in providing a framework for the nerve cells and their processes, establish numerous mutual contacts at which specialized cell junctions can be demonstrated by appropriate electron-microscopic procedures. Although thin-section electron microscopy is usually adequate to reveal distinct junctional appositions, freeze-fracture replicas are often required to define more precisely the features of the junctional membranes, thus providing a more reliable classification of the type of intercellular specialization(s) involved in the cell-to-cell contact. Freeze-fracture replicas are particularly useful in the classification of cell junctions in tissue culture.

As known, cell junctions are classified into three main categories: intermediate, or adhering, junctions; gap junctions; and tight junctions (reviewed by Simons and Fuller, 1985). Furthermore, cell junctions may be homocellular or heterocellular; i.e., they may occur between cells belonging to the same class or to different categories. All cell junctions strengthen the mechanical attachment of one cell to another well beyond the degree of cohesion provided by special adhering molecules present at the cell surface (Edelman, 1983; Hynes, 1985). Adhering junctions, however, are thought to represent the most sophisticated device for mechanical attachment, binding one cell to another in relation to the cytoskeletal framework of individual cells. Gap junctions, instead, are specialized in intercellular communication. Their specific proteinaceous components, the connexons, establish charged

Vernadakis, A. (ed.), *Model Systems of Development and Aging of the Nervous System*.

aqueous channels through which small solutes, including ions and many intermediates, can pass directly from the cytoplasm of one cell to that of its junctionally related neighbor. The role of tight junctions appears even more complex: their particulate components, possibly a protein of 120,000 daltons in molecular weight (Simons and Fuller, 1985), assemble into strands and join tightly with those in the adjacent cell, thus quilting the membranes together in a manner that obliterates the extracellular space enough to block paracellular movement of large molecules. Diffusion of ions is affected variably, depending on the extent and arrangement of the tight junction strands (Claude, 1978). The strands also form a barrier to lateral movement of intramembrane proteins within the lipid bilayer, thus contributing to the preservation of cell polarity (Mugnaini and Schnapp, 1974).

This account deals exclusively with gap junctions and tight junctions. Homocellular and heterocellular junctions are included. We review our published studies on the junctions between glial cells in the three systems: 1) the mature mammalian nervous system (Massa and Mugnaini, 1982; Mugnaini, 1986); 2) cultures of dissociated rat embryo brain cells enriched in oligodendrocytes, henceforth referred to as "mixed glial cell cultures" (Massa et al., 1982, 1983, and 1984a; Massa and Mugnaini, 1985); and 3) purified and cultured oligodendrocytes isolated from lamb white matter (Massa et al., 1983 and 1984b; Szuchet, 1986; Szuchet et al., 1986). Furthermore, reference will be made to additional, unpublished studies of glial cells both in situ and in vitro. Procedures and methodological details are provided in the specific references.

2. FEATURES OF FREEZE-FRACTURED PLASMA MEMBRANES OF ASTROCYTES AND OLIGODENDOCYTES

Identification of astrocytic and oligodendrocytic plasma membranes in freeze-fracture replicas was first carried out in the white matter of adult animals. In this tissue the vast majority of cells are either astrocytes containing bundles of intermediate filaments (fibrous astrocytes) or interfascicular oligodendrocytes, which lack intermediate filaments. The plasma membranes of cells whose cytoplasm is exposed by the fracturing process and contains (astrocytes) or lacks (oligodendrocytes) bundles of intermediate filaments appear to have distinctly different patterns of intramembrane particles (IMPs). Astrocytic P faces show scarcely distinctive individual round particles and characteristic orthogonal assemblies of small particles (figure 11-1). Oligodendrocytic P faces, in addition to scarcely distinctive round particles, have a large population of characteristic elongated particles that is absent in astrocytic membranes (figure 11-2). Furthermore, orthogonal assemblies are lacking in oligodendrocytic P faces. Complementary features are present in astrocytic and oligodendrocytic E faces, respectively.

Thus, astrocytic and oligodendrocytic plasma membranes can be differentiated on the basis not only of negative criteria (absence of orthogonal

Figure 11-1. This electron micrograph shows a replica of a freeze-fractured plasma membrane from a white matter astrocyte with the characteristic orthogonal assemblies and round particles, as seen on the protoplasmic aspect (P face). A gap junction is indicated by the *arrow*. × 78,000.

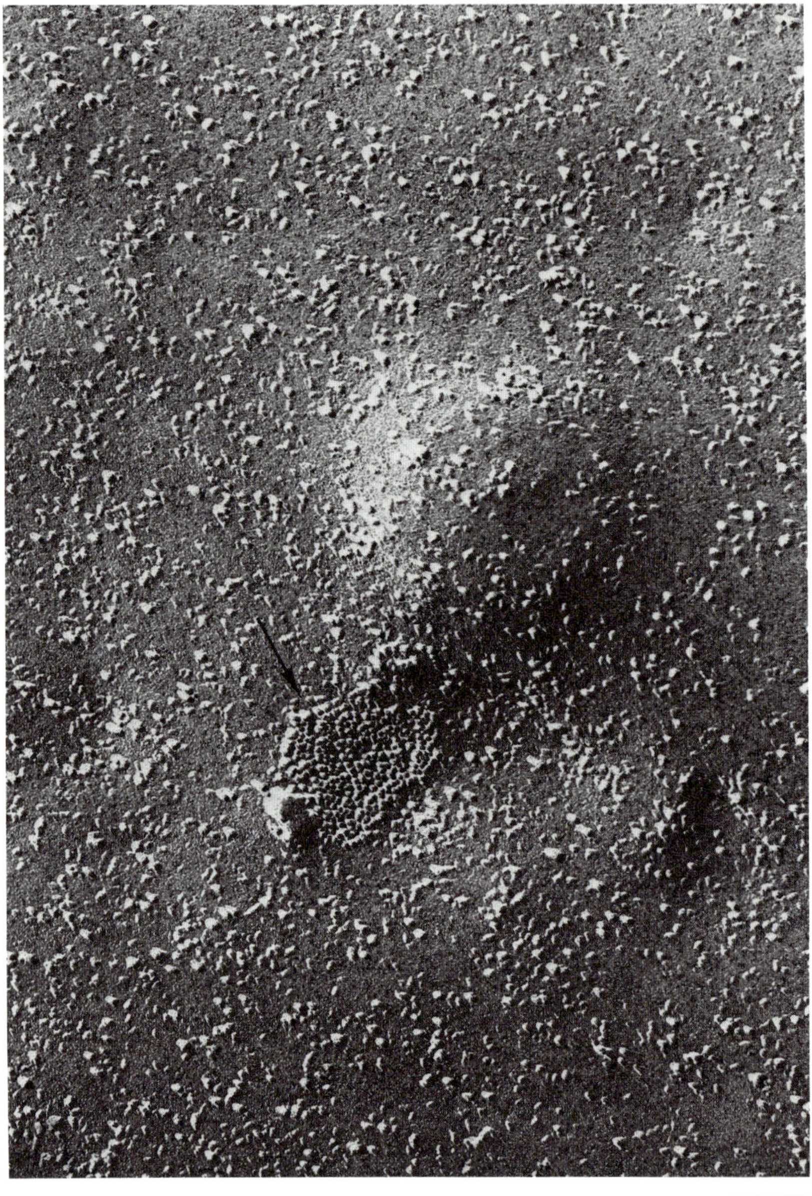

Figure 11-2. This replica of a freeze-fractured plasma membrane from the cell body of a white matter oligodendrocyte shows the characteristic elongated IMPs and the round IMPs, as seen on the protoplasmic aspect (P face). A gap junction is indicated by the *arrow*. Note the absence of orthogonal particle assemblies. × 105,300.

assemblies in oligodendrocytes; absence of individual elongated particles in astrocytes), but also of positive criteria (pressence of orthogonal assemblies, in astrocytes; presence of individual elongated particles in oligodentrocytes) (Massa and Mugnaini, 1982).

Successive, unpublished studies on glial cells in gray matter from various central nervous system (CNS) regions have shown that these criteria of identification of plasma membranes apply also to protoplasmic astrocytes and to gray matter oligodendrocytes. The oligodendrocytic pattern of IMPs is also found in the compact myelin membrane (figure 11-3) and in the loops of the myelin lamellae (Schnapp and Mugnaini, 1978). Thus the entire oligodendrocytic plasmalemma contains characteristic IMPs, although their density varies in different membrane regions.

The patterns of IMPs of astrocytic and oligodendrocytic plasma membranes, fortunately, are so uniformly peculiar that they can be used to identify even small plasmalemma fragments of profiles of cell bodies and processes whose cytoplasm has not been exposed by the fracturing process. It can be stated, therefore, that the distinctive IMPs are as good a criterion for identification of these cells as cell class-specific monoclonal antibodies. Obviously, this statement does not preclude the possibility that some immature cells might be present that cannot be safely classified on the basis of their IMPs, but express some of the antigens characteristic of astrocytes or oligodendrocytes.

Freeze-fracture replicas of dissociated rat embryo brain cells cultured in a serum-containing medium indicated that astrocytic and oligodendrocytic plasma membranes express the same classes of IMPs as their counterparts in situ (Massa and Mugnaini, 1985). Plasma membranes of oligodendrocytes purified from lamb brain and cultured in the presence of serum also show IMPs that resemble those of oligodendrocytes in situ (Massa et al., 1984b). Our culture conditions, thus, did not alter in a qualitatively significant way the expression of characteristic IMPs in glial cells.

In all the three systems analyzed, therefore, it was possible to identify reliably both partners in pairs of adjacent cells that form cell–cell junctional contacts. The criteria for identification of the adjoining cell membranes remained unchanged across the different cell systems.

3. CELL JUNCTIONS OF GLIAL CELLS IN SITU AND IN MIXED GLIAL CELL CULTURES

In situ, fibrous (Massa and Mugnaini, 1982) as well as protoplasmic (Mugnaini, 1986) astrocytes form gap junctions with each other (figure 11-4) (homocellular gap junctions) and with oligodendrocytes figures 11-5, 11-7, and 11-8) (heterocellular gap junctions). The density of glial gap junctions is very high (figure 11-6). In situ each glial cell was estimated to form more than a hundred gap junctions.

Oligodendrocytes form incomplete, linear tight junctions with each other

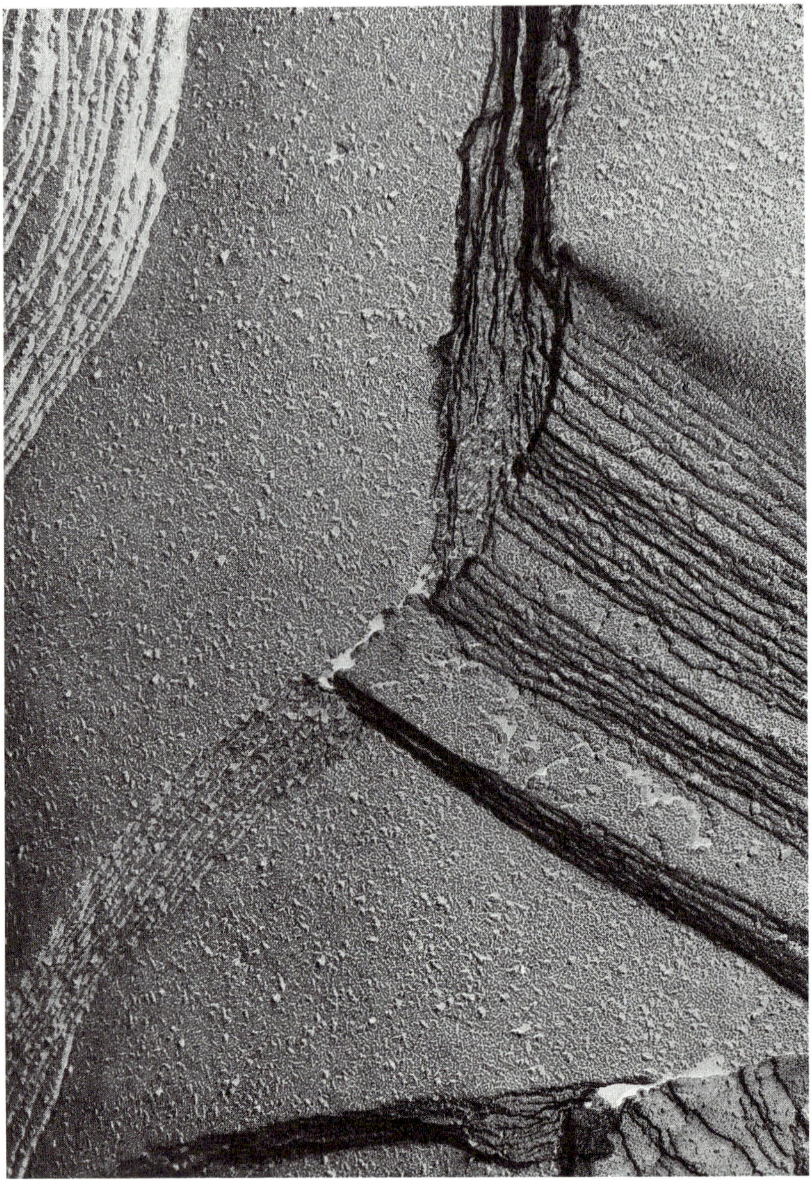

Figure 11-3. Freeze-fractured myelin lamellae of a thickly myelinated axon show elongated and round IMPs resembling those on the plasmalemma of oligodendrocytic cell body and processes. × 124,000.

Figure 11-4. These gap junctions (*GJ*) from the white matter are formed between two astrocytes, since orthogonal assemblies and their complementary pits (*arrows* and *crossed arrows*, respectively) are present in both membranes. × 128,700.

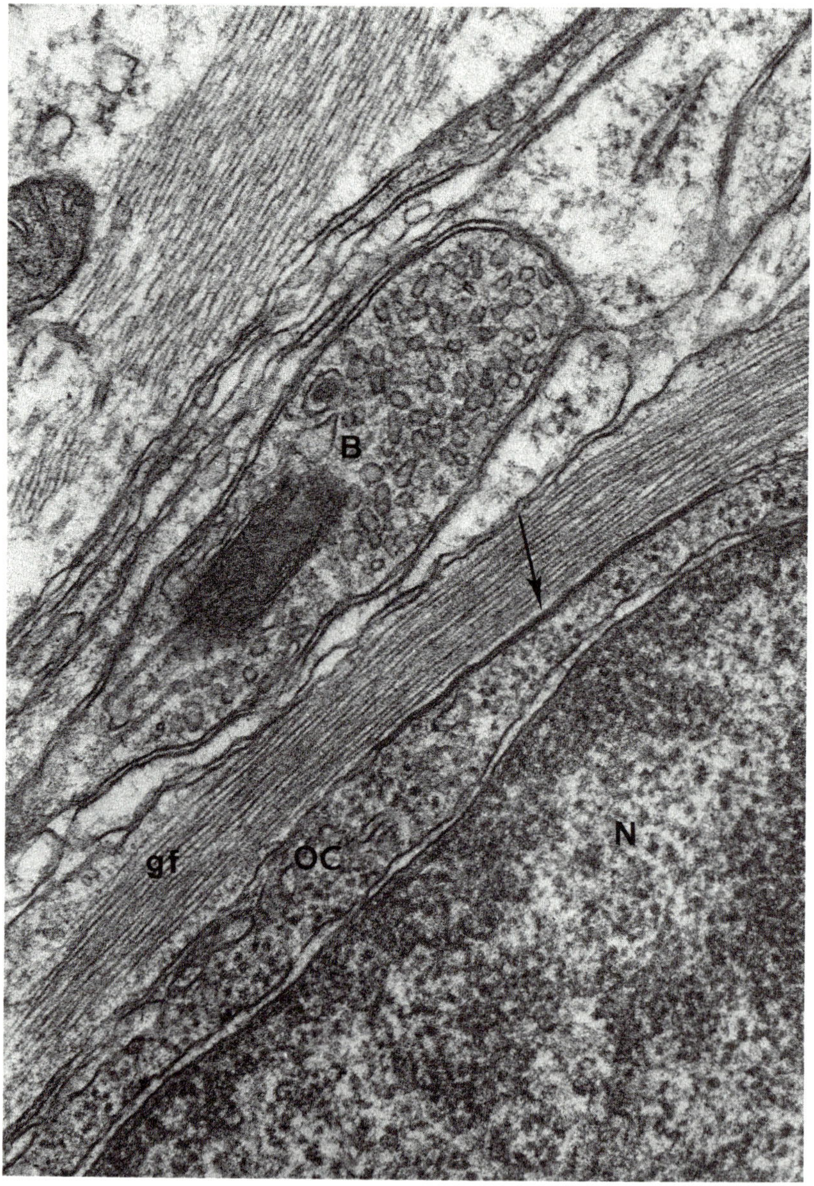

Figure 11-5. This detail of a thin section from the border region between white matter and gray matter contains a gap junction (*arrow*) between an astrocytic process filled with glial filaments (*gf*) and an oligodendrocytic cell body (*OC*). *N*, cell nucleus; and *B*, axonal varicosity filled with synaptic vescicles. × 70,500.

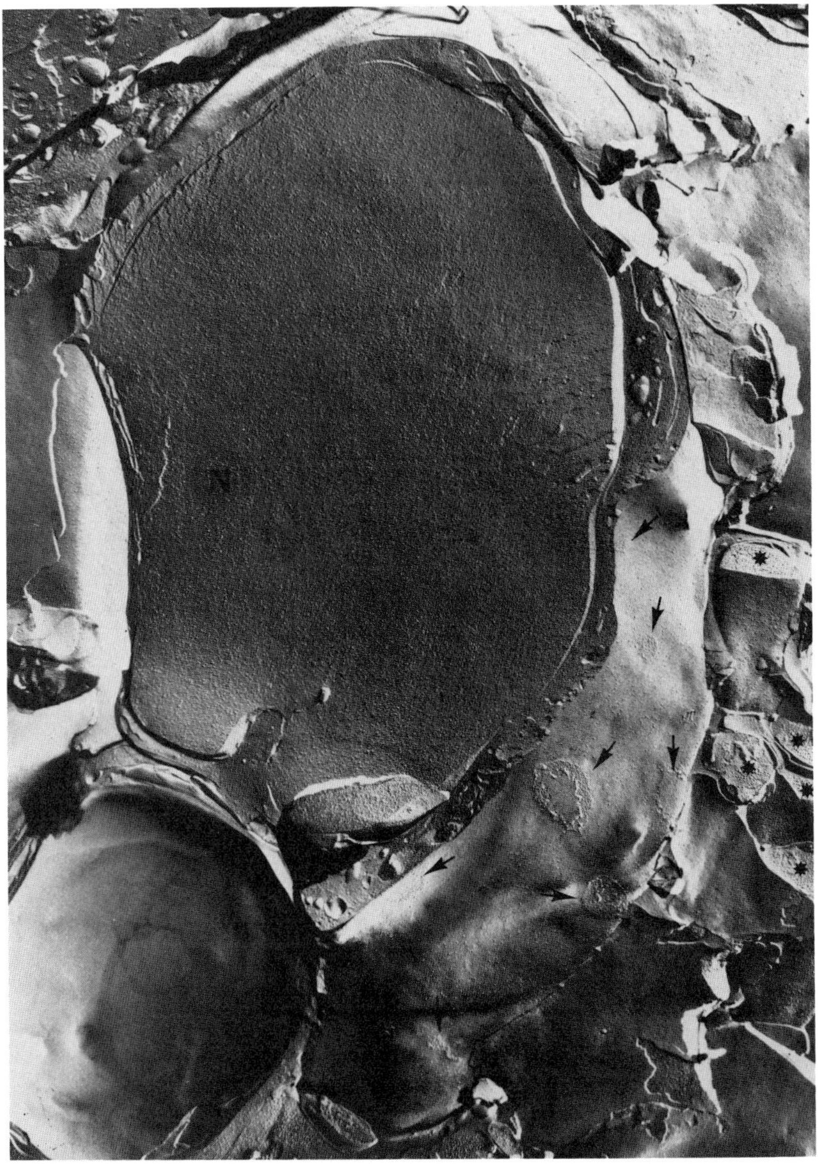

Figure 11-6. Replica of a freeze-fractured white matter oligodendrocyte that shows as many as seven gap junctions (*arrows*) in a small portion of its exposed plasmalemma. *N*, cell nucleus; and *asterisks*, astrocytic processes. × 21,000.

Figure 11-7. This micrograph shows a gap junction (*GJ*) between an oligodendrocyte (membrane with elongated particles) and an astrocyte (*AC*) identified by virtue of its intermediate filaments (*gf*) from the white matter. × 63,350.

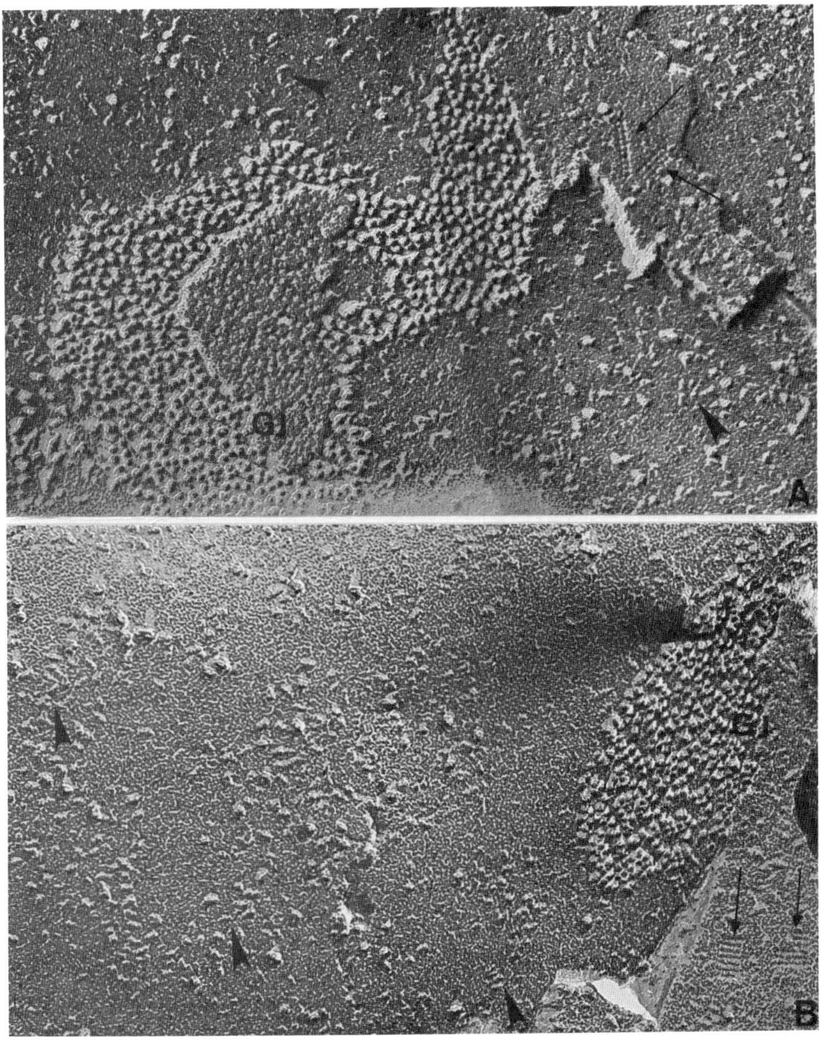

Figure 11-8. These micrographs show two gap junctions (*GJ*) between astrocytic and oligodendrocytic plasma membranes. These are identified by pits of orthogonal assemblies (*arrows*) and by elongated particles (*arrowheads*), respectively. × 156,000.

(figure 11-9) (homocellular tight junctions). Tight junctions between astrocytes were never observed. Interoligodendrocytic gap junctions, although tentatively (and rarely) identified in thin sections, could not be demonstrated with certainty in freeze-fracture replicas.

After approximately 10 days in vitro, dissociated cell cultures from brain of rat embryos contain a confluent bed layer of glial fibrillary acidic protein (GFAP)-positive astrocytes (figures 11-10A and 11), with oligodendrocytes laying scattered on top (figure 11-11) (McCarthy and De Vellis, 1980; Massa and Mugnaini, 1985). A varying number of neurons (not illustrated), identified by immunostaining with an antiserum against neuron-specific enolase (NSE) are included in the astrocytic bed layer. Oligodendrocytes at various stages of differentiation can be demonstrated by immunostaining with antisera to class-specific antigens such as myelin basic protein (MBP) (Massa et al., 1982 and 1983; Massa and Muganini, 1985) (figure 11-10B to D). After approximately 3–4 weeks in culture, mature oligodendrocytes produce empty tubes of compact myelin without physical contact with axons (Massa et al., 1984a) (figure 11-12A). After only 10 days in culture, agitation of the flasks at 250 rpm for 15 hours in an orbital shaker (a procedure developed by McCarthy and De Vellis, 1980) caused—in our cultures—the detachment of many oligodendrocytes and some astrocytes. The floating cells were replated in culture dishes and studied after 5 additional days in vitro. These replated glial cell cultures consisted of a nonconfluent layer of astrocytes with interspersed oligodendrocytes, most of which were clustered in aggregates attached partly to the dish itself and partly to the astrocytes (figure 11-13). Freeze-fracturing indicated that numerous gap junctions between astrocytes (figure 11-14) and between astrocytes and oligodendrocytes (figure 11-15) are also formed in this system. Similar to the situation in situ, oligodendrocytes develop tight junctions with one another (Massa and Mugnaini, 1985 and their figures 12–15) and interoligodendrocytic gap junctions could not be demonstrated.

The samples of freeze-fractured plasma membranes are relatively limited and may not constitute a truly random representation of all oligodendrocytic plasma membranes, especially in the cultures that tend to fracture near the surface of the culture dish. Clarification of the partners involved in gap junction formation is difficult also because, in most cases, the fracture process reveals a large portion of one membrane, but only a small fragment, if any, of the adjoining junctional membrane (figures 11-6 and 11-7). The findings in situ and in mixed glial cell cultures, therefore, do not prove that interoligodendrocytic coupling by gap junctions does not exist, but they indicated that, if such a coupling exists, it must be rare.

Coupling between presumed oligodendrocytes in explant cultures has been demonstrated with micropipette studies by other investigators (for references and discussion, see Mugnaini, 1986).

Figure 11-9. Linear tight junctions (*center*) between two interfascicular oligodendrocytes (OC_1 and OC_2). *N*, nuclear envelop with pores. ×49,500.

4. CELL JUNCTIONS OF PURIFIED AND CULTURED OLIGODENDROCYTES

Lamb oligodendrocytes were purified at a stage when, presumably, most of the cells had already formed myelin sheaths. In the process of mechanical and chemical dissociation, the oligodendrocytes are completely stripped of all their processes (figure 11-16) and become separated from one another in spite of their cell junctions. On the freshly isolated cells, gap junctions are still bound to the partner membrane that reseals, forming a vesicle (Massa et al., 1983 and 1984b). Analogous vesicles can be found on the tight junctional domains, although some of the tight junctions directly face the intercellular fluid and, therefore, must have separated neatly from the membrane of the adjoining cell. Presumably, both gap junctions and tight junctions are successively eliminated by internalization and digestion into the lysosomes, although it cannot be excluded that some of the component subunits may redistribute in the plane of the plasma membrane and become incorporated into newly formed junctions.

With time in culture, the cells adhere to one another, regrow extensive cell processes (figure 11-17A), and form lamellar flaps that attach to one another and to the dish, or are wrapped around oligodendrocytic cell bodies. Initially, and if plated at high density, the cells tend to form a confluent monolayer, but successively, as processes are extended, the cells begin to cluster (figure 11-17B). Such clustering is much less frequent and process extension more profuse if a relatively lower plating density is used. The purified oligodendrocytes are immunostained with antisera to MBP (figure 11-18A) and galactocerebroside. They are contaminated only by occasional GFAP-positive astrocytes (figure 11-18B) and by scattered, vacuolated cells identified as macrophages (not illustrated). Purified oligodendrocytes have been maintained in culture for 90 days and longer. Contaminating astrocytes may become more numerous in the older cultures. Cultures older than 3–4 weeks, however, are particularly interesting since they develop multilamellar plasma membrane arrays apposed to cell bodies and processes. These membranes have biochemical characteristics (e.g., they contain myelin-specific proteins such as MBP, PLP, and DM-20; Suchet et al., 1986) and ultrastructural features (e.g., periodicity; Suchet, 1986) similar to myelin (figure 12B). During the second or third week in culture, the purified oligodendro-

Figure 11-10. (**A**) GFAP-positive astrocytic cell bodies and processes form the astrocytic bed layer of a primary glial cell culture. × 105. (**B**) MBP-positive cell bodies and processes indicate oligodendrocytes atop the unstained bed layer of astrocytes in a primary glial cell culture. × 105. (**C**) MBP immunostaining shows oligodendrocytes at different stages of maturation in a 2-week-old primary culture. The pale (unstained) nuclei in the background belong to astrocytes in the underlying bed layer. × 315. (**D**) The processes of this MBP-positive oligodendrocyte from a 2-week-old primary culture are provided with flap-like expansions adhering to the substrate. × 630.

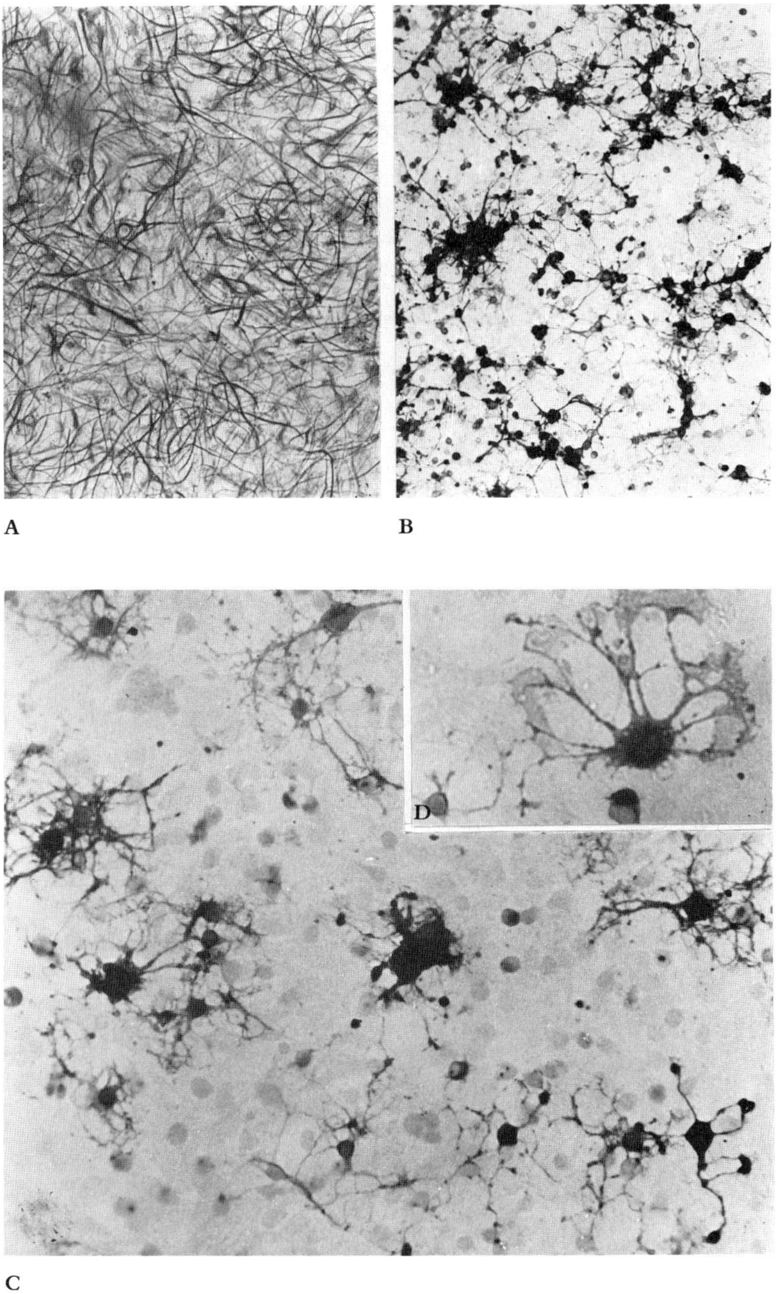

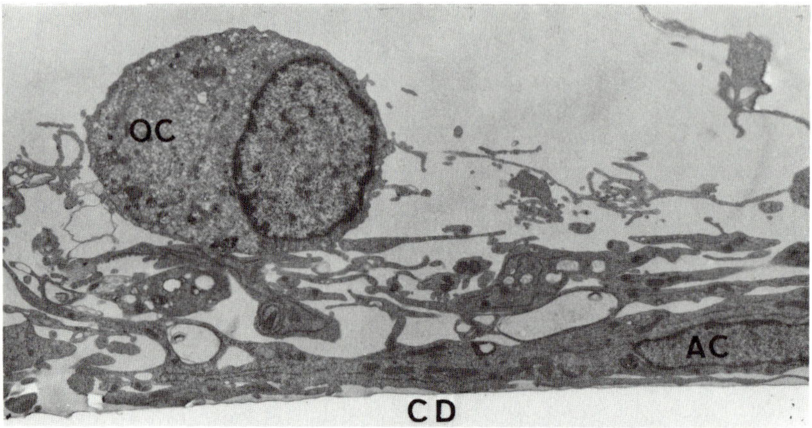

A

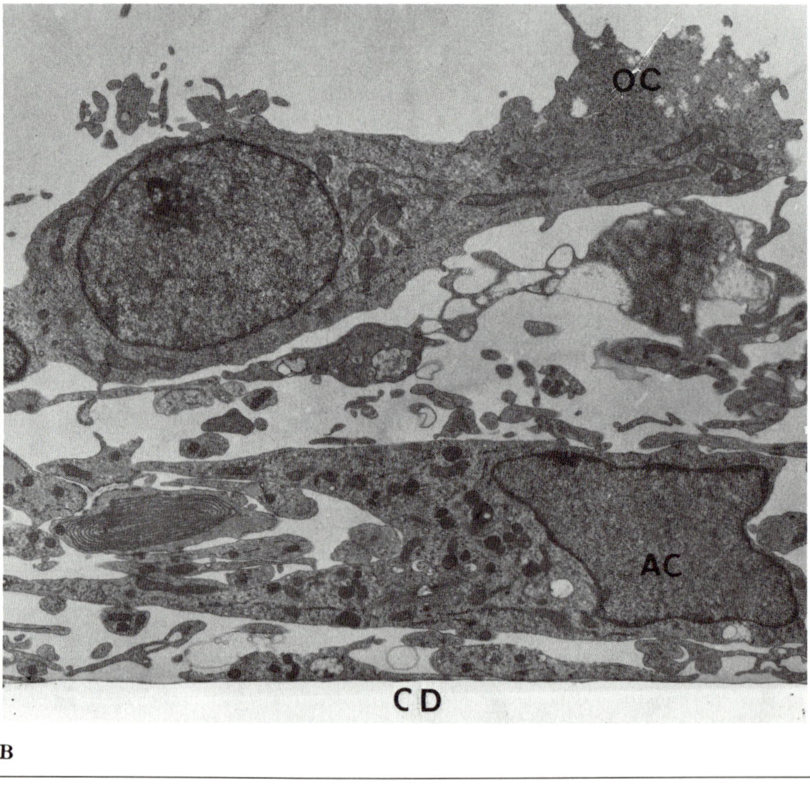

B

Figure 11-11. These electron micrographs are taken from thin sections cut perpendicular to the culture dish (*CD*) and show thinner (**A**) and thicker (**B**) portions of a 2-week-old primary glial cell culture. In both cases, astrocytes (*AC*) are attached to the dish and oligodendrocytes (*Oc*) are atop this bed layer. **A**, × 4700; and **B**, 6900.

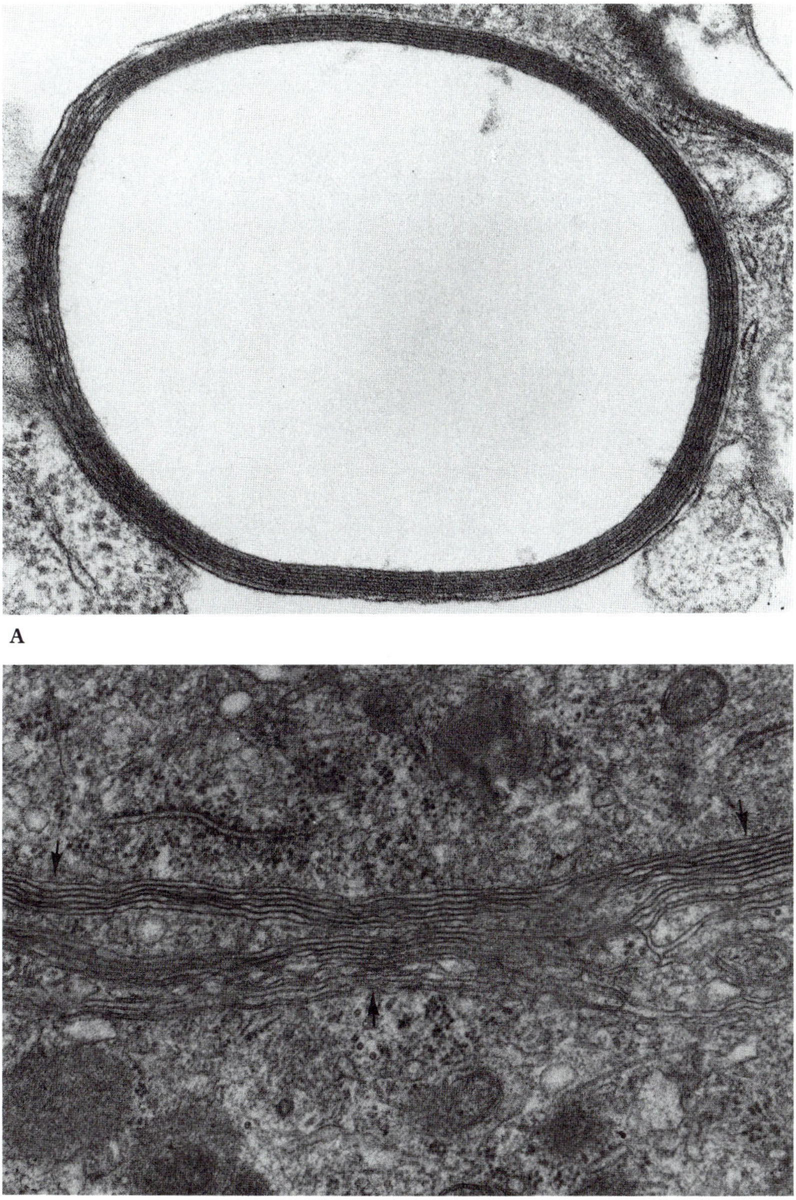

A

B

Figure 11-12. (**A**) This empty tube is formed by a myelin-like, spiraled, and compact membrane in a 5-week-old primary glial cell culture. × 86,800. (**B**) These myelin-like, compact lamellae (*arrows*) are situated between two cell bodies of purified ovine oligodendrocytes cultured for 4 weeks. × 39,500.

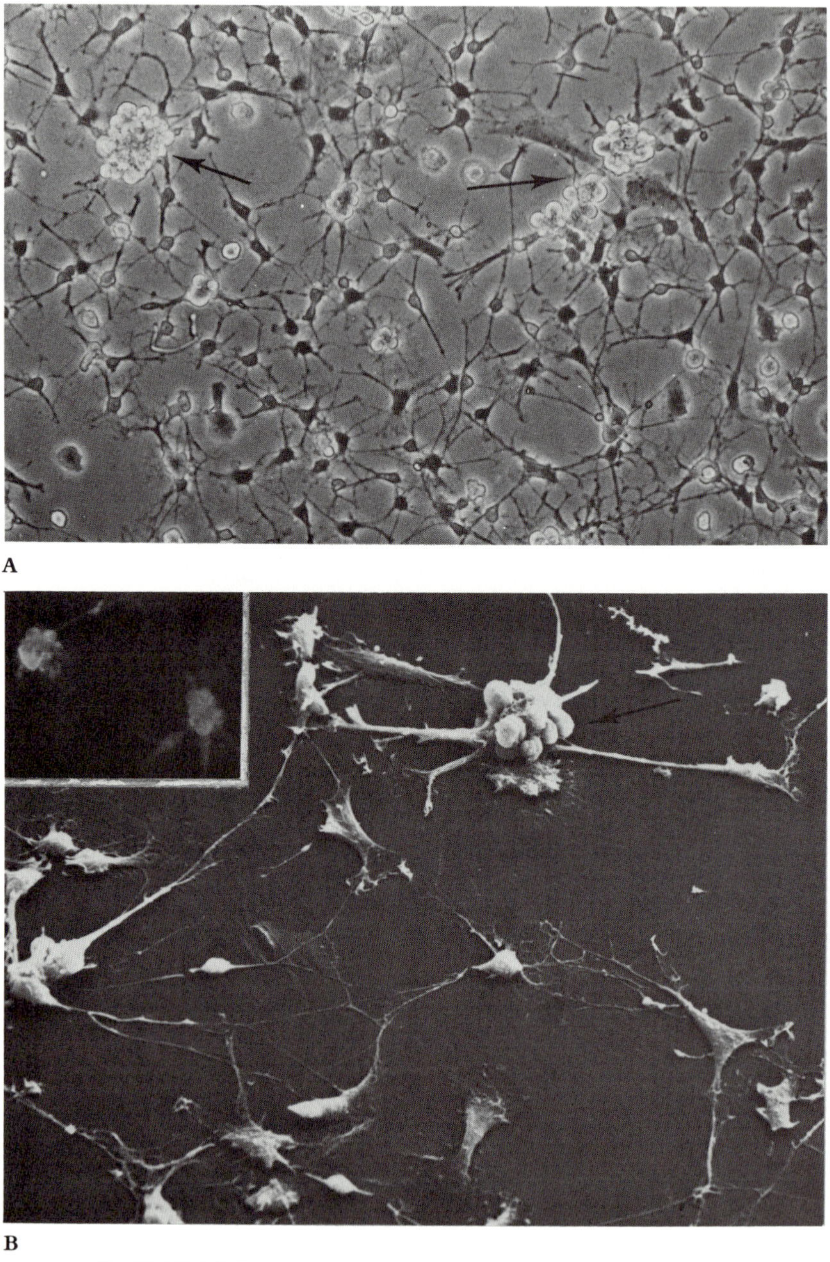

A

B

Figure 11-13. (**A**) Phase-contrast micrograph of replated floating cells obtained by vigorous agitation of primary glial cell cultures. These cultures are enriched of bright, clustered, small cell aggregates, some of which are indicated by *arrows*. × 175. (**B**) This scanning electron micrograph of a similar culture shows a small cell aggregate (*arrow*) attached partly to the dish and partly to a flat cell. × 410. (**Inset**) In this fluorescence micrograph, clustered small cell aggregates are immunostained with a MBP antiserum, indicating that they are oligodendrocytes. Isolated oligodendrocytes are also present. × 125.

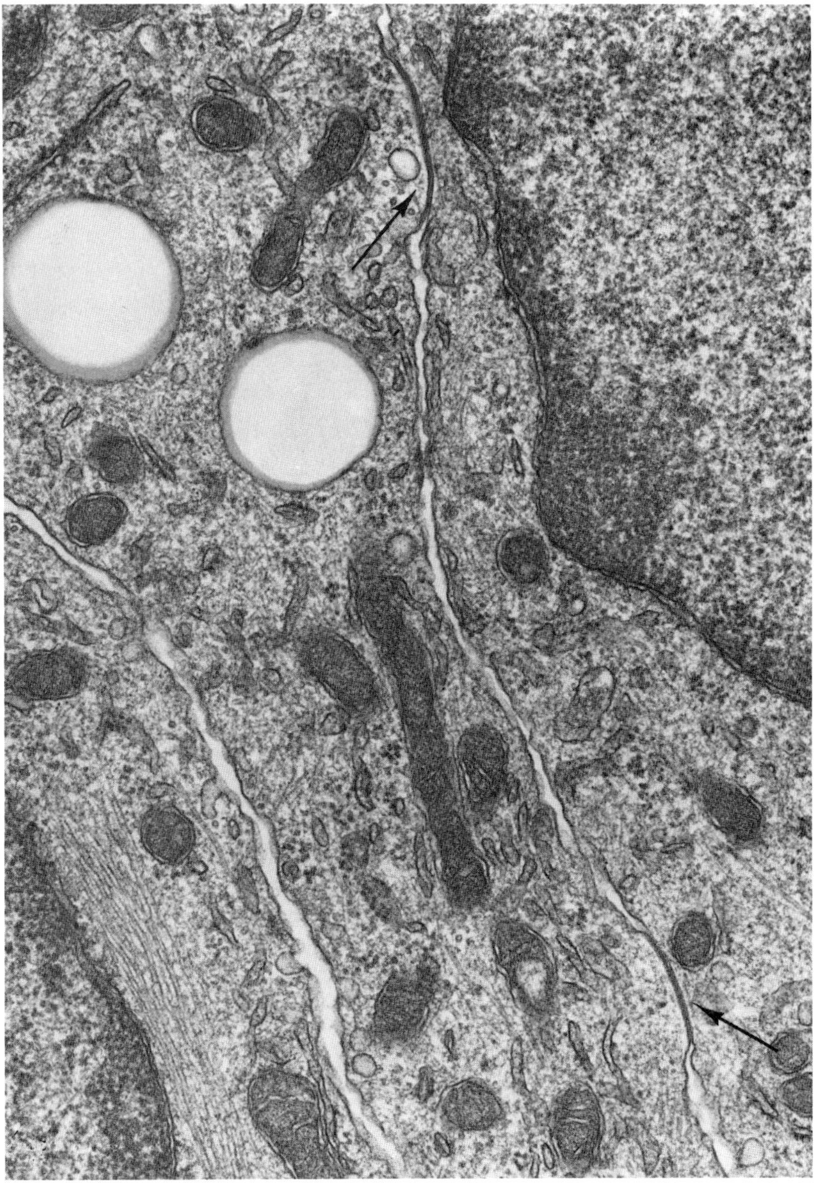

Figure 11-14. This electron micrograph of a mixed glial cell culture shows two gap junctions (*arrows*) between filament containing astrocytes in the bed layer. ×45,500.

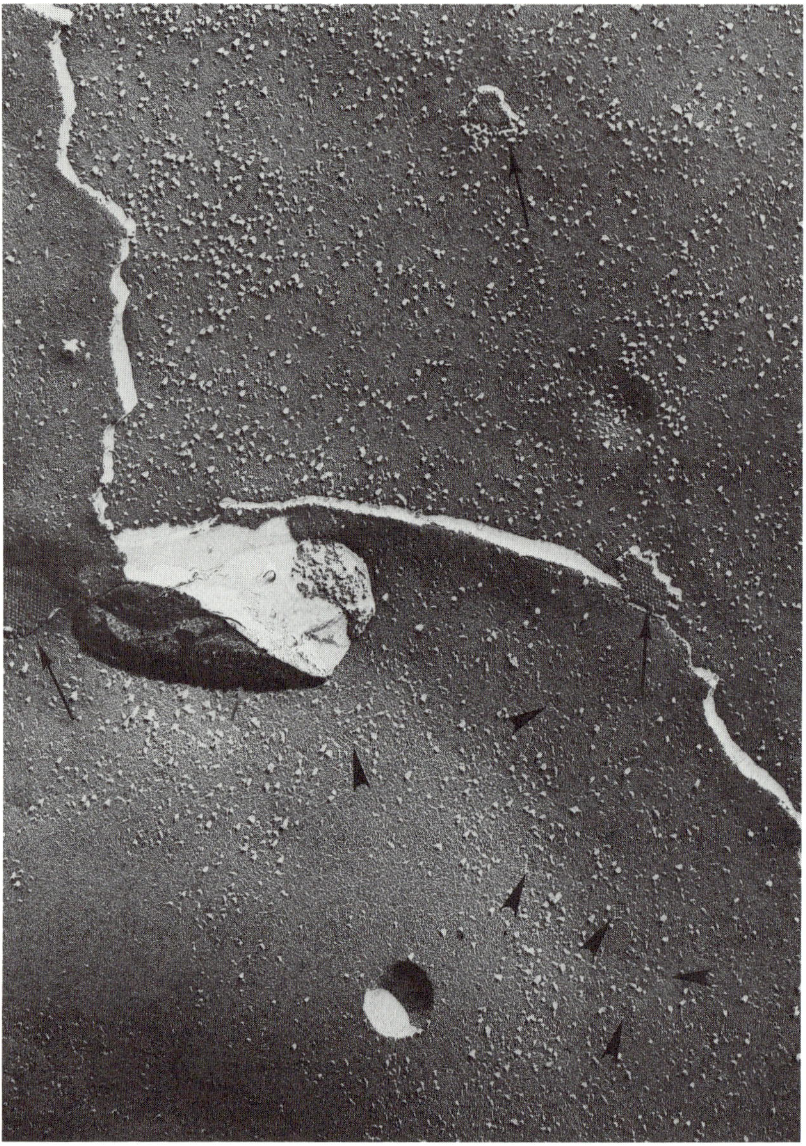

Figure 11-15. Gap junctions (*arrows*) are present at the site of contact between an oligodendrocytic and an astrocytic membrane. The former (P face) is identified by a mixture of round and elongated IMPs and the latter (E face) by pits complementary to orthogonal assemblies (*arrowheads*). × 123,000.

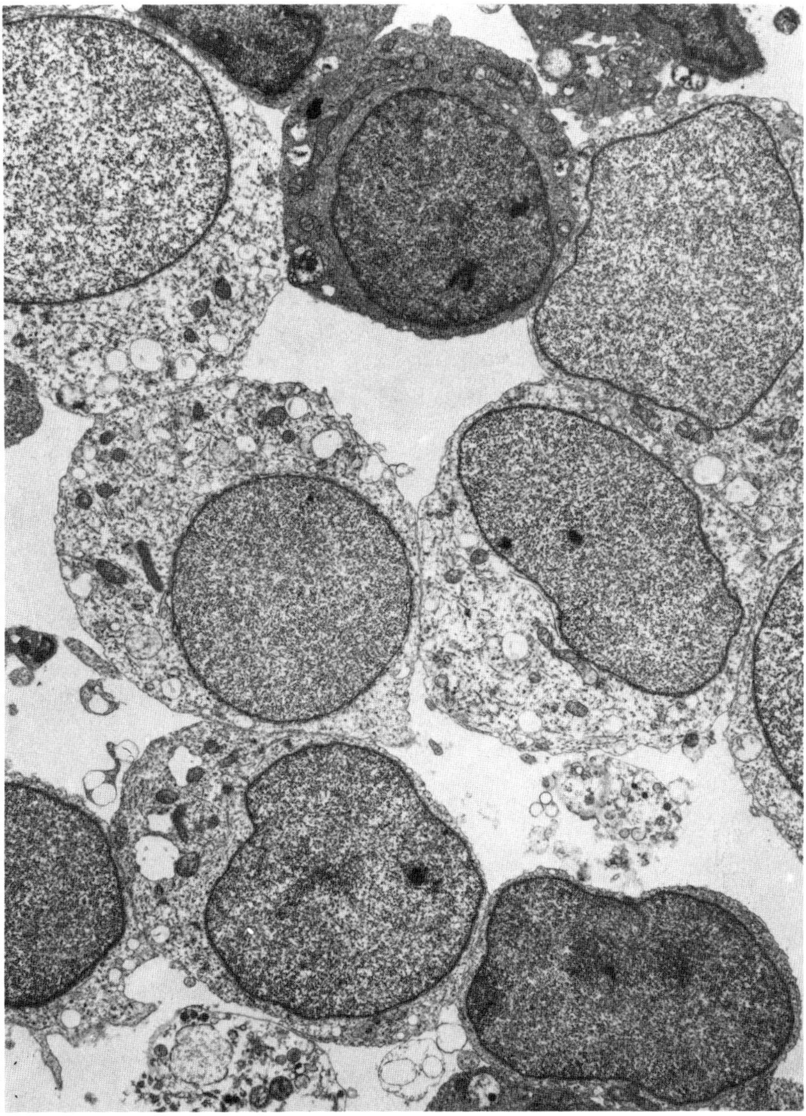

Figure 11-16. Freshly isolated, pelleted, ovine oligodendrocytes have lost all of their processes and contain several vacuoles, some of which represent internalized cell junctions. The electron density of these oligodendrocytes varies. ×5160.

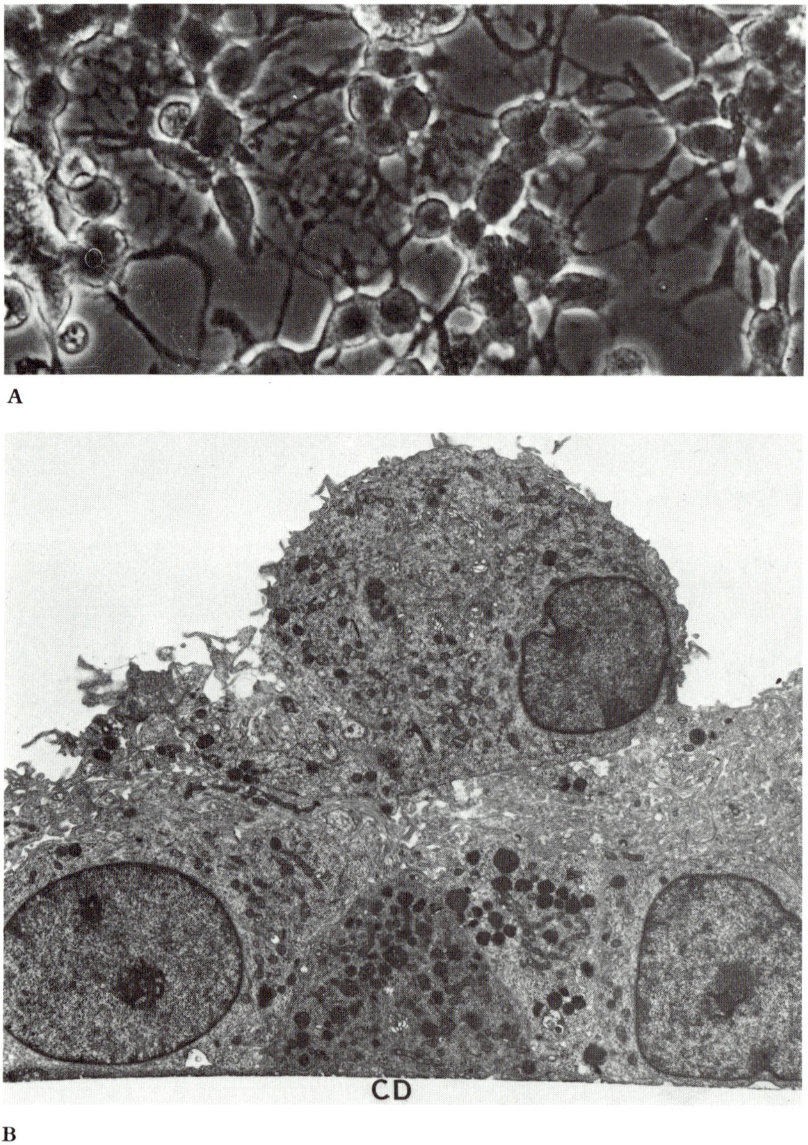

Figure 11-17. Phase-contrast micrograph (**A**) and electron micrograph (**B**) of purified ovine oligodendrocytes cultured for 11 days. *CD*, culture dish. **A**. × 800; and **B**, × 6000.

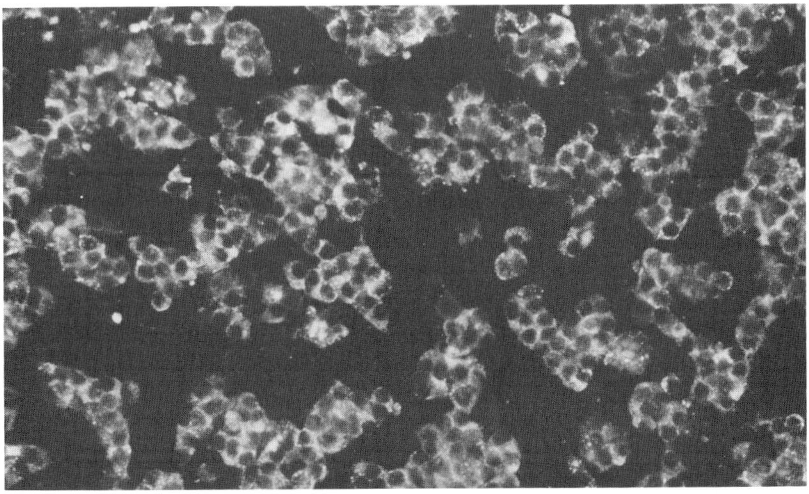

A

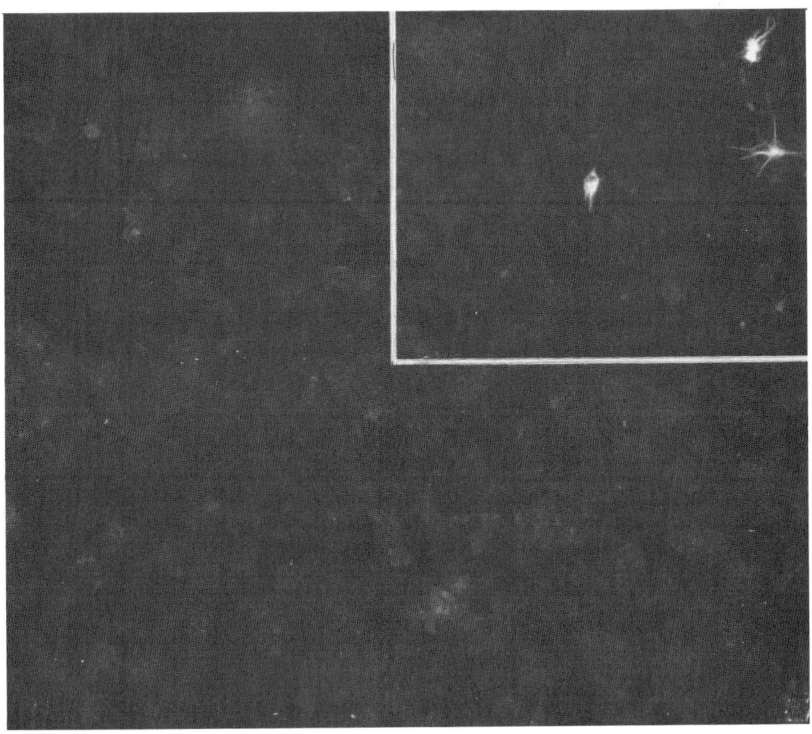

B

Figure 11-18. Immunofluorescent pictures of purified ovine oligodendrocytes, cultured for 7 days at high density indicate they are nearly homogeneously MBP positive (**A**) and GFAP negative (**B**). The **inset** shows a portion of a dish contaminated by three GFAP-positive astrocytes. **A**, × 320; **B**, × 130; and **inset**, × 130.

cytes, which are always provided with numerous microtubules, develop intermediate filaments (figure 11-19). The cell clusters can be redissociated and replated with reestablishment of a monolayer (Szuchet, unpublished observations). Notably, the purified oligodendrocytes can develop new myelin sheaths when cocultured with neurons (see Szuchet, 1986).

At the sites of apposition between cell bodies, between cell bodies and processes, and between cell processes, the membrane appositions show discontinuous tight junctional domains that increase in number and extent over time in culture, although a definite polarity is never attained. The amount of new membrane formed in vitro by the attached cells is so large that the vast majority of the tight junctions must be formed de novo. Different from the situation in situ, the tight junctions of purified and cultured oligodendrocytes show some anastomoses (figure 11-20). Occasional, small gap junctions occur in the corners of the anastomosed strands, but no free gap junctions are observed even after 3–4 weeks in vitro. The network of anastomosing oligodendrocytic tight junctions remains predominantly linear even in the older cultures and appears much less compact than that of endothelial cells in brain capillaries provided with a blood–brain barrier (figure 11-21).

5. FUNCTIONAL CONSIDERATIONS ON GAP JUNCTIONS IN GLIAL CELLS

Astrocytes are so extensively coupled by gap junctions to other astrocytes and to oligodendrocytes that the glial cell system must be considered a sort of functional syncythium. In the CNS, the extent of this syncythium depends on whether: 1) fibrous astrocytes and protoplasmic astrocytes are coupled to one another, 2) astrocytes in one region (for instance, one of the cortical layers or a nucleus in the thalamus) are coupled to astrocytes in neighboring regions, and 3) astrocytes are coupled to other cells in the astrocytic "family" such as ependymal cells. No specific data are available concerning the first two points. Nevertheless, since astrocytic gap junctions have been found in every region studied so far, it seems fair to suggest that all astrocytes may be coupled indiscriminately. Recently, it has been shown in two different brain regions that ependymal cells, which are known to be coupled to one another by gap junctions, also form gap junctions with protoplasmic and fibrous astrocytes (Mugnaini, 1986 and unpublished observations). One can hypothesize, therefore, that cells of the astrocytic family form a truly generalized functional syncythium extending from the ventricular lining to the membrana limitans gliosa (figure 11-22A) (Mugnaini, 1986). If oligodendrocytes do not form substantial homocellular gap junctions, as our studies indicate, such cells would represent blind-ended appendages on the functional astrocytic network. This situation is represented schematically in figure 22B, where the oligodendrocytes appear as grapes attached to a grape-stalk, the astrocytic syncythium. The notion of a generalized functional syncythium of glial cells does not preclude the possibility of topical specializations (for

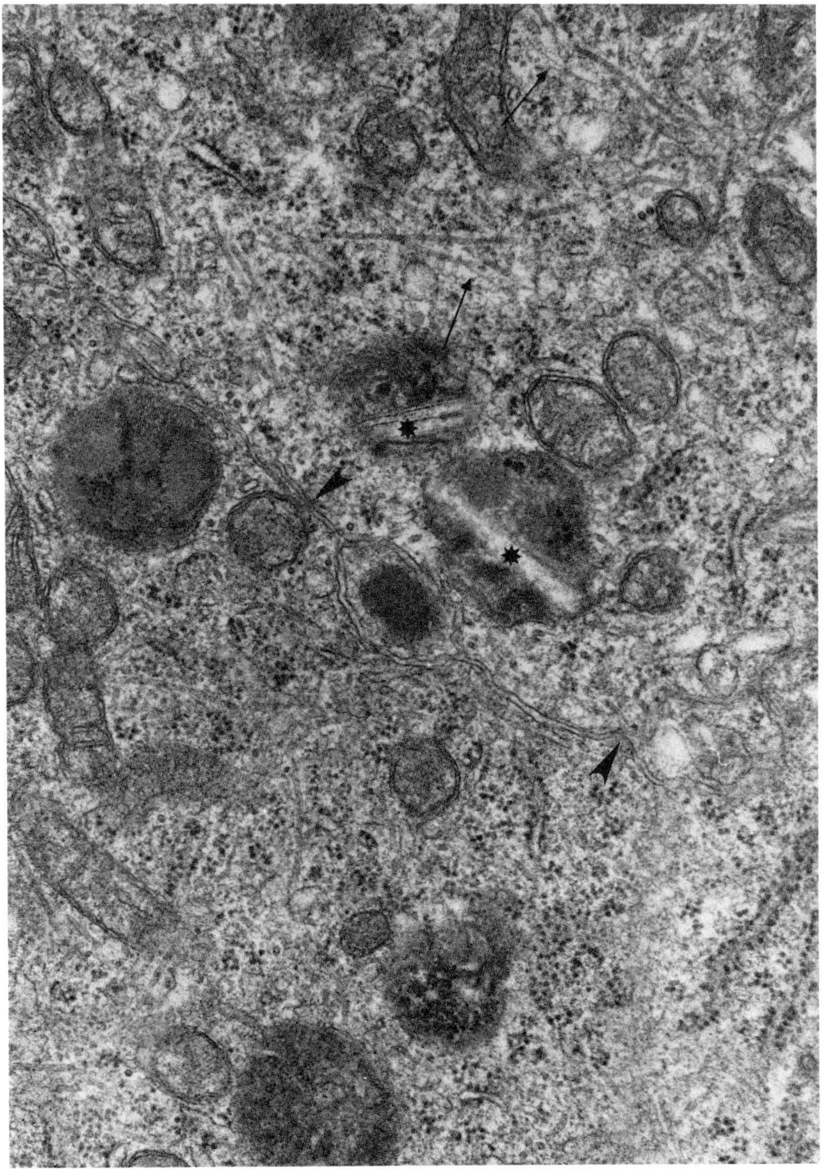

Figure 11-19. Electron micrograph of an area of contact between cell bodies and processes of purified ovine oligodendrocytes cultured for 3 weeks. Foci of close membrane apposition (presumably tight junctions, *arrowheads*) are present, but gap junctions are lacking. The *asterisks* mark dense bodies with electron-transparent inclusions characteristic of oligodendrocytes (see Mugnaini and Walberg, 1964). *Arrows* indicate groups of cytoplasmic filaments intermingled with mictotubles. × 44,550.

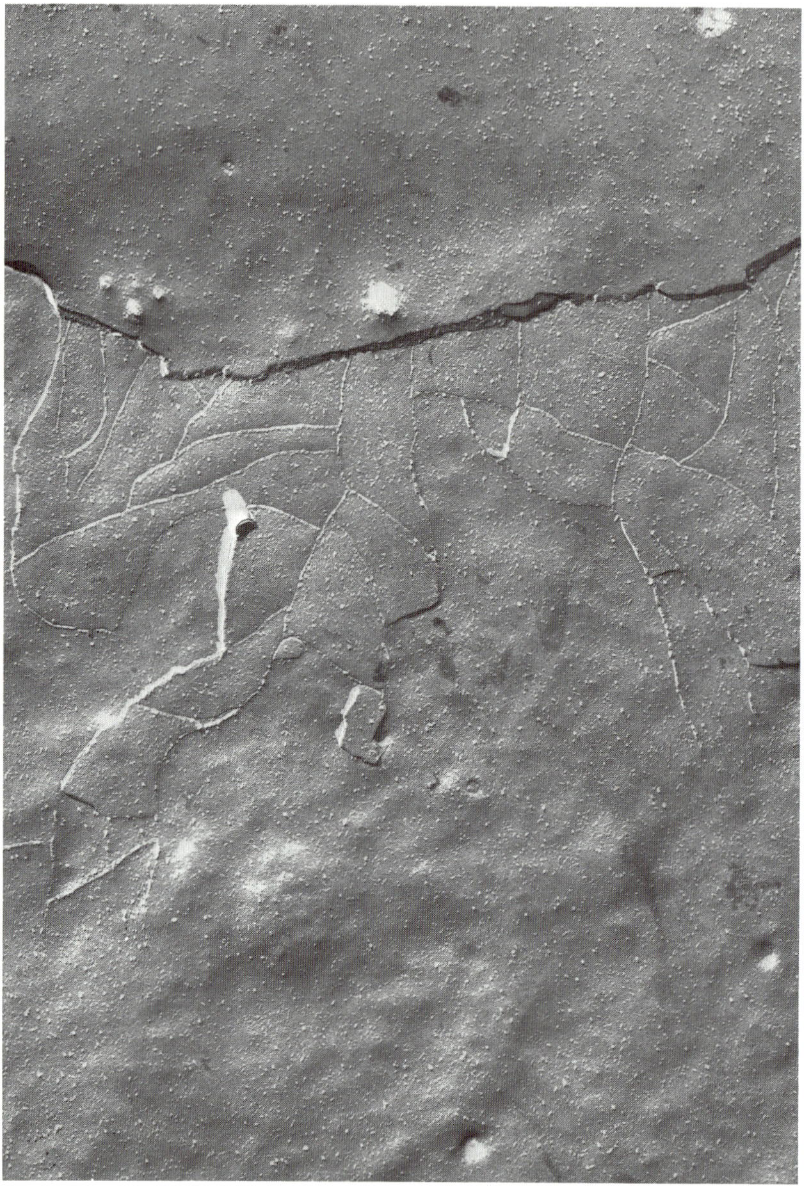

Figure 11-20. Tight junctions between adjoining purified ovine oligodendrocytes cultured for 3 weeks appear predominantly linear; although anastomoses of the strands are present, they are much less frequent than in the tight junctions of endothelial cells in brain capillaries (see figure 11-21). × 49,500.

Figure 11-21. This tight junctional complex of endothelial cells, in a cerebellar capillary that provides an effective blood–brain barrier, presents highly anastomosed strands over its whole extent (*arrows*). × 28,000.

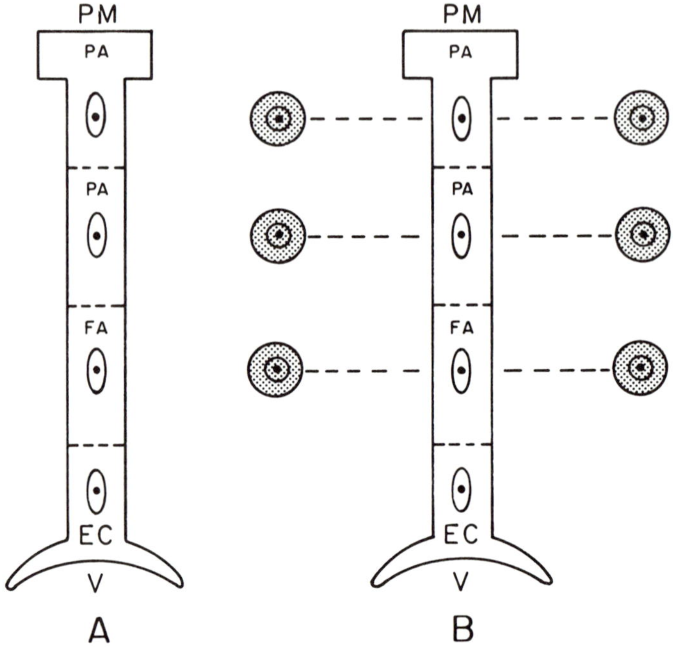

FIGURE 11-22. (**A**) This schematic diagram illustrates the hypothetical functional syncythium established between ependyma cells (*EC*), fibrous astrocytes (*FA*), and protoplasmic astrocytes (*PA*) as a result of communication through gap junctions (*dashed lines*). The syncythium extends from the ventricle (*V*) to the pia mater (*PM*). Adapted from Mugnaini (1986). (**B**) This schematic diagram illustrates the relation of oligodendrocytes (*shaded cells*) to the astroglial syncythium; it is based on the assumption that oligodendrocytes form gap junctions with astrocytes (*dashed lines*), but not with one another. Adapted from Massa and Mugnaini (1982).

instance, increased concentration of substrate-induced enzymes), nor the formation of gradients (for instance, with respect to concentration of K ions), as changes limited to a group of cells will tend to decrease with increasing distance from such a focus. Furthermore, since gating of the connexons is regulatable (see reviews in Bennett and Spray, 1985), it is possible that conditions might arise that functionally isolate one cell in the syncythium from the others, or even one group of cells from the rest of the network.

By virtue of the numerous gap junctions formed de novo in vitro, a generalized functional syncythium seems to form also in mixed glial cell cultures. In this situation, however, the tridimensional extent of the syncythium is altered, with the astrocytes tending to form a bed layer and the oligodendrocytes located on top of it, individually or in groups. On the contrary, the purified and cultured oligodendrocytes may show scarce cooperation, in view of the absence or rarity of homocellular oligodendrocytic gap junctions. It is interesting to note in this respect that enriched oligodendrocytic preparations separated from primary glial cell cultures synthesize myelin basic protein and other oligodendrocyte-specific products in lesser amounts per cell than when combined with astrocytes (Bhat et al., 1981).

6. EXPRESSION OF CELL JUNCTIONS BY GLIAL CELLS IN SITU AND IN VITRO

As mentioned, although oligodendrocytes form numerous gap junctions with astrocytes both in situ and in mixed glial cell cultures, they do not form any substantial number of gap junctions between each other in the three systems considered in this account. Thus, it appears that, although oligodendrocytes synthesize and insert connexons into their plasma membrane, they do not assemble them into junctional patches without contact with astrocytes, with few exceptions. Assembly of connexons into discrete patches may require factor(s) associated with the plasma membrane, such as membranoskeletal protein(s), that could be missing in oligodendrocytes. Lack of such factor(s) in these cells could be correlated with properties of their plasmalemmas differentiated for the formation of compact, spiraled myelin membranes. Alternatively, special properties of phospholipids in the oligodendrocytic plasma membrane may prevent clustering of connexons unless this process is initiated by contact with a neighboring astrocyte or facilitated by other unclear situations. The rare gap junctions observed in purified and cultured oligodendrocytes may originate by trapping of connexons, moving along the lipid bilayer, at the points of anastomosis of tight junctional strands. Occasional gap junctions have been observed also in vivo on the marginal cytoplasmic belt of the myelin sheath, between strands of the myelinic zonula occludens (reviewed by Schnapp and Mugnaini, 1978). Such gap junctions are of the reflexive type; i.e., they occur between portions of the same cell, and may lack functional significance. The presumed lack of some membranoskeletal elements in oligodendrocytes may accompany the well-known absence of intermediate filaments in these cells in situ and in mixed glial cultures and many bear some relation to the linear, rather than anastomosed, configuration of the tight junction strands observed in oligodendrocytic cell bodies, processes, and myelin membranes (Schnapp and Mugnaini, 1978; Massa and Mugnaini, 1982 and 1985). In this respect it should be noted that purified and cultured oligodendrocytes develop intermediate filaments after attachment, in the conditions used for our published studied (Massa et al., 1984b), and show a moderate degree of anastomosis in their tight junctional strands. Culturing conditions and media may exist that will allow experimental control of the expression of both intermediate filaments and membranoskeletal elements in these cells.

Tight junctions in monolayered epithelia are considered crucial in the maintenance of cell polarity. This is particularly evident in a cell line

(MDCK) that forms a monolayer in the culture dish and develops cell polarity, transepithelial resistance, and tight junctions with a similar time course (reviewed by Simons and Fuller, 1985). Oligodendrocytic tight junctions, with the exception of those forming the myelinic zonula occludens at the borders of the myelin sheath (Mugnaini and Schnapp, 1974), but including the tight junctions between compact myelin lamellae, do not seem to be related to polarity in situ. Likewise, in the monolayer formed by purified and cultured oligodendrocytes, polarity of tight junctions is absent, although some anastomosing of the strands occurs. This suggests that the lack of tight junctional polarity on the cell bodies of oligodendrocytes in situ may not be due to the fact that they become scattered in the tissue during growth and lose the capacity to form epithelial monolayers.

7. MODELS SYSTEMS OF GLIAL CELLS IN CULTURE

Primary glial cell cultures differ tridimensionally from the nervous tissue in situ, but the artificial situation of oligodendrocytes atop an astrocytic bed layer is reproducible. A degree of added complexity in this system is represented by continued proliferation of precursor cells at a rate higher than in mature nervous tissue. By vigorous shaking, most of the oligodendrocytes and some of the astrocytes detach from these cultures and can be replated one or more times, thus producing progressive enrichment of oligodendrocytes. Mixed glial cell cultures represent a good model system for the study of gap junctional communication in vitro (figure 11-23A).

The purified and cultured oligodendrocytes represent a unique material for in vitro studies on the formation of tight junctions (figure 11-23B), the synthesis of oligodendrocytic plasma membrane, and the regulation of expression of intermediate filaments by this kind of cell. Less myelin membrane is formed by purified oligodendrocytes, however, than by oligodendrocytes in mixed glial cultures. Since oligodendrocytes can synthesize gap junctional connexons, this model system may also be useful to study factors that promote assembly of gap junctional particles.

8. CONCLUSIONS

We have analyzed, in thin sections and in freeze-fracture replicas, the plasma membrane features and the specialized junctions between glial cells in three systems: 1) white matter of adult mammals; 2) dissociated rat embryo brain cells, cultured in a serum-containing medium and enriched of oligodendrocytes; and 3) oligodendrocytes purified from lamb white matter and cultured in a serum-containing medium. In all situations, plasma membrane features of astrocytes and oligodendrocytes are distinctive by virtue of characteristic intramembrane particles. In situ and in primary cultures, astrocytes form gap junctions with each other and with oligodendrocytes. At sites of contact between oligodendrocytes, there are numerous linear tight junctions but

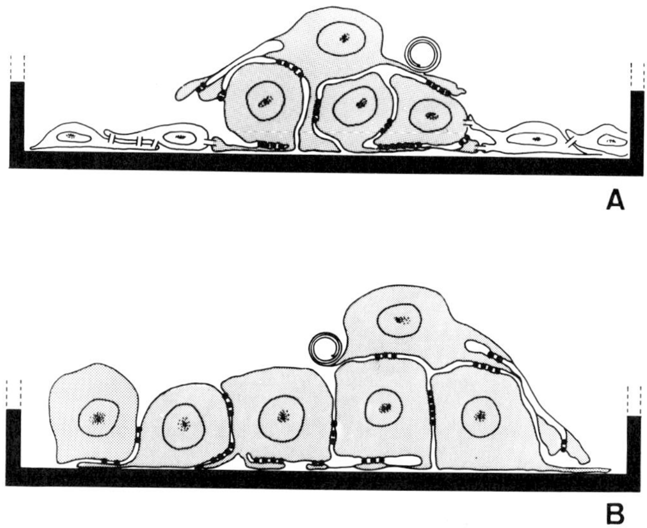

Figure 11-23. These diagrams show the junctional relationships of glial cells observed in two different in vitro systems, as seen in a cross section of the tissue culture dish (*black band*). Scheme **A** represents a mixed glial cell culture enriched of oligodendrocytes, derived from a primary culture of dissociated brain tissue of rat embryos; scheme **B** represents cultured oligodendrocytes purified from lamb white matter. Astrocytes are shown as *flattened, light cells* and oligodendrocytes as *rounded, shaded cells*. Gap junctions between astrocytes and between astrocytes and oligodendrocytes are indicated by *open bars*, and tight junctions between oligodendrocytes by *dark dots*. Purified oligodendrocytes are bound by tight junctions only. The membranous spirals represent empty, myelin-like tubes. Adapted from Massa et al. (1984b, and Massa and Mugnaini (1985).

usually no gap junctions. Purified oligodendrocytes regrow numerous cell processes and form confluent aggregates with tight junctions that become progressively more numerous with time in culture. It appears, thus, that although oligodendrocytes synthesize and insert into their plasma membrane the gap junctional connexons, usually they do not assemble them into junctional patches without contact with the astrocytes. In both in vitro model systems, oligodendrocytes form compacted or semicompacted, spiralled, or stacked processes, and thus maintain the capability of forming myelin-like membranes without physical contact with axons.

ACKNOWLEDGMENTS

Supported by USPHS grant 09904 and NMSS grant RG 1223-B3. The authors wish to thank Dr. A.W. Wachtel for the use of the scanning electron microscope, and Ms. Mary Jane Spring for artwork.

REFERENCES

Bennett, M.V.L., and Spray, D.C. (eds.) (1985) Gap junctions. Cold Spring Harbor, NY: Cold Spring Harbor Laboratory, pp. 1–409.

Bhat, S., Barbarese, E., and Pfeiffer, S.E. (1981) Requirement for non-oligodendrocytic cell signals for enhanced myelinogenic gene expression in long term cultures of purified rat oligodendrocytes. Proc. Natl. Acad. Sci. USA 78:1283–1287.

Claude, P. (1978) Morphological factors influencing transepithelial permeability: a model for the resistance of the zonula occludens. J. Membr. Biol. 39:219–232.

Edelman, G.M. (1983) Cell adhesion molecules. Science 219:450–457.

Hynes, R. (1985) Molecular biology of fibronectin. Annu. Rev. Cell Biol. 1:67–90.

Massa, P.T., and Mugnaini, E. (1982) Plasma membrane features and gap junctions of astyrocytes and oligodendrocytes. Neuroscience 7:523–538.

Massa, P.T., and Mugnaini, E. (1985) Cell-to-cell interactions and plasma membrane structure of cultured rat glial cells. J. Neuroscience 14:695–709.

Massa, P.T., Friedrich, V.L., Jr., and Mugnaini, E. (1982) Correlation between fine structural features and immunocytochemical labeling of phase dense cells in primary rat glial cell cultures. J. Cell Biol. 95:59a.

Massa, P.T., Szuchet, S., and Mugnaini, E. (1983) Maturation of cultured oligodendrocytes: an electron microscopic, immunocytochemical and freeze-fracture study. Anat. Rec. 205:125A.

Massa, P.T., Friedrich, V.L., Jr., and Mugnaini, E. (1984a) Oligodendrocytes and abortive myelin formation in primary cultures. In: Satellite myelin symposium. Am. Soc. Neurochem. Abstr., p. 271.

Massa, P.T., Szuchet, S., and Mugnaini, E. (1984b) Cell–cell interactions of isolated and cultured oligodendrocytes: formation of linear occludens junctions and expression of peculiar intramembrane particles. J. Neurosci. 4:3128–3139.

McCarthy, K.D., and De Vellis, J. (1980) Preparation of separate astroglial and olidendroglial cell cultures from rat cerebral tissue. J. Cell Biol. 85:890–902.

Mugnaini, E. (1986) Cell junctions of astrocytes, ependyma, and related cells in the mammalian central nervous system, with emphasis on the hypothesis of a generalized functional syncythium of supporting cells. In: Fedoroff, S., and Vernadakis, A. (eds.) The astrocytes, New York: Academic Press (in press).

Mugnaini, E., and Schnapp, B. (1974) Possible role of zonula occludens of the myelin sheath in demyelinating conditions. Nature 251:725–727.

Mugnaini, E., and Walberg, F. (1964) Ultrastructure of neuroglia. Rev. Anat. Embryol. Cell Biol. 37:195–236.

Schnapp, B., and Mugnaini, E. (1978) Membrane architecture of myelinated fibers as seen by freeze-fracture. In: Waxman, S.G. (ed.) Physiology and pathobiology of axons. New York: Raven, pp. 83–123.

Simons, K., and Fuller, S.D. (1985) Cell surface polarity in epithelia. Annu. Rev. Cell Biol. 1:243–288.

Szuchet, S. (1986) Myelin palingenesis: the reformation of myelin by mature oligodendrocytes in the absence of neurons. In: Althaus, H., and Siefert, W. (eds.) Glial–neuronal communication in development and regeneration. New York: Springer (in press).

Szuchet, S., Polak, P.E., and Yim, S.H. (1987) Mature oligodendrocytes cultured in the absence of neurons recapitulate the ontogenic development of myelin membranes. Dev. Neurosci. 8:208–221.

12. ELECTROPHYSIOLOGICAL AND IONIC TRANSPORT PROPERTIES OF GLIAL CELLS IN CULTURE

H. STEVE WHITE AND DIXON M. WOODBURY

1. INTRODUCTION

Over the past two decades, it has become increasingly clear that glial cells can no longer be considered as playing a passive role in the central nervous system (CNS). Numerous investigations have demonstrated that the glial cell, and particularly the astrocyte, is the unit that performs the function of removing depolarization products of neuronal firing such as CO_2, H^+, K^+, NH_3, and various neurotransmitters (for review, see Hertz, 1982; Walz and Hertz, 1983a; Hertz and Richardson, 1984). In addition, astrocytes possess an active anion transport system that is dependent in part on the glial-specific enzyme carbonic anhydrase and a bicarbonate-stimulated ATPase. It is through this action to regulate both potassium homeostasis and interstial pH in the CNS that glial cells are thought to be important determinants of brain excitability. As early as 1965, Hertz suggested that glial cells might be involved in K^+ homeostasis within the CNS and, in this respect, to be involved in regulating brain function (Hertz, 1965). This chapter reviews our current knowledge concerning the glial cell role in cation, anion, and acid–base homeostasis, and the effects of drugs on these various processes.

2. GLIAL CELL CULTURES

The studies conducted in this laboratory and described in this chapter have been derived from investigations carried out in vitro on glial cells isolated from the cerebral cortices of either newborn mice or rats and maintained

Table 12-1. Membrane potential, intracellular pH, and intracellular ion concentrations of astrocytes in primary culture.

	Range[a]	References
Membrane potential (mV)[b]	60–90	Hirata et al. (1983), Kimelberg et al. (1979b), Moonen et al. (1980), Moonen and Nelson (1978), Walz and Hertz (1983c), Walz et al. (1984), and present work
Intracellular pH[c]	7.0–7.5	Moonen and Nelson (1978) and White et al. (1986)
Intracellular ion concentration[d]		
Potassium (m*M*)	80–130	Kimelberg et al. (1979b), Moonen et al. (1980), Walz and Hertz (1983b and c), Walz et al. (1984), and present work
Sodium (m*M*)	8–30	Kimelberg et al. (1979b), and Walz and Hertz (1983b and 1984b)
Chloride (m*M*)	23–33	Kimelberg et al. (1979a) and Walz and Hertz (1983b and c)
Bicarbonate (m*M*)[e]	9.5–30	Kimelberg (1983) and White et al. (1986)

[a] Range of values represents those actually reported or those that could be calculated from the data provided.
[b] Range represents values obtained by standard electrophysiological techniques from primary astrocyte cultures isolated from both newborn rat and mouse brains. Membrane potentials were determined at $[K^+]_o$ between 3 and 10 m*M*.
[c] Calculated from steady-state distribution of ^{14}C-DMO and ^{3}H-antipyrine.
[d] Obtained from steady-state flux measurement and ion-selective microelectrodes.
[e] Calculated from the equation $pH = pK_a + \log [HCO_3^-]/[H_2CO_3]$ and assuming a P_{CO_2} of 40 Torr.

under standard culture conditions for varying lenths of time (Booher and Sensenbrenner, 1972; Kimelberg, et al., 1978; White et al., 1985). Cultures were determined to be primarily astrocytic in nature as identified by immunocytochemical staining with glial fibrillary acidic protein (Dahl and Bignami, 1976). The obvious advantage to utilizing such cultures is their relative cell homogeneity. Thus, it is possible to study ion transport processes of a single cell type by the use of radioisotopes and/or ion-selective microelectrodes (ISM).

The effects of brain extract on various aspects of anion and cation transport processes were also investigated. In these studies, brain extract was isolated from the whole brain of adult rats according to the method described by Pettman et al. (1981) and was added to the culture media at each media change beginning on culture day 4.

3. MECHANISMS OF CATION TRANSPORT

In this section, we discuss the primary mechanisms thought to be responsible for maintaining the intracellular concentrations of potassium ($[K^+]_i$) and sodium ($[Na^+]_i$), and the role of these two cations in the maintenance of the resting membrane potential. Table 12-1 summarizes the values reported in the literature for the membrane potential, $[K^+]_i$ and $[Na^+]_i$ of astrocytes in primary culture.

3.1 Potassium

Under conditions of normal neuronal firing, extracellular K^+ ($[K^+]_o$) concentration increases approximately 33% from 3 to 4 mM (Prince et al., 1978). Even larger shifts (from 3 up to 12 mM) in $[K^+]_o$ have been observed during periods of intense stimulation such as might be observed during a seizure or during spreading depression (Viskocil et al., 1972; Futamachi et al., 1974; Somjen et al., 1976; Somjen, 1979). The relatively narrow width of the extracellular cleft, necessitates that such excesses in $[K^+]_o$ be rapidly removed, and ultimately returned to the neuron. Over the years, several potential mechanisms have been advanced (Walz and Hertz, 1983a). Potassium might be expected to exit the brain by accumulation into adjacent blood vessels or by redistribution within brain tissue by diffusing within the extracellular space or via active removal by adjacent cells. The former possibility seems unlikely since redistribution is too slow to account for the actual observed rate of $[K^+]_o$ removal. The selective permeability of the astrocyte membrane to K^+ (Orkand et al., 1966; Kuffler, 1967; Orkand, 1977; Walz et al., 1984), the presence of a K^+-mediated C^- transport process (Gill et al., 1974), and the stimulation of an electrogenic Na^+/K^+ pump at high $[K^+]_o$ (Moonen et al., 1980) make the astrocyte particularly well suited for removing $[K^+]_o$.

In order for the glial cell to function as a spatial buffer for K^+, no ions other than K^+ should contribute significantly to membrane conductance. Although the membrane potential is dependent upon the $[K^+]_o$, the change in potential for a tenfold alteration in $[K^+]_o$ ranges from 51 to 57 mV for astrocytes (Kimelberg et al., 1979b; Moonen et al., 1980; Walz et al., 1984), and 52 mV for oligodendrocytes (Kettenmann et al., 1983), which is less than the 61 mV that is predicted by the Nernst equation for cells that behave as perfect K^+ electrodes. Three principal explanations have been advanced in an attempt to explain this disrepancy (for review, see Hertz, 1986). First, several investigators have demonstrated that, at $[K^+]_o$ above normal, there is an enhanced rate of K^+ uptake leading to an increase in the intracellular concentration of K^+ (Hertz, 1978; Moonen et al., 1980; Hertz and Chaban, 1982; Walz and Hertz, 1982, 1983b, and 1984a). The major mechanism of this increased uptake appears to be related to activation of both a ouabain-sensitive Na^+/K^+-ATPase at $[K^+]_o$ (Walz and Hertz, 1982) and a furosemide-sensitive K^+-mediated Cl^- uptake process (Hertz, 1978; Walz and Hertz, 1984a). Second, Moonen and coworkers (1980), who observed a 52-mV change per tenfold increase in $[K^+]_o$, have provided evidence that high $[K^+]_o$ activates an electrogenic component of the membrane potential.

Lastly, Kimelberg et al., (1979b) has reported a 57-mV slope over the range of 10–90 mM $[K^+]_0$, a slope considerably closer to the 61 mV predicted. Unlike other investigators, these authors report a significant (Na^+) permeability, a finding that possibly contributes to the lower slope of the membrane potential when $[K^+]_o$ is less than 10 mM.

3.2 Sodium

The large negative membrane potential and the low intracellular Na^+ concentration (table 12-1) of the astrocyte relative to the extracellular level provide a strong driving force for the large Na^+ influx rate that has been observed in these cells (Walz and Hertz, 1984b). In addition, Na^+ may gain entrance into the cell through selective activation of a cation-dependent furosemide-sensitive Cl^- transport process that transports both Na^+ and Cl^- into the cell (Walz and Hertz, 1983c; Kimelberg and Bourke, 1984), as well as via an amiloride-sensitive Na^+-in-for-H^+-out exchange (Kimelberg et al., 1979a; Walz and Hertz, 1984a and b). These latter processes appear to play an important role in both volume regulation and acid–base homeostasis (for review, see Kimelberg and Bourke, 1984; White et al., 1986).

Sodium is extruded from the cell in exchange for $[K^+]_o$ via activation of Na^+/K^+-ATPase, a process that requires ATP. Inhibition of this enzyme by the competitive inhibitor ouabain produces a marked and significant increase in $[Na^+]_i$ and decrease in $[K^+]_i$ (Latzkovits, et al., 1982; Walz and Hertz, 1984a; unpublished observations). As discussed above, one of the proposed mechanisms for the increased intracellular K^+ observed when $[K^+]_o$ is raised is through activation of Na^+/K^+-ATPase.

Numerous reports in the literature suggest that, despite the large driving force on Na^+ to enter the cell, the permeability of the astrocytic membrane to Na^+ is negligible (Moonen et al., 1980); Walz et al., 1984) and that Na^+ does not contribute significantly to maintenance of the resting membrane potential (Kimelberg et al., 1979b; Moonen et al., 1980; Walz et al., 1984). Similar findings have been reported by Kettenmann et al., (1983) for cultured oligodendroglia. In contrast to these findings, Kimelberg et al. (1979b) have suggested that Na^+ permeability may not be negligible, and Bowman et al. (1984) and Bowman and Kimelberg (1984) have reported depolarization of the astrocytic membrane when $[Na^+]_o$ was reduced. They have suggested that such an effect could result from inhibition of an electrogenic Na^+/K^+ pump or modification of intracellular pH. Bowman and Kimelberg (1984) have also demonstrated that L-glutamate, L- and D-aspartate, and kainic acid are all capable of directly depolarizing immunocytochemically identified astrocytes in culture in a Na^+-dependent manner. Similar findings have recently been reported by Kettenmann and Schachner for glutamate, asparate, and GABA (Kettenmann and Schachner, 1985). In addition to these findings, alpha-adrenergic agonists are capable of directly depolarizing astrocytes in culture (Hirata et al., 1983). Furthermore, a tetrodotoxin-sensitive, veratridine-induced Na^+ channel (Bowman et al., 1983 and 1984) has also been identified in astrocytes in primary culture. These findings suggest that the membrane potential of astrocytes, and other glial cells, can no longer be thought of as being responsive only to alterations in $[K^+]_o$.

4. MECHANISMS OF ANION TRANSPORT

This section summarizes the various processes that have been demonstrated to be responsible for maintaining the intracellular concentration of Cl^- and HCO_3^-. In addition, it examines the possible role of these two anions in maintenance of the resting membrane potential (for review, see Kimelberg and Bourke, 1982). Table 12-1 summarizes the range of intraceullar concentrations of Cl^- and HCO_3^- reported in the literature. This section also describes recent experiments conducted in this laboratory utilizing tracer concentrations of radiolabeled lodide ($^{125}I^-$). These studies suggest that astrocytes in culture are capable of concentrating $^{125}I^-$ in a manner similar to Cl^-.

4.1 Chloride and Bicarbonate

In view of the integral relationship that has been demonstrated for these two anions, they are discussed together. There is little disagreement in the literature with regared to the ability of cultured astrocytes to actively transport Cl^- into the cell against a large electrochemical gradient (Kimelberg et al., 1979a; Walz and Hertz, 1983a and 1984a; Kimelberg, 1981). As shown in table 12-1 and figure 12-1, estimates of $[Cl^-]_i$ in primary astrocyte cultures range from 23 to 53 mM, values that are several times greater than those predicted from passive distribution according to a membrane potential of approximately -75 mV.

Astrocytes in culture have a Cl^- exchange system similar to that described for erythrocytes (Dunham et al., 1980), glial cell lines (Gill et al., 1974; Wolpaw and Martin, 1984), in vitro brain slice preparations (Bourke, 1969) and cerebral cortex in vivo (Bourke and Nelson, 1972). It is comprised primarily of two active components: 1) a SITS-sensitive (4-acetamido-4′-isothiocyanostilbene-2, 2′-disulfonic acid) ion-exchanger that transports HCO_3^- and Cl^-; and 2) a furosemide-sensitive Na^+- and/or K^+-mediated Cl^- cotransporter (Kimelberg et al., 1979a; Kimelberg, 1981; Walz and Hertz, 1983c). The HCO_3^-/Cl^- exchange system is presumed to be important in regulation of acid–base homeostasis, and is dependent in part on the glial-specific enzyme carbonic anhydrase for its substrate (see below for discussion). As discussed above, Na^+- and/or K^+-mediated Cl^- cotransport may be important in removal of $[K^+]_o$ during intense seizure activity and may contribute to the astrocytic swelling observed under conditions of hypoxia, cerebral edema, and spreading depression (for review, see Kimelberg and Bourke, 1984).

As illustrated in figure 12-2, SITS significantly reduces the uptake of $^{36}Cl^-$ into primary astrocyte cultures. The degree of inhibition by SITS pretreat-

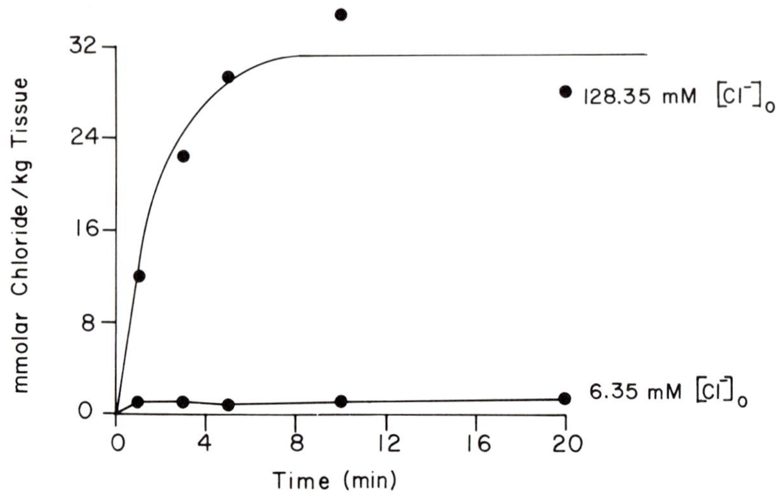

Figure 12-1. Time course of the uptake of $^{36}Cl^-$ into primary cultures of astrocytes measured in the presence of normal $[Cl^-]_o$ (128.4 m*M*) and low $[Cl^-]_o$ (6.4 m*M*). Growth medium was poured off the cells grown in 35-mm Falcon dishes, and the cells were washed three times with phosphate-buffered saline. They were then incubated at 37°C with 2 cc of Hepes-buffered salt solution (final concentration in m*M*: NaCl, 122; KCl, 3; $CaCl_2$, 1.3; $MgSO_4$, 1.4; KH_2PO_4, 1.2: Hepes, 20; and glucose, 10; pH adjusted to 7.4 with 1 *N* NaOH) for 60 minutes in a 5% CO_2/95% air incubator prior to addition of $^{36}Cl^-$. At time zero, 100 μl of $^{36}Cl^-$ (5.0 μCi/ml) was added to the buffer solution. At various times after addition of $^{36}Cl^-$, the buffer was pourd off and the cells washed three times with ice-bold buffer solution using a repipet (Labindustries). After the last wash, cells were removed by scraping into 0.5 ml of 1 *N* NaOH and placed into a scintillation vial and capped. After digesting for 1 hour at 60°C, the vial was vortexed and 50 μl of cell suspension was removed for protein determination (Lowry et al., 1951). The remaining suspension was prepared for scintillation counting by adding 10 cc of Scint A scintillation cocktail (United Technologies Packard). The effect of low $[Cl^-]_o$ on $^{36}Cl^-$ uptake was determined by substituting NaCl with Na isethionic acid. Each point represents the mean ± SEM of 4–5 determinations, which are previously unpublished results of H.S. White and D.M. Woodbury.

ment reported in the present study is similar to that reported by other laboratories (Kimelberg et al., 1979a; Kimelberg, 1981; Walz and Hertz, 1983c) and represents only a small fraction of the total pool of $[Cl^-]_i$. Similar findings have been observed utilizing double-barreled ion-selective microelectrodes (figure 12-3). Only slight changes in steady-state Cl^- levels are seen when cells are pretreated with SITS, a finding that possibly reflects the fact that Cl^- can gain entrance to the cell through a number of different processes. In contrast to these findings, pretreatment with furosemide markedly reduces both the influx and efflux as well as the steady-state Cl^- levels in primary astrocyte cultures (Walz and Hertz, 1983c) as well as in a glioma cell line (Wolpaw and Martin, 1984). These findings suggest that the major route through which astrocytes actively accumulate Cl^- is via the Na^+- and/or K^+-mediated Cl^- cotransport system.

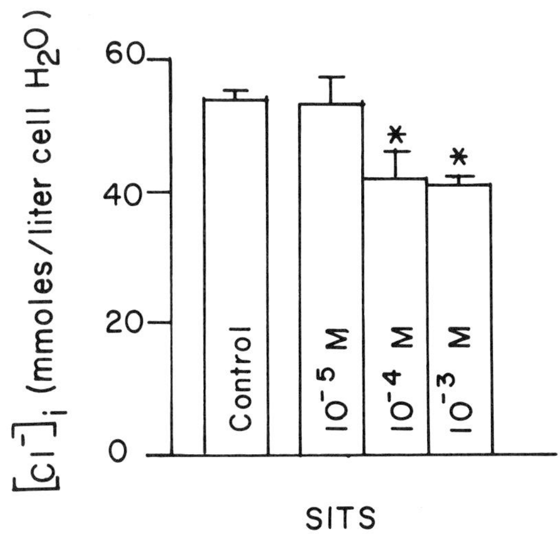

Figure 12-2. Effect of SITS on $^{36}Cl^-$ uptake. Cells were preincubated with isotope-free buffer for 60 minutes as described in figure 12-1. After 60 minutes, cells were incubated for an additional 15 minutes in the presence of SITS; 1 minute after addition of $^{36}Cl^-$, uptake was terminated as previously described (see the legend to figure 12-1). Each point represents the mean ± SEM of 4–5 determinations, which are previously unpublished results of H.S. White and D.M. Woodbury.

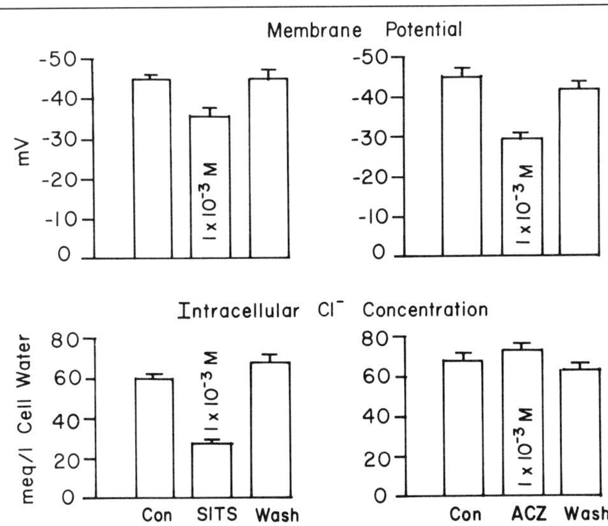

Figure 12-3. Effect of SITS and acetazolamide on membrane potential (mV) and $[Cl^-]_i$ of glial cells in culture. The membrane potential and $[Cl^-]_i$ of cultured glial were determined on populations of cells prior to (control), and 30 minutes after, exposure to either SITS or acetazolamide utilizing standard electrophysiological techniques (Chow et al., 1983 and 1984; White et al., 1985). At the completion of the experiment, cells were washed with drug-free buffer and the membrane potential and $[Cl^-]_i$ redetermined. Results represent the mean ± SEM of 20–40 determinations, which are previously unpublished data of H.S. White, S.Y. Chow, Y.C. Yen-Chow, and D.M. Woodbury.

Virtually all electrophysiological studies performed on glial cells in vitro agree that these cells act primarily as K^+ electrodes when the $[K^+]_o$ is increased and, with the exception of reports by Kimelberg et al. (1979a and b), agree that the permeability to Na^+ and Cl^- is negligible. It is also in keeping with these observations that replacement of Na^+ and Cl^- with nonpermeable ions has little effect on their resting membrane potential (Kimelberg et al., 1979b; Moonen et al., 1980; Walz et al., 1984). However, it has also been suggested by Hirata et al. (1983) that the norepinephrine-induced depolarization of cultured astrocytes could result from an increase in Cl^- conductance. Such an increase in the conductance of this ion could lead to depolarization since the $[Cl^-]_i$ is several fold higher than would be expected from the electrochemical gradient. With double-barreled ISMs, we have observed a marked and significant depolarization of cultured astrocytes under conditions in which HCO_3^-/Cl^- exchange is compromised (i.e., pretreatment with SITS) (figure 12-3). Membrane depolarization was simultaneously accompanied by a significant reduction in $[Cl^-]_i$. Both membrane potential and $[Cl^-]_i$ returned to normal when SITS was removed from the media. These findings, along with those of Bowman et al. (1983 and 1984), and Bowman and Kimelberg (1984), and those of Hirata et al. (1983), certainly suggest that, under certain conditions, additional conductance pathways besides K^+ might be activated.

As mentioned above, carbonic anhydrase is exclusively localized to the glial compartment in the CNS, and functions to hydrate CO_2 released from the neurons to carbonic acid, which then dissociates to HCO_3^- and H^+. Through hydration of CO_2, carbonic anhydrase provides the substrates for both HCO_3^-/Cl^- and Na^+/H^+ exchange and, in so doing, is directly involved in acid–base homeostasis (for review, see White et al., 1986). Although most authors agree that carbonic anhydrase is localized primarily in the oligodendrocyte (Chandour et al., 1979; Delaunoy et al., 1980), Kimelberg et al. (1982) have provided both immunocytochemical and biochemical evidence that primary astrocyte cultures contain carbonic anhydrase. Studies conducted in this laboratory have also found that primary astrocyte cultures contain carbonic anhydrase activity and that this enzyme is highly sensitive to inhibition ($Ki = 2 \times 10^{-10}$) by the carbonic anhydrase inhibitor acetazolamide (White et al., 1986). Acetazolamide significantly reduces intracellular pH (pH_i) of cultured astrocytes (White et al., 1986) as determined from dual labeled experiments with ^{14}C-DMO and ^{3}H-antipyrine (Chow et al., 1983, 1984, and 1985), or as determined by the acridine-orange fluorescence method that detects qualitative shifts in pH_i (Dell'Antone et al, 1972; Berglindh et al., 1980; Anderson et al., 1985). In so doing, acetazolamide significantly reduces the ability of the astrocyte and oligodendrocyte to buffer CO_2 that is continuously being produced, and it is through inhibition of this enzyme and subsequent buildup of CO_2 that acetazolamide exerts its anticonvulsant effect. As with SITS, acetazolamide significantly depolarized the glial

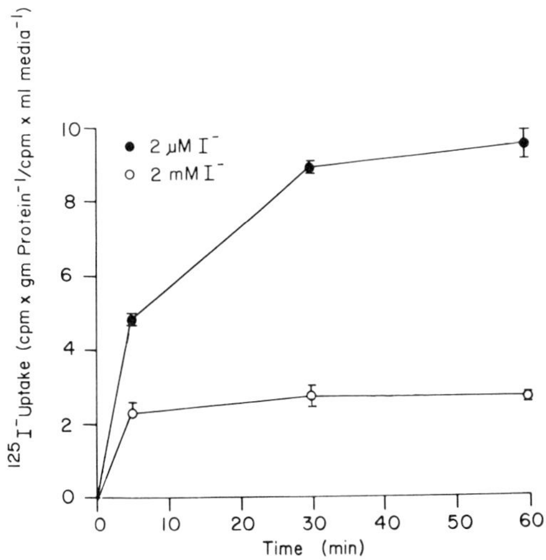

Figure 12-4. Time course of the uptake of $^{125}I^-$ in the presence of low (2 μ*M*) and high (2 m*M*) carrier-free iodide. Culture media was poured off, and the culture dish washed three times with phosphate-buffered saline. They were then incubated at 37°C with 2 cc of Hepes-buffered salt solution (final concentration in m*M*: NaCl, 122; KCl, 3; $CaCl_2$, 1.3; $MgSo_4$, 1.4; KH_2PO_4, 1.2; Hepes, 20; and glucose, 10; pH adjusted to 7.4 with 1 *N* NaOH) for 60 minutes in a 5% CO_2/95% air incubator prior to addition of $^{125}I^-$. At time zero, 100 μl of $^{125}I^-$ (2.0 μCi/ml) was added to the buffer solution. At various times after addition of isotope, an aliquot of buffer was removed for counting and the cells washed three times with ice-cold buffer solution using a repipet (Labindustries). After the last wash, cells were removed by scraping into 0.5 ml of H_2O and placed into a capped vial for gamma counting. Results are expressed as a ratio of c.p.m × gm protein^{-1}/c.p.m. × ml media^{-1}. Each point represents the mean ± SEM of five determinations. Unpublished results of H.S. White, S.Y. Chow, and D.M. Woodbury.

cell membrane (figure 12-3); however, this effect was not accompanied by any change in $[Cl^-]_i$ as determined with ISMs (figure 12-3).

4.2 Iodide

Astrocytes in culture actively accumulate inorganic iodide as determined from radiolabeled uptake studied with $^{125}I^-$ (figure 12-4). We have long used $^{125}I^-$ as a marker for anion transport, and extensive investigations suggest that I^- transport occurs by a mediated process that is similar in many respects to the Cl^- exchange process described above. Although the physiological significance of I^- uptake into astrocytes has yet to be determined, the localization in the CNS of a specific deiodinase enzyme, whose developmental profile is similar to that of glial cells, suggests that these cells may play a role in the uptake and metabolism of thyroid hormones within the CNS (Kaplan and Yaskoski, 1981).

As shown in figure 12-4, the cell–medium ratio of $^{125}I^-$ reaches steady state within 30 minutes after cells have been exposed to isotope, and that

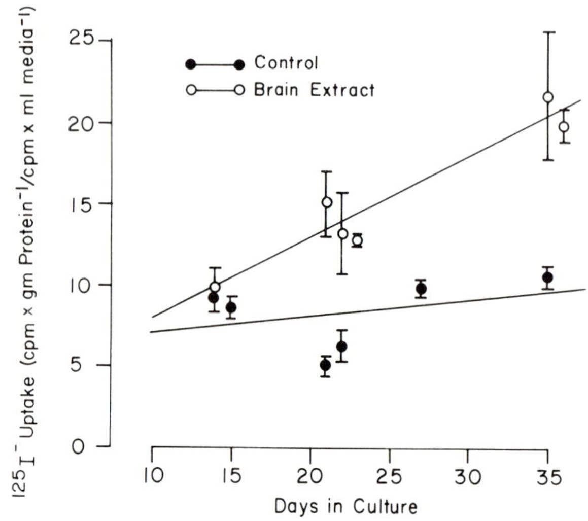

Figure 12-5. Effects of age and brain extract on radio-iodide uptake into cultured glial cells. Brain extract was isolated from the whole brain of adult rats according to the method of Pettmann et al., (1981) and added to the culture media at each media change beginning on culture day 4. Iodide uptake was determined on the days indicated according to the procedure outlined in figure 12-4. Results are expressed as a ratio of cell to medium $^{125}I^-$ and represent the mean ± SEM of 4–5 determinations. Unpublished results of H.S. White, S.Y. Chow, and D.M. Woodbury.

uptake is reduced to passive levels in the presence of high concentrations (2 mM) of carrier-free I^-. The data illustrated in figure 12-5 demonstrate that the ability of cultured glial cells to transport $^{125}I^-$ is dependent on both the culture conditions and the age of the cultures. As shown in this figure, there is little increase in I^- uptake with age into astrocytes grown in the absence of brain extract; however, cultures treated with brain extract from culture day 4 demonstrate a significant age-related increase in their ability to transport this anion. Brain extract contains a soluble protein(s) that, when added to the media, induces marked morphological and biochemical changes in cultured glia. Chronic treatment transforms normal flat polygonal cells into cells possessing abundant processes (Lim et al., 1973, 1976, and 1977; Sensenbrenner, 1977; Sensenbrenner et al., 1980a and b; Pettmann et al., 1981), and it has been suggested that these cells more closely resemble astrocytes in vivo. As can be seen in figure 12-6, I^- transport in brain-extract-treated cells appears to be Na^+ dependent, whereas uptake into nontransformed control cells exhibits little or no Na^+ dependency. These studies serve to illustrate that culture conditions may significantly influence the conclusions of such studies, and further underscore the importance of confirming the validity of the culture model with additional in vivo experimentation. It is of interest,

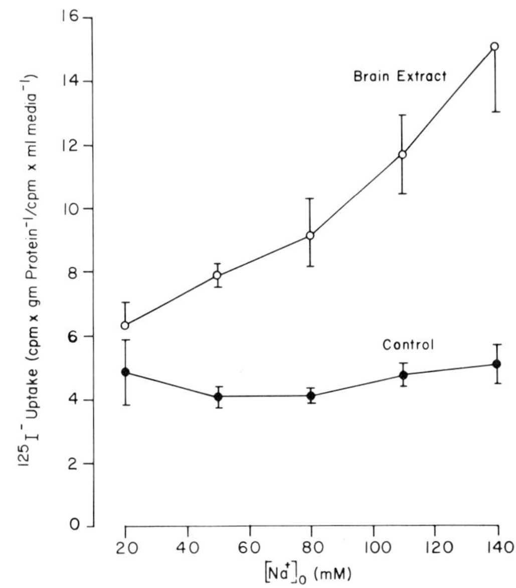

Figure 12-6. Radio-iodide uptake into primary cultures of astrocytes as a function of external sodium concentration. Sodium was replaced by substituting NaCl with choline Cl^-. The determination of $^{125}I^-$ uptake was carried out as described in figure 12-4. Results represent the mean ± SEM of 3–5 determinations, which are the unpublished observations of H.S. White, S.Y. Chow, and D.M. Woodbury.

however, that $[I^-]$ has been shown to be concentrated in glial cells in vitro by this laboratory.

5. EFFECTS OF DRUGS ON ANION AND CATION TRANSPORT PROCESSES

This section focuses on the physiological and pharmacological consequences associated with altering the various transport processes outlined above. Since one of the main roles proposed for the astrocyte in vivo is to regulate the extracellular milieu including acid–base homeostasis and, in this way, modulate neuronal excitability, it is of interest to examine the effect of agents whose principal effect is either to decrease or to increase CNS excitability. Specifically, this section focuses on the anticonvulsant drugs acetazolamide, phenytoin, and ethosuximide, and the convulsant drug pentylenetetrazol.

5.1 Acetazolamide

The effect of this drug on astrocytes in culture has been discussed above in relation to HCO_3^-/Cl^- transport. In addition to these effects, acetazolamide significantly reduces the pH_i of cultured astrocytes, which is quite high when compared to other cell types (table 12-1). Investigations in this laboratory have demonstrated that the pH_i of these cells is dependent on both culture age and culture conditions (White et al., 1986). With respect to acid–base homeostasis, it is also of interest that pH_i increases with age pari passu with the development of carbonic anhydrase activity.

The fact that glial cells are the sole possessors of carbonic anhydrase in the CNS, and that inhibition of this enzyme by acetazolamide protects against maximal electroshock seizures in laboratory animals and is effective in the therapy of tonic–clonic seizures in man, provides direct clinical evidence that these cells are important determinants of CNS excitability.

5.2 Phenytoin

The mechanism of action of phenytoin has yet to be completely elucidated (for review, see Woodbury, 1982, and Woodbury et al., 1983). We have previously reported that chronic treatment of astrocytes in primary culture with micromolar concentrations of the anticonvulsant drug phenytoin enhances the efficiency of glial Na^+/K^+-ATPase over a wide range of $[K^+]_o$ (White et al., 1985). This finding suggests that phenytoin, through an enhancement of glial Na^+/K^+-ATPase, may prevent the transition of interictal discharges to ictal episodes, an event that, as suggested by Dichter et al., (1972) and more recently by Rutecki et al. (1985), results from an increase in $[K^+]_o$.

Therapeutic concentrations of phenytoin have also been observed to hyperpolarize the astrocytic membrane (figure 12-7) (White et al., 1985). The marked membrane hyperpolarization is time dependent, with peak effect occurring between 2 and 4 hours. Results from two experiments utilizing separate cultures and double-barreled ISMs for the simultaneous measurement of membrane potential and ion (either K^+ or Cl^-) activity have suggested that $[K^+]_i$ decreases during the initial period of hyperpolarization (figure 12-6). This is somewhat perplexing, since it would be expected that a decrease in the $[K^+]_i$ of this magnitude would depolarize, rather than hyperpolarize, the membrane. What is clear from this study is that, after 1 hour of exposure to phenytoin, both $[Cl^-]_i$ and $[K^+]_i$ increase. Chloride remains increased for the duration of the experiment, whereas $[K^+]_i$ again is reduced after 3 hours. Preliminary flux studies with $^{36}Cl^-$ and $^{125}I^-$ suggest that phenytoin activates anion transport processes in astrocytes. This, coupled with the enhanced pumping of Na^+/K^+ (discussed above), would account for the observations reported here. Extensive investigations have determined that these effects of phenytoin are not related to any change in the pH_i of these cells (White et al., 1986).

These results suggest that, in addition to effects on neurons (MacDonald and McLean, 1986), therapeutic concentrations of phenytoin can produce marked effects on cultured astroytes. Through such as action, phenytoin provides the CNS with an enhanced ability to regulate $[K^+]_o$ homeostasis

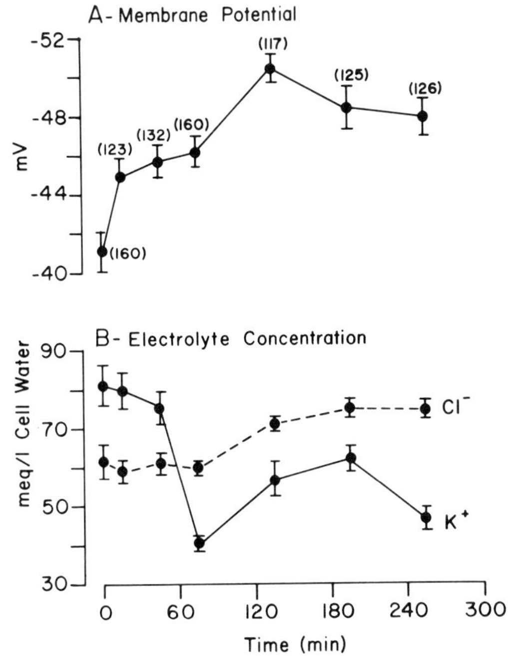

Figure 12-7. Effect of phenytoin ($5 \times 10^{-6}\,M$) on (**A**) membrane potential (mV) and on (**B**) $[K^+]_i$ and $[Cl^-]_i$ as determined with double-barreled ion-selective microelectrodes utilizing standard electrophysiological techniques (Chow et al., 1983 and 1984; White et al., 1985). Experiments were conducted at room temperature and at a $[K^+]_o$ of 5.4 m*M*. The results represent the mean ± SEM from two separate experiments and multiple cell impalements (*number in parentheses* represents number of cells impaled at each time point). Unpublished results of H.S. White, Y.C. Yen-Chow, S.Y. Chow, and D.M. Woodbury.

within the brain extracellular fluid, thereby limiting the spread of seizure activity.

5.3 Pentylenetetrazol

The most thoroughly documented and successful model of absence seizures in man is the pentylenetetrazol (PTZ)-induced threshold seizure. Drugs effective in preventing the clonic seizure induced by PTZ in lab animals almost always possess antiabsence activity in man (Woodbury, 1972). Although the exact mechanism by which PTZ produces seizures is not completely understood, it has been observed to reduce K^+ conductance and depolarize ganglion cells of the *Aplysia* (David et al., 1974; Wilson and Wachtel, 1974), to antagonize γ-aminobutyric acid (GABA)-mediated postsynaptic inhibition (Scholfied, 1982; Olsen, 1982), and to increase Ca^{2+} conductance across the membrane (Heinemann et al., 1977).

The effects of PTZ on membrane potential, $^{36}Cl^-$ and $^{125}I^-$ uptake are

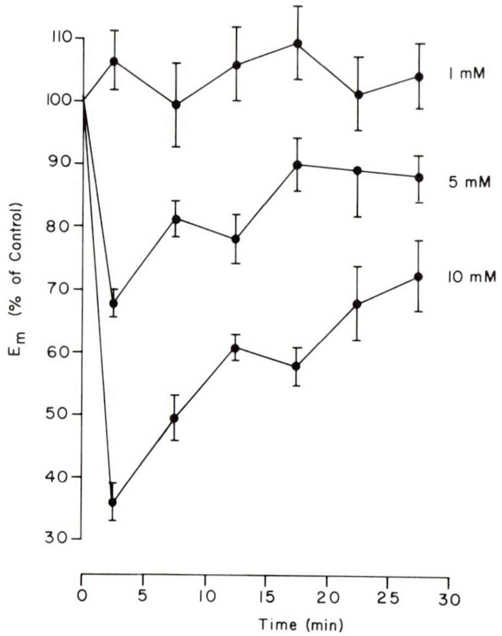

Figure 12-8. Effect of pentylenetetrazol (PTZ) on the membrane potential of cultured astrocytes. The average predrug membrane potential (mV) determined from a population of astrocytes utilizing standard electrophysiological techniques (Chow et al., 1983 and 1984; White et al., 1985) was found to be -47.8 ± 1.3mV ($n = 42$) for 1 m*M* PTZ, -59.7 ± 1.4 mV ($n = 21$) for 5 m*M* PTZ, and -58.9 ± 1.4 mV ($n = 18$) for 10 m*M* PTZ. After addition of PTZ to the bathing solution, membrane potential from several cells ($n = 5-10$) was determined at 5-minute intervals. Experiments were conducted at room temperature and at a $[K^+]_o$ of 5.4 m*M*. Results (mean ± SEM) are expressed as a percent of the predrug control and are previously unpublished results of H.S. White, Y.C. Chow, S.Y. Chow, and D.M. Woodbury.

summarized in figures 12-8 through 12-10. It is clear from the results shown in figure 12-8 through 12-10, that, in addition to an effect on neurons, PTZ is capable of producing significant alterations in both the electrophysiology and the anion transport systems of astrocytes in culture. Increasing concentrations of PTZ dose-dependently depolarized the astrocyte membrane (figure 12-8). The peak time of this effect was between 0 and 5 minutes, and parallels the time course observed in vivo for PTZ-induced seizures. Despite the continued presence of PTZ in the bathing solution, membrane potential returned toward control levels over the next 30 minutes. Although the ionic basis of this depolarization has yet to be completely elucidated, preliminary results obtained with ISMs suggest that PTZ decreases $[K^+]_i$, an effect that could result from either increased K^+ conductance across the membrane or decreased K^+ transport into the cell. The results of figure 12-9 would tend to suport the latter hypothesis. As can be seen from this figure, PTZ dose-

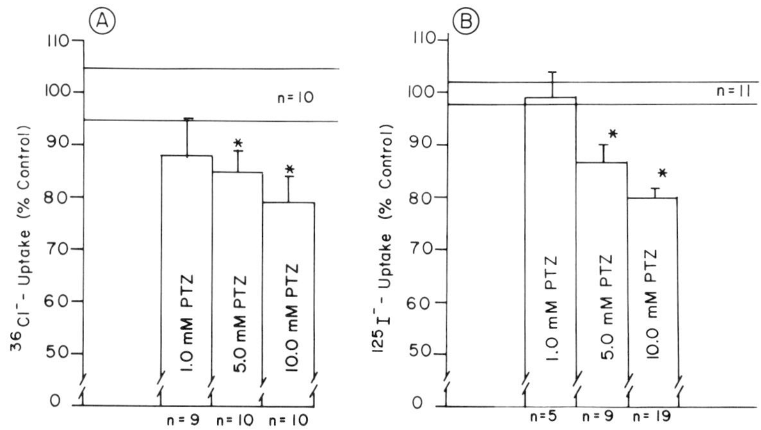

Figure 12-9. Effect of pentylenetetrazol on steady-state Cl^- concentrations (**A**), and on steady-state I^- concentrations (**B**), as determined from $^{36}Cl^-$ and $^{125}I^-$ uptake curves, respectively. Isotope ($^{36}Cl^-$ or $^{125}I^-$) was added to the culture dish of cells in equilibrium with culture media. After either 20 minutes ($^{36}Cl^-$) or 60 minutes ($^{125}I^-$), uptake was terminated by washing the cells in a manner identical to that previously described (see the legend to figure 12-1). Results represent the mean ± SEM of 9–12 determinations from two different experiments. $\star p < 0.05$ when compared to control. Unpublished results of S. Reinhold, H.S. White, and D.M. Woodbury.

dependently decreased the steady-state concentration of both $[Cl^-]_i$ and $[I^-]_i$. Experiments are currently being conducted to determine whether the transport of these two anions is coupled with K^+ via a K^+-mediated Cl^- and a K^+-mediated I^- cotransport system. Inhibition of these processes by PTZ would account for both membrane depolarization and reduced intracellular anion concentration that occur when cultured astrocytes are exposed to this agent. Furthermore, the actions of PTZ on anion and cation transport processes of cultured astrocytes reported herein may possibly contribute to, or account for, the clinical seizure observed in vivo with PTZ administration. The importance of the present findings is further emphasized by the data presented in figure 12-10.

5.4 Ethosuximide

The most characteristic action of ethosuximide (ETS) is its ability to protect against PTZ-induced seizures and the poor ability of ETS to block the tonic extension component of maximal electroshock seizures. Evidence that ETS affects membranes is provided by earlier observations from this laboratory that demonstrated that this drug inhibits the increase in short-circuit current and K^+ permeability induced by PTZ in toad bladder, an epithelial tissue (Woodbury, 1974). Effects on nonexcitable membranes (toad bladder) and lack of any demonstrable effect on nervous tissue suggest that the anticonvulsant action of ETS may in part be through an action on nonneuronal cells

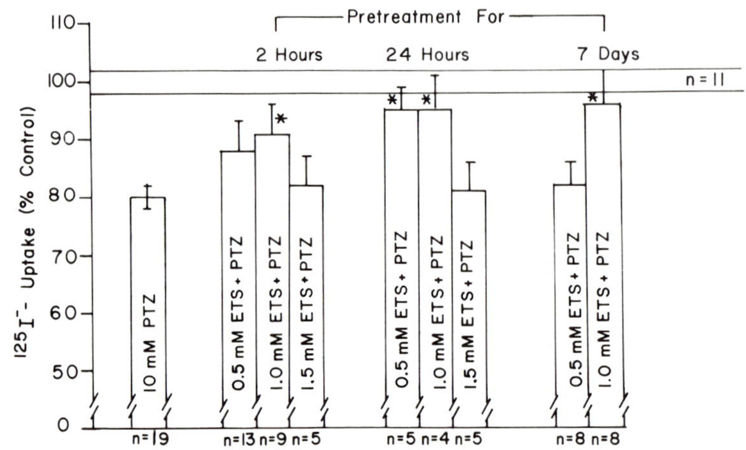

Figure 12-10. Effect of therapeutic concentrations of ethosuximide (*ETS*) on the effect of *PTZ* to reduce $^{125}I^-$ uptake into cultured astrocytes. Cells were exposed to increasing concentrations of ETS or drug-free buffer (*control*) for varying amounts of time. At zero time, PTZ (10 m*M*) was added to the culture dish and, 1 minute later, $^{125}I^-$ (0.2 μCi total) was added. Cells were then equilibrated at 37°C for 60 minutes in a 5% CO_2/95% air incubator. The experiment was then stopped by washing the culture dish with isotope-free media. Cells were then harvested as previously described (see the legend to figure 12-4). Results (mean ± SEM) are expressed as a percent of control. ⋆$p < 0.05$ when compared to untreated control. Unpublished results of S. Reinhold, H.S. White, and D.M. Woodbury.

such as glia, choroid plexus, or the endothelial cells that constitute the blood–brain barrier.

As shown in figure 12-10, pretreatment of primary astrocyte cultures for varying lengths of time with therapeutic concentrations of ethosuximide attenuates the effect of PTZ to lower $[I^-]_i$ at selected doses. Since ETS selectively blocks PTZ-induced seizures, a convulsant that in part acts by blocking GABA-mediated Cl^- conductance, it is possible that it is acting directly to block this action of PTZ on anion transport.

The above results confirm previous findings that astroyctes in culture respond to a number of pharmacologically active substances (for review, see Hertz, 1982; Walz and Hertz, 1983a; Hertz and Richardson, 1984). In addition, the demonstration that ETS, which is an anticonvulsant that selectively blocks PTZ-induced seizures in laboratory animals and which antagonizes the PTZ-effect in culture strongly suggests that astrocytes may directly contribute to the inherent ability of a drug to increase or decrease CNS excitability.

6. CONCLUSIONS

Glial cells in general, and astrocytes in particular, are no longer considered as playing a passive role in the CNS. Improved techniques for separating astrocytes from neurons through gradient centrifugation of whole brain tissue and the development and refinement of tissue culture procuedures have led to extensive investigations of the biochemistry and electrophysiology of both astrocytes and neurons.

In this chapter, we have summarized the burgeoning volumes of literature concerned with anion and cation transport processes and their role in maintenance of the resting membrane potential of astrocytes in primary culture. Furthermore, an attempt was made to identify areas of conflict in the literature so as to stimulate additional experimentation that will further our understanding of these processes and their regulation. The role of the astrocyte in buffering shifts in extracellular potassium observed during enhanced CNS excitability was briefly examined. Thus, $[K^+]_o$ is regulated via Na^+/K^+-ATPase, and a K^+-mediated Cl^- cotransport system. In addition, astrocytes maintain acid–base homestasis in the interstial fluid by Na^+/H^+ and HCO_3^-/Cl^- exchange systems that require carbonic anhydrase and possibly HCO_3^--ATPase. The demonstration that the carbonic anhydrase inhibitor, acetazolamide, depolarized glial cells, does not change $[Cl^-]_i$ or $[K^+]_i$, but decreases $[HCO_3^-]_i$ provides evidence that this anion system helps to maintain the resting membrane potential. In addition to active transport of Cl^- and HCO_3^-, astrocytes have also been shown to transport iodide.

Finally, some recent work has been presented that has focused on the pharmacology of astrocytes. These studies provide strong evidence that the astrocyte is directly involved in modulating the pharmacological effects of both anticonvulsant and convulsant agents in doses equivalent to therapeutic plasma concentrations in man. Thus, phenytoin increases the sensitivity of the astrocyte Na^+/K^+ transport system to $[K^+]_o$ and thereby prevents its accumulation in interstial fluid. Ethosuximide appears to prevent astrocytic membrane depolarization and inhibition of Cl^- and I^- transport by pentylenetetrazol and acetazolamide blocks CO_2 removal by astrocytes, which results in accumulation in neurons, thereby blocking neuronal conduction.

It is evident from this review that astrocyte cultures offer an elegant and homogeneous model for studying the effects of pharmacological agents on cation and anion transport processes. Such a system makes it possible to eliminate many of the interpretational problems associated with studies in whole brain homogenates, subcellular fractions, and brain slices. The results of such studies have demonstrated surprisingly that many of the CNS effects of anticonvulsants are exerted on glia.

ACKNOWLEDGMENTS

The authors express their gratitude to Dr. Doris Dahl for her generous gift of GFAP antisera and to Mrs. Toni Gillett for the artwork of this manuscript. This research was supported by grants from the Epilepsy Foundation of

America and grant 1-R01-22200-01 from NINCDS, NIH. D.M.W. is a recipient of Career Research Award 5-K06-NS-13838 from NINCDS, NIH.

REFERENCES

Anderson, R.E., Jee, W.S.S., and Woodbury, D.M. (1985) Stimulation of carbonic anhydrase in osteoclasts by parathyroid hormone. Calcif. Tissue Int. 37:646–650.

Berglindh, T., Dibona, D.R., Ito, S., and Sachs, G. (1980) Probes of parietal cell function. Am. J. Physiol. 238:G165–G176.

Booher, J., and Sensenbrenner, M. (1972) Growth and cultivation of dissociated neurons and glial cells from embryonic chick, rat and human brain in flask cultures. Neurology 2:97–105.

Bourke, R.S. (1969) Evidence for mediated transport of chloride in rat cerebral cortex in vitro. Exp. Brain Res. 8:219–231.

Bourke, R.S., and Nelson, K.M. (1972) Further studies on the K^+-dependent swelling of primate cerebral cortex in vivo: the enzymatic basis of the K^+-dependent transport of chloride. J. Neurochem. 19:663–685.

Bowman, C., Edwards, C., Kimelberg, H.K. (1983) Tetrodotoxin sensitive deploarizations induced by veratridine and alpha-scorpion toxin in primary astrocytes in culture. Biophys. J. 41:386a.

Bowman, C.L., and Kimelberg, H.K. (1984) Excitatory amino acids directly depolarize rat brain astrocytes in primary culture. Nature 311:656–659.

Bowman, C.L., Kimelberg, H.K., Frangakis, M.V. Berwald-Netter, Y., and Edwards, C. (1984) Astrocytes in culture have chemically activitated sodium channels. J. Neurosci. 4:1527–1534.

Chow, S.Y., Yen-Chow, Y.C., and Woodbury, D.M. (1983) Effects of thyrotropin, acetazolamide, 4-acetamido-4′-isopthiocyanostilbene-2-2′disulfonic acid, perchlorate, ouabain and furosemide of pH and HCO_3 concentrations in cells and luminal fluid of turtle thyroid as calculated from the distribution of 14C-dimethyloxazolidine-2, 4-dione. J. Pharmcol. Exp. Ther. 225:17–23.

Chow, S.Y., Yen-Chow, Y.C., Jee, W.S.S., and Woodbury, D.M. (1984) Electrophysiological properties of osteoblast-like cells from the cortical endosteal surfaces of rabbit long bones. Calcif. Tissue Int. 36:401–408.

Chow, S.Y., Yen-Chow, Y.C., and Woodbury, D.M. (1985) Water and electrolyte contents, cell pH and membrane potentials of cultured turtle thyroid cells. J. Endocrinol 104:45–52.

Dahl, D., and Bignami, A. (1976) Immunogenic properties of the glial fibrillary acidic protein. Brain Res. 116:150–157.

David, R.J. Wilson, W.A. Escueta, A.V. (1974) Voltage clamp analysis of pentylenetetrazol effects on Aplysia neurons. Brain Res. 67:549–554.

Delaunoy, J.P., Hog, F., Devilliers, G., Bansart, M., Mandel, P., and Sensenbrenner, M. (1980) Developmental changes and localization of carbonic anhydrase in cerebral hemispheres of the rat and in rat glial cell cultures. Cell. Mol. Biol. 26:235–240.

Dell'Antone, P., Colona, R., and Azzone, G.F. (1972) The membrane structure studied with cationic dyes. Eur. Acad. Sci. 24:553–565.

Dichter, M.A., Herman, C.J., and Selser, M. (1972) Silent cells during interictal discharges and seizures in hippocampal penicillin foci: evidence for the role of extracellular K^+ in the transition from the interictal state to seizures. Brain Res. 48:173–183.

Dunham, P.B., Stewart, G.W., and Ellory, J.C. (1980) Chloride-activated passive potassium transport in human erythrocytes. Proc. Natl. Acad. Sci. USA 77:1711–1715.

Futamachi, K.J., Mutani, R., and Prince, D.A. (1974) Potassium activity in rabbit cortex. Brain Res. 75:5–25.

Ghandour, M.S., Langley, O.K., Vincendon, G., and Gombos, G. (1979) Double labeling immunohistochemical technique provides evidence of the specificity of glial cell markers. J. Histochem. Cytochem. 27:1634–1637.

Gill, T.H., Young, O.M., and Tower, D.B. (1974) The uptake of 36 Cl into astrocytes in tissue culture by a potassium-dependent, saturable process. J. Neurochem. 23:1011–1018.

Heinemann, U., Lux, H.D., and Gutnick, M.J. (1977) Extracellular free calcium and potassium during paroxysmal activity in the cerebral cortex of the cat. Exp. Brain Res. 27:237–243.

Hertz, L. (1965) Possible role of neuroglia: a potassium-mediated neuronal–neuroglial–neuronal impulse transmission system. Nature (Lond.) 206:1091–1094.

Hertz, L. (1978) An intense potassium uptake into astrocytes, its further enhancement by high concentrations of potassium, and its possible involvement in potassium homeostasis at the cellular level. Brain Res. 145:202–208.

Hertz, L. (1982) Astrocytes. In: Laitha, A. (ed.) Handbook of neurochemistry vol. 1, 2nd edn. New York: Plenum, pp. 319–355

Hertz, L. (1986) Potassium transport in astrocytes and neurons in primary cultures. Ann. N.Y. Acad. Sci. (in press).

Hertz, L., and Chaban, G. (1982) Indications for an active role of astrocytes in potassium homeostasis at the cellular level: potassium uptake and metabolic effects of potassium. In: Pfeiffer, S.E. (ed.) Neuroscience approached through cell culture, vol. 1. Boca Raton, Fl: CRC Press, pp. 157–174.

Hertz, L., and Richardson, S.J. (1984) Is neuropharmacology merely the pharmacology of neurons or are astrocytes important too? Trends Pharmacol. 5:272–276.

Hirata, H., Slater, N.T., and Kimelberg, H.K. (1983) Adrenergic receptor-mediated depolarization of rat neocortical astrocytes in primary culture. Brain Res. 270:358–362.

Kaplan, M.M., and Yaskoski, K. A. (1981) Maturational patterns of iodothyronine phenolic and tyrosyl ring deiodinase activities in rat cerebrum, cerebellum, and hypothalamus. J. Clin. Invest. 67:1208–1214.

Kettenmann, H., and Schachner, M. (1985) Pharmacological properties of γ-aminobutyric acid-, glutamate-, and aspartate-induced depolarizations in cultured astrocytes. J. Neurosci. 5:3295–3301.

Kettenmann, H., Sonnhof, U., and Schachner, M. (1983) Exclusive potassium dependence of the membrane potential in cultured mouse oligodendrocytes. J. Neurochem. 3:500–505.

Kimelberg, H.K. (1981) Active accumulation and exchange transport of chloride in astroglial cells in culture. Biochim. Biophy. Acta 646:179–184.

Kimelberg, H.K. (1983) Primary astrocyte cultures: a key to astrocyte function. Cell. Mo. Neurobiol. 3:1–16.

Kimelberg, H.K., and Bourke, R.S. (1982) Anion transport in the nervous system. In: Lajtha, A. (ed.) Handbook of neurochemisty, 2nd edn. New York: Plenum, pp. 31–67.

Kimelberg, H.K., Narumi, S., and Bourke, R.S. (1978) Enzymatic and morphological properties of primary rat brain astrocyte cultures and enzyme development in vivo. Brain Res. 143–55:77.

Kimelberg, H.K., Biddlecome, S., and Bourke, R.S. (1979a) SITS-inhibitable Cl^- transport and Na^+-dependent H^+ production in primary astroglial cultures. Brain Res. 173:111–124.

Kimelberg, H.K., and Bourke, R.S. (1984) Mechanisms of astrocytic swelling. In: Bes, A., Braquet, P., Paoletti, R., and Siesjo, B.K. (eds.) Cerebral ischemia. Amsterdam: Elsevier, pp. 131–146.

Kimelberg, H.K., Bowman, C., Biddlecome, S., and Bourke, R.S. (1979b) Cation transport and membrane potential properties of primary astroglial cultures from neonatal rat brains. Brain Res. 177:533–550.

Kimelberg, H.K., Steig, P.E., and Mazurkiewicz, J.E. (1982) Immunocytochemical and biochemical analysis of carbonic anhydrase in primary astrocyte cultures from rat brain. J. Neurochem. 39:734–742.

Kuffler, S.W. (1967) Neuroglial cells: physiological properties and a potassium mediated effect of neuronal activity on the glial membrane potential. Proc. R. Soc. Lond. [B] 168:1–21.

Latzkovits, L., Rimanoczy, A., Juhasz, A., Torday, and Sensenbrenner, M. (1982) Control of cation transport in cultured glial cells by external Ca^{2+}: a possible signal in glial–neuronal interaction. Dev. Neurosci. 5:92–100.

Lim, R., Mitsunobu, K., and Li, W.K.P. (1973) Maturation-stimulating effect of brain extract and dibutyryl cyclic AMP in dissociated embryonic brain cells in culture. Exp. Cell. Res. 79:243–247.

Lim, R., Turriff, D.E., and Troy, S.S. (1976) Response of glioblasts to a morphological transferring factor: cinematographic and chemical correlations. Brain Res. 113:165–170.

Lim, R., Turriff, D.E., Troy, S.S., Moore, B.W., and Eng, L.F. (1977) Glia maturation factor: effect on chemical differentiation of glioblasts in culture. Science. 195:195.

Lowry, O.H., Rosebrough, N.J., Farr, A.L., and Randall, R.J. (1951) Protein measurement with the folin phenol reagent. J. Biol. Chem. 193:265–275.

MacDonald, R.L., and McLean, M.J. (1986) Anticonvulsant drug mechanisms of action. In: Delgado-Escueta. A.V., Ward, A.A. Woodbury, D.M., and Porter, R.J., (eds.) Basic mechanisms of the epilepsies; molecular and cellular approaches. New York: Raven, pp. 713–736.

Moonen, G., and Nelson, P.G. (1978) Some physiological properties of astrocytes in primary cultures. In: Schoffeniels, E., Franck, G., Hertz, L., and Tower, D.B. (eds.) Dynamic properties of glial cells. New York: Pegamon, pp. 389–393.

Moonen, G., Franck, G., Schoffeniels, E. (1980) Glial control of neuronal excitability in mammals. I. Electrophysiological and isotopic evidence in culture. Neurochem. Int. 2:299–310.

Olsen, R.W. (1982) Drug interactions at the GABA receptor–ionophore complex. Annu. Rev. Pharmacol. Toxicol. 22:245–277.

Orkand, R.K. (1977) Glial cells. In: Brockart, J.M., Mountcastle, V.B., Kandel, E.R., and Geiger, S.R. (eds.) Handbook of phsiology, vol. , sec. 1. The nervous system: cellular biology of neurons. Bethesda, MD: American Physiological Society, pp. 855–875.

Orkand, R.K., Nicholls, J.G., and Kuffler, S.W. (1966) Effect of nerve impulses on the membrane potential of glial cells in the central nervous system of amphibia. J. Neurophysiol. 29:788–806.

Pettman, B., Labourdette, G., Devilliers, G., and Sensenbrenner, M. (1981) Effects of brain extracts from chick embryo on the development of astroblasts in culture. Dev. Neurosci. 4:37–45.

Prince, D.A., Pedley, T.A., and Ransom, B.R. (1978) Fluctuations in ion concentrations during excitiation and seizures. In: Schoffeniels, E., Franck, G., and Hertz, L. (eds.) Dynamic properties of glial cells. Oxford: Pergamon, pp. 281–303.

Rutecki, P.A., Lebeda, F.J., and Johnston, D. (1985) Epileptiform activity induced by changes in extracellular potassium in hippocampus. J. Neurobiol. 54:1363–1374.

Scholified, C.N. (1982) Antagonism of γ-aminobutyric acid and muscimol by picrotoxin, bicuculline, strychnine, bemegride, leptazol, *d*-tubocurarine and theophylline in the isolated olfactory cortex. Arch. Pharm. (Weinheim) 318:274–280.

Sensenbrenner, M. (1977) Dissociated brain cells in primary culture. In: Federoff, S., and Hertz, L. (eds.) Cell, tissue and organ cultures in neurobiology. New York: Academic Press. pp. 191–213.

Sensenbrenner, M., Devilliers, G., Bock, E., and Porte, A. (1980a) Biochemical and ultrastructural studies of cultured rat astroglial cells: effect of brain extract and dibutyryl cyclic AMP on glial fibrillary acidic protein and glial filaments. Differentiation 17:51–61.

Sensenbrenner, M., Labourdette, G., Delaunoy, J.P., Pettman, B., Devilliers, G., Moonen, G., and Bock, E. (1980b) Morphological and biochemical differentiation of glial cells in primary culture. In: Giacobini, E., Vernadakis, A., and Shahar, A. (eds.) Tissue culture in neurobiology. New York: Raven, pp. 385–395.

Somjen, G.G. (1979) Extracellular potassium ion in the mammalian central nervous system. Annu. Rev. Physiol. 41:159–177.

Somjen, G.G., Rosentahal, M., Cordingley, G., Lamanna, J., and Lothman, E. (1976) Potassium, neuroglia and oxidative metabolism in central gray matter. Fed. Proc. 35:1266–1271.

Viskocil, F., Kriz, and N., Bures, J. (1972) Potassium selective microelectrodes used for measuring the extracellular brain potassium during spreading depression and anoxic depression in rats. Brain Res. 39:255–259.

Walz, W., and Hertz, L. (1982) Ouabain-sensitive and ouabain-resistant net uptake of potassium into astrocytes and neurons in primary cultures. J. Neurochem. 39:70–77.

Walz, W., and Hertz, L. (1983a) Functional interactions between neurons and astrocytes. II. Potassium homeostasis at the cellular level. Prog Neurobiol. 20:133–183.

Walz, W., and Hertz, L. (1983b) Intracellular ion changes of astrocytes in response to extracellular potassium. J. Neurosci. Res. 10:411–423.

Walz, W., and Hertz, L. (1983c) Comparison between fluxes of potassium and of chloride in astrocytes in primary cultures. Brain Res. 277:321–328.

Walz, W., and Hertz, L. (1984a) Intense furosemide-sensitive potassium accumulation into astrocytes in the presence of pathologically high extracellular potassium levels. J. Cereb. Blood Flow Metab. 4:301–304

Walz, W., and Hertz, L. (1984b) Sodium transport in astrocytes. J. Neurosci. Res. 11:231–239.

Walz, W., Wuttke, W., and Hertz, L. (1984) Astrocytes in primary cultures: membrane potential characteristics reveal exclusive potassium conductance and potassium accumulator properties. Brain Res. 292:367–374.

White, H.S., Yen-Chow, Y.C., Chow, S.Y., Kemp, J.W., and Woodbury, D.M. (1985) Effects of phenytoin on primary glial cell cultures. Epilepsia 26:58–68.

White, H.S., Honda, T., Chow, S.Y., and Woodbury, D.M. (1986) Distribution and regulation of hydrogen ion in astrocytes. In: Fedoroff, S., and Vernadakis, A. (eds.) Astrocytes, vol. 2. Orlando, FL: Academic Press, pp. 239–250.

Wilson, P.H., and Wachtel, H. (1974) Negative resistance characteristic essential for the maintenance of slow oscillations in bursting neurons. Science 186:932–934.

Wolpaw, E.W., and Martin, D.L. (1984) Cl^- transport in a glioma cell line: evidence for two transport mechanisms. Brain Res. 297:317–327.

Woodbury, D.M. (1972) Applications to drug evaluations. In: Purpura, D.P. Penry, J.K., Tower, D.B., Woodbury, D.M., and Walters, R.D., (eds.) Experimental models of epilepsy: a manual for the laboratory worker. New York: Raven, pp. 557–601.

Woodbury, D.M. (1974) Antiepileptic drugs: pharmacology and mechanisms of action. In: Harris, P., and Mawdsley, C. (eds.) Epilepsy: proceedings of the Hans Berger centenary symposium. Edinburgh: Churchill Livingstone, pp. 78–95.

Woodbury, D.M. (1982) Phenytoin: mechanisms of action. In: Woodbury, D.M., Penry, J.K., and Pippenger, C.E. (eds.) Antiepileptic drugs, 2nd edn. :269–281.

Woodbury, D.M., Kemp, J.W., and Chow, S.Y. (1983) Mechanism of action of antiepileptic drugs. In: Ward, A.A., Jr., Penry, J.K., Purpura, D.P. (eds.) Epilepsy, vol. 61. [Association for Research in Nervous and Mental Disease (ARNMD), research publications], pp. 179–223.

13. IN VITRO REGULATION OF DOPAMINERGIC MATURATION

ALAIN PROCHIANTZ

1. INTRODUCTION

For several years, in spite of the fact that, according to what was known from studies on the peripheral nervous system, the existence of growth factors specific to central subpopulations of neurons was postulated, most experiments have focused on the survival and differentiation of brain cells in general. The central concern related to specificity was mainly restricted to the establishment of the neuronal circuitry: target recognition and synapse formation.

More recently a new trend of research has focused not only on general trophic support necessary for all neurons, but also on more specific signals responsible for the survival, differentiation, and maturation of discrete neuronal populations in the central nervous system.

In this chapter I shall refer mainly to studies on central dopaminergic (DA) neurons as an illustration of one possible strategy to demonstrate the existence of specific factors involved both in target recognition and in the specific maturation of neuronal subpopulations as a means of their characterization.

2. STIMULATORY EFFECT OF TARGET CELLS ON DA MATURATION

The mesencephalic DA neurons constitute a rather well-defined cellular population. They are located in the mesencephalic flexure of the brain where they can be characterized by histochemical and immunocytochemical techniques as early as 13 days after fertilization in the mouse (ED13). When the

Vernadakis, A. (ed.), *Model Systems of Development and Aging of the Nervous System*.

mesencephalic flexure is removed, dissociated, and plated, most DA cells are able to survive and differentiate. Although, these cultures contain a very large proportion of non-DA cells, DA cells can be specifically followed due to their ability to take up and synthesize dopamine. Indeed these neurons can be visualized by immunocytochemistry using an antibody against tyrosine hydroxylase (TOH) or by autoradiography after ^{3}H-DA uptake. The biochemical maturation can also be assessed by measuring the uptake and release of ^{3}H-DA or the enzymic conversion of ^{3}H-tyrosine into ^{3}H-DA (Prochiantz et al., 1979; Daguet et al., 1980; Berger et al., 1982).

These DA neurons project massively into the striatum, a telencephalic subcortical structure. Since the striatum contains no DA cell bodies, the influence of striatal target cells on mesencephalic DA neurons behavior in vitro can be directly assessed in mixed cultures. When mesencephalic and striatal cells are mixed together in monolayer cultures, the ability of the DA cells to take up and synthesize DA is increased after 8 or 15 days of coculture. This happens without significant changes in the total number of surviving DA cells, indicating that coculturing DA neurons with their target cells enhances their biochemical maturation, but has no or little effect on their survival (Prochiantz et al., 1979).

Using an aggregate-culture technique, Hoffman et al. (1983) reported an increased synthesis of DA in coculture of mesencephalic and striatal cells. This increase was attributed in part to an enhanced survival of DA neurons. This discrepancy between results obtained in monolayer and aggregate cultures may reflect the different culture techniques. It may also be due to the relative sensitivities of the methods used to visualize the DA neurons (autoradiography after ^{3}H-DA uptake versus histofluorescence of accumulated DA) (Berger et al., 1982).

Injecting dissociated embryonic DA mesencephalic cells in the rat adult striatum previously deprived of its DA innervation allows correction of the motor deficiency of the animals as assessed by their rotatory behavior (Björklund et al., 1983). Recently Brundin et al. (1986) have reported that coinjection of mesencephalic and striatal embryonic neurons exerts stimulatory effects on DA cell development (morphologically and possibly functionally) with no modification in cell survival. Such a stimulatory effect decreases the delay necessary for behavioral recovery.

3. DIRECT CONTACTS BETWEEN DA CELLS AND THEIR TARGET ARE NECESSARY

The enhanced ^{3}H-DA uptake and synthesis observed in coculture experiments could not be elicited by adding striatal or coculture-conditioned medium to mesencephalic cultures. This raised the possibility that direct contacts between DA neurons and their target cells are necessary. To elucidate this point, an experimental protocol was established in which mesencephalic cells were cocultured with target and nontarget cells at low cell density (5×10^4 cells/cm^2) in short-term cultures (2–6 days), allowing no significant glial development. Under these conditions, almost no cellular aggregates could be formed and any DA growth cone as it progressed on the polyornithine substratum could meet target or nontarget neurons. After 2, 4, and 6 days, DA cells were visualized by ^{3}H-DA uptake and autoradiography, and the size of their arborization, mainly one long "axon-like" process was estimated. It was demonstrated that increasing the proportion of target versus nontarget cells results in a decrease in the length of the DA neuritic shaft (Denis-Donini et al., 1983).

This result indicates that, given the choice between polyornithine substratum and striatal neurons, DA "axonal" growth cones prefer to grow on target cells. This is not the case when the choice involves polyornithine and nontarget neurons. Whether this preferential adhesion to target neurons involves a specific recognition, arrest of growth, and synapse formation cannot be concluded from this type of experiment. However, these results seem to preducle the possibility that target cells act on DA development by secreting a diffusible compound that would promote fiber outgrowth. On the contrary it indicates that DA growth cones prefer target cell surfaces to any other growth substratum so far tested. It also suggestes that, when the cell density is increased in culture (from 5×10^4/cm^2 to 1.5×10^5/cm^2), the presence of striatal neurons stimulates the growth of axonal DA fibers. Indeed in the high-density cultures, growth cones did not stop on isolates target cells, but progressed from one to another as is the case in vivo.

Such a possibility is illustrated by experiments showing that mesencephalic cells grown with striatal membranes have an increased ability to take up ^{3}H-DA (Prochiantz et al., 1981). This increase is highly dependent on the age of the animal from which the membranes are prepared. Only 2- and 3-week-old rat striatal membranes are active, a period corresponding to the peak of innervation of the striatum by DA afferences. No stimulation is observed when DA neurons are cultured with membranes from nontarget regions. The ability of striatal target cell membranes to provide a substratum favorable to axonal outgrowth is also demonstrated by the work of Hemmendinger et al. (1981) and Heller et al. (1983), which strongly suggests that striatal target cells, as well as striatal membranes, specifically stimulate DA axonal growth. These authors also point out for the first time that in vitro axonal and dendritic outgrowth might be regulated separately.

4. NEURONAL CIRCUITS

In most of the experiments described above, no real distinction was made between the respective roles of neuronal and glial cells either from the mesencephalon or from the target. It is commonly assumed that regional specificity is linked to the presence of neuronal subpopulations. This assumption comes from the frequent confusion between specificity and synaptogenesis. Synaptic specificity requires a specific recognition between presynaptic and postsynaptic elements. Target recognition is likely to precede synaptogenesis and might involve the entire target cell population, neuron and glia sharing some molecular cues bearing positional information and "recognized" by the afferent fibers.

To study whether the enhanced maturation of DA neurons in coculture with striatal cells resulted from neuro-neuronal interactions, these cultures and cocultures were achieved in a chemically defined medium that reduces the numbers of glial cells (Di Porzio et al., 1980). In these conditions where more than 95% of the cells are neuronal in nature [as demonstrated by their specific decoration by antibodies against neurofilament proteins (Prochiantz et al., 1982)], the stimulatory effect observed in the presence of serum could be reproduced. It can therefore be proposed that neuro-neuronal interactions are sufficient to produce the enhanced maturation of the DA cells in presence of their target elements.

If we suppose that the effect is specifically neuro-neuronal and has nothing to do with glial elements, it becomes likely that specific connections between both populations of neurons are involved. Considering the anatomy of the nigrostriatonigral pathway, it is possible that circuits existing in vivo are reproduced in vitro with DA neurons establishing synaptic contacts with striatal neurons and, conversely, striatal neurons with DA neurons. Among these neurons, those using substance P (SP) as a neuromediator are of special interest in view of their excitatory activity. In fact, it has been recently reported that tyrosine hydroxylase activity is increased in the developing substantia nigra in reponse to depolarization and also in the presence of SP agonists (Friedman et al., 1985).

It can therefore be proposed that the in vitro establishment of specific connectivity might be reponsible for the enhanced maturation of the DA cells in vitro. In this case, it is also likely that differentiation of the striatal neurons is stimulated by the presence of the DA cells. Does that mean that the glial elements play no role in the development of these neurons and that the function of astrocytes is to foster neuronal development through the production of nonspecific trophic factors only?

5. ASTROCYTIC SPECIFICITY

Although it is conceivable that specific neuroneuronal interactions including synapse formation can occur in the quasi-absence of glial elements, it is clear from several in vitro studies that normal neuronal survival and development requires the presence of glial cells.

The classic function of astrocytes during development is to produce trophic compounds for the neurons and to serve as a surface favorable to neuritic outgrowth (Noble et al., 1984; Fallon, 1985; Mallat et al., 1986). More recently we have developed an experimental model showing that these functions might be specific and produce their full effect when the neuroastroglial cocultures are composed of neurons and astrocytes belonging to the same region of the brain (homotopic cocultures).

When embryonic mesencephalic DA neurons are grown on monolayers of striatal flat astroglial cells (G str.), they adopt after 2 days in culture a simple morphology, the cell body being surrounded by several short neurites with the exception of one that is very long (up to 20 times the cell body diameter), thin, and with no branching points. On mesencephalic flat astrocytes (Gmes.), this population can still be observed, but only accounts for one-third of the cells. The other two-thirds are more symmetrical with several branched and varicose neurites. The two morphological types are called poorly branched and highly branched (Denis-Donini et al., 1984). These two morphological configurations may represent the selective survival of one population (the highly branched one) on Gmes. This is, however, unlikely in view of the following facts. First, the total number of DA cells is identical on both glial substrata. Second, neurons grown on Gmes. or Gstr., trypsinized and replated, respectively, on Gstr. or Gmes., can reverse their morphological traits (Prochiantz, unpublished results). In view of the characteristics of the cells in light microscopy, it was suggested that highly and poorly branched neurites could, respectively, represent the dendritic and the axonal arborization (Denis Donini et al., 1984). Preliminary studies using electron microscopy and with specific dendritic and axonal markers agree with this hypothesis. Moreover, it seems that this phenomenon is not restricted to the DA cells nor only to the mesencephalic region, but that the development of a full dendritic arbor can only be obtained when locally matching neurons and astrocytes are cocultured together. Conversely heterotopic neuroastroglial cocultures give rise only to poorly mature cells with an arborization restricted to the axon, very similar (although more developed) to neurons grown for 2 days on artificial substrata.

The specific outgrowth of dendrites in homotopic neuroastroglial cocultures may be due to two mechanisms. First, astrocytes may secrete trophic factors that specifically stimulate matching neurons and the dendritic arboization is a result of such an enhanced maturation. Second, homotopic astrocytes may secrete, or present at their surfaces, factors that allow the outgrowth of dendrites of the matching neurons. Another version of this hypothesis is that heterotopic astrocytes inhibit dendritic outgrowth.

These experiments are reminiscent of those of Hemmendinger et al. (1981) and Heller et al. (1983), which demonstrate that the outgrowth of axonal-like DA fibers is triggered in the presence of striatal cells in aggregate coculture while dendrite-like fibers are predominant in their absence. Although our experiments tend to attribute this growth regulation to the astrocytic compartment, one cannot exclude that this property is shared by neuronal cells.

6. CONCLUSIONS

This chapter brings together two series of experiments directed to the questions of neuroneuronal and neuroastroglial interactions in the regulation of DA maturation. It seems that DA "axonal" growth cones specifically associate with target neurons and may also be target glia, allowing discrimination in vitro between target and nontarget areas. This is in principle similar to the results obtained on the retiotectal pathway (Bonhoeffer and Huf, 1980 and 1982; Halfter et al. 1981) or on the discrimination by sensory neurons or retina ganglion cells between tectal and spinal cord explants (Smalheiser et al., 1982).

In addition to this target recognition, the biochemical maturation of DA neurons is stimulated in coculture with striatal cells. Although this was one of the first systems in which such a phenomenon could be demonstrated, it appears to be no longer unique. The biochemical maturations in the presence of their targets of the noradrenergic locus coeruleus-hippocampus or -cerebellar and of the raphe-hippocampus neurons have been reported (Azmitia et al., 1983; Di Porzio and Estenoz, 1984; Dreyfus et al., 1985) However, none of the molecules responsible for these trophic effects have so far been isolated. The only defined molecule shown to have a trophic effect on a specific brain neuronal subpopulation is the nerve growth factor (NGF) known to stimulate the survival and differentiation of sympathetic and of some sensory peripheral cells. In the brain, several reports indicate that NGF stimulates choline acetyltransferase (CAT) activity of most cholinergic cells (Honegger and Lenoir, 1982; Hefti et al., 1985; Mobley et al., 1985; Martinez et al., 1985).

With the exception of NGF, which might be synthesized by glial or neuronal cells, the trophic interactions in neuron–target cocultures are believed to be due to neuroneuronal interactions. This has been directly demonstrated in the case of mesencephalic–striatal cocultures (Di Porzio et al., 1980), and further documented by the report that depolarizing agents or SP agonists increase TOH activity in substantia nigra explants (Friedman et al., 1985).

It cannot be precluded that target glial cells, mostly astrocytes, share with the neurons the ability to secrete and/or to expose at their surfaces molecular signals that induce the maturation of the afferent neurons. The influence can be direct, the afferent fibers interacting with these signals. It can also be indirect, the target astrocytes being necessary for the maturation of the target homotopic neurons, including the development of their dendritic arbors, and subsequently increasing the areas of contact between the afferent axon and the target neurons.

This latter possibility is raised by our recent experiments showing that a full development of neuronal dendritic arborization might be stimulated when DA neurons, but also other neuronal classes, are cultured on monolayers of homotopic astrocytes as opposed to heterotopic ones, a finding that points to the possibility that specific local factors may exist in the brain. Although these factors seem to be secreted or expressed at their surfaces by the astrocytes, they might also be expressed by other cell types, their main characteristic being their regional specificity.

REFERENCES

Azmitia, E., Whitaker, P.M., Lauder, J., and Privat, A. (1983) Primary culture of dissociated fetal mesencephalic raphe: differential stimulation of serotonergic growth by target tissue. Soc. Neurosci. Abstr. 9:10.

Berger, B., Di Porzio, U., Daguet, M.-C. Gay, M., Vigny, A., Glowinski, J., and Prochiantz, A. (1982) Long-term development of mesencephalic dopaminergic neurons of mouse embryos in dissociated primary cultures: morphological and histochemical characteristics. Neuroscience 7:193–205.

Björklund, A., Stenevi, U., Schmidt, R.H., Dunnett, S.B., and Gage, F.H. (1983) Intracerebral grafting of neuronal cell suspensions. I. Introduction and general methods of preparation. Acta. Physiol. Scand. Suppl. 522:1–7.

Bonhoeffer, F., and Huf, J. (1980) Recognition of cell types by axonal growth cones in vitro. Nature 288:162–164.

Bonhoeffer, F., and Huf, J. (1982) In vitro experiments on axon guidance demonstrating and anterior–posterior gradient in the tectum. EMBO J. 1:427–431.

Brundin, P., Isacson, O., Gage, F.H., and Björklund, A. (1986) Intracerebral grafting of dopamine-containing neuronal cell suspension: effect of mixing target and non-target cells. Dev. Brain Res. (in press).

Daguet, M.-C., Di Porzio, U., Prochiantz, A., Kato, A., and Glowinski, J. (1980) Release of dopamine from dissociated mesencephalic dopaminergic neurons in primary cultures in absence or presence of striatal target cells. Brain Res. 191:564–568.

Denis-Donni, S., Glowinski, J., and Prochiantz, A. (1983) Specific influence of striatal target cells on the in vitro outgrowth of mesencephalic dopaminergic neurites: a morphological quantitative study. J. Neuroscience 3:2292–2299.

Denis-Donini, S., Glowinski, J., and Prochiantz, A. (1984) Glial heterogeneity might define the three-dimensional shape of mouse mesencephalic dopaminergic neurons. Nature 307:641–643.

Di Porzio, U., and Estenoz, M. (1984) Positive control of target cerebellar cells on norepinephrine uptake in embryonic brainstem cultures in serum free medium. Dev. Brain Res. 16:147–157.

Di Porzio, U., Daguet, M.-C., Glowinski, J., and Prochiantz, A. (1980) Effects of striatal cells on in vitro maturation of mesencephalic dopaminergic neurons grown in serum-free conditions. Nature 288:370–373.

Dreyfus, C.F., Gillies, M.A., Goldstein, M., and Black, I.B. (1985) The target hippocampus increases tyrosine hydroxylase and NE uptake in locus coeruleus in culture. Soc. Neurosci. Abstr. 11:947.

Fallon, J.R. (1985) Preferential outgrowth of central nervous system neurites on astrocytes and Schwann cells as compared to non-glial cells in vitro. J. Cell Biol. 110:198–207.

Friedman, W.J., Dreyfus, C.F., McEwen, B.S., and Black, I.B. (1985) Depolarizing signals increase tyrosine hydroxylase development in cultured mouse substantia nigra. Soc. Neurosci. Abstr. 11:1142.

Halfter, W. Claviez, M., and Schwartz, U. (1981) Preferential adhesion of tectal membranes to anterior embryonic chick retina neurites. Nature 292:67–70.

Hefti, F., Hartikka, J., Eckenstein, F., Gnahn, H., Heumann, R., and Schwab, M. (1985) Nerve growth factor increases choline acetyl transferase but not survival or fiber outgrowth of cultured septal cholinergic neurons. Neuroscience 14:55–68.

Heller, A., Hoffmann, R.L., and Wainer, B. (1983) Enhancement of dopamine neuron survival and axonal proliferation by membranes from striatal target cells. Soc. Neurosci. Abstr. 9:10.

Hemmendinger, L.M., Garber, B., Hoffmann, P.L., and Heller, A. (1981) Target neuron-specific process formation by embryonic mesencephalic dopaminergic neurons in vitro. Proc. Natl. Acad. Sci. USA 78:1264–1268.

Hoffmann, P.C., Hemmendinger, L.M., Kotake, C., and Heller, A. (1983) Enhanced dopamine cell survival in reaggreates containing telencephalic target cells. Brain Res. 274:275–281.

Honegger, P., and Lenoir, D. (1982) Nerve growth factor (NGF) stimulation of cholinergic telencephalic neurons in aggregating cell cultures. Dev. Brain Res. 3:229–238.

Mallat, M., Moura Neto. V., Gros, F., Glowinski, J., and Prochiantz, A. (1986) Two simian virus 40 (SV40) transformed cell lines from the mouse striatum and mesencephalon presenting astrocytic characters. II. Interactions with mesencephalic neurons. Dev. Brain Res. 26:23–31.

Marinez, H.J., Dreyfus, C.F., Jonakait, J.M., and Black, I.B. (1985) Nerve growth factor promotes cholinergic development in brain striatal cultures. Proc. Natl. Acad. Sci. USA 82:7777–7781.

Mobley, W.C., Rutkowski, T.L., Tennekoon, G.I., Buchanan, K., and Johnston, M.V. (1985) Choline acetyl transferase in striatum of neonatal rats increased by nerve growth factor. Science 229:284–287.

Noble, M., Fok-Seang, J., and Cohen, J. (1984) Glia are a unique substrate for the in vitro growth of central nervous system neurons. J. Neurosci. 4:1892–1903.

Prochiantz, A., Di Porzio, U., Kato, A., Berger, B., and Glowinski, J. (1979) In vitro maturation of mesencephalic dopaminergic neurons from the mouse embryo is enhanced in the presence of their striatal target cells. Proc. Natl. Acad. Sci. USA 76:5387–5391.

Prochiantz, A., Daguet, M.-C., Herbet, A., and Glowinski, J. (1981) Specific stimulation of in vitro maturation of mesencephalic dopaminergic neurones by striatal membranes. Nature 293:570–572.

Prochiantz, A., Delacourte, A., Daguet, M.-C., and Paulin, D. (1982) Intermediate filament proteins in mouse brain cells cultured in the presence or in the absence of fetal calf serum. Exp. Cell Res. 134:404–410.

Smalheiser, N.R., Peterson, E.R., and Crain, S.M. (1982) Neurites from mouse retina and dorsal root ganglion explant show specific behaviour within co-cultured tectum or spinal cord. Dev. Brain Res. 2:362–382.

14. DO MOUSE CEREBELLUM ASTROCYTES PLAY A ROLE IN NEURONAL SURVIVAL AND DIFFERENTIATION?

BERNARD PESSAC AND FRANÇOISE ALLIOT

1. INTRODUCTION

The cerebellum is a region of the central nervous system (CNS) that has often been used as a model because of the well-studied development, circuitry, and properties of its few neuronal cell types (Eccles et al., 1967; Palay and Chan-Palay, 1974). It has been proposed that, in cerebellum, some astroglial cells have an architectonic role. Indeed, during late embryonic development, the Golgi epithelial cells, whose bodies are located in the Purkinje cell layer, extend radial astroglial processes, the Bergmann fibers, to the pial surface of the cerebellum cortex, probably to guide the migration of granule cell neurons from the ventricular zone to the external granular layer (Rakic, 1971; De Blas, 1984). In addition, the astroglia organizes postmitotic mature neurons into compartments; in particular there is evidence that the processes of the velate protoplasmic astrocytes form septa that wall off lamellar, veil-like compartments that contain groups of granule cell bodies (Palay and Chan-Palay, 1974).

Other investigations in tissue culture have show that nonneuronal, essentially astroglial, cells have a trophic effect on some CNS neurons, i.e., that either contact with or diffusible molecules from astroglial cells permit the survival and/or the neurite extension of neurons (Banker, 1980; Whatley et al., 1981; Barbin et al., 1974; Faltin et al., 1985; Muller et al., 1984). These data may be relevant to the question of the control of the final number of neurons in the adult cerebellum, i.e., the process of neuronal stabilization

Vernadakis, A. (ed.), *Model Systems of Development and Aging of the Nervous System*.

(Hamburger and Oppetuim, 1982; Cowan et al., 1984). Recent data also suggest that, conversely, neuronal cells may regulate astroglial morphology and proliferation in vitro (Hatten, 1984 and 1985).

These different types of experiments do not address the question of a role of astroglial cells in the choice of a differentiation pathway by neurons. Moreover, these in vitro data have been mainly obtained with primary cell cultures that are made of a mixture of the various kinds of astroglial cells and of a nonnegligible percentage of other cell types. In this chapter, we present an experimental system where the role of the different types of astroglial cells on the survival and the differentiation of CNS neurons could be investigated. As outlined above, the mouse cerebellum appears convenient to study these questions in an in vitro approach.

2. MODEL SYSTEM

Different methods have been proposed to obtain the neuronal and glial cells necessary for these experiments. Their yield, however, has been low and the different astroglial cell types are not separated. For these reasons, we have set up experiments whose goal was to obtain adequate quantities of homogeneous astrocytes through the derivation of permanent clonal cell lines. We started from explant cultures of 8-day postnatal mouse cerebella as the three main types of astroglial cells—the Golgi-Bergmann, the velate-protoplasmic, and the fibrous astrocytes—are already present at that stage and localized, respectively, in the granular layer, the molecular layer, and the white matter. Cells from these cultures acquired "spontaneously," i.e., without the addition of either carcinogens or oncogenic viruses, the capacity to develop into permanent cell lines from which many single cell clones could be derived (Alliot and Pessac, 1984). Some of these clones are neuronal-like, as they bind tetanus toxin and do not express neuronal markers; others have oligodendrocyte-like features such as the binding of antigalactocerebroside antibodies. Finally many clones synthetize glial fibrillary acidic protein (GFAP), (Alliot and Pessac, 1984) a molecule characteristic of astroglial cells in the CNS. In addition, they do not bind tetanus toxin or the monoclonal antibody A2B5 (Eisenbarth et al., 1979) and they have no oligodendroglial marker. These cerebellar astrocytic cell clones can be distinguished according to their morphology into three separate types that may be the in vitro counterparts of the Golgi-Bergmann, the velate, and the fibrous protoplasmic astrocytes (Alliot and Pessac, 1984).

The question we want to investigate are whether astrocytes play a role in the survival and/or the differentiation of neurons. We have therefore designed an experimental protocol where 14–15-day embryonic neurons from mouse cerebella are cocultured on top of layers of each of the astrocytic cell clones. At that stage the only neurons that are about to complete their last round of mitosis are Purkinje cells and neurons of the deep nuclei while the other neurons, i.e., the Golgi, basket, stellate, and granule cells originate at later times (Altman and Bayer, 1978). Single cell suspensions from embryonic day-14–15 mouse cerebella adhered readily and equally well to the layers of the three astroglial clones as 80%–90% of the cells were firmly attached after 3 hours. The neurons identified by the binding of tetanus toxin with the indirect immunofluorescence method represent the vast majority of the seeded cerebellar embryonic cells (Hatten and Liem, 1981; and manuscript in preparation).

After 24 hours of coculture the same number of neurons was present on the three types of astrocytic monolayers. However, the survival of neurons was strikingly dependent upon the type of astroglial cells on which they were cocultured. On the "Golgi-Bergmann" or the "velate protoplasmic" astrocytes, about 50% of the neurons that had initially adhered were still present after 1 week. In contrast, the number of neurons seeded on the "fibrous" astrocytes began to drop after 3 days in coculture and only $<5\%$ of these neurons could be detected after 1 week. Various hypotheses could account for these effects of astrocytes on neuronal survival; on one hand, the "Golgi-Bergmann" and the "velate protoplasmic" astrocytes, but not the "fibrous" astrocytes, might produce a neurotrophic factor(s). On the other hand, the "fibrous" astrocytes might release a substance(s) toxic for these cerebellar neurons. Preliminary data suggest that indeed the toxic effect of fibrous astrocytes is mediated through a soluble factor(s), as their conditioned medium is also deleterious to cerebellar embryonic neurons. It should be pointed out that the presence of toxic factors in their culture media does not imply that the "fibrous" astrocytes do not also synthesize neurotrophic substances. Further experiments should clarify the specificity and the mechanisms of these trophic and toxic effects of the astrocytic clones on the survival of neurons.

The acquisition of differentiation markers by the cerebellar embryonic neuroblasts was also dependent on the type of astroglial cells on which they were cocultured. We have investigated three distinct sets of markers, namely, neuronal morphology, the synthesis of neurotransmitters, and the expression of epitopes that are specific for granule cells.

2.1 Morphological pattern

On the "velate protoplasmic" astrocytes, two morphologically distinct types of tetanus toxin-positive cells could already be distinguished after 2 days in coculture. One type of neuron was bipolar with a small soma and long bifurcating processes while the other type had a large soma with few short processes. The small soma neurons represented the majority of tetanus toxin-positive cells after 1 week in coculture with the "velate astrocytes;" they aggregated to each other and formed small clusters; many of their numerous processes tended to remain parallel (figure 14-1a). On the "Golgi-Bergmann" layers, most neurons had a small soma that gave rise to few long

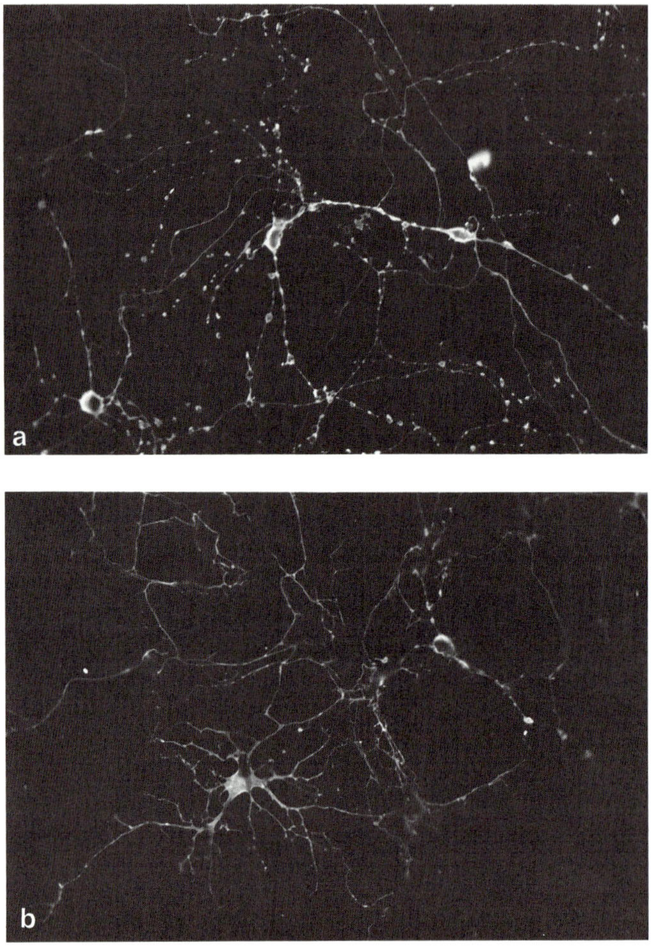

Figure 14-1. Cocultures of cell suspensions from 15-day mouse embryonic cerebella upon layers of (**a**) "velate-protoplasmic" astrocytes or (**b**) "Golgi-Bergmann" cells. The neurons are visualized after 5 days by tetanus toxin binding. (Indirect immunofluorescence technique, × 320)

thin extensions, which sometimes followed the Bergmann-like fibers, but most often formed a spiderweb network (figure 14-1b). The other tetanus toxin-positive cells had only one process or a large flat soma. On the "fibrous" astrocytes the neurons extended few long processes that retracted and began to disappear after 3 days of coculture.

2.2 Presence of neurotransmitters

The two major neurotransmitters present in the neurons of adult mouse cerebellum are glutamate in the granule cells (Valcana et al., 1972; Young et al., 1979) and γ-aminobutyric acid (GABA) in most other neurons (Tran Vinh Tan and Snyder, 1979). To detect the presence of these neurotransmitters in the neurons cocultured on the astrocytic layers, we have used antisera that are specific for GABA (Seguela et al., 1984) or glutamate (M. Geffard, personal communication). In preliminary experiments, we have verified that the "Golgi-Bergmann" and the "velate protoplasmic" astrocytes are not immunoreactive to the antiglutamate or anti-GABA antisera. We have investigated the presence of immunoreactive (IR) cells in the cocultures. About 20%–35% of the neurons were GABA IR on both astrocytic layers as early as 2 days after the start of cocultures, although their labeling pattern was different. In contrast, there was a marked difference in the number of glutamate IR neurons between the two astrocytic layers, as the antiglutamate antiserum labeled three times more neurons cultured on the "velate protoplasmic" astrocytes than on the "Golgi-Bergmann" cells.

2.3 Expression of granule cell-specific epitopes

To define more rigorously the neuronal cell types that differentiate on the astrocytic layers, specific reagents able to recognize the various neurons of cerebellum, such as monoclonal antibodies, are necessary. Indeed, a set of monoclonals that appear to recognize cerebellar granule cells has been recently described (Webb and Woodhams, 1984). Two of these monoclonals, 78D2 and 8.20.1, bind to the granular and molecular layers of postnatal rat cerebella and to cells with a granule-like morphology in primary cultures. In preliminary experiments, we have verified that 78D2 and 8.20.1 also label microneurons in cultures from postnatal *mouse* cerebellum, but do not bind to the "Golgi-Bergmann" and "velate protoplasmic" astrocytic cell clones. When 15-day embryonic cerebellum neurons were cultured on the "velate protoplasmic" layers, after 3 days, tetanus toxin-positive cells were already labeled with either 78D2 or 8.20.1; after 1 week in coculture, 50%–80% of the embryonic neurons were strongly stained with either monoclonal antibody. Although it is not possible to double label the neurons with both 78D2 and 8.20.1, the morphology of the neurons labeled by either antibody appeared similar. In contrast, no more than 15% of the tetanus toxin-positive cells were stained by 78D2 or 8.20.1 when the embryonic neurons were seeded on the "Golgi-Bergmann" astrocytes. Taken together these data show that, when 15-day embryonic cerebellar neurons are cocultured on top of monolayers of the "velate protoplasmic" astrocytes, they acquire the capacity to synthesize glutamate and to express epitopes that are found in postnatal granule cells.

The results reported here show that it is possible to derive, from mouse postnatal cerebellum, permanent cell clones that express differentiated characteristics. In particular, three of these clones appear to be the in vitro counterpart of the three main types of astroglial cells present in the adult cerebellum. The availability of pure populations of stable astrocytes that display morphological properties comparable to those of the Golgi-Bergmann glia, the velate protoplasmic or the fibrous astrocytes, has given us the opportunity to begin a study on the possible role of each of these astrocytes on the survival and the differentiation of neurons.

2.4 Neuronal survival

We have observed a striking difference in the capacity of the various astrocytic clones to support the survival of embryonic (and also postnatal) neurons from cerebellum or cortex. When these neurons are cocultured at very low densities on the "Golgi-Bergmann" or "velate protoplasmic" astrocytes, they survive well, as 30%–50% of their initial number are still present after 1 week. In contrast, when these same neuronal suspensions are seeded on the "fibrous" astrocytes or cultured on polyornithine in the absence of astrocytes, they die rapidly so that at most only 5% of the neurons survive in these conditions. These data show for the first time that the role of astrocytes in the survival of neurons is not equivalent and that one type of astrocyte does not support neuronal survival. As pointed out above, the data available now suggest that this toxic effect is mediated by a soluble factor(s). Concerning the trophic effect of the astrocytes, our present data do not allow one to discriminate between cell contact- or soluble factor-mediated effects.

Previous studies on the role of astrocytes in the survival of neurons have been carried out mainly with primary cultures that are made of a mixture of protoplasmic and fibrous astrocytes; in such experimental conditions the trophic effect might be dominant and therefore the toxic effect of fibrous astrocytes might not appear. It is difficult to elaborate on the significance of this toxic effect on neurons during the development, although one possible physiological role might be to induce the neuronal death in particular in the white matter where most fibrous astrocytes are located.

The effect of "velate protoplasmic" astrocytes on the differentiation of embryonic cerebellum neuroblasts raises many questions. First, the cocultured neurons acquire markers that are those of granule cells. In contrast, the "Golgi-Bergmann" cells do not induce the appearance of granule cell markers in the same neuronal suspensions. In this context, it might be recalled that the velate astrocytes surround the granule cells that they compartment (Palay and Chan-Palay, 1974). Second, the granule cell markers appear rapidly in these 15-day embryonic cerebellum cells, as they are already clearly present after 3 days of coculture and are maximally expressed after 6 days, i.e., at a stage equivalent to embryonic day 21. It is important to notice that the 78D2 and 8.20.1 epitopes do not appear to be expressed in cerebellum neurons at that age during in vivo development. The acceleration of the differentiation process in vitro might be due to the fact that neuroblasts receive a "signal" from astrocytes that are derived from 8-day postnatal cerebellum. It is interesting that 8.20.1 could be detected in vitro in neurons that were chronologically younger than the earliest age at which it could be found in vivo (Webb and Woodhams, 1984). It is therefore possible that, in the tissue culture conditions, the neurons differentiate more rapidly than in vivo, perhaps because they stop dividing earlier than in the normal developing cerebellum.

Thus the velate protoplasmic astrocytes play an important role in the appearance of the granule cell markers, i.e., glutamate reactivity and expression of the 78D2 and 8.20.1 epitopes; however, one may ask whether the astrocytes are required for the appearance of these markers or if they only accelerate a differentiation process that can occur in their absence. To try to answer this question, we have cultured 15-day embryonic cerebellum neurons in tissue culture conditions that prevent the appearance of astrocytes, and we have investigated the expression of various markers in these astrocytic-free neuronal cultures. It is not possible to obtain quantitative data from such cultures, as the high density of seeded neurons, necessary to permit survival, does not allow cell counting. Nevertheless, it is clear that neurons are, in great proportion, strongly GABA and glutamate positive after 2–3 days in culture, while they appear very faintly or unlabeled by the 78D2 and 8.20.1 monoclonals. Thus the two sets of markers—neurotransmitter immunoreactivity and monoclonal binding—appear to be expressed independently in the "pure neuronal" cultures from 15-day embryonic cerebella. Synthesis of the neurotransmitters does not seem to require the presence of astrocytes, while the granule cell epitopes are only expressed in the presence of a differentiated type of astrocyte, the "velate protoplasmic." Indeed, in primary cultures from 15-day embryonic cerebella made under conditions where flat cells, which are probably the precursors of astrocytes, can spread and multiply, the 78D2 and 8.20.1 epitopes are very weakly (or not) expressed, even after 5–6 days.

3. CONCLUSIONS

Evidence discussed in this chapter strongly suggests that the cerebellar astroglia have a role in neuronal survival and differentiation. They also point to the fact that each type of astroglial cell may have a distinct function. Although a role of astroglial cells in neuronal survival has long been suspected, a "nontrophic" or toxic effect of a given type of astrocyte had not been reported previously. In addition, our data support the view that the astroglial cells have a distinct ability to induce or promote the differentiation program of neurons.

ACKNOWLEDGMENTS

We thank Mr. Geffard for the anti-GABA and antiglutamate sera and Mr. Webb for the 78D2 8.20.1 antibodies. This work was supported by INSERM, CNRS, and the *Fondation pour la Recherche Médicale Française*.

REFERENCES

Alliot, F., and Pessac, B. (1984) Astroglial cell clones derived from established cultures of 8-day post natal mouse cerebella. Brain Res. 306:283–291.

Altman, J., and Bayer, S.A. (1978) Prenatal development of the cerebellar system of the rat. I. Cytogenesis and histogenesis of the deep nuclei and the cortex of the cerebellum. J. Comp. Neurol. 179:23–48.

Banker, G.A. (1980) Trophic interactions between astroglial cells and hippocampal neurons in culture. Science 209:809–810.

Barbin, G., Selak, I., Manthorpe, M., and Varon, S. (1984) Use of central neuronal cultures for the detection of neurotrophic agents. Neuroscience 123:33–43.

Cowan, W.M., Fawcett, J.W., O'Leary, P.D.M., and Stanfield, B.B. (1984) Regressive events in neurogenesis. Science 225:1258–1265.

De Blas, A.L. (1984) Monoclonal antibodies to specific astroglial and neuronal antigens reveal the cytoarchitecture of the Bergmann glia fibers in the cerebellum. J. Neurosci. 4:265–273.

Eccles, J.C., Ito, M., and Szentagothai, J. (1967) The cerebellum as a neuronal machine. Berlin: Springer.

Eisenbarth, G.S., Walsh, F.S., and Nirenberg, M. (1979) Monoclonal antibody to a plasma membrane antigen of neurons. Proc. Natl. Acad. Sci. USA 76:4913–4917.

Faltin, J., Lodin, Z., Hartman, J., Gombos, G., and Sensenbenner, M. (1985) Morphological maturation and survival of chicken and rat embryonic neurons on different culture substrates. Int. J. Dev. Neurosci. 3:111–121.

Hamburger, V., and Oppenheim, R.W. (1982) Naturally occurring neuronal death in vertebrates. Neurosci. Commun. 1:39–55.

Hatten, M.E. (1984) Embryonic cerebellar astroglia in vitro. Dev. Brain Res. 13:309–313.

Hatten, M.E. (1985) Neuronal regulation of astroglial morphology and proliferation in vitro. J. Cell Biol. 100:384–396.

Hatten, M.E., and Liem, R.K. (1981) Astroglial cells provide a template for the positioning of developing cerebellar neurons in vitro. J. Cell Biol. 90:622–630.

Muller, H.W., Beckh, S., and Seifert, W. (1984) Neurotrophic factor for central neurons. Proc. Natl. Acad. Sci. USA 81:1248–1252.

Palay, S., and Chan-Palay, V. (1974) Cerebellum cortex. Berlin: Springer.

Rakic, P. (1971) Neuron–glia relationship during granule cell migration in developing cerebellar cortex: a Golgi and electron microscopic study in *Macacus* Rhesus. J. Comp. Neurol. 141:283–312.

Seguela, P., Geffard, M., Buijs, R.M., and Le Moal, M. (1984) Antibodies against gamma aminobutyric acid: specificity studies and immunocytochemical results. Proc. Natl. Acad. Sci. USA 81:3888–3892.

Tran Vinh Tan, and Snyder, S.H. (1979) Aminoacid neurotransmitter candidates in rat cerebellum: selective effects of kainic acid lesions. Brain Res. 167:345–353.

Valcana, T., Hudson, D.B., and Timiras, P.S. (1972) Effects of x-irradiation on the content of aminoacids in the developing rat cerebellum. J. Neurochem. 19:2229–2232.

Webb, M., and Woodhams, P.L. (1984) Monoclonal antibodies recognizing cell surface molecules expressed by rat cerebellar interneurons. J. Neuroimmunol. 6:283–300.

Whatley, S.A., Hall, C., and Lim, L. (1981) Hypothalamic neurons in dissociated cell culture: the mechanism of increased survival times in the presence of neuronal cells. J. Neurochem. 36:2052–2056.

Young, A., Oster-Granite, M., Herndow, R., and Snyder, S. (1974) Glutamic acid: selective depletion by viral induced granule cell loss in hamster cerebellum. Brain Res. 73:1–13.

15. PROLIFERATION AND MATURATION OF RAT NEUROBLASTS IN CULTURE: INFLUENCE OF MENINGEAL CELLS

CLAIRE GENSBURGER, GÉRARD LABOURDETTE
AND MONIQUE SENSENBRENNER

1. INTRODUCTION

It is well known that the development of the neural cell population is influenced by the surrounding environment, such as cellular contacts as well as hormones and soluble factors. Interactions between neural and mesodermal tissues have been demonstrated to occur at early developmental stages (Landesman, 1967; Toivenen and Saxen, 1968). Recent in vivo studies have indicated that meningeal membranes, which are of mesodermal origin, may play an important role during neuronal development, and it has been reported that they influence the proliferation of granule cells in rat cerebellum (Allen et al., 1981; Pehlemann et al., 1985).

To study the effect of cellular interactions and trophic factors, dissociated brain cells in primary monolayer culture offer several advantages. Our first in vitro investigations on neuroblasts derived from brain of young chick embryos have shown that meningeal cells or extracts stimulate the proliferation and maturation of these cells (Cam et al., 1975; Barakat et al., 1981 and 1982). More recently, we have described stimulatory effects on rat neuronal cells under the influence of meningeal cells (Gensburger et al., 1986).

In this chapter, we review the work from our laboratory on the development of rat brain neuronal stem cells in primary culture and on the influence of meningeal cells on the proliferation as well as on the maturation of these cells.

Vernadakis, A. (ed.), *Model Systems of Development and Aging of the Nervous System*.

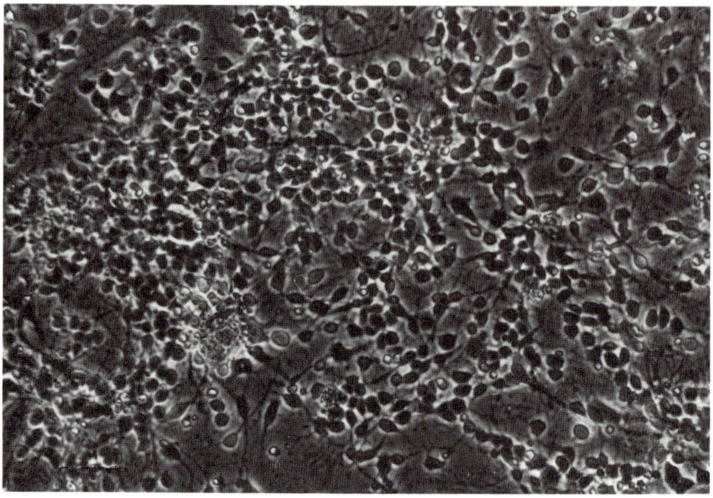

Figure 15-1. Phase-contrast micrographs of brain cells from a 13-day-old rat embryo cultured for 7 days on a polylysine substrate in a serum-free defined medium showing many process-bearing cells. Bar = 50 μm.

2. INFLUENCE OF MENINGEAL CELLS ON THE PROLIFERATION OF RAT NEUROBLASTS

We have been interested in the study of the proliferation of neuronal stem cells from rat brain in primary culture. Subsequently, the effect of meningeal cells on neuroblast multiplication was investigated.

To undertake these studies we had to establish a culture system highly enriched in primitive neuroblasts. Such cultures could be obtained by seeding cells dissociated from brain of 13-day-old rat embryos on a polylysine substrate (Yavin and Yavin, 1974) and by culturing them in a serum-free, chemically defined medium (Gensburger et al., 1986). Indeed, in vivo studies have shown that brain cells at this embryonic age are essentially immature and most of them are still dividing (Korr, 1980).

Cerebral hemispheres from 13-day-old rat embryos were either cleaned entirely or only partially of their meningeal membranes and were mechanically dissociated. The dissociated cells (0.25×10^6 cells/ml) were cultured for 2 hours in a serum-containing medium [Dulbecco's modified Eagle's medium (DMEM) plus 20% fetal calf serum] and then switched to a serum-free defined medium, which consisted of DMEM plus insulin (5 μg/ml), transferrin (50 μg/ml), progesterone (20 m*M*), putrescine (100 μ*M*), and sodium selenite (30 n*M*). The medium contained 50 U/ml penicillin and 50 μg/ml streptomycin. The culture medium was replaced by only half the volume and only once a week. In these conditions the culture consists essentially of process-bearing cells between 2 and 8 days in vitro (figure 15-1).

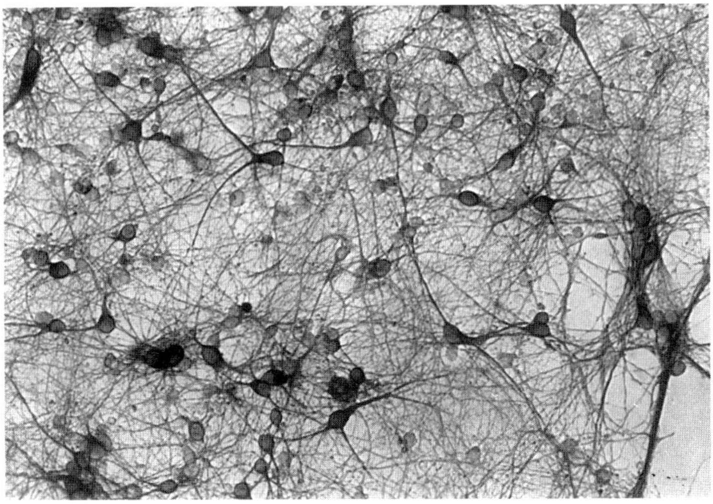

Figure 15-2. Neuronal cells immunostained for neurofilament protein* in an 8-day-old culture. (The NF antiserum was kindly provided by Dr. J.F. Leterrier.) Bar = 50 μm.

Most of these cells have been identified as neuronal cells by immunostaining against neurofilament (NF) proteins (figure 15-2).

The proliferative activity of the neuroblasts was evaluated by autoradiography after ^{3}H-thymidine (^{3}H-TdR) incorporation in combination with immunocytochemical staining for NF proteins. Labeling experiments were performed during the first 3 days after culture initiation. Since it has been shown that proliferating cells are more radiosensitive than differentiating cells (Hirayama et al., 1984; Kriegstein and Dichter, 1984), we have tested two different levels of ^{3}H-TdR: 0.1 μC/ml and 0.5 μC/ml. We observed that, by using a concentration of 0.5 μC/ml, many cells underwent degeneration, while a level of 0.1 μC/ml showed no detectable effect on the neuroblasts. A concentration of 0.1 μC/ml was therefore chosen for use throughout our autoradiographic studies.

Cultures were exposed for 24 hours to ^{3}H-TdR at 2, 24 and 48 hours after culture initiation. The radioactive medium was then removed, and cultures were extensively washed and supplied with fresh nutrient medium. After 8 days, cultures were submitted to the immunocytochemical (Gensburger et al., 1986) and autoradiographic (Korr, 1981) procedures.

The autoradiograms showed labeled neuroblasts after all three ^{3}H-TdR exposure conditions in control as well as in the presence of meningeal cells (figure 15-3). This observation indicates that neuronal cells from brains of 13-day-old rat embryos are capable of dividing under our culture conditions.

A quantitative analysis has shown that, in the presence of meningeal cells, the level of proliferation of the neuroblasts is always higher compared to

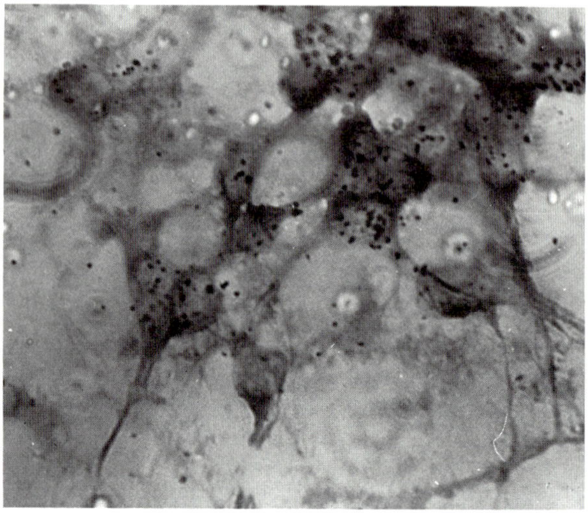

Figure 15-3. Autoradiograph showing labeled neurons exposed to ^{3}H-TdR for 24–48 h and grown for 8 days in the presence of meningeal cells. Neuronal cells are visualized by immunostaining for neurofilament proteins. Bar = 10 μm.

control cultures. The highest proliferative activity was observed between 24 and 48 hours in both culture conditions: 24% and 50% of labeled NF-positive cells to total NF-positive cells were found in control and experimental cultures, respectively.

We have previously shown that brain neuroblasts from chick embryo proliferate in monolayer cultures (Sensenbrenner et al., 1980). Other authors have reported that neuroblasts from rat brain divide in dispersed cell cultures (Kriegstein and Dichter, 1984; Asou et al., 1985). When cultures were derived from 15- and 18-day-old rat embryos, about 10%–13% and 6%–8% of neurons were labeled, respectively. Another research group (Fedoroff et al., 1982; Juurlink and Fedoroff, 1982) demonstrated that, in explant cultures from chick and mouse spinal cord, neuronal precursors were able to multiply. About 50%–60% of neurons were labeled in cultures derived from 9- to 10-day embryonic mouse neural tube.

Our observations that the proliferation of rat neuronal cells is stimulated by meningeal cells agree with our previous findings showing that meningeal cells influence the multiplication of chick embryo neuroblasts (Barakat et al., 1981 and 1982).

In further experiments, we tested the effects of crude extracts from meningeal membranes or from meningeal cell cultures as well as of conditioned medium. Meninges from brains of 13-day-old rat embryos or newborn rats were mechanically dissociated and seeded on a polylysine substrate in serum-containing medium (DMEM plus 10% fetal calf serum). After 24 hours, the medium was changed, and the same defined medium as described above was added and renewed after 5 days. Extracts from meninges and from meningeal cell cultures grown for 10 days were prepared in defined medium. The supernatant conditioned medium (CM) was collected at day 10 and was diluted (1:2) with fresh defined medium before use.

We found that the extracts added at day 1 had a toxic effect and many cells underwent dengeneration. In contrast, the addition of the CM exerted a strong stimulatory effect. Large cellular clumps were present in the treated culture and more labeled neuronal cells were seen. A quantitative analysis indicated that about 47% of neuroblasts had incorporated ^{3}H-TdR in CM-treated culture while, in control culture, only 24% were labeled when the radioactive precursor was given between 24 and 48 hours.

All these results indicate that meningeal cells affect the proliferation of brain neuroblasts from 13-day-old rat embryos. Moreover, the presence of meningeal cells as well as CM from these cells give rise to similar stimulatory effects. This strongly suggests that, to exert their influence, cellular contacts between meningeal cells and neuroblasts are not necessary, but that the effect is mediated by a soluble factor or factors released from the meningeal cells.

3. INFLUENCE OF MENINGEAL CELLS ON THE MATURATION OF RAT NEUROBLASTS

As already mentioned above, in cultures derived from dissociated brain cells of 13-day-old rat embryos, mainly process-bearing cells develop during the first week. Afterward, these cells extend long processes that form a dense fiber network. In control cultures, the fibers are relatively thin while, in the presence of meningeal cells, both fine and thick processes are found. In control cultures, process-bearing cells undergo degeneration between 2 and 3 weeks while, in experimental cultures, most of these cells survive for 21 days. The cell somata as well as the processes are stained for NF proteins (figure 15-4). In most of these cells, the immunostaining is stronger in the presence of meningeal cells or of CM.

From day 7 on, flat and fibrous astroglial cells develop, which have been characterized by a positive immunoreaction with glial fibrillary acidic protein (GFAP). Between 14 and 21 days, the cultures are still enriched in neurons; the astroglial cells multiply slowly and never form a continuous layer as found in cultures grown in serum-containing medium. Oligodendroglial cells are present, but in small numbers as revealed by electron microscopy and by a low enzyme activity of 2′, 3′-cyclic-nucleotide 3′-phosphodiesterase.

At the ultrastructural level the neuronal cells have an immature appearance in control cultures. The processes contain many microtubules, but few neurofilaments. Synaptic contacts are seldom seen and contain very few small clear vesicles. In contrast, in the presence of meningeal cells, the synapses are abundant, with characteristic symmetrical membrane thickenings and clusters of clear vesicles (figure 15-5).

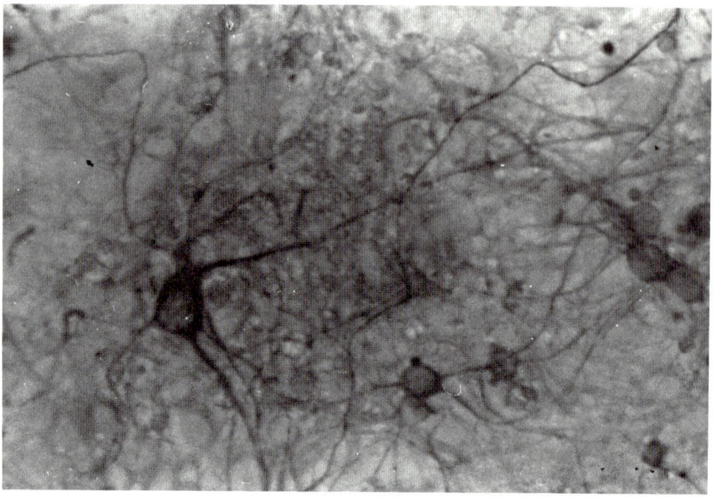

Figure 15-4. Neuronal cells cultured for 18 days in the presence of meningeal cells and immunostained for neurofilament proteins. Bar = 25 μm.

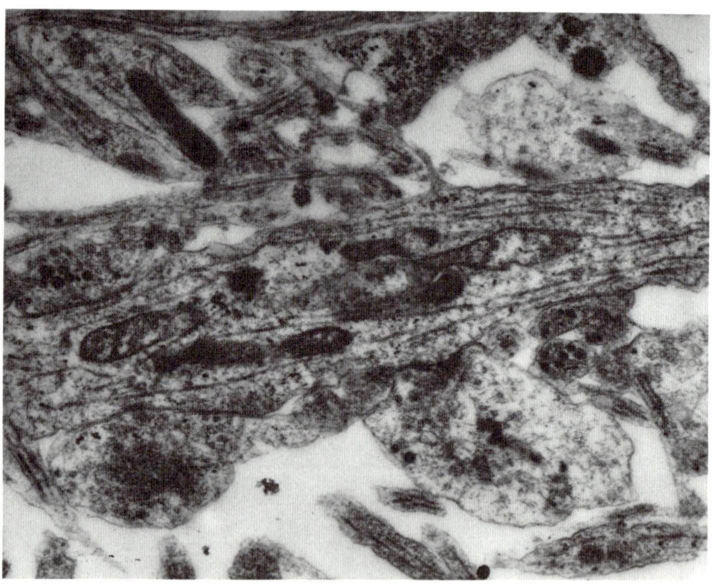

Figure 15-5. Electron micrograph of neurons after 15 days of culture in the presence of meningeal cells showing a synapse with clear vesicles. Bar = 0.5 μm.

Previously, we presented evidence that meningeal extract enhances the morphological maturation of chick embryo neuroblasts (Cam et al., 1975). Now we show that meningeal cells stimulate the maturation of neuroblasts from young rat embryos. An influence on the development of synaptic formations was especially observed.

4. CONCLUSIONS

Neuroblasts derived from cerebral hemispheres of 13-day-old rat embryos and cultured in the absence of meningeal cells on a polylysine substrate in a serum-free defined medium are capable of proliferation. The proliferative activity increases during the first 2 days of culture and then declines.

In this culture condition, most of the cells reach only a limited stage of maturation with time in culture.

In the presence of meningeal cells or meningeal-conditioned medium, the proliferative activity of these neuroblasts is enhanced. The morphological maturation of the cells is also stimulated.

Cellular contact between neuronal and meningeal cells seems not to be necessary to obtain a stimulatory effect. Rather, it may be mediated by soluble substances released from meningeal cells.

REFERENCES

Allen, C., Sievers, J., Berry, M., and Jenner, S. (1981) Experimental studies on cerebellar foliation. II. A morphometric analysis of cerebellar fissuration defects and growth retardation after neonatal treatment with 6-OHDA in the rat. J. Comp. Neurol. 203:771–783.

Asou, H., Iwasaki, N., Hirano, S., and Dahl, D. (1985) Mitotic neuroblasts in dissociated cell cultures from embryonic rat cerebral hemispheres express neurofilament protein. Brain Res. 332:355–357.

Barakat, I., Wittendorp-Rechenmann, E., Rechenmann, R.V., and Sensenbrenner, M. (1981) Influence of meningeal cells on the proliferation of neuroblasts in culture. Dev. Neurosci. 4:363–372.

Barakat, I., Sensenbrenner, M., and Vincendon, G. (1982) The importance of cell contact for the proliferation of neuroblasts in culture and its stimulation by meningeal extract. Neurochem. Res. 7:287–300.

Cam, Y., Sensenbrenner, M., and Mandel, P. (1975) A comparative study of the effects of brain extracts and mesodermal membrane extracts on nerve cell differentiation. Experientia 31:1430–1431.

Fedoroff, S., Krukoff, T.L., and Fisher, K.R.S. (1982) The development of chick spinal cord in tissue culture. III. Neuronal precursor cells in culture. In Vitro 18:183–195.

Gensburger, C., Labourdette, G., and Sensenbrenner, M. (1986) Influence of meningeal cells on the proliferation and maturation of rat neuroblasts in culture. Expt. Brain Res. 63:321–330.

Hirayama, M., Eccleston, P.A., and Silberberg, D.H. (1984) The mitotic history and radiosensitivity of developing oligodendrocytes in vitro. Dev. Biol. 104:413–420.

Juurlink, B.H., and Fedoroff, S. (1982) The development of mouse spinal cord in tissue culture. II. Development of neuronal precursor cells. In Vitro 18:179–182.

Korr, H. (1980) Proliferation of different cell types in the brain. Adv. Anat. Embryol. Cell Biol. 61:1–71.

Korr, H. (1981) Light microscopical autoradiography of nervous tissue. In: Heym, Ch., and Forssmann, W.G. (eds.) Techniques in neuroanatomical research. Berlin: Springer, pp. 218–244.

Kriegstein, A., and Dichter, M.A. (1984) Neuron generation in dissociated cell cultures from fetal rat cerebral cortex. Brain Res. 295:184–189.

Landesman, R. (1967) Neural–mesodermal interactions subsequent to neural induction in Ambystoma. Dev. Biol. 16:341–367.

Pehlemann, F.W., Sievers, J., and Berry, M. (1985) Meningeal cells are involved in foliation, lamination, and neurogenesis of the cerebellum: evidence from 6-hydroxydopamine-induced destruction of meningeal cells. Dev. Biol. 110:136–146.

Sensenbrenner, M., Wittendorp, E., Barakat, I., and Rechenmann, R.V. (1980) Autoradiographic study of proliferating brain cells in culture. Dev. Biol. 75:268–277.

Toivenen, S., and Saxen, L. (1968) Morphogenetic interaction of presumptive neural and mesodermal cells mixed in different ratio. Science 159:539–540.

Yavin, E., and Yavin, Z. (1974) Attachment and culture of dissociated cells from rat embryo cerebral hemispheres on polylysine-coated surface. J. Cell Biol. 62:540–546.

II. NEUROMODULATION, SYNAPSE AND RECEPTORS

16. NEUROTRANSMITTERS AS MORPHOGENETIC SIGNALS AND TROPHIC FACTORS

JEAN M. LAUDER

1. INTRODUCTION

This chapter summarizes evidence for the hypothesis that neurotransmitters play key roles in different phases of ontogenesis in both the developing nervous system and in other tissues where morphogenesis, cell proliferation, or differentiation is active. The possible ontogenetic influences of neurotransmitters discussed here include effects on: 1) neural tube formation; 2) morphogenetic cell movements in the head ectoderm, otocyst, heart, gut, visceral arches, brain, and palate; 3) cell proliferation and the onset of neuronal differentiation in the neural tube; 4) postnatal germinal cell proliferation; 5) neurogenesis and gliogenesis; and 6) axonal growth and synaptogenesis. These studies, together with information gathered from the sea urchin and other simple organisms, provide evidence that, in both a phylogenetic and ontogenetic sense, neurotransmitters and neuronhumoral substances may exert important epigenetic influences prior to their use in neurotransmission or other functional aspects of the adult nervous system. Thus it would appear that these substances may have evolved and developed into ever more specific signaling devices to achieve their status as mediators of chemical neurotransmission and neuronal function in the vertebrate nervous system.

Vernadakis, A. (ed.), *Model Systems of Development and Aging of the Nervous System*.

2. EARLY NEUROGENESIS

The hypothesis that neurotransmitters might have other, nontransmitter functions during development is derived from the early presence of these substances during embryogenesis, prior to the onset of synaptogenesis or neurotransmission, together with the ability of pharmacologic agents that affect the synthesis, uptake, receptor interaction, or metabolism of these neurotransmitters to cause defects in embryologic development. For example, neurotransmitters such as serotonin (5-HT), norepinephrine (NE), dopamine, and acetylcholine have been measured biochemically and localized histochemically in the fertilized egg of a number of animal species during early phases of embryogenesis, such as cleavage and gastrulation, where they have been implicated in modulating cell movements and cell shape changes. Moreover, these substances have been found in lower organisms, where they have been proposed to play a role in morphogenesis and regeneration occurring at this time (see Burden and Lawrence, 1973; Buznikov, 1984; Buznikov and Shmukler, 1981; Gustafson and Toneby, 1970; Markova et al., 1985).

During early phases of chick embryogenesis, such as neural tube formation and closure, 5-HT and NE have been localized in areas of active morphogenesis, suggesting that they might play a role in neurulation or other aspects of early neurogenesis. For example, NE and dopamine have been measured biochemically in whole embryo extracts of chicks as early as day 1 of incubation (Ignarro and Shideman, 1968a and b), and extracts of the chick notochord have revealed the presence of 5-HT as well as catecholamines (Strudel et al., 1977a and b). Using fluorescence microscopy, Allan and Newgreen (1977), Lawrence and Burden (1973), Kirby and Gilmore (1972), and Newgreen et al. (1981) have reported sites of NE and 5-HT uptake in both the notochord and neural tube, and in other sites of morphogenesis outside the nervous system, following incubation of early chick embryos in these substances. Wallace (1979 and 1982) has provided a detailed description of sites of 5-HT uptake in both the neural tube and notochord of the chick embryo, and has observed that these sites are primarily nonoverlapping with sites for NE uptake. Wallace et al. (1985) have also characterized these sites pharmacologically and found that they are blocked by the 5-HT uptake inhibitors amitriptyline, fluoxetine, imipramine, and chlorimipramine.

In order to ascertain whether these sites of neurotransmitter uptake are devleopmentally meaningful, it is important to determine where these substances might be synthesized during early embryogenesis. In the chick embryo, one possibility is the yolk itself, since it has been found to contain NE (Ignarro and Shideman, 1968a), 5-HT, and its precursor L-tryptophan (L-trp; Emanuelsson, 1974). Moreover, 5-HT-interactive drugs that cause malformations also produce delayed degradation of yolk granules (Palen et al., 1979), which are taken up into the neural plate and neural tube during neurulation (Santander and Cuardrado, 1976). In addition, the notochord has been reported to contain detectable levels of 5-HT following incubation of chick embryos in L-trp or 5-hydroxytryptophan (5-HTP; Kirby and Gilmore, 1972; Wallace, 1979 and 1982). This structure also contains catecholamines following incubation with their precursor L-DOPA (Allan and Newgreen, 1977; Kirby and Gilmore, 1972). Thus, the notochord could also be a site of synthesis for 5-HT and NE in the chick embryo. In the rat, the visceral yolk sac may be a source of monoamines for the early embryo, since Schlumpf and Lichtensteiger (1979) have reported catecholamine histofluorescence in this structure.

Recently, we have undertaken a series of experiments using the method of whole mammalian embryo culture (New et al., 1973; Sadler and Kochar, 1976; Sadler and New, 1981) to explore the mouse as a model for studying morphogenetic roles of neurotransmitters in a system more analogous to the human for the study of drug-related birth defects. Using immunocytochemistry, a different pattern of 5-HT uptake was observed, as compared to the chick embryo, but these sites also appear to coincide with locations of morphogenetic cell movements, such as the developing heart, hindgut, otocyst, tongue, roof of the mouth, palate, pharyngeal and mandibular arches, and brain (epiphysis and roof of the third ventricle). These sites have been studied at three different stages: embryonic (E) days 9.5–9.75, 12–12.5, and 14 (figures 16-1 and 16-2). We have also begun to characterize these sites pharmacologically in terms of their capacities for 5-HT uptake and/or synthesis (tables 16-1 and 16-2), and have obtained evidence that these sites are primarily involved in the uptake of 5-HT rather than its synthesis. Possibilities for endogenous sources of 5-HT for these uptake sites are 1) mast cells (which are abundant in the embryonic mouse brain, and mesenchyme of the head and face), 2) the visceral yolk sac, and 3) the fetal circulation. The only possible exogenous source of 5-HT in the whole embryo culture system is the culture medium itself, since it contains serum, which is known to contain varying amounts of 5-HT, depending on the species from which it is derived (Agrez et al., 1984).

In the palate (E14), the distribution of 5-HT uptake sites found in our studies correlates with the demonstrated regional effects of 5-HT on reorientation of the palatal shelves in vitro (Zimmerman and Wee, 1984), suggesting a functional link between the uptake of 5-HT and morphogenetic cell movements in this tissue. As in the notochord, neural tube, and brain, sites of 5-HT uptake may be related to malformations caused by treatment of chick embryos (Palen et al., 1979), pregnant mice (Jurand, 1980), hamsters (Guram et al., 1980 and 1982), or humans (Idänpään-Heikkila and Saxen, 1973) during neurulation phases of embryogenesis with the tricyclic antidepressants amitriptyline, imipramine, chlormipramine, trimipramine, or chlorpromazine, all of which inhibit 5-HT and/or NE uptake. It should also

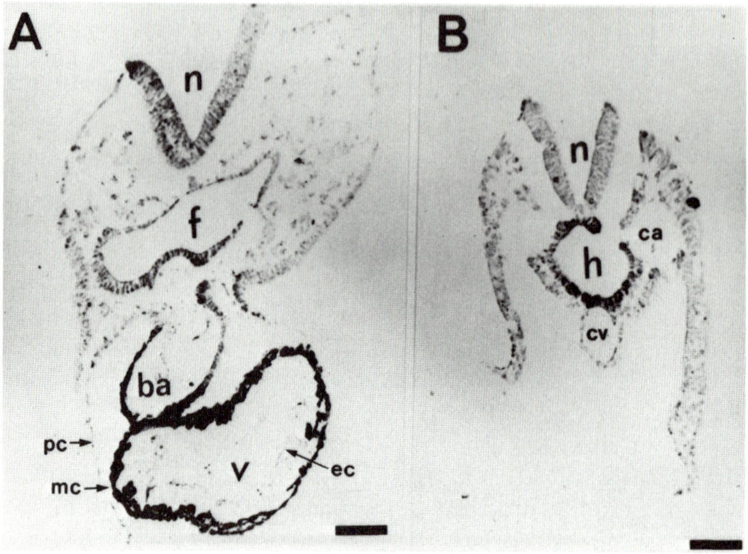

Figure 16-1. Sites of uptake of 5-HT in the seven somite mouse embryo (E9.5) cultured according to the method of Sadler (1979) and stained using anti-5-HT immunocytochemistry: (**A**) heart, and (**B**) hindgut. Embryos were cultured for 3 hours in the presence of 5-HT (10^{-5} *M*) preceded by a 1 hour of incubation in nialamide + L-cysteine (10^{-5} *M*). These same sites were seen with 5-HTP (10^{-5} *M*), and when 5-HT or 5-HTP were used at 10^{-7} *M*. L-trp (10^{-5} *M*), however, only labeled the heart, suggesting that only this site could contain the entire 5-HT pathway. *n*, neural tube; *f*, foregut; *ba*, bulbus arteriosus; *v*, ventricle; *pc*, pericardium; *mc*, myocardium; *ec*, endocardium; *h*, hindgut; *cv*, caudal vein; and *ca*, caudal artery. Bars = 100 μm. From Lauder and Sadler (in prep.)

be noted that excess dietary L-trp during pregnancy has been reported to cause adverse effects on embryogenesis in the hamster (Meier and Wilson, 1983). Therefore, we are now investigating the capacity of 5-HT, NE, their precursors, and appropriate drugs to produce malformations in cultured mouse embryos, and to determine whether they are related to sites of uptake and/or synthesis for these neurotransmitters.

In addition to the tricyclic antidepressants, the benzodiazepines also appear to cause malformations in the mouse embryo (Guram et al., 1982; Jurand, 1980), which raises the possibility that γ-aminobutyric acid (GABA) or its receptors might also play a role in morphogenesis. This observation is noteworthy since GABA has been measured biochemically in the mouse neural tube during closure (McCandless and Scott, 1981), and has been implicated in palate morphogenesis, based on the finding that it inhibits palatal shelf reorientation in vitro (Wee and Zimmerman, 1983). Moreover,

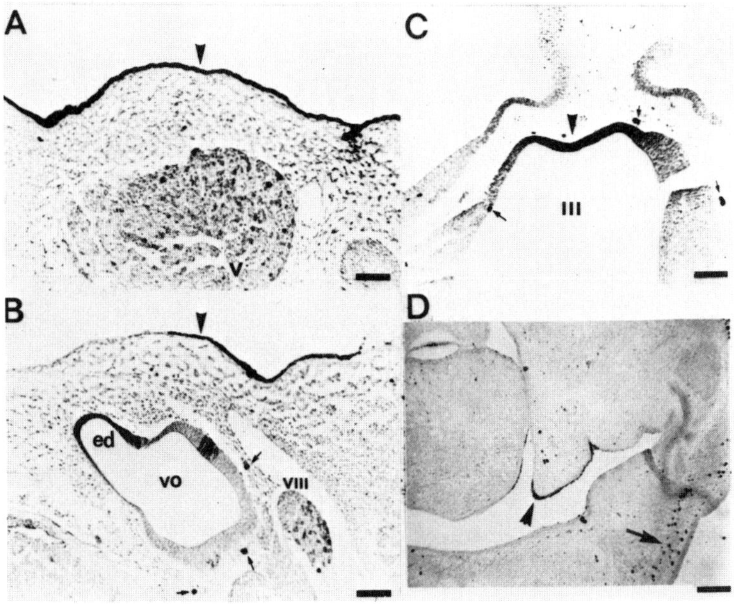

Figure 16-2. Sites of uptake of 5-HT (10^{-5} M) in the 45 somite E12.5 (**A–C**) and E14 (**D**) in the cultured mouse embryo stained with anti-5-HT immunocytochemistry. (**A**) Head ectoderm spanning the mesencephalic brain flexure (*arrowhead*): *v*, ganglion fifth cranial nerve. (**B**) Head ecotderm (*arrowhead*) and otocyst: *ed*, endolympatic sac; *vo*, vestibular part of otocyst; *arrows*, mast cells; and *VIII*, ganglion eighth cranial nerve. (**C**) Roof of third ventricle/epiphysis (*arrowhead*): *arrows*, mast cells. (**D**) Anterior palate (*arrowhead*): *arrows*, mast cells. All sites are blocked by fluoxetine (10^{-5} M). 5-HTP (10^{-5} M) or 5-HT (10^{-7} M) labels all sites, except **C**. Bars = 100 μm. A–C: From Lauder and Sadler (in prep.); D from Lauder and Zimmerman (in prep.)

diazepam has been reported to cause cleft palate in mice and to mimic the effects of GABA in palatal cultures (Zimmerman and Wee, 1984). Thus, sites for GABA uptake and/or synthesis may also exist in the early mouse embryo, a possibility that we will explore in future studies.

In summary, drugs that act via neurotransmitter-related mechanisms may cause teratogenic effects because of their ability to interfere with the uptake, synthesis, or metabolism of these substances at precise morphogenetic sites.

3. CELL PROLIFERATION

The possibility that neurotransmitters influence cell division in the early embryo stems from the work of Buznikov and coworkers, who have reported the presence of acetylcholine, dopamine, epinephrine, norepinephrine,

Table 16-1. Sites of uptake and/or synthesis of 5-HT in the day-12 to day-12.5 mouse embryo.

Treatment	Location and relative intensity of immunoreactivity					
	Head ectod.	Otocyst	Roof III/ epiphysis	Tongue	Mouth r.	Phary. a.
None (cult)	0	0	0	0	0	0
Nial + Cys (10^{-5} *M*)	+	+	0	+	+	+
5-HT ($10^{-5}/10^{-7}$ *M*)	++/+	++/+	++/0	++/+	++/+	++/+
L-trp ($10^{-5}/10^{-6}$ *M*)	+	+	0	+	+	+
5-HTP ($10^{-5}/10^{-7}$ *M*)	++/+	++/+	0	++/+	++/+	++/+
5-HT + Fluox ($10^{-5} + 10^{-5}$ *M*)	0	0	0	0	0	0
L-trp + Fluox ($10^{-5} + 10^{-5}$ *M*)	0	0	0	0	0	0
Nial + Cys + Fluox (10^{-5} *M*)	0	0	0	0	0	0

Embryos were preincubated for 1 hour in nialamide (Nial) + L-cysteine (cys) (10^{-5} *M*) ± fluoxetine, followed by incubation in 5-HT or precursors for 3 hours. Culture medium consisted of 25% Tyrode's solution + 75% heat-inactivated rat serum. L-trp, L-trptophan; 5-HTP, 5-hydroxytryptophan; Fluox, fluoxetine; Head ectod., ectoderm spanning mesencephalic flexure; Roof III, roof of third ventricle; Mouth r., roof of mouth; and Phary. a., ectoderm of pharyngeal and mandibular arches. Intensity of immunoreactivity was rated on a relative scale of 0 (none) to ++ (very intense). Values represent the range of intensities from a number of different experiments and are, therefore, only meant as relative indices rather than as absolute values.

and 5-HT in fertilized eggs and early zygotes of sea urchins and fish (Buznikov et al., 1964, 1968, 1970 and 1974; Manukhin et al., 1981), the levels of which appear to fluctuate with the cell division cycle. The effects of these neurotransmitters or their pharmacologic analogs on cleavage divisions in the eggs of sea urchins and other species have been studied in detail by the Buznikov group and have been comprehensively reviewed recently (Buznikov, 1984).

These studies have shown that analogs of neurotransmitters (receptor agonists or antagonists) are more potent than the transmitters themselves in affecting cleavage furrow formation (Buznikov et al., 1974), which appears to relate to the fact that they are more lipophilic and thus cross the cell membrane better (Landau et al., 1981). The effects of these analogs on cleavage furrow formation appear mainly to be inhibitory, an effect that can be reversed by addition of the transmitter itself in most cases (Rakic et al., 1978a and b). These findings imply the presence of neurotransmitter receptors in the cleaving egg, possibly localized to the cleavage furrow (Renaud et al., 1983).

Table 16-2. Sites of uptake and/or synthesis of 5-HT in the day-14 mouse embryo.

	Location and intensity of immunoreactivity			
Treatment	Roof of mouth	Tongue	Ant. palate	Post. Palate
None (not cult.)*	0	0	0	0
None (cultured)*	+	++/+	++	0
Nial + Cys (10^{-5} *M*)	+	++/+	++/+	0
5-HT ($10^{-5}/10^{-7}$ *M*)	+++/+	+++/+	+++/++	0
L-trp (10^{-5} *M*)	+	++	++	0
L-trp (10^{-5} *M*)*	+	++	++	0
5-HT (10^{-7} *M*) + Fluox (10^{-5} *M*)	0	0	0	0
L-trp (10^{-5} *M*) + Fluox (10^{-5} *M*)*	0	0	0	0
Nial + Cys (10^{-5} *M*) Fluox (10^{-5} *M*)	0	0	0	0

Except where indicated (*), embryos were preincubated for 1 hour in nialamide and L-cysteine ± fluoxetine. Incubation in 5-HT or precursors was for 3 hours. Culture medium consisted of modified Dulbecco's medium + 10% fetal calf serum. Intensity of immunoreactivity was rated on a relative scale of 0 (none) to +++ (most intense). Abbreviations as in table 16-1.

Studies with neurotropic drugs in whole rodent brain and slices have provided evidence that catecholamines and 5-HT may be involved in controlling cell proliferation during postnatal development (Lewis et al., 1977; Patel et al., 1977, 1979 and 1981; see review by Patel et al., 1983), since drugs that exert their effects via these transmitter systems inhibit ^{3}H-thymidine incorporation into forebrain and cerebellum, whereas cholinergic drugs do not. However, the direct effects of these drugs on cell proliferation must also be considered. Catecholamines have also been implicated in controlling cell division in hepatocytes from regenerating liver (Cruise et al., 1985; reviewed by Morely and Royse, 1981), where they appear to act via α1-adrenergic receptors to stimulate DNA synthesis.

Peptides, such as the endogenous opioids, may also play roles in cell proliferation in developing brain. For example, β-endorphin has been detected biochemically in rat brain as early as E13 (Shoemaker et al., 1984), and studies by Zagon and McLaughlin (1984) have show that treatment of prenatal or postnatal rats with opiates causes a variety of behavioral and morphological deficits, including decreased head circumference and brain size as well as retarded somitic growth. These workers have also reported that treatment of infant rats with the opiate antagonist naltrexone produces increased brain size and cellular content (Zagon and McLaughlin, 1983). Moreover, the localization of opioid peptides in proliferating cells has been suggested by their recent study in which they report the presence of enkephalin immunoreactivity within the external granular layer of the cerebellum, during the period when many of these cells are still actively proliferating (Zagon et al., 1985).

In addition to the opiates, other neurohumoral substances such as taurine and vasopressin may also influence cell proliferation during postnatal neurogenesis, as suggested by disturbances of cell acquisition in the cerebellum, forebrain, and olfactory bulbs in animals with deficiencies in these substances (Boer and Patel, 1983; Sturman et al., 1985).

Thus it appears that neurotransmitters and neurohumors may play roles in controlling cell proliferation in both nervous and nonnervous tissues during key phases of ontogenesis. As discussed below, these substances may also influence the differentiation of neurons and glia during brain development.

4. NEURONAL AND GLIAL DIFFERENTIATION

4.1 Serotonin

Our studies concerning the development of monoamine neurons in the embryonic rat brain have demonstrated that these are early developing neurotransmitter systems that begin to differentiate just after neural tube closure (Lauder and Bloom, 1974). Since most other cells are still dividing, this raises the possibility that neurons containing these transmitters might influence the differentiation of other neurons or precursor cells to which they project. Based on these temporal correlations and the results of pharmacologic studies (described below), it is proposed that 5-HT acts as a differentiation signal for specific populations of neurons in the embryonic brain.

In pharmacologic experiments (Lauder and Krebs, 1976 and 1978), we have tested the hypothesis that 5-HT neurons influence certain populations of neuroepithelial cells as they stop dividing and begin to differentiate as neurons. In these studies, the drug *p*-chlorophenylalanine (pCPA), which inhibits 5-HT synthesis (Aghajanian et al., 1973), was administered to pregnant rats beginning on embryonic day 8 (E8). This maternal treatment significantly inhibited 5-HT synthesis in the embryo (Lauder et al., 1981), particularly in developing 5-HT neurons (Lauder et al., 1985), and produced prolonged proliferation (delayed differentiation) of specific populations of neuroepithelial cells in the embryonic brain, all of which give rise to neurons innervated by 5-HT axons in the adult (5-HT target cells). These results have been correlated with the spatiotemporal patterns of 5-HT axonal growth in the embryonic brain (Lauder et al., 1982a; Wallace and Lauder, 1983). They indicate that the effects of pCPA on embryonic neuronal differentiation are dependent on the presence of developing 5-HT axons, suggesting that these pCPA effects are specific to 5-HT neurons and their target cells.

In this context, interactions between developing 5-HT neurons and proliferating neuroepithelial cells in the neural tube have been examined (Lauder et al., 1982a and b). Such interactions exist in a variety of different configurations. For example, developing 5-HT axons pass through the marginal zone of the neural tube where they are able to contact the apical processes of dividing neuroepithelial cells. Likewise, varicose processes of 5-HT neurons pass directly into the neuroepithelium where they appear to contact dividing cells. Sometimes even the cell bodies of 5-HT neurons are in contact with proliferating cells in the immediately adjacent neuroepithelium. In addition, certain populations of 5-HT neurons appear to contact the cerebrospinal fluid directly, raising the possibility that 5-HT could also act as a diffuse humoral signal in the embryonic brain (Lauder et al., 1982a).

Recently, we completed collaborative studies (Whitaker-Azmitia and Lauder, 1985) that have demonstrated changes in the number of 5-HT S1 receptors measured postnatally in the forebrain and brain stem of offspring of mothers treated with either pCPA or the 5-HT agonist 5-methoxytryptamine (5-MT) from E8 to E17. In the case of pCPA, which depletes 5-HT in the embryonic brain, 5-HT receptors were elevated after birth, whereas with 5-MT, which activates 5-HT receptors, the number of receptors was decreased. These results suggest that 5-HT may influence the development of their receptors and that the effects of 5-HT neurons on the differentiation of their target cells could be receptor mediated.

After birth, axons of 5-HT neurons continue to develop and grow into most regions of the neonatal rat brain. At this time, however, most cell proliferation is related to the genesis of glia or small interneurons. In these regions (e.g., cerebrellum, hippocampus), 5-HT axons are frequently seen in association with such neuronal and glial precursor cells, raising the possibility that they could exert an epigenetic influence on postnatal neurogenesis or gliogenesis. In the opossum, 5-HT axons have also been found in the intermediate zone of the cerebellar plate, where they have been proposed to influence neuronal migration and differentiation (Bishop et al., 1985).

Recently, we have begun to explore the possible influences of 5-HT on the in vitro differentiation of neurons and glia from developing rat brain using dissociated cell cultures of embryonic 5-HT neurons and their target cells. The dopaminergic (DA) neurons of the substantia nigra appear to be one of the earliest targets of 5-HT axons in the developing brain, as shown by double immunocytochemistry with antisera to 5-HT and tyrosine hydroxylase. Using dissociated cell cultures of 5-HT and DA neurons from embryonic day 14 (E14) rat brain, the effects of coculture on the morphological and biochemical differentiation of these neurons were examined by using immunocytochemistry and uptake of 3H-5-HT or 3H-DA to label 5-HT and DA neurons, respectively. Coculture had a stimulatory effect on the morphological differentiation of both DA and 5-HT neurons (König et al.,

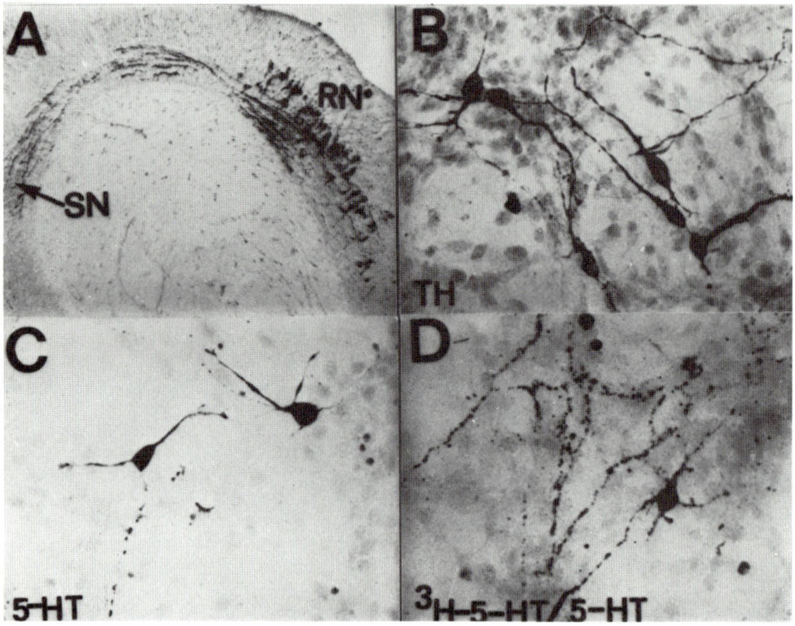

Figure 16-3. Serotonergic (5-HT) and dopaminergic (DA) neurons from the raphe (*RN*) and substantia nigra (*SN*) of the E14 rat embryo (**A**, sagittal section) grown in dissociated cell culture for 4 days (*B–D*). (**B**) SN neurons immunocytochemically stained with antityrosine hydroxylase (*TH*). (**C**) RN neurons stained with anti-5-HT. (**D**) 5-HT neuron stained with anti-5-HT and labeled with ^{3}H-5-HT (combined immunocytochemistry and autoradiography). A: From Lauder et al. (in prep.)

1987). This could be interpreted to mean that 5-HT neurons in these cultures influence the differentiation of DA neurons, as predicted by our hypothesis and their in vivo ontogenic relationship (figure 16-3A), although other possibilities must also be considered, such as indirect effects via glial cells. No effect of coculture on the biochemical differentiation of these DA neurons has yet been demonstrated, however, as measured by the amont of uptake of 3H-DA/DA neuron, even though there are clearly effects on the morphological complexity of these cells. We are also examining the effects of 5-HT added directly to the culture medium on the differentiation of DA neurons.

In addition, we are exploring the possible influence of 5-HT on glial cell proliferation in monolayer cultures of astrocytes from postnatal rat brain. In studies where we used combined ^{3}H-thymidine autoradiography (to label proliferating glial cells) and combined anti-5-HT/anti-GFAP (glial fibrillary acidic protein) immunocytochemistry, we were able to demonstrate that 5-HT neurons readily contact proliferating astrocytes in these cultures. We are now engaged in studies where 5-HT is added to the culture medium and quantitative measurements of ^{3}H-thymidine incorporation are made to assess the effects of 5-HT on glial cell proliferation.

Thus, it appears that these in vitro model systems may be useful for exploring effects of 5-HT on the proliferation and differentiation of target neurons and glial cells. These systems may also prove useful for examination of the direct effects of various drugs, hormones, and humoral agents on the differentiation of 5-HT neurons and their interactions with other neurons or glia.

In addition to our work, other in vitro studies have recently appeared that lend support to the working hypothesis that 5-HT plays roles in development prior to its use in neurotransmission. For example, Gromova et al. (1983) added 5-HT to the medium of organotypic hippocampal cultures from newborn rats that were grown for 2–35 days in vitro and reported that 5-HT stimulated neuropil development, axon myelination, and synaptogenesis. In addition, more neurons in these cultures exhibited spontaneous activity and did so earlier than in untreated cultures. As in our own studies discussed above, however, it is not clear where these effects are direct or indirect. One reason for considering the possibility of indirect effects is the results of studies by Kater and coworkers (Haydon et al., 1984; also see chapter 17) on cultured leech neurons, which have directly demonstrated inhibitory effects of 5-HT on growth-cone motility and synaptogenesis in 5-HT neurons and their target cells. Similar results have recently been obtained by Azmitia and Whitaker-Azmitia (1984) and Davila-Garcia et al. (1985), who have reported that growth of cultured 5-HT neurons from embryonic rat brain is inhibited either by 5-HT itself or by the 5-HT receptor agonist 5-MT, presumably by the stimulation of autoreceptors on these 5-HT neurons. Thus it appears that, when applied directly to responsive cells, 5-HT appears to exert an inhibitory rather than stimulatory influence.

Further studies are in progress by the Kater group to examine the effects of other transmitters such as GABA and norepinephrine on growth-cone motility and synaptogenesis in snail neuronal cultures. Preliminary evidence (Kater, personal communication) suggests that these substances each have their own specific effects on different types of snail neurons such that some inhibit, while others stimulate, growth-cone motility. This suggests that different neurotransmitters, released from growth cones, could produce a mutual orchestration of cell–cell interactions during neurite outgrowth and synaptogenesis, thus participating in the construction of those neuronal circuitries in which they will later function in neurotransmission.

4.2 Catecholamines

Studies employing catecholamine-depleting drugs such as 6-hydroxydopamine (6-HODA) and reserpine have examined the possibility that catecholamines may also play roles in the differentiation of their target cells.

For example, 6-HODA administration to neonatal rats and mice has been reported to produce changes in cerebellar development, including defects in granule cell migration, decreased numbers of granule cells, and disorientation of Purkinje cell dendritic trees (Berry et al., 1980 and 1981; Lovell, 1982; König et al., 1983; Sievers et al., 1981). Although it is not clear to what extent these results can be correlated with the absence of noradrenergic axons rather than to toxic side effects of the drug itself, one study (Lovell, 1982), in which a relatively low dose of the drug was used, reported similar effects with no evidence of cytotoxicity or cell death. Further evidence for an influence of catecholamines on cell migration in the developing cerebellum was found in a study by Rosenstein and Brightman (1981), who transplanted fragments of the adrenergic superior cervical ganglion into the rat cerebellum postnatally and observed that granule cells were either arrested at the brain surface or migrated aberrantly to invade the graft as if they were being "attracted" there.

With regard to cortical development, Schlumpf et al. (1980) have suggested that the early developing catecholaminergic innervation in cortex could exert a trophic influence on cell differentiation. Berger and Verney (1984) further suggested that norepinephrine and dopamine may play separate roles in cortical development due to their different spatiotemporal patterns of development. In support of a role for catecholamines in cortical development, Felten et al. (1982) reported dose-dependent changes in pyramidal cell dendritic arborizations following neonatal 6-HODA administration. Their results further suggest that, in order to demonstrate such influences of noradrenergic neurons on their target cells, it may be necessary to destroy this innervation completely. This finding, together with the existence of compensatory mechanisms (Bear and Daniels, 1983; Bear et al., 1983), could explain the discrepancies between earlier studies of possible neurotrophic influences of noradrenergic neurons on cortical development (see review by Rakic and Goldman-Rakic, 1982). In addition, 6-HODA treatment of neonates has been reported to cause changes in cortical ganglioside development (Segler-Stahl et al., 1982) and cell density (Onteniente et al., 1980).

Particularly interesting are the studies by Kasamatsu and coworkers (Kasamatsu and Pettigrew, 1976 and 1979; Kasamatsu et al., 1979 and 1985; Jonsson and Kasamatsu, 1983; see review by Kasamatsu, 1984), in which norepinephrine has been suggested to play a role in the development and plasticity of ocular dominance in the visual system, acting via β-adrenergic receptors (Kasamatsu and Shirokawa, 1985). Equally interesting is the work of Tennyson et al. (1982 and 1983), who reported significant differences in the numbers of both presynaptic and postsynaptic elements in the putamen of fetal rabbits treated chronically with reserpine, which depletes monamines. These data suggest that changes in the differentiation of target elements in the putamen have occurred subsequent to depletion of transmitter and destruction of presynaptic elements containing these substances.

Prenatal exposure of developing catecholaminergic neurons to psychotropic drugs, such as the neuroleptic haloperidol, or to drugs that deplete them of their transmitter content, such as α-methyltyrosine, 6-HODA, or DSP-4, causes changes in behavior (König et al., 1985) and in the development of catecholamine receptors (Rosengarten and Friedhoff, 1979; see Brown and Fishman, 1984, and Friedhoff and Miller, 1983, for reviews) that persist into the postnatal period. The sensitive, "critical," period for these effects coincides with the period of active neurogenesis of these neurons (Rosengarten et al., 1983), suggesting that receptors for these transmitters are present that can be affected by changes in levels of transmitter in developing catecholaminergic neurons or by interaction with other receptor ligands. Recently, autoradiographic studies of catecholaminergic receptor development have provided evidence supporting this view (Schlumpf et al., 1983). These findings are consistent with the results we have obtained regarding the effects of alterations in the embryonic 5-HT system on 5-HT receptors, discussed above. Thus there is evidence that drug exposure in pregnancy may have long-lasting consequences for functional brain development, possibly due to the presence of receptors for neurotransmitters during the prenatal period.

4.3 GABA

Prenatal development of the GABAergic system in the rat brain has recently been studied using an antiserum to GABA–glutaraldehyde–hemocyanin conjugates, which is specific for GABAergic neurons (Lauder et al., 1985). In this study, the GABA system was found to differentiate prior to most other transmitters, including the monoamines. Thus, by embryonic day 13 (E13) a well-developed ascending fiber system exists in the brain stem, mesencephalon, and diencephalon, including a large projection in the posterior commissure and adjacent areas on the surface of the mesencephalon and tectum. GABAergic cell bodies begin to appear on E14 in the frontal pole of the cerebral cortex. By E16, they are visible in all regions of cortex, where they are located above and below the cortical plate and in the outer part of layer I. Also contained in the outer part of layer I is a dense fiber plexus that stains intensely for GABA. These fibers may be part of the first contingent of cortical afferents to invade the telencephalic vesicle, an event that may be a stimulus for the beginning of neuronal differentiation in this region (König and Marty, 1981; Marin-Padilla, 1984).

Interestingly, the trajectories taken by growing GABAergic fibers in the brain stem, mesencephalon, and diencephalon at E13 and at subsequent stages of development coincide with regions of both monoaminergic and peptidergic differentiation and appear to correspond to recently reported patterns of benzodiazepine receptor development (Schlumpf et al., 1983). These receptors, which are thought to be functionally linked to the GABA receptor (Johnson et al., 1984), appear to develop coincidentally with the GABAergic innervation in these same regions (Lauder et al., 1986). This relationship may be relevant to the finding that prenatal exposure to the benzodiazepine diazepam results in long-lasting changes in GABA uptake in certain brain regions (Frieder and Grimm, 1985) as well as apparently permanent behavioral changes, including sleep disturbances. Moreover, these behavioral changes are accompanied by long-lasting reductions in the numbers of benzodiazepine receptors in particular brain regions (Livezey et al., 1985).

Evidence from other studies suggests that GABA 1) can stimulate neuronal differentiation and receptor development in vitro (Eins et al., 1983; Hanson et al., 1984; Meier et al., 1985; Spoerri and Wolff, 1981), 2) is released from growth cones, a process that is inhibited by receptor antagonists (Gordon-Weeks et al., 1984; Lockerbie and Gordon-Weeks, 1985), and 3) may be involved in the plasticity of neuronal connections in adulthood (Houser et al., 1983; Wolff et al., 1978). Thus, the early differentiation of the GABAergic system may indicate a trophic role for GABA in brain development, possibly involving receptors for this neurotransmitter or related substances.

ACKNOWLEDGMENTS

This work was supported by the following grants from the NIH: NS-13481, NS-15706, NS-06196, NS-00507 (RCDA), and HD-17381.

REFERENCES

Aghajanian, G.K., Kuhar, M.J., and Roth, R.H. (1973) Serotonin-containing neuronal perikarya and terminals: differential effects of *p*-chlorophenylalanine. Brain Res. 54:85–101.

Agrez, M.V., Ames, M.M., Lieber, M.M., and Tyce, G.M. (1984) The presence of serotonin in animal sera used in cell culture. Biogenic Amines 1:223–228.

Allan, I.J., and Newgreen, D.F. (1977) Catecholamine accumulation in neural crest cells and the primary sympathetic chain. Am. J. Anat. 149:413–421.

Bear, M.F., and Daniels, J.D. (1983) The plastic response to monocular deprivation persists in kitten visual cortex after chronic depletion of norepinephrine. J. Neurosci. 3:407–416.

Bear, M.F. Paradisio, M.A., Schwartz, M., Nelson, S.B., Carnes, K.M., and Daniels, J.D. (1983) Two methods of catecholamine depletion in kitten visual cortex yield different effects on plasticity. Nature 302:245–247.

Berger, B., and Verney, C. (1984) Development of the catecholamine innervation in rat neocortex: Morphological features. In: Monoamine innervation of cerebral cortex. New York Alan R. Liss, pp. 95–121.

Berry, M., McConnell, P., and Sievers, J. (1980) Dendritic growth and the control of neuronal form. Curr. Top. Dev. Biol. :67–101.

Berry, M., Sievers, J., and Baumgarten, H.G. (1981) Adaptation of the cerebellum to deafferentation. Prog. Brain Res. 53:65–92.

Bishop, G.A., Ho, R.H., and King, J.S. (1985) An immunohistochemical study of serotonin development in the opossum cerebellum. Anat. Embryol. 171:325–338.

Boer, G.J., and Patel, A.J. (1983) Disorders of cell acquisition in the brain of rats deficient in vasopressin (Brattelboro mutant). Neurochem. Int. 5:463–469.

Brown, R.M., and Fishman, R.H.B. (1984) An overview and summary of the behavioral and neural consequences of perinatal exposure to psychotic drugs. In: Yanai, J. (ed.) Neurobehavioral teratology. Amsterdam: Elsevier, pp. 3–54.

Burden, H.W., and Lawrence, I.F. (1973) Presence of biogenic amines in early rat development. Am. J. Anat. 136:251–257.

Buznikov, G.A. (1984) The action of neurotransmitters and related substances on early embryogenesis. Pharmacol. Ther. 25:23–59.

Buznikov, G.A., and Shmukler, Y.B. (1981) Possible role of "prenervous neurotransmitters" in cellular interactions of early embryogenesis: a hypothesis. Neurochem. Res. 6:55–68.

Buznikov, G.A., Chudakova, I.V., and Znezdina, N.D. (1964) The role of neurohumors in early embryogenesis. I. Serotonin content of developing embryos of sea urchin and leech. J. Embryol. Exp. Morphol. 12:563–573.
Buznikov, G.A., Chudakova, I.V., Berdysheva, L.V., and Vyazmina, N.M. (1968) The role of neurohumors in embryogenesis. II. Acetylcholine and catecholamine content in developing embryos of sea urchin. J. Embryol. Exp. Morphol. 20:119–128.
Buznikov, G.A., Kost, A.N., Kucherova, N.F., Mndzhoyan, A.L., Suvorov, N.N., and Berdysheva, L.V. (1970) The role of neurohumors in early embryogenesis. II. Pharmacological analysis of the role of neurohumors in cleavage divisions. J. Embryol. Exp. Morphol. 27:339–351.
Buznikov, G.A., Rakic, L., Turpaev, T.M., and Markova, L.N. (1974) Sensitivity of sea urchin embryos to antagonists of acetycholine and monoamines. Exp. Cell Res. 86:317–324.
Cruise, J.L., Houck, K.A., and Michalopoulos, G.K. (1985) Induction of DNA synthesis in cultured rat hepatocytes through stimulation of α1-adrenoreceptor by norepinephrine. Science 227:749–751.
Davila-Garcia, M.I., Alvarez, P., Whitaker-Azmitia, P.M., Towle, A.C., and Azmitia, E.C. (1985) Effects of serotonin and 5-methoxytrpytamine on GABA and 5-HT immunocytostained cells from hippocampus and mesencephalic raphe grown in tissue culture. Soc. Neurosci. Abstr. 11:601.
Eins, S., Spoerri, P.E., and Heyder, E. (1983) GABA or sodium-bromide-induced plasticity of neurites of mouse neuroblastoma cells in culture. a quantitative study. Cell. Tissue Res. 229:457–460.
Emanuelsson, H. (1974) Localization of serotonin in cleavage embryos of *Ophryotrocha labronica* La Greca and Racci. Wilhelm Roux Archiv. 175:253–271.
Felten, D.L., Hallman, H., and Jonsson, G. (1982) Evidence for a neurotrophic role of noradrenaline neurons in the postnatal development of rat cerebral cortex. J. Neurocytol. 11:119–135.
Frieder, B., and Grimm, V.E. (1985) Some long-lasting neurochemical effects of prenatal or early postnatal exposure to diazepam. J. Neurochem. 45:37–42.
Friedhoff, A.J., and Miller, J.C. (1983) Prenatal psychotropic drug exposure and the development of central dopaminergic and cholinergic neurotransmitter systems. In: Schlumpf, M., and Lichtensteiger, W. (eds.) Drugs and hormones in brain development. S. Basel: S. Karger, pp. 91–98.
Gordon-Weeks, P.R., Lockerbie, R.O., and Pearce, B.R. (1984) Uptake and release of [^{3}H] GABA by growth cones isolated from neonatal rat brain. Neurosci. Lett. 52:205–210.
Gromova, H.A., Chubakova, A.R., Chumasov, E.I., and Konovalov, H.V. (1983) Serotonin as a stimulator of hippocampal cell differentiation in tissue culture. Int. J. Dev. Neurosci. 1:339–349.
Guram, M.S., Gill, T.S., and Geber, W.F. (1980) Teratogenicity of imiprimine and amitriptyline in fetal hamsters. Res. Coo. Psychol. Psychiatr. Behav. 5:275–282.
Guram, M.S., Gill, T.S., and Geber, W.F. (1982) Comparative teratogenicity of chlordiazepoxide, amitriptyline, and a combination of the two compounds in the fetal hamster. Neurotoxicology 3:83–90.
Gustafson, T., and Toneby, M. (1970) On the role of serotonin and acetylcholine in sea urchin morphogenesis. Exp. Cell Res. 62:102–117.
Hanson, G.H., Meier, E., and Schousboe, A. (1984) GABA influences the ultrastructural composition of cerbellar granule cells during development in culture. Int. J. Dev. Neurosci. 2:247–257.
Haydon, P.G., McCobb, D.P., and Kater, S.B. (1984) Serotonin selectively inhibits growth-cone motility and synaptogenesis of specific identified neurons. Science 226:561–564.
Houser, C.R., Lee, M., and Vaughn, J.E. (1983) Immunocytochemical localization of glutamic acid decarboxylase in normal and de-afferented superior colliculus: evidence for reorganization of gama-aminobutyric acid synapses. J. Neurosci. 3:2030–2042.
Idänpään-Heikkila, J., and Saxen, L. (1973) Possible teratogenicity of imipramine/chloropyramine. Lancet 2:282–284.
Ignarro, L.J., and Shideman, F.E. (1968a) Appearance and concentrations of catecholamines and their biosynthesis in embryonic and developing chick. J. Pharmacol. Exp. Ther. 159:38–48.
Ignarro, L.J., and Shideman, F.E. (1968b) Norepinephrine and epinephrine in the embryo and embryonic heart of the chick: uptake and subcellular distribution. J. Pharmacol. Exp. Ther. 159:49–58.

Johnson, G.A.R., Allan, R.D., and Skerritt, J.H. (1984) GABA receptors. In: Lajtha, A. (ed.) Handbook of neurochemistry, vol. 6. New York: Plenum, pp. 213–237.

Jonsson, G., and Kasamatsu, T. (1983) Maturation of monoamine neurotransmitters and receptors in cat occipital cortex during postnatal critical period. Exp. Brain Res. 50:449–458.

Jurand, A. (1980) Malformations of the central nervous system induced by neurotropic drugs in mouse embryos. Dev. Growth Diff. 22:61–78.

Kasamatsu, T. (1984) The role of the central noradrenaline system in regulating neuronal plasticity in the developing neocortex. In: Marois, M. (ed.) Prevention of physical and mental congenital defects, part C. New York: Alan R. Liss, pp. 369–373.

Kasamatsu, T., and Pettigrew, J.D. (1976) Depletion of brain catecholamines: failure of ocular dominance shift after monocular occlusion in kittens. Science 194: 206–209.

Kasamatsu, T., and Pettigrew, J.D. (1979) Preservation of binocularity after monocular deprivation in the striate cortex of kittens treated with 6-hydroxydopamine, J. Comp. Neurol. 185:139–162.

Kasamatsu, T., and Shirokawa, T. (1985) Involvement of β-adrenoreceptors in the shift of ocular dominance after monocular deprivation. Exp. Brain Res. 59:507–514.

Kasamatsu, T., Pettigrew, J.D., and Ary, M. (1979) Restoration of visual cortical plasticity by local microperfusion of norepinephrine. J. Comp. Neurol. 185:163–182.

Kasamatsu, T., Watabe, K., Heggelund, P., and Scholler, E. (1985) Plasticity in cat visual cortex restored by electrical stimulation of the locus coeruleus. Neurosci. Res. 2:365–386.

Kirby, M.L., and Gillmore, S.A. (1972) A fluorescence study on the ability of the notochord to synthesize and store catecholamines in early chick embryos. Anat. Rec. 173:469–478.

König, N., Han, V., Lieth, E. and Lauder, J. (1987) Effects of co-culture on the morphology of identified raphe and substantia nigra neurons from the embryonic rat brain. J. Neurosci. Res. (in press).

König, N., and Marty, R. (1981) Early neurogenesis and synaptogenesis in cerebellar cortex. Bibl. Acta 19:152–160.

König, N., Serrano, J.-J., Szafarczyk, A., and Dememes, D. (1983) Role of catecholamines in nervous system development: effects of prenatal 6-hydroxydopa treatment on behaviour and brain structures. In: Schlumpf, M., and Lichtensteiger, W. (eds.) Drugs and hormones in brain development. Basel: S. Karger, pp. 11–19.

König, N., Serrano, J.-J., Jonsson, G., Malaval, F., and Szafarczyk, A. (1985) Prenatal treatment with 6-hydroxyDOPA and DSP-4: biochemical, endocrinological and behavioural effects. Int. J. Dev. Neurosci. 3:501–509.

Landau, M.A., Buznikov, G.A., Kabankin, A.S., Teplitz, N.A., and Chernilovskaya, P.E. (1981) The sensitivity of sea urchin embryos to cytotoxic neuropharmacological drugs: the correlation between activity and lipophility. Comp. Biochem. Physiol. 69C:359–366.

Lauder, J.M., and Bloom, F.E. (1974) Ontogeny of monoamine neurons in the locus coeruleus, raphe nuclei and substantia nigra of the rat. I. Cell differentiation. J. Comp. Neurol. 155:469–482.

Lauder, J.M., and Krebs, H. (1976) Effects of *p*-chlorophenylalanine on time of neuronal origin during embryogenesis in the rat. Brain Res. 107:638–644.

Lauder, J.M., and Krebs, H. (1978) Serotonin as a differentiation signal in early neurogenesis. Dev. Neurosci. 1:15–30.

Lauder, J.M., Han, V.K.M., Henderson, P., Verdoorn, T. and Towle, A.C. (1986) Prenatal ontogeny of the GABAergic System in the rat brain: An immunocytochemical study. Neurosci. 19:465–493.

Lauder, J.M., Petrusz, P., Wallace, J.A., Di Nome, A., Wilkie, M.B., and McCarthy, K. (1982b) Combined immunocytochemistry of serotonin and ^{3}H-thymidine autoradiography: in vivo and in vitro methods. J. Histochem. Cytochem. 30:788–793.

Lauder, J.M., Sze, P.Y., and Krebs, H. (1981) Maternal influences on tryptophan hydroxylase activity in the developing brain. Dev. Neurosci. 4:291–295.

Lauder, J.M., Towle, A.C., Patrick, K., Henderson, P., and Krebs, H. (1985) Decreased serotonin content of embryonic raphe neurons following maternal administration of *p*-chlorophenylalanine: a quantitative immunocytochemical study. Dev. Brain Res. 20:107–114.

Lauder, J.M., Wallace, J.A., Krebs, H., Petrusz, P., and McCarthy, K. (1982a) In vivo and in vitro devleopment of serotonergic neurons. Brain Res. Bull. 9:605–625.

Lawrence, I.E., Jr., and Burden, H.W. (1973) Catecholamines and morphogenesis of the chick neural tube and notochord. Am. J. Anat. 137:199–208.

Lewis, P.D., Patel, A.J., Bendek, G., and Balazs, R. (1977) Effect of reserpine on cell proliferation in the developing rat brain: a quantitative histological study. Brain Res. 129:299–308.
Livezey, G.T., Radulovacki, M., Isaac, L., and Marczynski, T.J. (1985) Prenatal exposure to diazepam results in enduring reductions in brain receptors and deep slow wave sleep. Brain Res. 334:361–365.
Lockerbie, R.O., and Gordon-Weeks, P.R. (1985) γ-Aminobutyric$_A$ acid (GABA$_A$) receptors modulate [^{3}H]GABA release from isolated growth cones in the rat. Neurosci. Lett. 55:273–277.
Lovell, K.L. (1982) Effects of 6-hydroxydopamine-induced norepinephrine depletion on cerebellar development. Dev. Neurosci. 5:359–368.
Manukhin, B.N., Volina, E.V., Markova, L.N., Rakic, L., and Buznikov, G.A. (1981) Biogenic monoamines in early embryos of sea urchins. Dev. Neurosci. 4:322–328.
Marin-Padilla, M. (1984) Neurons of layer I: a development analysis. In: Peters, A., and Jones, E.G. (eds.) Cerebral cortex. New York: Plenum, pp. 447–478.
Markova, L.N., Buznikov, G.A., Kovacevic, N., Rakic, L., Salimova, N.B., and Volina, E.V. (1985) Histochemical study of biogenic monoamines in early ("prenervous") and late embryos of sea urchins. Int. J. Dev. Neurosci. 3:493–499.
McCandless, D.W., and W.J. Scott (1981) The effect of 6-aminonicotinamide on energy metabolism in rat embryo neural tube. Teratology 23:391–395.
Meier, A.H., and J.M. Wilson (1983) Tryptophan feeding adversely influences pregnancy. Life Sci. 32:1193–1196.
Meier, E., Drejer, J., and Schousboe, A. (1985) The trophic effect of GABA on cerebellar granule cells is mediated by GABA receptors. Int. J. Dev. Neurosci. 3:401–407.
Morely, C.G.D., and Royse, V.L. (1981) Adrenergic agents as possible regulators of liver regeneration. Int. J. Biochem. 13:969–973.
New, D.A.T., Coppola, P.T., and Terry, S. (1973) Culture of explanted rat embryos in rotating tubes. J. Reprod. Fertil. 35:135–138.
Newgreen, D.F., Allan, I.J., Young, H.M., and Southwell, B.R. (1981) Accumulation of exogenous catecholamines in the neural tube and non-neural tissues of the early fowl embryo: correlation with morphogenetic movements. Wilhelm Roux Arch. 190:320–330.
Onteniente, B., König, N., Sievers, J., Jenner, S., Klemm, H.P., and Marty, R. (1980) Structural and biochemical changes in rat cerebral cortex after neonatal 6-hydroxydopamine administration. Anat. Embryol. 159:245–256.
Palen, K., Thorneby, L., and Emanuelsson, H. (1979) Effects of serotonin antagonists on chick embryogenesis. Wilhelm Roux Arch. 187:89–103.
Patel, A.J., Bendek, G., Balazs, R., and Lewis, P.D. (1977) Effect of reserpine on cell proliferation in the developing rat brain: a biochemical study. Brain Res. 129:283–297.
Patel, A.J., Barochovsky, O., and Lewis, P.D. (1981) Psychotropic drugs and brain development: effects on cell replication in vivo and in vitro. Neuropharmacology 20:1243–1249.
Patel, A.J., Bailey, P., and Balazs, R. (1979) Effect of reserpine on cell proliferation and energy stores in the developing brain. Neuroscience 4:139–143.
Patel, A.J., Barochovsky, O., Borges, S. and Lewis, P.D. (1983) Effects of neurotropic drugs on brain cell replication in vivo and in vitro. In: Schlumpf, M., and Lichtensteiger, W. (eds.) Drugs and hormones in brain development. Basel: S. Karger, pp. 99–110.
Rakic, P., and Goldman-Rakic, P.S. (eds.) (1982) Development and modifiability of the cerebral cortex. Neurosci. Res. Prog. Bull. 20.
Rakic, L.M., Manukhin, B.N., Buznikov, G.A., and Turpaev, T.M. (1978a) Binding of neuropharmacological drugs by sea urchin embryos. I. Early sea urchin embryos as experimental models in neuropharmaca binding studies. Jugoslav. Physiol. Pharmacol. Acta 14:445–454.
Rakic, L.M., Buznikov, G.A., Manukhin, B.N., and Turpaev, T.M. (1978b) Binding of neuropharmacological drugs by sea urchin embryos. II. Influence of some substances on the binding of and the sensitivity to neuropharmaca. Jugoslav. Physiol. Pharmacol. Acta 14:455–463.
Renaud, F., Parisi, E., Capasso, A., and De Prisco, P. (1983) On the role of serotonin and 5-methoxy-tryptamine in the regulation of cell division in sea urchin eggs. Dev. Biol. 98:37–46.
Rosenstein, J.M., and Brightman, M.W. (1981) Anomalous migration of central nervous tissue to transplanted autonomic ganglia. J. Neurocytol. 10:387–409.

Rosengarten, H., and Friedhoff, A.J. (1979) Enduring changes in dopamine receptor sites of pups after drug administration to pregnant or nursing rats. Science 203:1133–1135.

Rosengarten, H., Friedman, E., and Friedhoff, A. (1983) Sensitive periods for the effect of haloperidol on development of striatal dopamine receptors. Birth Defects 19:511–513.

Sadler, T.W., and Kochar, D.M. (1976) Biosynthesis of DNA, RNA and proteins by mouse embryos cultured in the presence of a teratogenic dose of chlorambucil, J. Embryol. Exp. Morphol. 36:273–281.

Sadler, T.W., and New, D.A.T. (1981) Cultures of mouse embryos during neurulation. J. Embryol. Exp. Morphol. 66:109–116.

Santander, R.G., and Cuadrado, G.M. (1976) Ultrastructure of the neural canal closure in the chicken embryo. Acta Anat. 95:368–383.

Schlumpf, M., and Lichtensteiger, W. (1979) Catecholamines in the yolk sac epithelium of the rat. Anat. Embryol. 156:177–187.

Schlumpf, M., Shoemaker, W.J., and Bloom, F.E. (1980) Innervation of embryonic rat cerebral cortex by catecholamine-containing fibers. J. Comp. Neurol. 192:361–376.

Schlumpf, M., Richards, J.G., Lichtensteiger, W., and Mohler, H. (1983) An autoradiographic study of the prenatal development of benzodiazepine-binding sites in rat brain. J. Neurosci. 3:1478–1487.

Segler-Stahl, K., Rahmann, H., and Rosner, H. (1982) Effects of neonatal 6-hydroxydopamine treatment on cortical development in mice and rats as monitored by developmental changes of gangliosides. Dev. Neurosci. 5:554–566.

Shoemaker, W.J., Baetge, G., and Bayon, A. (1984) Pro-opiomelanocortin peptides in early CNS development. In: Caciagli, F., Giacobini, E. and Paoletti, R. (eds.) Developmental neuroscience: physiological, pharmacological and clinical aspects. Amsterdam: Elsevier, pp. 59–65.

Sievers, J., Berry, M., and Baumgarten, H. (1981) The role of noradrenergic fibres in the control of postnatal cerebellar development. Brain Res. 207:200–208.

Spoerri, P.E., and Wolff, J.R. (1981) Effect of GABA administration on murine neuroblastoma cells in culture. I. Increased membrane dynamics and formation of specialized contacts. Cell Tissue Res. 218:567–579.

Strudel, G., Meinel, P., and Gateau, G. (1977a) Recherches d'amines flurigenes dans les chordes d'embryons de Poulet traites par des cholinergiques. C.R. Acad. Sci. Paris 284:1097–1100.

Strudel, G., Recasens, M., and Mandel, P. (1977b) Identification de catecholamines et de serotonine dans les chordes d'embryons de Poulet. C.R. Acad. Sci. Paris 284:967–969.

Sturman, J.A., Moretz, R.C., French, J.H., and Wisniewski, H.M. (1985) Postnatal taurine deficiency in the kitten results in a persistence of the cerebellar external granular layer: correction by taurine feeding. J. Neurosci. Res. 13:521–528.

Tennyson, V.M., Budinikas-Schoenebeck, M., and Gershon, P. (1982) Effects of chronic reserpine treatment on development of maturity of the putamen in fetal rabbits. Brain Res. Bull. 9:651–662.

Tennyson, V.M., Gershon, P., Budininkas-Schoenebeck, M., and Rothman, T.P. (1983) Effects of extended periods of reserpine and α-methyl-*p*-tyrosine treatment on the development of the putamen in fetal rabbits. Int. J. Dev. Neurosci. 1:305–318.

Wallace, J.A. (1979) Biogenic amines in the development of the early chick embryo. PhD thesis, University of California, Davis.

Wallace, J.A., (1982) Monoamines in the early chick embryo: demonstration of serotonin synthesis and the regional distribution of serotonin-concentrating cells during morphogenesis. Am. J. Anat. 165:261–276.

Wallace, J.A., and Lauder, J.M. (1983) Development of the serotonergic system in the rat embryo: an immunocytochemical study. Brain Res. Bull. 10:459–479.

Wallace, J.A., Lilly, S., and Maez, R.R. (1985) Characterization of serotonin uptake mechanisms in the region of the caudal neuropore of the early chick embryo. Soc. Neurosci. Abstr. 11:605.

Wee, E.L., and Zimmerman, E.F. (1983) Involvement of GABA in palate morphogenesis and its relation to diazepam teratogenesis in two mouse strains. Teratology 28:15–22.

Whitaker-Azmitia, P.M., and Lauder, J.M. (1985) Effects of maternal treatment with pCPA on the development of high affinity serotonin receptors in rat brain. Soc. Neurosci. Abstr. 11:601.

Wolff, J.R., Joo, F., and Dames, W. (1978) Plasticity in dendrites shown by continuous GABA administration in superior cervical ganglion of adult rat. Nature 274: 72–74.

Zagon, I.S., and McLaughlin, P.J. (1983) Increased brain size and cellular content in infant rats treated with an opiate antagonist. Science 221:1179–1180.

Zagon, I.S., and McLaughlin, P.J. (1984) An overview of the neurobehavioral sequelae of perinatal opioid exposure. In: Yanai, J. (ed.) Neurobehavioral teratology. Amsterdam: Elsevier, pp. 197–234.

Zagon, I.S., Rhodes, R.E., and McLaughlin, P.J. (1985) Distribution of enkephalin immunoreactivity in germinative cells of developing rat cerebellum. Science 227:1049–1051.

Zimmerman, E.F., and Wee, E.L. (1984) Role of neurotransmitters in palate development. Curr. Top. Dev. Biol. 19:37–63.

17. MULTIFUNCTIONAL ROLES FOR NEUROTRANSMITTERS: THE REGULATION OF NEURITE OUTGROWTH, GROWTH-CONE MOTILITY AND SYNAPTOGENESIS

STANLEY B. KATER AND PHILIP G. HAYDON

1. INTRODUCTION

There seems to be an ever-increasing list of what might be aptly termed *multifunctional bioactive compounds*. These are particular molecules known to have a variety of widely divergent roles ranging from affecting development through the regulation of fully established cellular ensembles. Certainly among these molecules, calcium and cyclic nucleotides have received significant attention because of their profound and obvious general intracellular regulatory roles throughout the animal kingdom. For some time now there has been a question as to whether classic neural transmitters may also be such multifunctional bioactive compounds, but acting as intercellular messengers.

Experiments performed in our laboratory now indicate that the neurotransmitters serotonin and dopamine may be involved in the control of the expression of behavioral repertoires, in regulating neurite outgrowth, and in controlling the establishment of functional neuronal circuitry. In particular, our most recent work demonstrates a clear effect of the neurotransmitters serotonin and dopamine on the dynamic behavior of neuronal growth cones of specific identified neurons. We believe that, in addition to acting as intercellular signals determining states of electrical activity, they may also serve important roles in the establishment and maintenance of neuronal morphology and connectivity.

We have employed the nervous system of the pond snail *Helisoma trivolvis* for the study of developmental neurobiology (Murphy and Kater, 1978 and

Vernadakis, A. (ed.), *Model Systems of Development and Aging of the Nervous System*.

1980a and b; Bulloch et al., 1980; Wong et al., 1981; Bulloch and Kater, 1982; Hadley et al., 1982; Hadley and Kater, 1983; Haydon et al., 1984 and 1985; Murphy et al., 1985). The primary reason we have focused our work on this system is the fact that its nervous system is made up of neurons that are uniquely identifiable at the levels of position, size, and coloration of their cell bodies within the ganglia; of their electrophysiological characteristics (i.e., inputs, outputs, and intrinsic membrane properties); of their characteristic morphologies; and of their transmitter phenotypes. Within the ganglion, intercellular dye injection readily reveals characteristic forms of different identified neurons. For instance, among the buccal ganglian neurons, neuron 5 is a simple monopolar neuron resembling a balloon on a string, whereas neuron 19 is a bipolar neuron with a characteristic unidirectional dendritic morphology extending thoroughout both buccal ganglia. Furthermore, since it is possible to identify individual neurons visually, we are able to isolate specific neurons from the nervous system and plate them into cell culture as individuals, where they can be studied as identified neurons either alone or as parts of experimentally constructed neural circuits (Hadley et al., 1983).

For nearly 20 years, our laboratory has focused its studies on the neurobiology of the buccal ganglia of *Helisoma* that are involved in producing the feeding behavior of this organism. Among our findings relevant to the present chapter, immunocytochemical staining of *Helisoma*, using an antiserum to serotonin, has revealed that a single pair of neurons, previously identified as neurons LC1 and RC1 (Granzow and Kater, 1977), provide the total serotonergic innervation of the buccal ganglia (Murphy et al., 1985). Activity of these serotonergic neurons causes an excitation of the buccal feeding motor program and addition of exogenous serotonin similarly activates this feeding behavior (figure 17-1). This, then, is a case of what is considered a classic role for a neurotransmitter. The remainder of this review describes experiments which indicate that this neurotransmitter plays additional roles from embryogenesis up to and including maturity in regulating the development of neuronal architecture and connectivity.

2. NEURON-SPECIFIC GROWTH-CONE PROPERTIES

By studying identified neurons in situ, and then the same neurons in cell culture, we have addressed questions concerning the intrinsic differences among these neurons and the environmental factors that influence the expression of such intrinsic properties to result in the discretely different morphologies and connectivity characteristics of each. When plated in cell culture, growth cones arise from the spherical isolated cell body and elaborate an extensive network of neurites over the course of several days. We have found (Haydon et al., 1985) that, under these conditions, one can identify, simply on the basis of growth pattern, the unique identities of neurons 4, 5, and 19. Clearly, given the careful analysis by others (especially

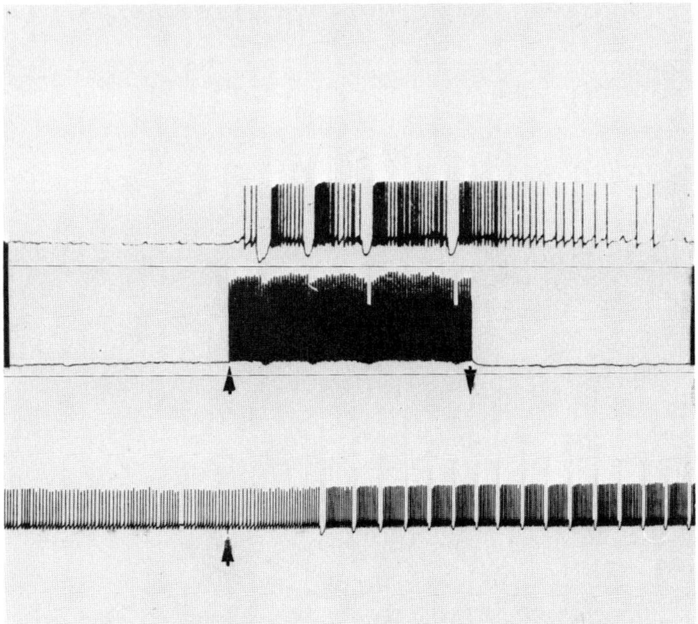

Figure 17-1. Serotonin elicits motor program activity for the feeding behavior of *Helisoma*. (*Top*) Intracellular stimulation of the identified serotonergic neuron C1 causes an activation of the cyclic activity of the feeding behavior as recorded in an identified motoneuron. (*Bottom*) Addition of exogenous serotonin (10^{-6} M) to the buccal ganglia similarly causes an activation of this motor program activity. From Granzow and Kater (1977).

Bray, 1973), the final morphology of a neuron is largely influenced by the individual growth cones of those neurons. We wondered whether one could indeed see differences in the qualities of growth cones between different identified neurons.

Figure 17-2 demonstrates the basis of our initial observations wherein neuron 5 is clearly shown to have distinctly different growth-cone morphologies from those of neuron 19. There are striking differences between the growth cones of neurons 5 and 19 simply based on morphological observations. While there are no significant differences in growth cone surface area, there are three major differences between growth-cone filopodia of these neurons: 1) the growth cones of neuron 5 contain a significantly greater number of filopodia per growth cone as compared with those of neuron 19. This is not simply because the growth cones are larger, since 2) the distance between points of origin of individual filopodia of neuron 5 growth cones is significantly less than the interfilopodial distance of neuron 19. 3) The average length of filopodia from growth cones of neuron 19 is significantly greater than the length of filopodia from neuron 5. Thus there are significant

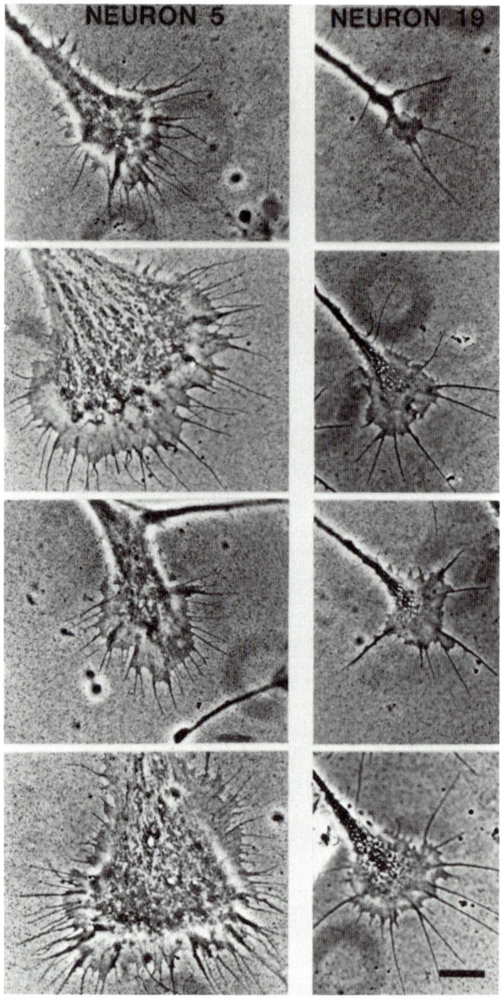

Figure 17-2. Photomicrographs showing examples of the growth cones of neurons 5 and 19. The growth cones of neuron 19 have fewer filopodia, longer fiolpodia, and a greater interfilopodial distance than the growth cones of neuron 5. Bar = 10 μm. From Haydon et al. (1985).

differences in the structural features of the growth cones of different neurons just as there are characteristic differences in the in situ and in vitro morphology of their parent neurons (Cohan et al., 1985). It was with this realization—that individual neurons were different even at the finest level of these motile organelles—that we began our analysis of the extrinsic factors that could regulate the behavior of these intrinsically different motile elements.

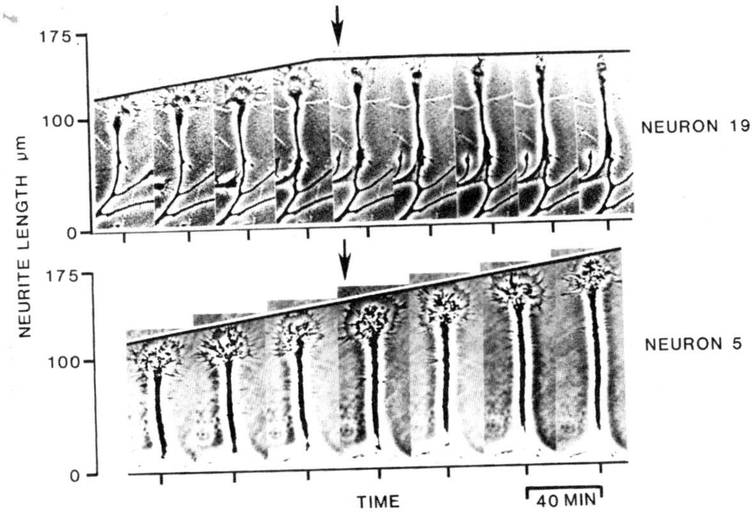

Figure 17-3. Photomicrographs of the behavior of intact growth cones from identified neurons in cell culture. Intact growth cones connected to the remainder of the neuron (not shown) are displayed at frame intervals of 40 minutes. Intact growth cones produce neurite elongation at a constant linear rate before serotonin treatment (*left of arrow*). Application of serotonin (*arrows*) inhibits the elongation of neuron 19 (*top*) but has no effect on neuron 5 (*bottom*). From Haydon et al. (1984).

3. SEROTONIN INHIBITS NEURITE ELONGATION AND GROWTH-CONE MOTILITY

Isolated neuronal somata of *Helisoma* display a characteristic sequence of outgrowth activity in cell culture. Many growth cones arise from a single cell body. As these growth cones advance over the substratum, each leaves in its trail a single stable neurite. An extensive network of neurites is elaborated over a period of 3–4 days (figure 17-9) whereupon a morphological steady state is attained (Hadley et al., 1985). During this period, neurons extend neurites at a reliably constant rate (figure 17-3; Haydon et al., 1985). This constant background rate of extension allows us to ascertain the actions of exogenous agents on neurite extension. As a first approach to determining the actions of serotonin on growth cone motility, a 40-µl dose of the neurotransmitter was added to the culture medium and its effects determined on the rate of neurite elongation and on growth-cone structure. Serotonin (threshold final concentration of $10^{-7}M$) caused a complete inhibition of the motile activity of neuron 19's growth cones and an inhibition of neurite extension (figures 17-3 and 17-4; Haydon et al., 1984). Accompanying this inhibition of neurite elongation, serotonin caused structural changes to the

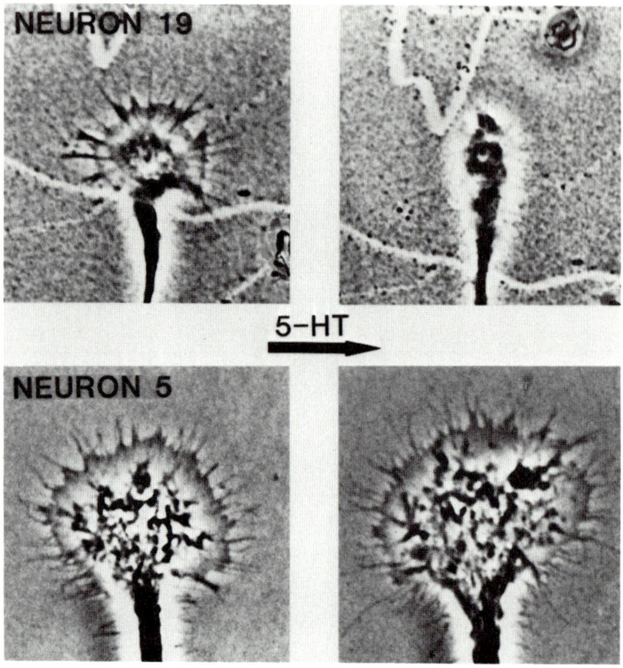

Figure 17-4. Serotonin causes a neuron-specific inhibition of growth-cone motility Exposure of the growth cones of neuron 19 to serotonin (5-HT; threshold concentration $10^{-7}\,M$) causes a retraction of growth-cone filopodia and an inhibition of growth-cone motility (*top panels*). By contrast, the growth cones of neuron 5 are unaffected by serotonin at concentrations as high as $5 \times 10^{-5}\,M$; these growth cones retain their filopodia and motile activity. Bar = 20 μm. From Haydon et al. (1985).

growth cones of neuron 19. The motile growth cone is a flattened structure with actively probing filopodia (microspikes) and a mobile veil-like structure, the lamellipodium. Serotonin caused a retraction of both of these motile structures and consequently the growth cone assumed the appearance of a terminal club-like structure (figure 17-4). Exposure to the carrier medium for serotonin or to medium adjusted to the pH of the serotonin solution never caused any of the growth-inhibitory effects. Similarly, both the precursor of serotonin, 5-hydroxytryptophan (5-HTP) and serotonin's metabolite 5-hydroxyindoleacetic acid (5-HIAA), had no effect on either the neurite outgrowth or growth-cone motility of neuron 19 (table 17-1). Thus, the inhibitory action on outgrowth is specific to the presence of serotonin. An individual neuron has many dozens of neurites and associated growth cones. The inhibitory actions of serotonin were seen on all neurites of individual neuron 19, and persisted for a period of at least 24 hours (the limit of our observation time).

The inhibitiory actions of serotonin on neurite elongation are neuron

Table 17-1. The effect of various agents on neurite outgrowth of cell-cultured identified neurons.

	5-HT	DA	Forskolin	5-HTP	5-HIAA
19	★	★	★	0	0
5	0	0	★	—	—
P5	★	0	—	—	—

★ Inhibition of neurite elongation.
0, no effect

specific. In contrast to its inhibitory effects on neuron 19, serotonin has no inhibitory action either on neurite extension or on growth-cone motility of neuron 5. At concentrations of serotonin as high as $5 \times 10^{-5} M$, the growth cones of neuron 5 retain their filopodia and lamellipodia (figure 17-4) and continue to advance over the substrate (figure 17-3). Similary, the rate of elongation of neuron 5's neurites is totally unaffected by $5 \times 10^{-5} M$ serotonin (Haydon et al., 1984). Thus, serotonin selectively inhibits the neurite outgrowth of specific identified neurons; the outgrowth of neuron 19 is totally inhibited by serotonin whereas the outgrowth of neuron 5 is unaffected.

We eventually hope to test the effects of a battery of neurotransmitters on an array of different *Helisoma* neurons. For instance, the response of neuron P5 to serotonin is different from that of either neuron 19 or neuron 5. Of 12 neurons examined, serotonin ($10^{-6} M$ to $5 \times 10^{-5} M$) caused either a permanent ($n = 4$) or a transient inhibition of P5's motile activity ($n = 8$; Haydon, McCobb, and Kater, in preparation). When the elongation of neuron P5's neurites was completely inhibited, its response to serotonin was similar to that of neuron 19. P5's growth-cone filopodia and lamellipodia retracted. When the motile activity of P5's growth cones was only transiently inhibited, however, these growth-cone filopodia and lamellipodia reappeared within 20–60 minutes. Accompanying the transient structural changes of P5's growth cones was also a transient inhibition of neurite elongation. In some cases, however, we were not able to determine whether serotonin caused a concurrent transient inhibition of neurite elongation, since the duration of the transient retraction of growth-cone filopodia and lamellipodia was similar to our sampling period. Nonetheless, serotonin has a spectrum of action that ranges from total inhibition of outgrowth (neurons 19 and P5) through transient inhibition (neuron P5) to no effect whatsoever (neuron 5) and, just as in chemical synaptic transmission, the particular effect of a neurotransmitter is specified by the target neuron it acts upon.

Having shown that serotonin can inhibit neurite outgrowth, we determined the locus of action of this monoamine (Haydon et al., 1984). There are several loci where serotonin could act in order to inhibit neurite outgrowth: at the cell body, the neurite, or, alternatively, directly at the growth cone. To determine its locus of action, the application of serotonin was focused to specific areas of the membrane of neuron 19 while monitoring the structure and motile activity of its growth cones. Focal application of serotonin was achieved by causing a pressure-evoked efflux of the neurotransmitter from a fine-tipped micropipette. In these experiments it was important to apply serotonin for short periods of time (less than 50 minutes) to minimize dispersal of serotonin throughout the culture medium and thus effectively retain the focal nature of application. Due to this time constraint, we analyzed the rapid time-course inhibitory actions of serotonin by observing the accompanying structural changes of neuron 19's growth cones. A growth-inhibitory response to serotonin was detected only when the serotonin-containing pipette was adjacent to the growth cone itself. Application of serotonin in this manner caused retractions of growth-cone filopodia and lamellipodia that are characteristic of the growth-inhibitory response. Growth-inhibitory responses were not duplicated when serotonin was omitted from the pipette medium. While pipette application of serotonin caused an inhibition of the growth cone adjacent to the source of serotonin, other growth cones attached to the same neuron may still retain their normal motile structural features, indicating that the behavior of individual growth cones is controlled locally and that they can act in a relatively autonomous fashion. Furthermore, serotonin applied to the cell body or neurite of a neuron 19 never caused an inhibition of growth-cone motile behaviors. Taken together, these results indicate that serotonin's growth-inhibiting action is exerted directly at the growth cone and that the individual growth cones of a neuron can act independently, since these organelles are directly responsive to the signals contained in their local microenvironment.

It is possible to show more conclusively the autonomous reaction of the growth cone to serotonin by direct isolation of these organelles from the cell proper. By severing the interconnecting neurite with the tip of a glass micropipette, one produces a viable, isolated growth cone. Isolated growth cones continue to exhibit normal motile behaviors similar in character to the activity of intact growth cones, for periods in excess of 2 hours. Exposure of such isolated growth cones to serotonin causes an inhibition of motile activity. Serotonin applied from a micropipette caused a retraction of growth-cone filopodia and lamellipodia, with a latency to onset of 10 minutes (figure 17-5). Following removal of the pipette, this inhibitory effect was reversed over a time course of apporximately 15 minutes with the first filopodium reextending within 1 minute of pipette removal (figure 17-5). These experiments conclusively demonstrate that the growth cone itself can detect serotonin and transduce this response into a growth-inhibitory effect. While this does not exclude additional contributions from the rest of the cell, it seems reasonable to regard the growth cone, in this respect, as a rather autonomous organelle.

It is as yet unclear what the precise spatial nature of the transduction

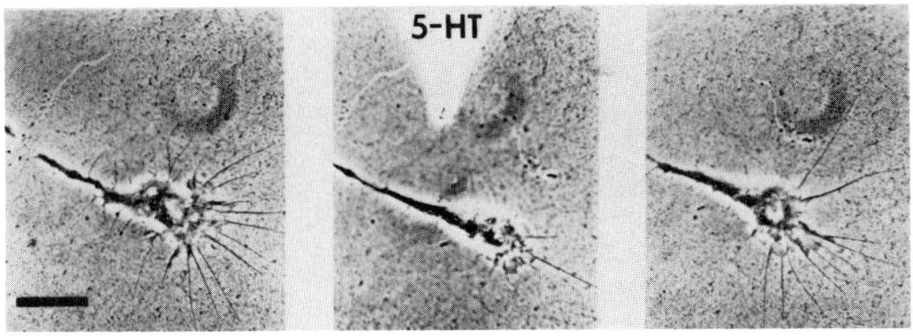

Figure 17-5. Serotonin's growth-inhibitory actions are mediated directly by the growth cone. Physically isolated growth cones of neuron 19 (**left**) retain normal motile activities for many hours in culture. Pipette application of serotonin to such isolated growth cones causes a reversible inhibition of growth-cone motility, as evidenced by the retraction of growth-cone filopodia and lamellipodia (**middle**). Upon removal of the serotonin-containing pipette, growth-cone filopodia and lamellipodia rapidly reextend (**right**). Thus, serotonin acts locally at the growth cone to cause growth-inhibitory actions. Cal. bar 10 μm. From Haydon et al. (1984).

mechanism is. In particular, it will be of significant importance to understand the sensory receptors that allow a growth cone to identify a neural transmitter as an environmental signal worthy of responding to. At this point our observations lead us to speculate that the filopodia may play important roles along these lines. Their long extension into the environment allows them to act as antennae at a significant distance from the growth cone itself and also allows a large spatial separation of the sensing area in order to identify gradients present in the local milieu. The ability to identifiy spatial gradients would not only allow agents such as serotonin to inhibit neurite outgrowth, but perhaps even more exciting, as suggested by Trinkaus (1985), might allow them to guide the direction of outgrowth. One could envision that a serotonergic neuron's release site could establish a significant spatial gradient of this neurotransmitter that would cause a responding growth cone to grow down the gradient in the opposite direction to the source of this growth-inhibiting agent.

4. IN SITU REGULATION OF NEURITE EXTENSION BY SEROTONIN

It is important to establish whether the phenomena that we have described under the experimental conditions of cell culture have any bearing on the normal control of neural plasticity and neuronal development. Two separate lines of evidence reinforce our belief that we are talking about processes actually utilized by the nervous system. The first of these comes from experiments performed on the extension of neurites in situ and the second from experiments performed on embryonic specimens of *Helisoma trivolvis*. Adult buccal ganglia, containing the identified neurons 5 and 19, were isolated from the animal and cultured in a defined medium. Individual nerve trunks containing the axons of the neurons under study were axotomized to evoke neurite outgrowth. Over a period of several days, neurites elongate at a characteristic rate within nerve roots toward the denervated target tissue of the respective neurons. Since neurite extension occurs at a reliable rate, it is possible to predict the extent achieved at any given time after axotomy. In control cultures, numerous neurites arose from the proximal axon stumps and extended for a few hundred microns into the left and right heterobuccal and ventrobuccal nerves. When cultured in the presence of serotonin, however, only growth cones or small neurites were seen. The extent of outgrowth was significantly reduced by the presence of serotonin (Murphy, Haydon, and Kater, in preparation). In contrast to the serotonin-induced growth inhibition of neuron 19, there was no detectable effect of serotonin on the outgrowth of neurites from axotomized neuron 5. Neuron 5 grew profusely following axotomy, independently of whether serotonin was added to the culture medium (figure 17-6). Thus, within the environment of the neurvous system exogenously applied serotonin not only inhibits neurite outgrowth, but is still selective in this process.

The bilateral cerebral ganglion neurons C1 are being added to our matrix of identified *Helisoma* neurons studied with respect to the action of neurotransmitters on neurite outgrowth (Murphy, Haydon, and Kater, in preparation). Serotonin inhibits neurite outgrowth of these serotonin-containing identified neurons. When cultured for 2 days in the presence of serotonin, axotomy only evokes either the formation of a small swelling indicative of a growth cone or possibly a retraction club at the proximal axon stumps of neuron C1 or alternatively a few small neuritic sprouts. In contrast, after 2 days of culture under control conditions, neurons C1 exhibit profuse neuritic sprouting (figure 17-6) that may extend several hundred microns so that, within this short period of time, growth cones are ready to enter the buccal ganglia. Thus, exogenously applied serotonin is capable of inhibiting the outgrowth of the serotonergic neuron C1 as well as neuron 19.

The inhibition of outgrowth of the serotonergic neuron C1 by its own neurotransmitter points to the possibility that this neuron may self-regulate its outgrowth by the orchestrated release of serotonin at its terminals; that is, in the presence of the appropriate receptor mechanism, the functional release of serotonin will cause a feedback inhibition of it's own neurite outgrowth whereas, in the absence of either the receptor or of the release of this neurotransmitter, outgrowth will continue. Similarly there is a preliminary report (Azmitia and Whitaker-Azmitia, 1984) that serotonin can suppress the outgrowth of serotonergic neurons from fetal rats in cell culture. This demonstration also raises the possibility that, in addition to regulating neurite outgrowth in adult nervous systems, this neurotransmitter may be involved in organizing the development of the neuronal arbor and connectivity during development. It is tempting to speculate that, during the development of *Helisoma*, the extension of neuron 19's dendrites ceases when the axon of neuron C1 invades the buccal ganglia and locally releases serotonin. In an

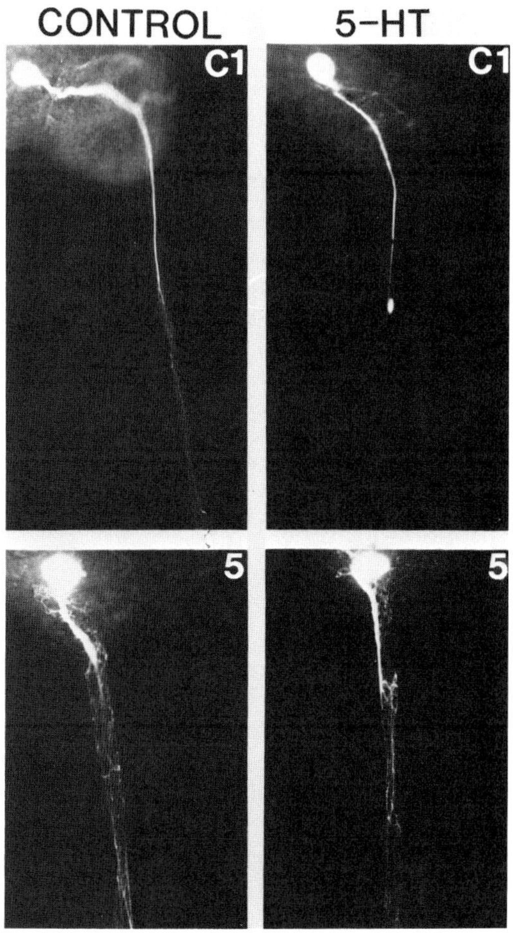

Figure 17-6. Lucifer yellow stains of neurons C1 (*upper panel*) and 5 (*lower panel*) cultured in standard culture media (*left column*) or in culture media plus 5×10^{-5} *M* serotonin (*right column*) for 2 days following axotomy. Growth of neuron C1 is strongly inhibited by serotonin, whereas serotonin has no effect on the growth of neurites from neuron 5. From Murphy et al. (in preparation for submission to J. Comp. Neurol.).

attempt to test this hypothesis, Goldberg and Kater (1985) treated embryonic specimens of *Helisoma* with the serotonin analog 5, 7-dihydroxytryptamine in order to lower the serotonin content of C1. After rearing these treated animals into adults, the structure of neuron 19 was determined by the intracellular injection of Lucifer yellow. As would be predicted by our cell culture data, this treatment caused major alterations in the structure of neuron 19, supporting the hypothesis that the release of serotonin from neuron C1 inhibits the continued growth of dendrites from neuron 19.

Furthermore, the continued serotonergic innervation of neuron 19 by C1 may exert a stabilizing action on its dendrites and prevent them from sprouting in the intact functional nervous system.

This potential stabilizing action of serotonin is supported by experiments described by Amaral et al. (1980). They have demonstrated in young adult animals that the neonatal treatment with 6-hydroxydopamine, to lower the brain content of noradrenaline, promotes a greater expansion of the commisural projection to the dentate gyrus following lesions to the entorhinal cortex. One possible explanation of this result, as pointed out by the authors, is that noradrenaline exerts as outgrowth-inhibitory influence on the commisural afferent inputs to this region of the brain. Thus, the demonstrated inhibitory action of serotonin on neurite outgrowth, combined with the proximity of serotonin release sites to the location of the responsive neuron 19, clearly suggests a novel role for this neurotransmitter in regulating neurite outgrowth and growth-cone motility.

5. THE ROLE OF SECOND MESSENGERS AND OTHER TRANSMITTERS IN REGULATING NEURITE OUTGROWTH

There are a variety of mechanisms that might mediate serotonin's growth-inhibitory action. Serotonin might exert its action by evoking a change in transmembrane ion fluxes that causes a change in intracellular ionic activity by changing the electrical activity of the cell or by causing a change in the concentration of cyclic AMP. As our first approach to examining the potential role of AMP in regulating neurite outgrowth, we determined the effect of forskolin on the outgrowth of cell-cultured neurons (Kater et al., 1985). Forskolin is a potent activator of adenylate cyclase, which consequently evokes an increase in the intracellular concentration of cAMP. Thus, a forskolin-induced inhibition of neurite outgrowth would implicate cAMP as a second messenger mediating serotonin's growth-inihibtory action. The addition of forskolin (final concentration $10^{-4} M$) to culture media caused a rapid retraction of growth-cone filopodia and a complete inhibition of neuron 19's neurite outgrowth. This growth-inhibitory action of forskolin is indistinguishable from that of serotonin on neuron 19. Whereas serotonin affects neither the neurite outgrowth nor motile growth-cone behavior of neuron 5, the addition of forskolin to culture media causes a rapid retraction of growth-cone filopodia and an inhibition of neurite elongation of this neuron (figure 17-7). This indicates that neurons 5 and 19 may have a common second messenger system that is able to regulate neurite outgrowth; conclusive proof, however, will require further experimentation. Nonetheless, that the outgrowth of neuron 5's neurites is inhibited by forskolin suggests that neurotransmitters other than serotonin may be capable of regulating the outgrowth of this cell if the appropriate receptor is coupled to the second messenger system (figure 17-8).

Concordant with this theme, we have recently demonstrated that the neurotransmitter dopamine inhibits neurite outgrowth and growth-cone

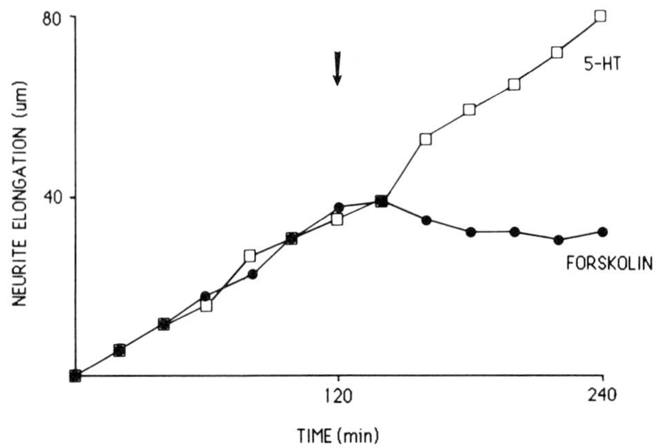

Figure 17-7. Forskolin inhibits neurite elongation of neuron 5. The addition of 10^{-4} *M* forskolin to culture medium inhibits the extension of neuron 5's neurites and causes a retraction of growth-cone filopodia and lamellipodia, whereas 5×10^{-5} *M* serotonin has no effect on either growth-cone motility or neurite elongation of this neuron. This indicates that a cyclic AMP second-messenger system may regulate the inhibition of neurite outgrowth in neuron 5 and that other neurotransmitters could gain access to this control if their receptor is coupled to adenylate cyclase. From Haydon et al. (in preparation for J. Neurosci.).

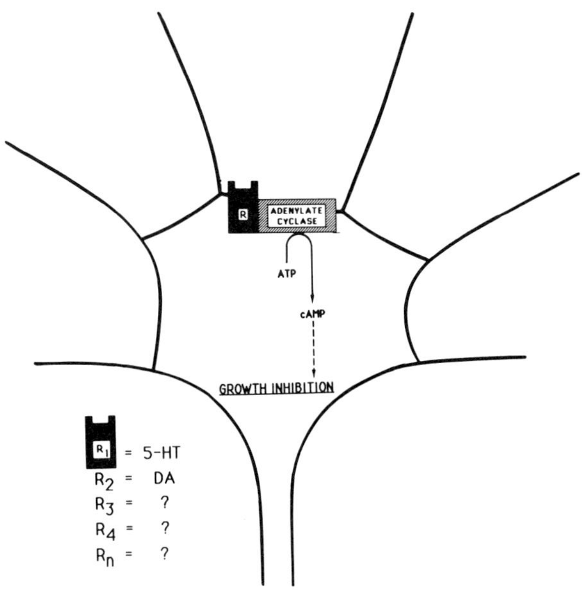

Figure 17-8. Hypothetical diagrammatic representation of a control mechanism for growth-cone inhibiton. Experiments utilizing forskolin indicate that the second-messenger cyclic AMP may mediate a neurotransmitter-evoked inhibition of neurite elongation and growth-cone inhibition. Neurotransmitters such as serotonin and dopamine could gain access to this control process if their receptor can activate adenylate cyclase.

motility of specific identified neurons. Cell culture experiments have demonstrated that the capability to inhibit neurite outgrowth is not restricted to serotonin. The addition to the culture medium of dopamine (10^{-6} to $5 \times 10^{-5} M$) specificially inhibits the outgrowth of neuron 19's neurites and causes retractions of the associated growth-cone filopodia and lamellipodia without affecting neurite extension of neuron 5 (McCobb et al., 1985). While serotonin and dopamine share this common outgrowth-inhibitory action on neuron 19, dopamine has no affect on the outgrowth of identified neuron P5, whose neurite extension is inhibited by serotonin (see Table 17-1). Thus, serotonin and dopamine inhibit the neurite outgrowth of distinct target neurons.

It is important to realize that certain neurons are unaffected by agents like serotonin. This implies a high degree of specific action of this neurotransmitter. Similarly, it is important to realize that not all neurotransmitters may be effective neurotrophic agents. We have performed experiments in which we have ascertained whether the neurotransmitter acetylcholine may exert similar growth-inhibitory action on particular neurons (McCobb, Haydon, and Kater, unpublished). To date, we have been unable to detect any action of this agent on neurite outgrowth, even though we know that this neurotransmitter can evoke potential changes in tested neurons. This result may be important. Acetylcholine is thought to function in the buccal ganglia as a *local* neurotransmitter involved in synaptic transmission. By contrast, both serotonin and dopamine act in a much more global sense since both agents are able to evoke entire behavioral repertoires. Thus these data lead us to speculate that neurotransmitters involved in activating whole neuronal ensembles may also have neurotrophic roles, whereas those involved in synaptic transmission between pairs of neurons probably do not have an additional role in regulating neurite outgrowth. This distinction, if correct, would further refine the degree of specificity and action that the release of specific neurotransmitters can have at propitious times in the life time of the organism.

6. SEROTONIN'S EFFECTS ON NEURITE OUTGROWTH HAVE PROFOUND EFFECTS ON THE FORMATION OF CONNECTIVITY PATTERNS

The processes of neurite outgrowth and synaptogenesis are intimately intertwined. Neurons that do not contact one another cannot form monosynaptic connections. This relationship goes further than simply spatial overlap. Even when contact is made between the neurites of *Helisoma* neurons, synaptic connections do not necessarily form. In order to form electrical synaptic connections, *Helisoma* neurons must be in an active state of neurite outgrowth. Pairs of neurons will form electrical connections only if both potential partners have a period of temporally coincident and spatially overlapping *active* neurite outgrowth (Hadley and Kater, 1983; Hadley et al., 1983 and

1985). In light of serotonin's inhibitory action on neurite outgrowth, we reasoned that it may prevent the formation of specific electrical synapses.

As a simple test, neurons 5 and 19 were plated in culture under conditions known to result in the formation of electrical synapses; then, 24 hours later, serotonin ($10^{-6}\,M$) was added to test cultures specifically to inhibit the continued neurite outgrowth of neuron 19 prior to an interaction with the growing neurites of neuron 5 (Haydon, et al., 1984) As predicted from the experiments heretofore mentioned, this manipulation prevented the continued outgrowth of neuron 19 while permitting the outgrowth of neuron 5, which eventually resulted in an extensive overlap of neurites between neurons 5 and 19. On days 3 and 4, the resultant connectivity was determined in this test and in parallel control cultures by pairwise intracellular recording. Our criteria for identifying an electrical connection between cell-cultured neurons are: 1) injection of direct hyperpolarizing and depolarizing current should cause a concomitant voltage change in the postsynaptic cell, 2) action potentials should cause short-latency postsynaptic potentials that follow 1:1 with each presynaptic spike, and 3) the ability to pass direct current and to evoke short-latency postsynaptic potentials should be shared by both potential partner neurons. In control cultures, neurons 19 and 5 always formed electrical synaptic connections (coupling coefficient 0.18 ± 0.04, mean ± SEM; $n = 9$; figure 17-9). By contrast, however, treatment with serotonin always prevented neuron 19 from forming electrical connections with neuron 5 (mean coupling coefficient $= 0.00 \pm 0.00$; $n = 7$; figure 17-9). When compared with control cultures, this effect of serotonin is significant, both with respect to the strength of the interconnecting synapse ($p < 0.002$; t-test), and with respect to the number of zero coupling values obtained ($p < 0.01$; chi-squared test). Thus, serotonin prevents neuron 19 from forming electrical synapses with neuron 5. In contrast to its inhibitory action on electrical synaptogenesis between neurons 19 and 5, serotonin did not prevent the formation of electrical connections between pairs of neuron 5 precisely as would be predicted from this neuron's resistance to serotonin's outgrowth effects. Pairs of neuron 5 formed electrical connections both in control cultures ($0.59 \pm n = 5$) and in test culture conditions (0.31 ± 0.09; $n = 5$; figure 17-9). Thus, serotonin regulates the formation of specific synaptic connections such that neuron 19 is prevented from forming electrical connections with other neurons that are themselves competent to interconnect.

7. CONCLUSIONS

Serotonin (as well as dopamine) plays a variety of roles: the regulation of formation of *Helisoma*'s nervous system, determining whether particular motor programs are activated and, consequently, which behaviors are elicited. In guiding the development of the central nervous system, serotonin

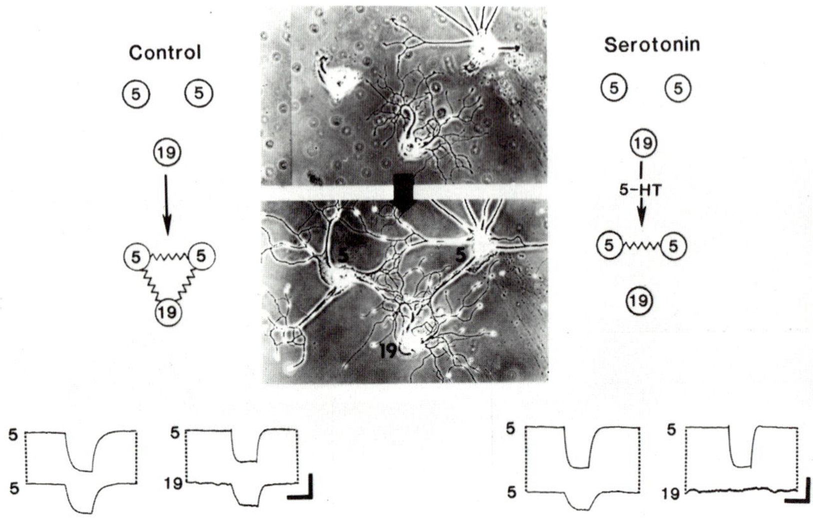

Figure 17-9. Serotonin prevents the formation of specific electrical synaptic connections in cell culture. Isolated, identified neurons, when plated in cell culture, grow new neurites. After 24 hours in culture, the neurites of separate neurons have not normally interacted with one another (*top photomicrograph*). Addition of serotonin ($10^{-6}\,M$) at this time prevents the further outgrowth of neurites from neuron 19. By day 3 (*bottom photomicrograph*), there was little additional outgrowth from neuron 19 compared with the major elaboration of the arbor of neuron 5. This continued outgrowth of neuron 5 caused an extensive overlap of neurites between neurons 5 and between neuron 5 and neuron 19. Injection of direct hyperpolarizing current into neuron 5 reveals that the unaffected neurons 5 have formed an electrical synapse with one another, but that neuron 19 has been excluded from forming such electrical connections (*right*). Under control conditions, in the absence of serotonin, however, neurons 5 and 19 do form electrical synaptic connections (*left*). Thus, the presence of serotonin prevents the formation of specific electrical synaptic connections. Cal. bars: (*vertical left*) upper 20 mV, lower 10 mV; (vertical right) upper 20 mV, lower 5 mV; and (horizontal) 2 seconds.

has prominent roles in regulating the formation of neuronal architecture and connectivity matrices. Clearly, these intercellular regulatory roles qualify this, and likely other agents we commonly refer to as "neurotransmitters," as multifunctional bioactive compounds.

REFERENCES

Amaral, D.G., Avendano, C., and Cohen, W.M. (1980) The effects of neonatal 6-hydroxydopamine treatment on morphological plasticity in the dentate gyrus of the rat following entorhinal lesions. J. Comp. Neurol. 194:171–191.

Azmitia, E.C., and Whitaker-Azmitia, P.M. (1984) Receptor mediated autoregulation of serotonergic growth in primary dissociated cell culture. Soc. Neurosci. Abstr. 10:663.

Bray, D. (1973) Branching patterns of individual sympathetic neurons in culture. J. Cell Biol. 56:702–712.

Bulloch, A.G.M., and Kater, S.B. (1982) Neurite outgrowth and selection of new electrical connections by adult *Helisoma* neurons. J. Neurophysiol. 48:569–583.

Bulloch, A.G.M., Kater, S.B., and Murphy, A.D. (1980) Connectivity changes in an isolated molluscan ganglion during in vivo culture. J. Neurobiol. 11:531–546.

Cohan, C.S., Haydon, P.G., and Kater, S.B. (1985) Single channel activity differs in growing and nongrowing growth cones of isolated identified neurons of *Helisoma*. J. Neurosci. Res. 13:285–300.
Goldberg, J.L., and Kater, S.B. (1985) Experimental reduction of serotonin content during embryogenesis alters morphology and connectivity of specific identified *Helisoma* neurons. Soc. Neurosci. Abstr. 11:158.
Granzow, B., and Kater, S.B. (1977) Identified higher-order neurons controlling the feeding motor program of *Helisoma*. Neuronscience 2:1049–1063.
Hadley, R.D., Kater, S.B., and Cohan, C.S. (1983) Electrical synapse formation depends on interaction of mutually growing neurites. Science 221:446–468.
Hadley, R.D., and Kater, S.B., (1983) Competence to form electrical connections is restricted to growing neurites in the snail, *Helisoma*. J. Neurosci, 3:924–932.
Hadley, R.D., Wong, R.G., Kater, S.B., Barker, D.L., and Bulloch, A.G.M. (1982) Formation of novel central and peripheral connections between molluscan central neurons in organ cultured ganglia. J. Neurobiol. 13:217–230.
Hadley, R.D., Bodnar, D.A., and Kater, S.B. (1985) Formation of electrical synapses between isolated, cultured *Helisoma* neurons requires mutual neurite elongation. J. Neurosci. 5:3145–3153.
Haydon, P.G., McCobb, D.P., and Kater, S.B. (1984) Serotonin selectively inhibits growth cone motility and synaptogenesis of specific identified neurons. Science 226:561–564.
Haydon, P.G., Cohan, C.S., Miller, H.R., McCobb, D.P., and Kater, S.B. (1985) Neuron-specific growth cone properties as seen in identified neurons of *Helisoma*. J. Neurosci. Res. 13:135–147.
Kater, S.B., Haydon, P.G., Murphy, A.D., and McCobb, D.P. (1985) A novel role for neurotransmitters: the control of neurite outgrowth. Soc. Neurosci. Abstr: 11:761.
McCobb, D.P., Haydon, P.G., and Kater, S.B. (1985) Dopamine: an additional regulator of neurite outgrowth in *Helisoma*. Soc. Neurosci. Abstr. 11:761.
Murphy, A.D., and Kater, S.B. (1978) Specific reinnervation of a target organ by a pair of identified molluscan neurons. Brain Res. 156:322–328.
Murphy, A.D., and Kater, S.B. (1980a) Sprouting and functional regeneration of an identified neuron in *Helisoma*. Brain Res. 186:251–272.
Murphy, A.D., and Kater, S.B. (1980b) Differential discrimination of appropriate pathways by regenerating identified neurons in *Helisoma*. J. Comp. Neurol. 190:395–403.
Murphy, A.D., Barker, D.L., Loring, J.F., and Kater, S.B. (1985) Sprouting and functional regeneration of an identified serotonergic neuron following axotomy. J. Neurobiol. 16:137–151.
Trinkaus, J.P. (1985) Further thoughts on directional cell movement during morphogenesis. J. Neurosci. Res. 13:1–19.
Wong, R.G., Hadley, R.D., Kater, S.B., and Hauser, G.C. (1981) Neurite outgrowth in molluscan organ and cell cultures: the role of conditioning factor(s). J. Neurosci. 1:1008–1021.

18. PEPTIDE MODULATION OF NEURONAL DIFFERENTIATION IN CULTURE

DOUGLAS E. BRENNEMAN AND PHILLIP G. NELSON

1. INTRODUCTION

In attempts at the analysis of the highly complex process of development, particularly of the mammalian central nervous system (CNS), there are two equal and opposite imperatives: complexity (to capture what are undoubtedly complicated interactions between the multiple components of the intact system) and simplicity (to permit coherent and comprehensible results). Clearly, a clever choice of preparation is required such that the level of questions and their resolution is matched to the possibilities inherent in the system. For some questions, only a behaviorally intact animal will be an appropriate preparation, while for others a single, cloned gene may be best. We have chosen dissociated cell cultures prepared from the fetal mammalian central nervous system (Ransom et al., 1977) as a useful model for study of mechanisms of central synaptic action and the role of electrical activity in neuronal development. In this chapter, we deal in a general way with the properties of such system, including some of the advantages and disadvantages, and then present in some detail the results of using the dissociated spinal cord cultures in analyzing an important role that vasoactive intestinal peptide (VIP) has in mediating activity-dependent regulation of neuronal development. In addition, we examine general considerations for investigating developmental effects of neuropeptides and finally present a model for activity-dependent influences on neuronal survival.

2. CELL CULTURE MODELS AND THE STUDY OF DEVELOPMENT

A full analysis of the developmental trajectory of a cell in the CNS (or elsewhere) involves tracing the lineage involved in generation of the cell and an understanding of the regulatory factors influencing the cell's development at each phase of that lineage. Typically, dissociated cell preparations are made from tissue in which a substantial portion of this developmental sequence has occurred already, and is, therefore, not available for manipulation and analysis in vitro. This development, antecedent to establishment of the culture system, may include such crucial steps as cell phenotypic commitment, for example, although the substantial degree of plasticity in this regard even relatively late in development is becoming evident. Despite this early period of highly significant developmental events not captured by the dissociated cell cultures, an enormous range of developmental phenomena undoubtedly occur and are, to a substantial degree, accessible in these systems. Morphological complexity increases rapidly and a variety of characteristic neuronal forms can be distinguished over time. Neurotransmitter-related enzymes and receptors increase with time and in some cases this reflects the developmental increases observed in vivo (see Vernadakis and Arnold, 1980; Brenneman and Nelson, 1985). Synapse formation is vigorous and diverse and, in combination with the development of dendrites and axons, quickly results in the formation of highly functional, spontaneously active circuits of synaptically coupled neurons (Westbrook and Brenneman, 1984). However, some aspects of neuronal function may deviate from the in vivo counterparts. Hippocampal pyramidal neurons in cell cultures, for example, do not depolarize after application of acetylcholine (Segal, 1983), while responses are quite marked in the intact hippocampus (Ben Ari et al., 1981) or in slice preparations (Dodd et al., 1981). The nature of spinal cord neuron responses to catecholamine was quite different in cultures that included explants of locus coeruleus than in "spinal-cord-alone" cultures (Pun et al., 1985).

Cultures prepared by dissociating a complex, heterogeneous tissue such as the mouse spinal cord into single cells, plating these cells in culture dishes, and providing a culture medium that supports survival, are themselves complex and heterogeneous. Their cellular composition is related to, but not necessarily the same as, the composition of the source tissue. As asserted above, the cellular complexity does allow for developmental interactions between different types of cells, which may bear some useful relationship to developmental interaction in vivo. A logical approach to analyzing these interactions would be to isolate pure populations of different cell types and undertake a stepwise cellular reconstitution of the original cellular system. A good deal of effort has gone into various cell purification procedures, and there have been some encouraging results, but such procedures are expensive and highly labor-intensive. An alternative approach is to use cell-type enrichment procedures in combination with cell-specific identification methods to study interactions within only partially purified populations (McCarthy and Partlow, 1976; Bottenstein et al., 1980; Eccleston and Silberberg, 1984). Neurons and glia can be distinguished reasonably well with antibodies to neurofilament proteins (Raju et al., 1981), neuron-specific enolase (Pickel et al., 1975), glial fibrillary acid protein (Bignami et al., 1972), and galactocerebroside (Raff et al., 1978). For our developmental studies, both neurofilament and neuron-specific enolase immunohistochemistry are employed for neuronal cell counts (figure 18-1). We utilize the neuron-specific enolase assay most often because there is less labeling of neuronal processes, thus making aggregated neurons more distinguishable. The further identification of subpopulations of cell types will certainly be aided with the use of monoclonal or polyclonal antibodies. Different classes of glial cells have been partially isolated (McCarthy and De Vellis, 1980; Hirayama et al., 1983) and various types of neurons (peptidergic, GABAergic, cholinergic) can be demonstrated with immunocytochemical (Neale et al., 1982, 1983 and 1985) or other methods (see Vernadakis and Arnold, 1980). Regional dissection and dissociation of portions of the brain can be carried out and culture conditions modified to support selectively the survival of either neurons or glia. Finally, the cell culture methodology lends itself to approaching mechanisms of interaction. These can be divided into those requiring cell contact and those that are mediated by a diffusible material. If an effect on one cell type by a second type may be duplicated by medium conditioned by contact with the second cell type, the isolation and characterization of the *factors* responsible can be undertaken. A number of such factors are under intense investigation at the present time.

3. ELECTRICAL ACTIVITY AND NEURONAL SURVIVAL

The experimental context for our studies with neuropeptides is the investigation of the molecular basis of the relationship between electrical activity and the development of neurons. A brief review of these studies provides the background for understanding the effects of a specific peptide—vasoactive intestinal peptide (VIP)—to be described later. In considering the effects of activity on neurodevelopment, a natural first step is to evaluate the effects of blocking all action potentials of the neurons in a developing culture. Of course, the normal course of culture development should be established by a variety of indicators as a baseline. When the number of living neurons are taken as a quantitative measure of culture development, it can be seen that a gradual but large decrease in the number of neurons occurs (figure 18-2), rather similar in extent to the decrease in the number of motoneurons in the spinal cord (Hamburger, 1975) and to the decreases in ciliary ganglion neurons (Landmesser and Pilar, 1974) that occur during development in vivo. When the effects of impulse blockade are examined, there is an apparent sharp distinction between the result in vivo and in vitro. Blockade of neuro-

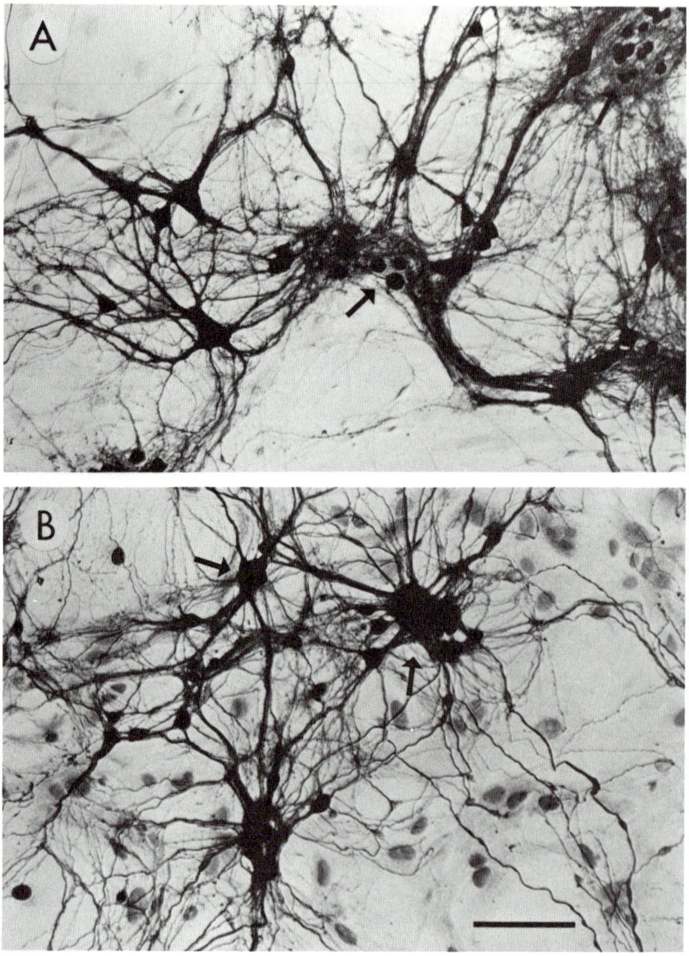

Figure 18-1. Comparison of 3-week-old dissociated spinal cord cultures stained by immunohistochemistry for neuron-specific enolase (**A**) and for neurofilament protein (**B**). Immunoreactivity was visualized by an avidin–biotin immunoperoxidase method (Hsu et al., 1981). The neurofilament antibody reacted with all three of the neurofilament polypeptides (Dahl and Bignami, 1977). Aggregated neurons are the points of comparison as indicated by the *arrows*. Cultures stained with anti-neuron-specific enolase showed less background cell labeling and a greater degree of resolution of somal bodies than did cultures stained with antineurofilament. The calibration bar = 100 μm.

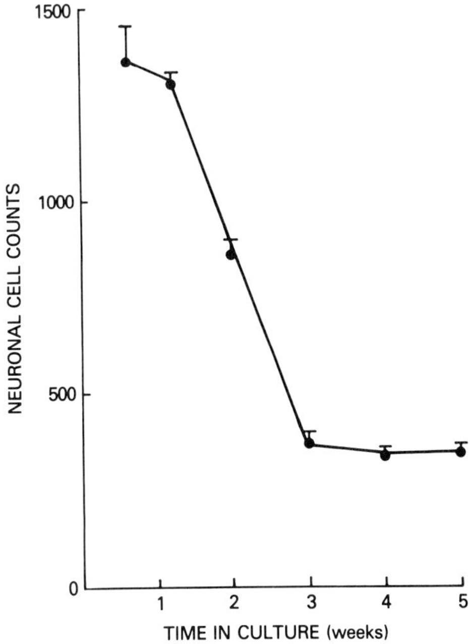

Figure 18-2. Decreases in neuronal cell counts during development in culture. Counts were obtained from 100 microscopic fields. Each value is the mean ± SEM of five cultures. From Brenneman et al. (1985a).

muscular activity in the intact system results in a decrease in cell death (Pittman and Oppenheim, 1978). Similarly, injections of α-bungarotoxin prevent neuronal cell death in the trochlear nucleus (Creazzo and Sohal, 1979). By contrast, cell death is increased by impulse blockade in dissociated spinal cord cultures (Brenneman et al., 1983). However, an additional manipulation of the culture system produces a result more consonant with the results obtained in vivo (Brenneman et al., 1984). Medium conditioned by contact with normally active neurons can be applied to the electrically silent cultures. Whereas there is little effect of the conditioned medium (CM) treatment on electrically active cultures, the same CM when applied to electrically inactive cultures not only reverses the deficit associated with inactivity, but also prevents the "normal" decrease in neuronal number that occurred during the experimental period in control cultures (figure 18-3); that is, essentially no neuronal death occurred in cultures that were 1) electrically silent and 2) exposed to material released from active cells (i.e., CM from normal cultures). We discuss the implications of these findings in a later section dealing with possible mechanisms and summarize here some other features of electrical activity–development interactions. A number of these features suggest a degree of specificity in this interaction: 1) Not all cell types

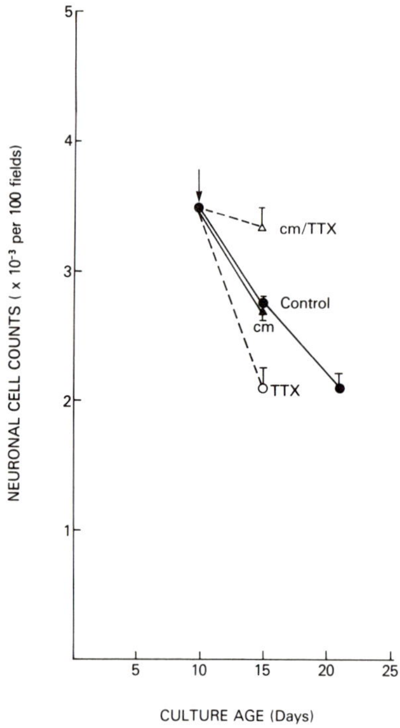

Figure 18-3. Neuronal cell counts after electrical blockade with TTX: effect of conditioned medium (*cm*) from electrically active spinal cord cultures. Treatment was started on day 9 as indicated by the *arrow*. Conditioned medium was obtained from spontaneously active spinal cord cultures between days 30 and 34. This conditioned medium prevented the decrease that occurred as a result of TTX treatment (1 μM) in addition to that which occurs "normally" during development. Each value is the mean ± SEM of three dishes.

are affected. Neurons that take up GABA, for instance, are not affected while cholinergic cells and the total number of neurons are affected (Brenneman et al., 1983). 2) There is a "critical period" of maximal sensitivity or vulnerability of neurons to the effects of blockade of electrical activity. No effect is seen from days 0–6 in vitro or later than about day 21. 3) Active donor cultures release maximal amounts of supporting material *outside* a restricted period [roughly from day 7 to day 21—the critical period during which neuronal death naturally occurs, (Brenneman et al., 1985c)]. The reduction in the detectable survival-promoting material during the critical period may play a role in the process of cell death that occurs during this period.

4. VASOACTIVE INTESTINAL POLYPEPTIDE (VIP) AS NEURODEVELOPMENTAL REGULATOR

Further progress in analyzing the activity–development relationship has been based on two areas of study: 1) the effect of a peptide (VIP) and 2) the use of

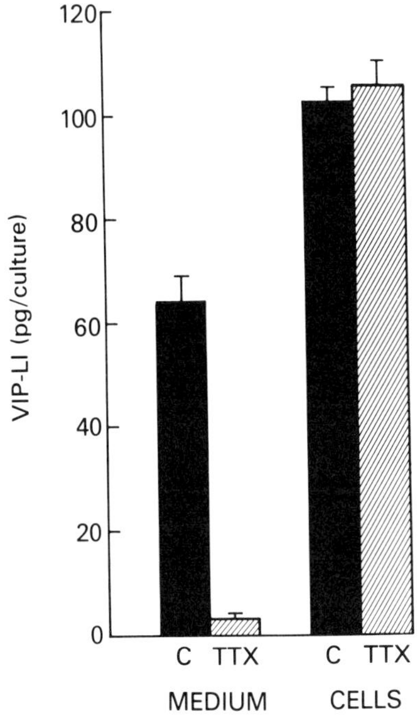

Figure 18-4. Effect of electrical blockade on VIP-like immunoreactivity of cultured cells and conditioned medium. One-month-old cultures were given a complete change of medium and, 4 days later, VIP radioimmunoassays were conducted on the culture medium and cells. Treatment with 1 μM TTX had no effect on the amount of VIP in the cells, but dramatically reduced the amount of VIP that was spontaneously released into the culture medium. Each value is the mean of four determinations ± SEM. From Brenneman et al. (1985b).

separate cultures purified or highly enriched for either neurons or glia. The experiments with spinal cord CM indicated the presence of endogenous substances that can influence neuronal survival and that are themselves influenced by electrical activity. The subsequent goals have been to isolate these substances and to characterize their mechanism of action. Two strategies were adopted in hopes of reaching these goals. The first strategy was to test substances that were known to be present in our model system, and the second was to isolate neurotrophic substances in CM biochemically. The peptidergic influences to be summarized are the result of the first strategy.

VIP was screened for its potential effects on neuronal survival for the following reasons: 1) VIP has been shown to be present in the spinal cord (Loren et al., 1979) and in some dorsal root ganglion neurons (Go and Yaksh, 1980). Previous work (Brenneman et al., 1985b) has shown that VIP-immunoreactive neurons are present in dissociated spinal cord cultures and that the spontaneous release of VIP is prevented by tetrodotoxin (figure 18-4). Examples of VIP-positive neurons in culture are presented in figure

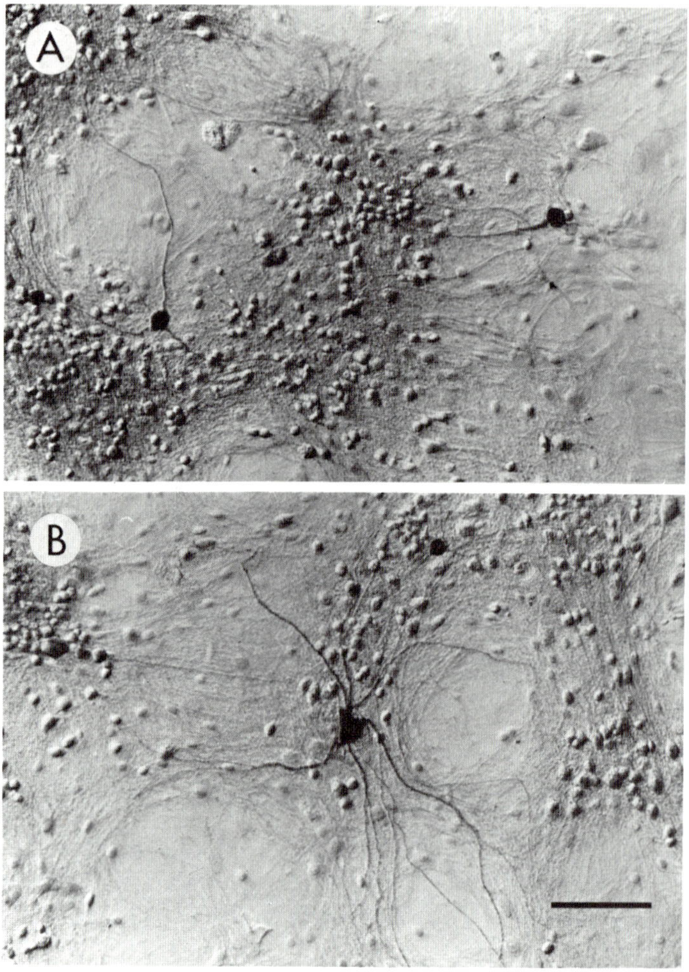

Figure 18-5. VIP-immunoreactive neurons in dissociated spinal cord cultures. The most abundant of the VIP-positive neurons were the small cells shown in **A**. These cells comprised 3%–5% of the total neurons present in the cultures. The large VIP neuron in **B** was very rare (<5 per 35-mm dish). All cultures were treated with colchicine for 24 hours prior to fixation and staining. The antiserum employed (Eiden et al., 1982) showed no reactivity with secretin or PHI-27. Photomicrographs were taken with Hoffman modulation contrast optics. The calibration bar is 100 μm.

18-5. 2) Cyclic AMP derivatives have been shown to increase neuronal survival in spinal cord cultures that are electrically blocked (Brenneman et al., 1985c). VIP has been reported to produce significant increases in cAMP levels in a variety of test systems (Rougon et al., 1983; Koh et al., 1984). 3) VIP has been shown to be colocalized with choline acetyltransferase in both

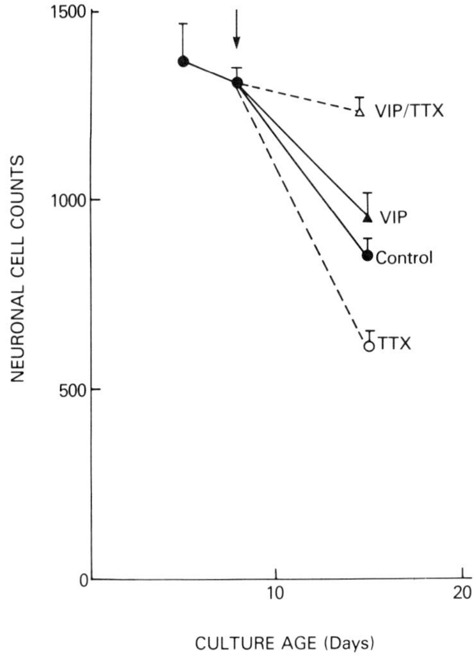

Figure 18-6. Effect of VIP and TTX on the survival of cultured spinal cord neurons. Treatment was started on day 8 as indicated by the *arrow*. Addition of 0.1 n*M* VIP and 1 μ*M* TTX prevented the decrease in neuronal cell counts that was produced by TTX treatment alone and that which occurred in control cultures during development. These results are similar to that presented for conditioned medium, as shown in figure 18-3. Addition of VIP alone had no significant effect on the number of surviving neurons. Each value is the mean of four dishes ± SEM.

the CNS (Eckenstein and Baughman, 1984) and the peripheral nervous system (PNS) (Lundberg et al., 1981). Since a population of cholinergic neurons appears to be among those that were affected by electrical blockade, we thought that these cholinergic neurons could be releasing VIP during electrical activity and thereby influence their own survival as well as that of neighboring neurons. 4) A number of polypeptide growth factors have been shown to be important for the regulation of a wide variety of cell types (see James and Bradshaw, 1984) including glia (Giulian et al., 1985). 5) Low molecular weight substances that possess neurotrophic activity have been reported to influence neurodevelopment (Muller et al., 1984; Varon et al., 1984; McManaman et al., 1985). These substances have not been fully characterized, but it appears that some of these are peptides.

Direct evidence of the effects of VIP on neuronal survival is presented in figure 18-6. Addition of 0.1 n*M* VIP to TTX-treated cultures prevented the neuronal deficits produced by electrical blockade and apparently also the

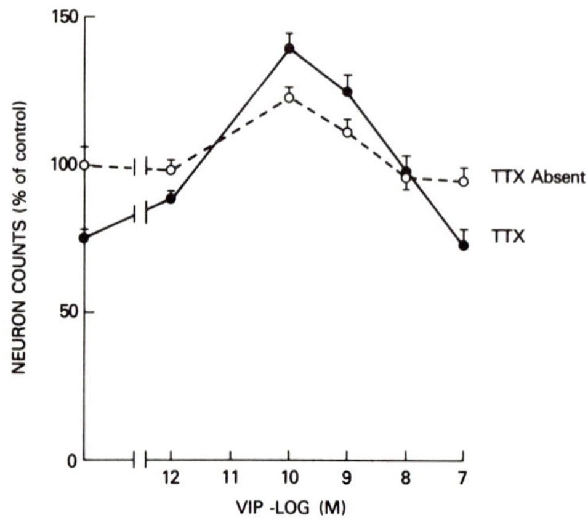

Figure 18-7. Concentration–effect study of VIP on neuronal cell counts in spinal cord cultures. VIP was added on day 9 to either electrically active (TTX absent) or TTX-treated cultures. After 5 days, the cultures were fixed and counted. Each point is the mean of 6–8 determinations from two experiments. From Brenneman and Eiden (1986).

Table 18-1. Comparison of VIP antiserum and tetrodotoxin on neuronal cell death.

Treatment[a]	Neurons per 100 fields	% of control
Control	1132 ± 38	100
VIP antiserum	890 ± 14*	79 ± 1*
TTX	902 ± 24*	80 ± 2*
VIP antiserum and TTX	859 ± 16*	76 ± 2*
Normal rabbit serum (0.25%)	1145 ± 62	101 ± 5

[a] VIP antiserum was added to dissociated spinal cord cultures 24 hours after plating. Sufficient antiserum was added to bind 300 pg of VIP. Cultures were fixed after 12 days of treatment. Each value is the mean ± SEM. From Brenneman and Eiden (1986).
*Significantly different ($p < 0.01$) from control.

death that occurred naturally. Concentration–effect studies (figure 18-7) indicated that 0.1 n*M* VIP was the optimum concentration, with higher amounts producing no sparing of neurons.

The role of VIP in determining neuronal survival has been shown also with the use of substances that block the action of VIP. Exposure of the cultures to VIP antiserum was found to reduce the number of neurons to the same extent as treatment with TTX (table 18-1). Thus, by binding VIP with antibodies throughout development in culture, neuronal death appears to be accelerated. However, it is possible that the group of neurons that are affected by TTX may not be precisely the same population affected by VIP. In addition to the VIP antiserum effects, a carboxyl fragment of VIP was used to study structure–activity relationships. A fragment of the VIP mole-

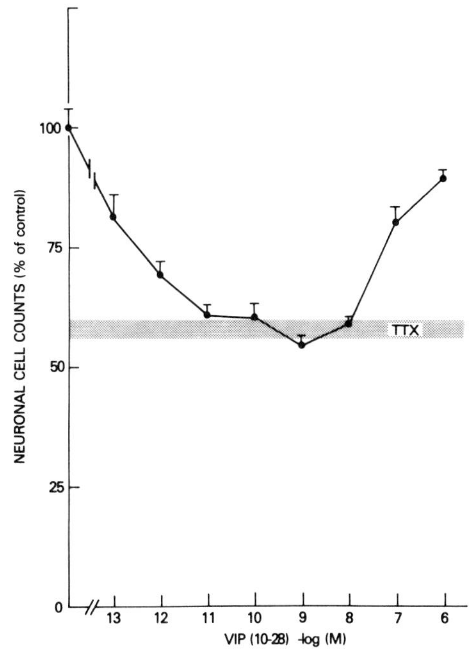

Figure 18-8. Neuronal cell counts after chronic treatment with VIP (10-28) during development. The carboxyl fragment of VIP was added to the cultures 24 hours after plating. Tratment was continued until day 13, when the cultures were fixed for counting. The *shaded area* indicates the neuronal cell counts from cultures treated with TTX only. Each point is the mean of 4–6 dishes ± SEM.

cule, VIP(10-28) has been shown to inhibit VIP stimulation of adenylate cyclase (Watling and Dowling, 1983; Bissonnette et al., 1984). This fragment of VIP does not have any demonstrable effect itself on cAMP formation. More recent evidence in the HT 29 human colonic adenocarcinoma cell line indicated that VIP(10-28) inhibits VIP binding in both intact cells and in membranes (Turner et al., 1985). In addition, the peptide fragment was found to inhibit VIP-stimulated adenylate cyclase, but it did not have any effect on basal activity or increases produced with forskolin. Thus, it appears that VIP(10-28) has the properties of a receptor antagonist for VIP. Addition of this putative VIP receptor antagonist to dissociated spinal cord cultures produced a concentration-dependent decrease in neuronal cell counts (figure 18-8). It is interesting that, when higher concentrations of VIP(10-28) were used, less cell death was produced. It may be that, at higher concentrations, VIP(10-28) has some agonist activity. Thus, the developmental effects of the native peptide are perhaps reversed by the presence of a fragment of the same peptide. Such a relationship could have biological importance if VIP were degraded to a fragment similar to VIP(10-28) in vivo. Such a mechanism could be a control point for the regulation of the action of this peptide during development. Other peptides that are closely homologous to VIP have been tested for trophic activity (Brenneman and Eiden, 1986). Secretin and PHI-27 (peptide, histidylisoleucine amide) did not increase the number of neurons in TTX-treated cultures. Therefore a high degree of structural specificity is apparent for the effects of VIP on neuronal survival during electrical blockade.

The mechanism by which VIP increases the number of surviving neurons in dissociated spinal cord cultures is unknown. Our current working hypothesis is that VIP acts to increase neuronal survival by stimulating the release of neurotrophic substances synthesized by astroglia (Brenneman et al., 1985a). Several lines of evidence have led us to this hypothesis. Functional VIP receptors have been observed on cultured astrocytes (Magistretti et al., 1983). Substances that affect neuronal survival of neurons derived from the CNS have been shown to be present in CM from glial cultures (Banker, 1980; Eagleson et al., 1985; Rudge et al., 1985). We would propose that VIP stimulates the release of trophic substances that may be important for the survival and maintenance of neurons throughout development and aging. Preliminary studies with CM obtained from glial cultures of dissociated spinal cord support this indirect-action hypothesis for VIP (Brenneman et al., 1985a). A comparison of CM obtained from VIP-stimulated glial cultures versus those without VIP has shown that the CM from VIP-treated cultures has more survival-promoting activity. These CM were tested on spinal cord cultures that had reduced number of astrocytes and that did not respond to VIP alone. The nature of these glial-derived substances that are stimulated by VIP is unknown.

5. PRACTICAL CONSIDERATIONS: NEUROPEPTIDES AND DEVELOPMENT

Our investigation of the biological activity of neuropeptides during development has revealed a number of characteristics that should be considered in experimental design. The following technical aspects of neuropeptide pharmacology will serve as examples of the experimental variables that must be considered in this area: effective concentration, developmental sensitivity, cell interactions, endogenous interference, and degradation.

Often the natural inclination in exploring the properties of a new biological agent is to expose the test system to a high concentration of the substance. Whereas this may be appropriate with some agents, for neuropeptides it may produce misleading negative results. This problem is evident from other studies of peptides that have shown concentration–effect curves that have an inverted-U shape. Investigation of the electrophysiological effects of somatostatin on cultured cortical neurons showed that 100-p*M* concentrations were effective in producing depolarizations (Delfs and Dichter, 1983), whereas increasing the amount of peptide attenuated this response. This gradual loss of effect progressed until at micromolar concentrations, no response was observed. This loss of biological activity at high concentrations has been observed in VIP effects on serotonin (5-HT) binding in hippocampus (Rostene et al., 1983) and in VIP-mediated increases in choline acetyltransferase (CAT) activity (Luine et al., 1984; Brenneman, 1986). The effects of VIP on neuronal cell counts during development also exhibit this type of dose–response relationship (Brenneman and Eiden, 1986). The reason for this apparent generalized phenomenon is not known. The possibilities that have been considered include receptor desensitization and the activation of other receptors that have regulatory roles. Clearly, peptides should be tested over a broad range of concentrations, including amounts that are subnanomolar.

Another aspect of neuropeptide research that became evident during the course of our studies with VIP was the issue of developmental sensitivity. Many developmental studies have indicated the existence of critical periods during which a system is vulnerable or accessible to pharmacological agents. For any process that is suspected of being affected by neuropeptides, the period during which that process is changing during development should be determined. Our studies on the effects of VIP on CAT activity (Brenneman, 1986) showed that, when VIP was applied to cultures before or after a "critical period" in development, during which neuronal cell survival and CAT activity were rapidly changing, no effect was observed. Therefore it is essential to know the developmental characteristics of the test system being studied before evaluating the effects of a neuropeptide.

Possible cell interactions should be considered when exploring the biological effects of neuropeptides. Responses to peptides may involve more than one cell type. When peptides are involved in an interaction between developing neurons and glia, both cell types must be present in order to demonstrate such an effect. Our studies with VIP suggest that, when the number of nonneuronal cells is too low in the test system, no survival-promoting effects of VIP are observed (Brenneman et al., 1985a). This possibility encourages the use of complex test systems like the dissociated spinal cord culture system. In this model system, a great variety of neuronal and nonneuronal cell types are present. Even though most of these cell types have not been fully characterized, this complexity allows for a variety of cellular interactions that may be required to reveal the role of neuropeptides. Once evidence of biological activity is found, the model system can be simplified to allow for more detailed mechanistic description.

The presence of endogenous peptides in the test system may interfere with the screening for biological effects of exogenous neuropeptides. When the substances to be tested is spontaneously released into the test system from endogenous sources, the addition of exogenous peptide could lead to no effect or perhaps to an opposite action from the physiological effect because of the unusual inverted-U shape of the concentration–effect curves. This problem was encountered in our studies of VIP and neuronal survival (Brenneman and Eiden, 1986). The problem of endogenous interference was approached by either preventing the release of VIP with TTX or by blocking its action with antibodies or receptor antagonists.

The last technical consideration to be discussed is the issue of peptidergic degradation. Inactivation of exogenous neuropeptides has been observed (Keltz et al., 1980; Kessler et al., 1983). Protease inhibitors such as bacitracin or captopril are usually included in the systems to prevent degradation. However, this is not always necessary. Our studies on the developmental effects of VIP did not require the use of such inhibitors (Brenneman and Eiden, 1986). This is particularly worthy of note since biological effects were observed at picomolar concentrations of VIP.

6. MODEL OF ACTIVITY-DEPENDENT NEURONAL DEVELOPMENT

We would like to propose a model incorporating the findings described above as well as other observations and some inferences derived from them. Apart from providing a convenient and concrete summary of results, such models may serve a heuristic purpose in suggesting new approaches and experiments.

The model shown in figure 18-9 indicates a conceptual framework incorporating these results. These observations seem to require two sorts of trophic interaction, both influenced by electrical activity. One is clear-cut and involves the activation of glial cells by VIP, which is released from neurons during spontaneous electrical activity. The other neurotrophic factor we have speculatively labeled retrophin, in analogy with a factor that is target tissue-derived and that acts *retrogradely* on neurons (Hendry and Hill, 1980). The elaboration of this factor is diminished by electrical activity. The action of this factor is revealed by the fact that, in the presence of VIP (or glia-released factor, GRF), survival is enhanced when electrical activity is blocked.

The two factors act together in a highly nonlinear, synergistic fashion. Retrophin by itself has essentially no effect. We reach this conclusion from the fact that TTX-treated cultures (high retrophin activity, but no VIP effect) are no different from electrically active cultures treated with VIP blockers (low retrophin activity, no VIP effect). Similarly, VIP-stimulated trophic activity is more effective in the presence of retrophin. This conclusion is based on the observations of a large effect of VIP on TTX-treated cultures as compared to the intermediate effect of VIP in the absence of retrophin. Thus, control cultures (VIP action but low retrophin) have more neurons than electrically active. VIP-blocked cultures (neither VIP nor retrophin action), but some cell death does occur. This is in contrast to the results with TTX-treated cultures in the presence of exogenous VIP (both VIP and retrophin factors present) in which cell death does not occur. This model has utility in suggesting additional experiments.

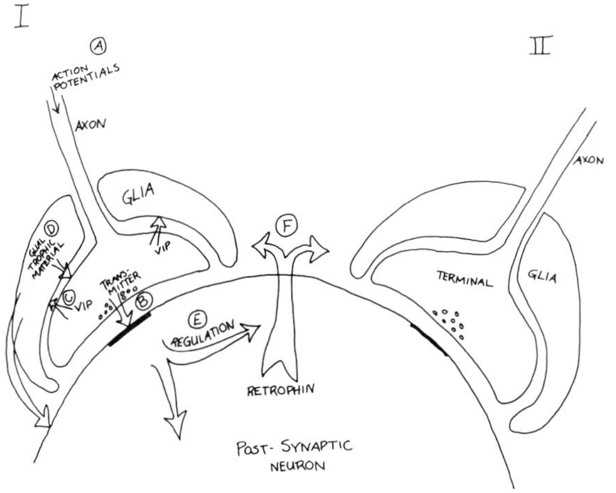

Figure 18-9. Model of activity-related trophic interactions between neurons and glia. The sequence of events during propagation of an action potential in one of the presynaptic fibers is as follows: (*A*) An action potential invades axon I. (*B*) Neurotransmitter is released from the nerve terminal and stimulates the postsynaptic cell. (*C*) Vascattive intestinal peptide (VIP) is simultaneously released and activates receptors on glia. (*D*) VIP-stimulated glial cells release a neurotrophic substance (glial-released factor, GRF) (*E*) An action potential, generated in the postsynaptic cell, regulates the synthesis and/or release of a retrograde trophic substance, retrophin. We assume that the regulatory effect of the action potential is to decrease the elaboration of retrophin. (*F*) The effect of retrophin may be either positive and supportive of neuronal and axonal development or negative and inhibitory for such development. We assume here a positive character for retrophin, and we further assume that the regulatory effect of the postsynaptic action potential is to *decrease* the elaboration of retrophin. GRF acts synergistically with retrophin to provide neuronal development and synaptic stabilization.

1) What is the site of action of GRF? Does it act on neuronal cell bodies, on axons, or perhaps on both? Experiments analogous to those done by Campenot (1981) with NGF should be possible.
2) Direct experiments are required to document the positive or negative neuronotrophic nature of retrophin.
3) More precise experiments should be done dealing with the interaction between GRF and retrophin.

The model presented in figure 18-10 explores the consequences of one assumption concerning the action of both trophic materials, namely, that the time course of the modulation of their elaboration by electrical activity may be rapid and coordinated; that is, we hypothesize that the release of GRF is increased following a neuronal action potential and declines with a time course of less than a second. Retrophin we suppose to be decreased following a postsynaptic action potential and recovers again with a time course of about a second or less. The critical interaction here is the temporal coordination between activity-mediated decreases in retrophin and activity-dependent in-

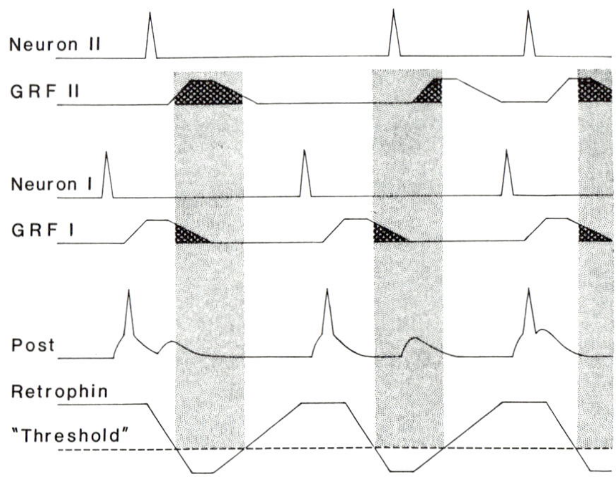

Figure 18-10. Scheme illustrating the consequences of relatively rapid kinetics for activity-related regulation of the glial-released factor (GRF, see steps *C* and *D* from figure 18-9) and retrophin (see steps *E* and *F* from figure 18-9). Neurons I and II propagate action potentials at approximately the same rate, but not synchronously (*top trace* is for neuron II and the *third trace* for neuron I). Following the action potential in each of these neurons, there is a transient local increase in GRF. Neuron I excites the postsynaptic cell (*Post*) and elicits an action potential while the input to the postsynaptic cell from neuron II is subthreshold. A decrease in retrophin is produced by the suprathreshold input to the postsynaptic cell from neuron I, but no such decrease is caused by the subthreshold input from neuron II.

The *stippled bands* show the periods of diminished retrophin levels during which the GRF is inactive (see text for discussion of synergistic interaction between GRF and retrophin). *Cross-hatched areas* on GRF traces represent inactive GRF. Note that the total area of inactive GRF is larger for neuron II than for neuron I. Subthreshold inputs receive less positive trophic activity than do suprathreshold inputs. The time course of retrophin "recovery" and GRF release provides for periods during which both factors are present at relatively high concentrations in the local environment of inputs capable of eliciting an action potential.

creases in GRF release. An appropriate duration of depressed retrophin levels after an action potential is the essential element of this model. The time course of depression in retrophin elaboration following a postsynaptic action potential must be of the general magnitude of the interval between presynaptic action potentials. A decrease in retrophin levels after an action potential would decrease the probability of sufficient levels of this trophic substance being present when a subthreshold input occurs at another postsynaptic site. Presynaptic neurons that produce an action potential would have more retrophin when GRF is maximal because the activity-mediated decrease in retrophin levels would be less than that occurring with subthreshold inputs. If these hypotheses were correct, then the model would operate in such a manner as to provide the "selective stabilization" of synapses as suggested by Changeux and colleagues (1973), which must be an important mechanism in the use-dependent development of functionally appropriate synaptic networks. The key feature of selective stabilization is that synapses that become capable of activating a postsynaptic element serve to reduce the effectiveness of, and ultimately eliminate, synapses that are functional but that do not produce spike activity in the postsynaptic cell. The system shown in figures 18-9 and 18-10 would have this property (given the temporal assumptions made above):

1) An action potential in neuron I ensures an increase in GRF, which interacts with the ambient retrophin to give an increment in survival-promoting action affecting neuron I.
2) Input from neuron I produces an action potential in the postsynaptic cell, which results in a corresponding decrement in retrophin.
3) At some later period, an action potential in neuron II causes a local increase in GRF, which interacts with a somewhat depressed (although recovering) level of retrophin to give a lesser increment in survival-promoting action for neuron II than in (1) above.
4) Neuron II does not elicit an action potential in the postsynaptic cell so that retrophin levels continue to recover during and after the action potential in neuron II.
5) When another action potential occurs in neuron I, retrophin levels will have recovered more than they had when neuron II fired. This situation will on average necessarily favor neuron I over neuron II and tend to make the neuron, which first successfully activates the postsynaptic cell, become increasingly dominant over other initially unsuccessful inputs. Those inputs activated together will tend to produce postsynaptic activation and to reduce the effectiveness of other noncoordinately activated inputs. This is, essentially, the process of "selective stabilization".

We have proposed an integrative model of both glial-derived and neuron-derived trophic support that may help explain neuronal development. The glial-released trophic substances may be triggered by peptide signals such as VIP. Dissociated cell cultures provide a useful model for studies of complex interactions because the system has the cellular components in an accessible form. Furthermore, by manipulating the cellular composition, mechanisms important in developmental regulation can be investigated.

ACKNOWLEDGMENTS

We wish to thank Dr. George Foster and Dr. Elaine Neale for providing immunocytochemistry expertise and Linda Bowers for help with the photography. Antibody against neuron-specific enolase was generously provided by Dr. Paul Marangos, and that against neurofilament protein, by Dr. Doris Dahl.

REFERENCES

Banker, G.A. (1980) Trophic interactions between astroglial cells and hippocampal neurons in cultures. Science 209:809–810.

Ben Ari, Y., Krnjevic, K., Reinhardt, W., and Ropert, N. (1981) Intracellular observations on the disinhibitory action of acetylcholine in the hippocampus. Neuroscience 6:2475–2484.

Bignami, A., Eng. L.F., Dahl, D., and Uyeda, C.T. (1972) Localization of the glial fibrillary acidic protein in astrocytes by immunofluorescence. Brain Res. 43:429–435.

Bissonnette, B.M., Collen, M.J., Adachi, H., Jensen, R.T., and Gardner, J.D. (1984) Receptors for vasoactive intestinal peptide and secretin on rat pancreatic acini. Am. J. Physiol. 246:G710–G717.

Bottenstein, J., Skaper, S., Varon, S., and Sato, G. (1980) Selective survival of neurons from chick embryo sensory ganglionic dissociates utilizing serum-free supplemented medium. Exp. Cell Res. 125:183–190.

Brenneman, D.E. (1986) Role of electrical activity and trophic factors during cholinergic development in dissociated cultures. Can. J. Physiol. Pharmacol. 64:356–362.

Brenneman, D.E., and Eiden, L.E. (1986) Vasoactive intestinal peptide and electrical activity influence neuronal survival. Proc. Natl. Acad. Sci. USA 83:1159–1162.

Brenneman, D.E., and Nelson (1985) Neuronal development in culture: role of electrical activity. In: Bottenstein, J.E., and Sato, G. (eds.) Cell culture in the neurosciences. New York: Plenum, pp. 289–316.

Brenneman, D.E., Neale, E.A., Habig, W.H., Bowers, L.M., and Nelson, P.G. (1983) Developmental and neurochemical specificity of neurochemical deficits produced by electrical impulse blockade in dissociated spinal cord cultures. Dev. Brain Res. 9:13–27.

Brenneman, D.E., Fitzgerald, S., and Nelson, P.G. (1984) Interaction between trophic action and electrical activity in spinal cord cultures. Dev. Brain Res. 15:211–217.

Brenneman, D.E., d'Autremont, S.W., and Neale, E.A., (1985a) Non-neuronal cells mediate VIP stimulation of neuronal survival during development. Neurosci. Soc. Abstr. 11:1148.

Brenneman, D.E., Eiden, L.E., and Siegel, R.E. (1985b) Neurotrophic action of VIP on spinal cord cultures. Peptides 6:35–39.

Brenneman, D.E., Fitzgerald, S., and Litzinger, M.J. (1985c) Neuronal survival during electrical blockade is increased by 8-bromo cyclic adenosine 3′, 5′ monophosphate, J. Pharmacol. Exp. Ther. 233:402–408.

Campenot, R.B. (1981) Regeneration of neurites in long-term cultures of sympathetic neurons deprived of nerve growth factor. *Science* 214:579–581.

Changeux, J.-P., Courrege, P., and Danchin, A. (1973) A theory of the epigenesis of neuronal networks by selective stabilization of synapses. Proc. Natl. Acad. Sci. USA 70:2974–2978.

Creazzo, T.L., and Sohal, G.S. (1979) Effects of chronic injections of α-bungarotoxin on embryonic cell death. Exp. Neurol. 66:135–145.

Dahl, D., and Bignami, A. (1977) Preparation of antisera to neurofilament protein from chicken brain and human sciatic nerve. J. Comp. Neurol. 176:645–658.

Delfs, J.R., and Dichter, M.A. (1983) Effects of somatostatin on mammalian cortical neurons in culture: physiological actions and unusual dose response characteristics. J. Neurosci. 3:1176–1188.

Dodd, J., Dingledine, R., and Kelly, J.S. (1981) The excitatory action of acetylcholine on hippocampal neurons of the guinea pig and rat maintained in vitro. Brain Res. 207:109–127.

Eagleson, K.L., Raju, T.R., and Bennett, M.R., (1985) Motoneurone survival is induced by immature astrocytes from developing avian spinal cord. Dev. Brain Res. 17:95–104.

Eccleston, P.A., and Silberberg, D.H. (1984) The differentiation of oligodendrocytes in a serum-free hormone-supplemented medium. Dev. Brain Res. 16:1–9.

Eckenstein, F., and Baughman, R.W. (1984) Two types of cholinergic innervation in cortex, one co-localized with vasoactive intestinal polypeptide. Nature 309:153–155.

Eiden, L.E., Nilaver, G., and Palkovits, M. (1982) Distribution of vasoactive intestinal polypeptide (VIP) in the rat brain stem nuclei. Brain Res. 231:472–477.

Giulian, D., Tomozawa, Y., Hindman, H., and Allen, R.L. (1985) Peptides from regenerating central nervous system promote specific populations of macroglia. Proc. Natl. Acad. Sci. USA 82:4287–4290.

Go, V.L.W., and Yaksh, T.L. (1980) Vasoactive intestinal peptide (VIP) and cholecystokinin CCK-8 in cat spinal cord and dorsal root ganglion: release from cord by peripheral nerve stimulation. Regul. Pept. 1:43.

Hamburger, V. (1975) Cell death in the development of the lateral motor column of chick embryo. J. Comp. Neurol. 160:535–546.

Hendry, I.A., and Hill, C.E. (1980) Retrograde axonal transport of target tissue-derived macromolecules. Nature 287:647–649.

Hirayama, M., Silberberg, D.H., Lisak, R.P., and Pleasure, D.E. (1983) Long-term culture of oligodendrocytes isolated from rat corpus callosum by Percoll density gradient: lysis by polyclonal antigalactocerebroside serum. J. Neuropathol. Exp. Neurol. 42:16–28.

Hsu, S.-M., Raine, L., and Fanger, H. (1981) Use of avidin–biotin–peroxidase complex (ABC) in immunoperoxidase technique: a comparison between ABC and unlabeled antibody (PAP) procedures. J. Histochem. Cytochem. 29:577–580.

James, R., and Bradshaw, R.A. (1984) Polypeptide growth factors. Annu. Rev. Biochem. 53:259–292.

Keltz, T.N., Straus, E., and Yalow, R.S. (1980) Degradation of vasoactive intestinal polypeptide by tissue homogenates. Biochem. Biophys. Res. Commum. 92:669–674.

Kessler, J.A., Adler, J.E., and Black, I.B. (1983) Substance P and somatostatin regulate sympathetic noradrenergic function. Science 221:1059–1061.

Koh, S.-W.M., Kyritsis, A., and Chader, G.J. (1984) Interaction of neuropeptides and cultured glial (Muller) cells of the chick retina: elevation of intracellular cyclic AMP by vasoactive intestinal peptide and glucagon. J. Neurochem. 43:199–203.

Landmesser, L., and Pilar, G. (1974) Synaptic transmission and cell death during normal ganglionic development. J. Physiol. 241:737–749.

Loren, I., Emson, P.C., Fahrenkrug, J., Björklund, A., Alumets, J., Hakanson, R., and Sundler, F. (1979) Distribution of vasoactive intestinal polypeptide in the rat and mouse brain. Neuroscience 4:1953–1976.

Luine, V.N., Rostene, W., Rhodes, J., and McEwen, B.S. (1984) Activation of choline acetyltransferase by vasoactive intestinal peptide. J. Neurochem. 42:1131–1134.

Lundberg, J.M., Anggard, A., Emson, P., Fahrenkrug, J., and Hokfelt, T. (1981) Vasoactive intestinal polypeptide and cholinergic mechanisms in cat nasal mucosa: studies on choline acetyltransferase and release of vasoactive intestinal peptide. Proc. Natl. Acad. Sci. USA 78:5255–5259.

Magistretti, P.J., Manthorpe, M., Bloom, F.E., and Varon, S. (1983) Functional receptors for vasoactive intestinal polypeptide in cultured astroglia from neonatal rat brain. Regul. Pept. 6:71–80.

McCarthy, K.D., and De Vellis, J. (1980) Preparation of separate astroglial and oligodendroglial cell cultures from rat cerebral tissue. J. Cell Biol. 85:890–902.

McCarthy, K., and Partlow, L. (1976) Separation of pure neuronal and non-neuronal cultures from embryonic chick sympathetic ganglia: a new method based on both differential cell adhesion and the formation of homotypic neuronal aggregates. Brain Res. 114:391–414.

McManaman, J.L., Smith, R.G., and Appel, S.H. (1985) Low-molecular-weight peptide stimulates cholinergic development in ventral spinal cord cultures. Dev. Biol. 112:248–252.

Muller, H.W., Beckh, S., and Seifert, W. (1984) Neurotrophic factor for central neurons. Proc. Natl. Acad. Sci. USA 81:1248–1252.

Neale, E.A., Matthew, E., Zimmerman, E.A., and Nelson, P.G. (1982) Substance P-like immunoreactivity in neurons in dissociated cell cultures of mammalian spinal cord and dorsal root ganglia. J. Neurosci. 2:169–177.

Neale, E.A., Oertel, W.H., Bowers, L.M., and Weise, V.K. (1983) Glutamate decarboxylase immunoreactivity and γ-(^{3}H) aminobutyric acid accumulation within the same neurons in dissociated cell cultures of cerebral cortex. J. Neurosci. 3:376–382.

Neale, E.A, Bruce, G., Matthew, E., d'Autremont, S.W., Strauss, W., Hersh, L.B., and Nelson, P.G. (1985) Cholinergic neurons demonstrated by immunohistochemistry in mouse spinal cord cultures. Neurosci Soc. Abstr. 11:862.

Pickel, V.M., Reis, D.J., Marangos, P.J., and Zomzely-Neurath, C. (1975) Immunocytochemical localization of nervous system-specific protein (NSP-R) in rat brain. Brain Res. 105:184–187.

Pittman, R.H., and Oppenheim, R.W. (1978) Neuromuscular blockade increases motoneurone survival during normal cell death in the chick embryo. Nature 271:364–366.

Pun, R.Y.K., Marshall, K.C., Hendelman, W.J., Guthrie, P.B., and Nelson, P.G. (1985) Noradrenergic responses of spinal neurons in locus coeruleus–spinal cord co-cultures. J. Neurosci. 5:181–191.

Raff, M.C., Mirsky, R., Fields, K.L., Lisak, R.P., Dorfman, S.H., Silberberg, D.H., Gregson,

N.A., Leibowitz, S., and Kennedy, M. (1978) Galactocerebroside is a specific cell-surface antigenic marker for oligodendrocytes in culture. Nature 274:813–816.

Raju, T.R., Bignami, A., and Dahl, D. (1981) In vivo and in vitro differentiation of neurones and astrocytes in the rat embryo: immunofluorescence study with neurofilament and glial filament antisera. Dev. Biol. 85:344–357.

Ransom, B.R., Neale, E., Henkart, M., Bullock, P.N., and Nelson, P.G. (1977) Mouse spinal cord in cell culture. I. Morphology and intrinsic neuronal electrophysiological properties. J. Neurophysiol. 40:1132–1150.

Rostene, W.H., Fischette, C.T., and McEwen, B.S. (1983) Modulation by vasoactive intestinal peptide (VIP) of $serotonin_1$ receptors in membranes from rat hippocampus. J. Neurosci. 3:2414–2419.

Rougon, G., Noble, M., and Mudge, A.W. (1983) Neuropeptides modulate the β-adrenergic response of purified astrocytes in vitro. Nature 305:715–717.

Rudge, J.S., Manthorpe, M., and Varon, S. (1985) The output of neuronotrophic and neurite-promoting agents from rat brain astroglial cells: a microculture method for screening potential regulatory molecules. Dev. Brain Res. 19:161–172.

Segal, M. (1983) Rat hippocampal neurons in culture: responses to electrical and chemical stimuli. J. Neurophysiol. 50:1249–1264.

Turner, J.T., Ray-Prenger, C.J., and Bylund, D.B. (1985) VIP(10–28) is an antagonist of VIP adenylate cyclase stimulation of HT29 cells. Neurosci. Soc. Abstr. 11:414.

Varon, S., Skaper, S.D., Barbin G., Selak, I., and Manthorpe, M. (1984) Low molecular weight agents support survival of cultured neurons from the central nervous system. J. Neurosci. 4:654–658.

Vernadakis, A., and Arnold, E.B. (1980) Age-related changes in neuronal and glial enzyme activites. In: Fedoroff, S., and Hertz, L. (eds.) Advances in cellular neurobiology. New York: Academic Press, pp. 229–283.

Watling, K.J., and Dowling, J.E. (1983) Effects of vasoactive intestinal peptide and other peptides on cyclic AMP accumulation in intact pieces and isolated horizontal cells of the teleost retina. J. Neurochem. 41:1205–1213.

Westbrook, G.L., and Brenneman, D.E. (1984) The development of spontaneous electrical activity in spinal cord cultures In: Caciagli, E., Giacobini, E., and Paoletti, R. (eds.) Developmental neurosciences: physiological, pharmacological and clinical aspects. Amsterdam: Elsevier, pp. 11–17.

19. TROPHIC AND TOXIC MECHANISMS IN NEURONAL SURVIVAL

SILVIO VARON, STEPHEN D. SKAPER AND MARSTON MANTHORPE

1. INTRODUCTION

Forty years of investigations of the nerve growth factor (NGF) phenomenon have spurred a more recent search for other neuronotrophic factors (NTFs)—i.e., specific proteins required for the survival, functional maintenance, and general growth capability of selected populations of nerve cells (for reviews, see Varon and Adler, 1981; Varon, 1985). The search, thus far, has led to the isolation of three new NTFs: a pig brain-derived neuronotrophic factor (BDNF), addressing dorsal root ganglionic neurons (Barde et al., 1982); a chick eye-derived ciliary neuronotrophic factor (chick CNTF), addressing the cholinergic neurons of ciliary ganglia (Barbin et al., 1984a; Manthorpe and Varon, 1985), but also neurons of dorsal root and sympathetic ganglia—the traditional targets of NGF; and a similar factor (rat CNTF) obtained from adult rat nerve (Manthorpe et al., 1986a). In vitro cultures of dissociated neurons have provided the best tools for the recognition, quantitative evaluation, and further investigation of NTFs. The test neurons were collected at early *developmental* ages and mainly from peripheral neural tissues.

The likelihood that similar NTFs may play a role with regard to *adult* neurons in the central (CNS) as well as the peripheral (PNS) nervous systems has encouraged a "trophic hypothesis" for CNS neurons, which proposes an involvement of NTFs in neural regeneration (e.g., Varon, 1977; Björklund and Stenevi, 1981; Crutcher and Collins, 1982; Varon et al., 1983 and 1984a; Gage et al., 1984), certain neurodegenerative disorders (Appel, 1981; Varon

Vernadakis, A. (ed.), *Model Systems of Development and Aging of the Nervous System*.

et al., 1982 and 1984a), and brain aging processes (Varon, 1975; Appel, 1981). Several recent in vivo findings in the adult rat support such a hypothesis. Mechanical injury causes a rise in CNTF, and possibly NGF, in the injured tissue and surrounding wound fluids in both PNS (Longo et al., 1983a and b) and CNS (Nieto-Sampedro et al., 1982 and 1983; Manthorpe et al., 1982 and 1983a). Local administrations of CNTF-containing fluids improve cholinergic sprouting from injured adult septal neurons (Schonfeld et al., 1981 and 1985) and from embryonic striatal grafts (Nieto-Sampedro et al., 1984) Most recently, Hefti (1985) has reported that NGF promotes *survival* of septal cholinergic neurons after fimbria–fornix lesion. Another quantitative study of a similar adult rat model has further demonstrated that continuous infusions of NGF (either intraparenchymally or intraventricularly) protect both cholinergic and noncholinergic septal neurons against axotomy-induced death, and may also promote intraseptal cholinergic sprouting (Varon et al., 1986a; Williams et al., 1986).

Neuronal cell cultures have shown that neuronal survival requires not only the appropriate neuronotrophic factors, but also several other components of the culture medium and substratum. In addition, some materials examined for putative trophic activities have revealed the occurrence of *neuronotoxic* agents as well (Manthorpe et al., 1982 and 1983a; Nieto-Sampedro et al., 1983). The emphasis on survival and survival-promoting (trophic) agents has yet to be matched by an interest in the converse situation, neuronal cell death and "death-promoting" agents. Developmental neuronal death is a well-recognized event (e.g., Cowan, 1973; Varon and Adler, 1980; Hamburger and Oppenheim, 1982; Cunningham, 1982; Cowan et al., 1984). In the adult animal, neuronal death may occur as a consequence of mechanical trauma (Feringa and Vahlsing, 1985; Gage et al., 1986), pathological situations (Appel, 1981; Varon et al., 1984a) and/or aging (Varon, 1975; Appel, 1981; Whitehouse et al., 1982). One of the challenges in modern cellular neurobiology will be to determine whether neuronal death is the consequence of deficient survival support (with the emphasis on trophic agents) or conversely results from the active involvement of neuronotoxic agents or conditions (with the trophic agents providing protection against them). The problem is discussed in this chapter in the light of recent findings concerning the roles of pyruvate and the peroxide-degrading enzyme catalase as agents promoting neuronal survival.

2. MULTIPLE REQUIREMENTS FOR NEURONAL SURVIVAL IN VITRO

In vivo, nerve cells are subject to influences, both stimulatory and inhibitory, derived from their physical environment (extracellular fluid, extracellular matrix) and their cellular environment (glial cells, presynaptic neurons, postsynaptic innervation targets). The same convergence of multiple influences occurs in vitro, as illustrated by the diagram in figure 19-1. Not only are neurons affected by culture medium, culture substratum, and cocultured

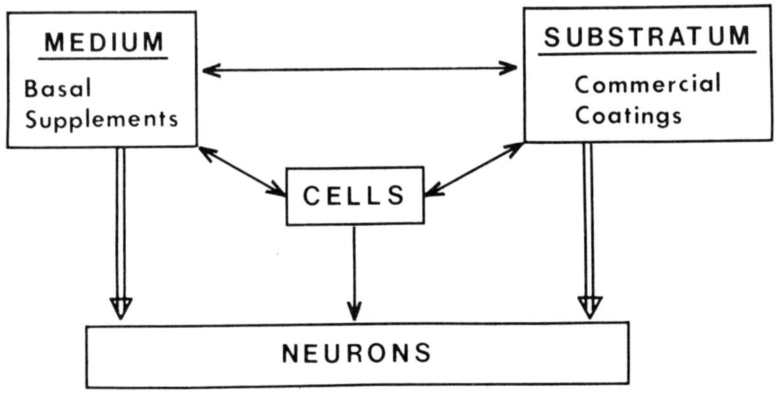

Figure 19-1. Neurons in vitro (and in vivo) are subject to multiple extrinsic influences from humoral, substratum-bound, and cellular sources.

cells, but medium and cultures are continuously modified by the activities of the living cells themselves. This means that, even in the best in vitro studies, we can at most define only the initial composition of medium and substratum and the initial number and nature of the cells seeded, but not the true, ever-changing environment in which the neurons under investigation are operating. In addition, since most investigators use prenatal neurons in their test cultures, the neurons themselves will undergo developmental changes as they age in vitro, and these changes may include the loss and/or the acquisition of neuronal requirements for survival and functional performance. It is important, when evaluating extrinsic influences on neuronal behaviors, to distinguish between responsiveness (i.e., the ability to react to extrinsic agents with a change in cell behavior) and dependence (i.e., the inability to perform a given task unless the extrinsic agent is available).

Some of these points are illustrated in figure 19-2, with regard to the dependence of neuronal survival on exogenous neuronotrophic factors (Skaper and Varon, 1985). Dorsal root ganglia (DRG) can be collected from chick embryos at different embryonic ages (E6–E15), dissociated, and preplated in serum-supplemented medium on tissue culture plastic to remove most of the nonneuronal cells. The enriched neuronal suspension is reseeded on polyornithine–laminin-coated substrata in serum-free, Dulbecco's modified Eagle's medium (DMEM) supplemented with the defined N1 mixture (Bottenstein et al., 1980). *With NGF* added to the medium, only about 40% of the 1000 neurons seeded (per 6-mm well) from E6 DRG will survive for even 24 hours, whereas 80%–90% of them do so from E8–E15 ganglia. Thus, some of the younger DRG neurons have survival requirements that are not met by NGF, at least under the culture conditions tested. If another neuronotrophic factor, *CNTF*, is supplied instead of NGF, neuronal survival remains negligible until E8 and then sharply increases, so that by E10 the

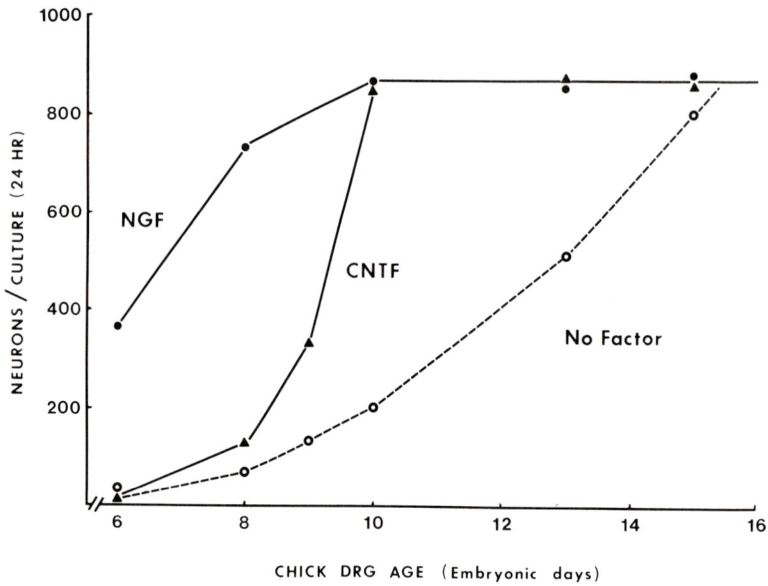

Figure 19-2. Chick embryo DRG neurons change the dependance of their 24-hour survival on exogenous neuronotrophic factors with increasing developmental age. *NGF*, nerve growth factor; and *CNTF*, ciliary neuronotrophic factor.

same maximal numbers of neurons (hence the same neurons) survive as with NGF. Such a behavior suggests that, between E7 and E10, DRG neurons acquire the ability to respond to CNTF while retaining their responsiveness to NGF and their dependence on either exogenous trophic factor for their survival in vitro. If *no trophic factor* is supplied, neuronal survival is negligible at the early ages, but progressively increases to reach by E15 the same maximal level as with NGF or CNTF. It appears, therefore, that the *dependence* on exogenous trophic factors for short-term survival by DRG neurons declines with developmental age, even though *responsiveness* to those factors may well persist for neuronal behaviors other than survival. The basis for this age-related decline in dependence on exogenous trophic factors for survival has not yet been elucidated. It is intriguing to speculate, among other possibilities, that the DRG neurons become increasingly resistant to some environmental toxic agent and, thus, lose their need for the protection against it, which neuronotrophic factors might be able to provide.

Neurons require extrinsic support for their in vitro survival by other extrinsic agents beside trophic factors, and in the past some such support was provided by supplementation of the culture medium with serum or other biological fluids. Bottenstein and Sato (1979) have designed a chemically defined supplement (designated as N1 or N2, depending on its transferrin concentration) capable of substituting for serum in the promotion of neuro-

Table 19-1. Neurons from different tissues and ages have different requirements for individual constituents of the serum-replacing N1 supplement.

	N1 ingredients		
Neuronal sources	Insulin	Transferrin	Selenium
Sympathetic ganglia			
Chick E8–10	×	—	×
Chick E11–15	×	×	×
Dorsal root ganglia			
Chick E8–15	×	×	—
Mouse neonate	×	×	—
Ciliary ganglia			
Chick E8	×	—	—
CNS			
Chick E8 forebrain	×	—	—
Rat E18 hippocampus	×	—	—
Rat E18 septum	×	—	—
Rat E18 striatum	×	—	—

blastoma cell *proliferation*. The N1 supplement has since then been shown also to replace serum with regard to the *maintenance* of postmitotic, primary neurons from peripheral (e.g., Bottenstein et al., 1980; Skaper et al., 1982 and 1984; Selak et al., 1983) and central (e.g., Skaper et al., 1979; Longo et al., 1982; Barbin et al., 1984b) neural tissues. Of the five N1 ingredients (insulin, transferrin, selenite, putrescine, and progesterone), only the first three appear to be important for short-term neuronal survival. Table 19-1 shows that different neurons have different requirements for even those three components, although insulin is needed by all neurons tested thus far. Also shown in table 19-1 is another example of developmental changes in survival requirements for exogenous agents. Neurons from chick sympathetic ganglia require insulin and selenium, but not transferrin, before E11. After E11 is reached, either in vivo or in vitro (e.g., by E8 cells maintained for 3 days in culture), the neurons will die in 24 hours if transferrin is not provided exogenously to them (Selak et al., 1983).

Beside appropriate media and medium supplements, culture substrata also have an important bearing on neuronal maintenance. Polyornithine or other polycationic substances confer considerable adhesiveness to a plastic substratum, and promote neuronal attachment (e.g., Yavin and Yavin, 1984; Letourneau, 1975; Varon, 1979). Further treatment of polyornithine with *laminin* or related polyornithine-binding neurite promoting factors (PNPFs) produces a substratum that is particularly conducive to neuritic outgrowth from the cultured neurons (e.g, Collins, 1978; Adler and Varon, 1980; Manthorpe et al., 1981 and 1983b; Wewer et al., 1983; Davis et al., 1985a–c). PNPFs and laminin also apparently promote survival of both central (Longo et al., 1982; Barbin et al., 1984b) and peripheral neurons (Edgar and

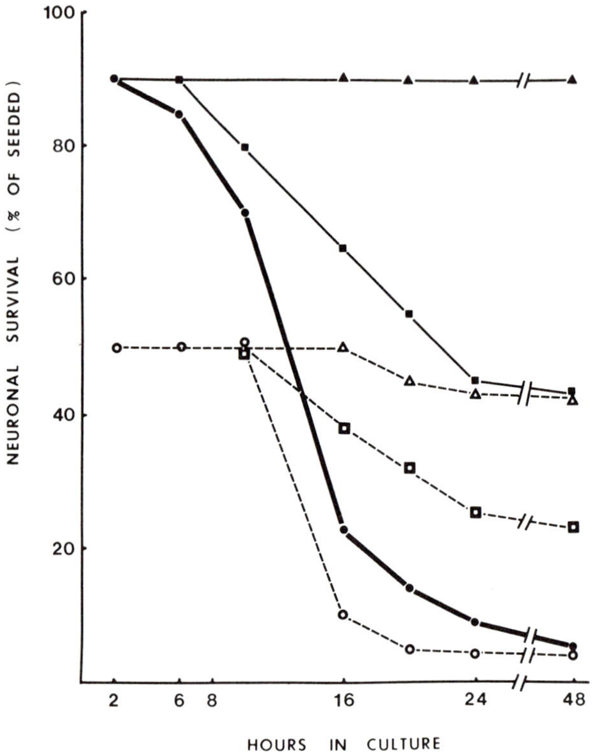

Figure 19-3. Effects of substrata, basal media, and media supplements on the attachment and survival of purified E8 chick DRG neurons in monolayer cultures: *Solid lines* and *solid symbols*, laminin-coated polyornithine substratum; *broken lines* and *open symbols*, uncoated polyornithine substratum; ▲, △, DMEM + N1 + NGF; ■, □, EMB + N1 + NGF; and ●, ○, either basal medium + N1, but no NGF.

Thoenen, 1982; Skaper and Varon, 1986). Figure 19-3 shows a temporal analysis of E8 chick DRG neurons cultured in different media on either polyornithine alone or polyornithine plus laminin (Skaper and Varon, 1986). The laminin-coated substratum (solid lines) supports maximal neuronal survival in DMEM/N1 medium supplemented with NGF, partial support when Eagle's basal medium (EBM) rather than DMEM is used with the same supplements, and only transient support when NGF is omitted. The same medium influences on survival are observed on plain polyornithine substrata, except that all survival levels are reduced by about 50%. Since this reduction is apparent as early as 2 hours after seeding, one may conclude that the survival promotion by laminin reflects a promotion of *initial attachment* rather than of the subsequent maintenance of the attached neurons. However, one should not exclude the possibility of differential effects of the laminin substratum on neuronal behaviors other than the initial attachment.

Table 19-2. Subpopulations of PNS neurons require exogenous pyruvate for their in vitro survival.

	24-hour neuronal survival (% of maximal)[a]		
	cCG8 +CNTF	cDRG8 +NGF	cDRG10 +NGF or CNTF
EBM			
+ —	0	0	0
+ Serum	100	60	60
+ Dialyzed serum	35	40	40
+ N1	35	60	60
DMEM			
+ —	0	0	10
+ N1	100	100	100
EBM/N1			
+ Fe^{3+}, SER, GLY	35	60	60
+ PYR	70	100	100
+ PYR + Fe^{3+} + SER	100	100	100

[a]cCG8, E8 chick ciliary ganglia; and cDRG8 and cDRG10, E8 or E10 chick dorsal root ganglia.

3. IMPORTANCE OF PYRUVATE FOR NEURONAL SURVIVAL IN VITRO

The above-mentioned difference in 24-hour survival (but not attachment) by DRG neurons cultured in EBM rather than DMEM (see figure 19-3) is one example of yet another extrinsic regulation, in this case by regular basal media ingredients rather than by media supplements such as trophic factors or N1 components. A series of studies has determined that *pyruvate* is the main constituent responsible for the differences among basal media with regard to both PNS and CNS neuronal support. The following is a brief review of such studies and their conclusions.

Table 19-2 summarizes some of the observations made with chick embryo PNS ganglionic neurons (Skaper et al., 1984; Skaper and Varon, 1986). E8 ciliary ganglionic (CG) neurons, cultured with their trophic factor CNTF and on a laminin substratum, fail to survive for even 24 hours in EBM alone. Serum supplementation provides for maximal survival. However, dialyzed serum only supports about one-third of the neurons, and the same partial support is achieved when N1 is used to replace the serum, indicating that N1 can replace the nondialyzable serum components but dialyzable serum constituents are separately required by a majority of CG neurons. Full survival is restored when N1 is used in conjunction with a different basal medium, DMEM. DMEM contains four ingredients that are absent in EBM: Fe^{3+}, serine, glycine, and pyruvate. Addition of the first three to EBM/N1 does not improve survival. Pyruvate, on the other hand, doubles the number of surviving neurons when added alone and restores full survival when added together with both Fe^{3+} and serine. Thus, the E8 chick CG neuronal population may comprise three subsets with different survival requirements. One is

Table 19-3. Exogenous pyruvate is also required by CNS neurons, and is responsible for the trophic activity of astroglia-conditioned media (ACM).

	24-hour neuronal survival (% of maximal)[a]			
A)	cFB	rHIPP	rSP	rSTR
EBM/N1	0	0	0	0
DMEM/N1	100	100	100	100
EBM/N1 + PYR	80	100	100	100
EBM/N1 + PYR + SER	100	100	100	100
ACM/N1	100	100	100	100
ACM-F_2/N1	100	100	100	100

B)	PYR nmol/ml	Activity TU/ml	PYR ED_{50} nmol/TU
DMEM/N1	910	27	33
EBM/N1 + 10^{-3} *M* PYR	960	28	34
ACM/N1	350	11	33
ACM/N1 + enzym. degradation (PDH)	44	1.2	37

cFB, E8 chick forebrain; rHIPP, rSP, and rSTR, E18 rat hippocampus, septum, and striatum. Data in B are derived from Selak et al (1985).

independent of pyruvate while the other two require it and, of those two, one also needs Fe^{3+} plus serine. Unlike CG neurons, neurons from chick embryo DRGs reveal, through a similar analysis, only two subsets of about equal sizes: one supported by EBM/N1 and thus pyruvate independent and the other requiring either DMEM/N1 or EBM/N1 supplemented with exogenous pyruvate. The same partial dependence on pyruvate is found with neurons from E8 DRGs supported by NGF, from E10 DRGs supported by NGF or CNTF, or from E15 DRGs unsupported by any exogenous trophic factor. The pyruvate requirement, then, appears to be unrelated to developmental age and to the need (or lack of a need) for an appropriate neuronotrophic protein factor.

Pyruvate dependence for neuronal survival is an even more prominent feature with neurons derived from CNS tissues (Selak et al., 1985). As illustrated in table 19-3, neurons from E8 chick forebrain, or E18 rat hippocampus, septum, and striatum—among others tested—fail to survive altogether in EBM/N1 but achieve maximal survival when cultured with DMEM/N1 or with EBM/N1 supplemented with pyruvate (the chick, but not the rat, CNS neurons include a small subset that needs serine in addition to pyruvate). Previous work (Manthorpe et al., 1982; Barbin et al., 1984b; Varon et al., 1984b; Rudge et al., 1985) had shown that the survival of CNS neurons in vitro can be supported by media conditioned over glial cell cultures. Table 19-3A illustrates this feature with regard to rat astroglia-conditioned medium (ACM), and shows that the trophic activity of ACM resides with low molecular weight substances that pass through ultrafiltration membranes having a 1000-Dalton cutoff (ACM-F_2). The main trophic substance in ACM was again identified as pyruvate (Selak et al., 1985). Table 19-3B compares pyruvate content, trophic activity, and their ratio (i.e., the pyruvate ED_{50} for CNS trophic support) in: 1) DMEM (which contains pyruvate as a normal ingredient), 2) EBM supplemented with pyruvate (at the same $10^{-3} M$ concentration as in DMEM), 3) ACM, and 4) ACM after treatment with the pyruvate-degrading enzyme, pyruvate dehydrogenase. Conditioning by the astroglial culture has provided an initially pyruvate-free EBM with a pyruvate content and a CNS trophic activity equal to about one-third those of DMEM, hence with the same ED_{50}. Pyruvate dehydrogenase treatment reduces by about 90% both pyruvate content and trophic activity, leaving unaffected the ED_{50} of the undegraded pyruvate.

Several questions arise from the findings that pyruvate is essential to the survival in vitro of all prenatal CNS and many embryonic PNS neurons tested thus far. Why are these neurons restricted in their capacity to either generate or effectively utilize endogenous pyruvate from the glucose abundantly available in their medium (e.g., EBM)? Does a similar pyruvate deficit occur with adult neurons and with in vivo situations? Is the contribution of pyruvate or pyruvate-like substances by glial cells to nerve cells an integral component of the glia–neuron functional unit concept, already proposed by Hydèn several years ago (c.f. reviews by Varon and Somjen, 1979; Manthorpe et al., 1986b)? We find it intriguing to speculate (Varon et al., 1986b and c; Manthorpe et al., 1986b) that adult neurons in situ may exceed their endogenous pyruvate capacity when taxed by excessive demand (e.g., functional overload) or restrictive circumstances (e.g., pathological situations) and may then signal to the surrounding astroglia their need for exogenous supplementation. It has been shown that astroglial cells store energy in the form of glycogen and can mobilize it (glycogenolysis) under external stimuli such as K^+, neurotransmitters, and neuropeptides—all signals of putatively neuronal origin (Magistretti et al., 1983).

Another important set of questions concerns the neuronal metabolic needs that must be met by exogenous pyruvate. We have shown that exogenous pyruvate can be replaced by α-ketoglutarate or oxaloacetate, but not by any other compounds related to the tricarboxylic acid cycle—a main source of cellular ATP (Facci et al., 1985). α-Keto acids other than pyruvate, α-ketoglutarate, and oxaloacetate can also substitute for those three compounds, but only when accompanied by certain α-amino acids, specifically those (and only those) that can transaminate with an α-keto acid to yield pyruvate, α-ketoglutarate, or oxaloacetate (Facci et al., 1985 and 1986). When ^{14}C-glutamate is used as a source of α-ketoglutarate in conjunction with an α-keto acid (e.g., phenylpyruvate) as an amino acceptor, $^{14}CO_2$ is released by CNS neurons in the same dose–response relationships as those needed for neuronal survival, thereby encouraging the view that pyruvate or its surrogates may exert their "trophic" role by meeting energy metabolism needs of the neurons (Facci et al., 1986). An additional observation was that the same $^{14}CO_2$ measurements apply to freshly generated CNS cell suspensions as to 24-hour monolayer CNS neuronal cultures, which suggests that the "pyruvate deficit" is a feature already intrinsic to the dissociated neurons rather than imposed on them by the culture conditions.

4. CNS-DIRECTED NEURONOTROPHIC PROTEINS AND CATALASE

One puzzling difference in the pyruvate needs of PNS and CNS neurons is that PNS neurons (table 19-2) require their appropriate protein neuronotrophic factors in order to benefit from exogenous pyruvate, whereas CNS neurons (table 19-3) survive with pyruvate in the absence of any recognized neuronotrophic factor (other than the insulin component of the N1 supplement, which is required even in the absence of glucose). Previous work with E4 chick spinal cord neurons has shown that these CNS neurons could be equally well supported by either the low molecular weight components of a glia-conditioned medium or by certain protein-like neuronotrophic agents (Manthorpe et al., 1982). One source of such CNS-directed neuronotrophic proteins was the fluid accumulating in vivo in experimental lesions of young or adult rat brains (Manthorpe et al., 1983a; Nieto-Sampedro et al., 1983), and similarly active proteins for rat CNS neurons were detectable in the cerebrospinal fluid from human head-trauma patients (Longo et al., 1984). Both types of would fluids also contain neuronotoxic proteins, which can be readily detected and measured by use of CNS neuronal microcultures supported by a constant and independent supply of low molecular weight (i.e., pyruvate) trophic agents (Manthorpe et al., 1982). While little research has yet been carried out with regard to those neuronotoxic factors, the ability of the same CNS neurons to be supported either by pyruvate or by some neuronotrophic protein independently from each other pointed to some common need by the neurons that either trophic agent was able to meet.

Since traumatic brain lesions are accompanied by considerable local hemorrhage, blood itself might conceivably be a source of the neuronotrophic proteins detected in a wound fluid. This line of investigation led to the examination of several blood samples and the recognition of neuronotrophic proteins residing in the red blood cells (RBCs) of all species tested (Williams et al., 1985). Human RBCs were chosen for purification of the active protein by successive steps of ion-exchange chromatography at two different pHs, ultrafiltration concentration, and preparative Immobiline isoelectric focusing (Walicke et al., 1986). The physicochemical properties of the purified trophic protein were then compared with those of known RBC-soluble proteins. The human RBC trophic protein was found to be nearly identical to *catalase* by isoelectric point (pH 6.6), native molecular weight (260 kd), and

Table 19-4. Catalase can substitute for exogenous pyruvate in the support of CNS and PNS neuronal survival in EBM/N1 medium.

		24-hour neuronal survival (% of maximal)		
		No supplement	Pyruvate (1 mM)	Catalase 30 μg/ml
cForebrain	—	0	100	105
rHippocampus	—	0	100	93
rSeptum	—	0	100	100
rStriatum	—	0	100	100
rCerebral cortex	—	0	100	100
cCG8	—	0	0	0
	+ CNTF	36	69	65
	+ CNTF + SER + FE^{3+}	36	100	98
cDRG8	—	0	0	0
	+ NGF	50	96	100
cDRG10	—	10	20	21
	+ NGF	60	100	95
	+ CNTF	62	100	98
	+ NGF + CNTF	100	100	100
cDRG15	—	60	100	100

c, chick embryo; and r, rat fetus.

subunit M_r (60 kd) on SDS–gel electrophoresis. The human RBC trophic protein displays catalase-like enzymatic activity, a commercial preparation of purified catalase conversely exhibits trophic activity for CNS neurons, and both proteins lose their trophic competence when treated with irreversible inhibitors of the enzymatic competence.

As shown in table 19-4, catalase faithfully reproduces the survival effects of pyruvate in all the culture situations in which CNS and PNS neurons have been examined thus far, including the concurrent requirement for neuronotrophic factors by PNS neurons and the recognition of neuronal subsets in CG and DRG populations. It is appropriate, therefore, to combine information and suggestions derived from either pyruvate or catalase in an attempt to speculate about the fundamental nature of neuronal survival (or death) regulation.

5. THE "OXYGEN RADICAL" HYPOTHESIS: A BASIS FOR NEURONOTOXIC AND NEURONOTROPHIC MECHANISMS

The only known function of catalase is the enzymatic breakdown of hydrogen peroxide (and, possibly, other peroxides). We can thus view this point as the main clue around which to organize our perceptions of neuronotoxic and neuronotrophic mechanisms and orient their future investigations. We *hypothesize* the following propositions, summarized diagrammatically in figure 19-4:

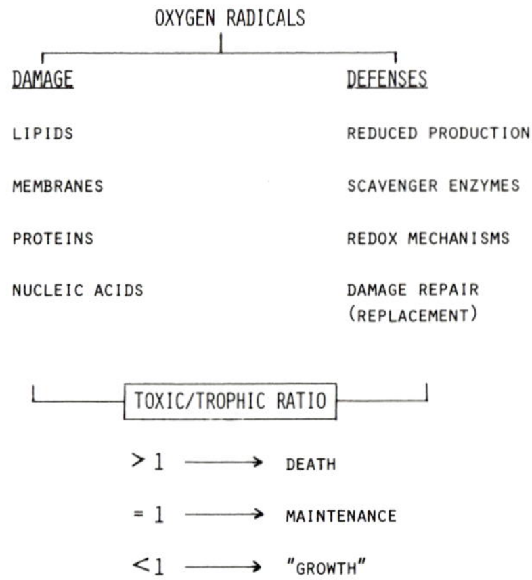

Figure 19-4. The balance between damages from and defenses against oxygen radicals (toxic-to-trophic ratio) may determine the fate of a nerve cell.

1) Cellular damage caused by *oxygen radicals* are at the core of neuronotoxic events. We speculate that a variety of noxious situations lead to neuronal damage via a common "second message" for cell death, the oxygen radicals.
2) *Oxygen radical-derived damage* may be increased by either an increase in oxygen radicals or a decrease in cellular defense mechanisms. Conversely, peroxidation damage would be reduced by either a decrease in oxygen radicals or an enhanced cellular tolerance to them.
3) *Neuronotoxic and neuronotrophic agents* may operate by altering the balance between oxygen radicals and cellular defenses to favor the former (toxic) or the latter (trophic), respectively. For example, catalase might conceivably reduce oxygen radicals in the medium, or at the neuronal cell surface (or even intracellularly); pyruvate may reduce the production of oxygen radicals by the cells themselves; neuronotrophic factors may enhance other defense processes, including anabolic activities to replace damaged cell constituents.

5.1 Oxygen radical damage

Oxygen radicals (and the cellular damage they may cause) are the topic of a vast literature (e.g., Pryor, 1982; Fridovich, 1983; Bors et al., 1984; Cutler, 1985), concerning such diverse areas as mutagenesis and cancer, environmental and radiation pollution, nutritional deficits, macrophage-mediated inflammation, and cellular aging, among many others. Free radicals also appear involved in neural tissue damage, including specific neuropathological situations such as retinal phototoxicity, postischemic brain injury, and brain aging processes (e.g., Vitorica et al., 1984; Halliwell and Gutteridge, 1985).

Free radicals are molecular species containing one (or more than one) unpaired orbital electron, and endowed with a wide range of reactive half-lives (from nanoseconds to nearly inert states). Molecular oxygen (O_2) has two relatively unreactive unpaired electrons, but may become the highly reactive free radical O_2^- when it acquires one electron, mostly from electrons leaking out of mitochondrial and microsomal electron-transport chains (e.g., ubiquinone $\rightarrow$ semiquinone $\rightarrow$ quinone). Interacting with biomolecules in the aqueous phases and membrane lipid core of the cell, O_2^- triggers a chain of organic free radicals (e.g., lipid peroxidation chain) that propagates their damage to membrane molecules, generates insoluble protein and lipid deposits (e.g., lipofuscin), and causes DNA breakage. Transition metal ions (e.g., Fe^{2+}, Cu^{2+}, and Mn^{2+}) can readily transfer single electrons, hence their role in metallo-oxidases. Metalloproteins such as hemoglobin, transferrin, and the metallo-oxidases keep the electrons from escaping to free solution. In intracellular and extracellular low molecular weight metallo-complexes, however, electrons donated by transition metals (e.g., $Fe^{3+} \rightarrow Fe^{2+} + e^-$) play major roles in peroxidation reactions, including those involving aromatic compounds (tyrosine and tryptophan, catecholamines), hydroquinones, and sulfhydromolecules. The following three basic reactions appear to be most important:

$$\text{I.}\quad \boxed{O_2^-} + 1\,e^- + H_2O \xrightarrow{\text{dismutases}} H_2O_2 + \tfrac{1}{2}O_2$$

$$\text{II.}\quad H_2O_2 + 1\,e^- \xrightarrow[Fe^{2+} \frown Fe^{3+}]{} HO^- + \boxed{\cdot OH}$$

$$\text{III.}\quad O_2^- - 1\,e^- \xrightarrow[Fe^{3+} \frown Fe^{2+}]{} O_2$$

Hydrogen peroxide, H_2O_2 (reaction I), is by itself relatively inert and can be readily degraded to H_2O and O_2 by the enzyme catalase. Its major involvement in free radical damage lies in its susceptibility to conversion (reaction II) into the hydroxyl radical, $\cdot OH$, probably the most reactive radical in aqueous phases. Generation of $\cdot OH$ depends on the availability of transition metals in their reduced form (e.g. Fe^{2+}), and is favored by O_2^- not only via the production of H_2O_2 (reaction I) but also via the ability of O_2^- to reduce Fe^{3+} into the electron-donor Fe^{2+} form (reaction III). In the latter capacity, O_2^- can be replaced by other reducing agents (e.g., ascorbic acid, when present at low concentrations).

5.2 Cellular defenses against radical damage

One main line of defense is provided by *scavanger enzymes*, which tend to minimize the accumulation of free radicals. *Superoxide dismutases* accelerate the shift from O_2^- to H_2O_2 (reaction I) and thereby also limit the opportunity for O_2^- to generate the reduced form of metal ions (reaction III) needed for the conversion of inert H_2O_2 into highly reactive ·OH radicals (reaction II). *Catalase* cooperates with the dismutases by disposing of the excess H_2O_2. Brain tissue is reported to be rich in dismutases, but poor in catalase.

Another main defense line consists of *antioxidant* action, aimed at reducing the free radicals themselves. *Vitamin E* (α-tocopherol) is a highly liposoluble molecule well suited to attack free radicals along the lipid peroxidation chain in cell membranes. By donating one electron to the radical, it becomes itself a poorly reactive free radical that degrades easily or is reduced back to the original form (for example, by ascorbic acid). *Glutathione* (GSH) is the most abundant and ubiquitous sulfhydrocompound in animal cells. Free radical reduction is effected by GSH oxidation to GSSG, catalyzed by the enzyme glutathione peroxidase, which requires selenium as its cofactor. GSSG is, in turn, reduced back to GSH by a glutathione reductase, driven by the oxidation of NADPH to NADP. Glutathione and its two enzymes are found in all neural tissues (c.f. Halliwell and Gutteridge, 1985). A decline in brain GSH has been observed as a consequence of catecholamine oxidation in the Parkinson-like syndrome induced by the MPTP drug (Bannon et al., 1984). Lastly, *ascorbic acid*, at high concentrations, also acts as an antioxidant by converting to dehydroascorbate under the action of an ascorbic acid oxidase. Active transport systems for ascorbic acid are present both in the choroid plexus and in neural cells, securing high ascorbate levels in cerebrospinal and intracellular fluids, respectively (Spector and Eells, 1984).

5.3 Neuronal cultures and neuronotrophic agents

Traditional culture media contain ingredients (e.g., aromatic amino acids, riboflavin, phenol red, and Hepes) that may be subject to auto-oxidation under the action of light (e.g., Wang and Nixon, 1978; Spierenburg et al., 1984). One possible mechanism for the neuronal survival protection by *catalase* (table 19-4) would be the scavanging of radicals accumulating in the medium from auto-oxidized medium components or from generation and release by the cells. *Vitamin E*, added to the medium in place of catalase, has a similar neuronotrophic effect in neuronal monolayer cultures (Walicke et al., 1985) and reaggregate cultures (Halks-Miller et al., 1982). A variety of nonneural cells have also been reported to increase their survival in cultures supplemented with catalase (e.g., Darfler and Insel, 1983; Sagone et al., 1984) or vitamin E (e.g., Hornby, 1982). It is interesting also to note that one of the N1 ingredients found critical for certain neurons has been *selenite* (table 19-1), the cofactor required for an effective use of glutathione as an antioxidant against free radicals.

In the absence of antioxidant supplements to the culture medium, *pyruvate* is required by most neurons for their in vitro survival (table 19-2 and 19-3).

It is reasonable to assume, then, that exogenous pyruvate permits the neurons to build up other defenses against free radical damage. One possibility is that pyruvate, by sustaining a healthy performance of the tricarboxylic acid (TCA) cycle, reduces the escape of electrons from electron transport chains, and thus the generation of free radicals by the neurons themselves. Alternatively, pyruvate may favor indirectly the generation of NADPH to sustain the GSH cycle of antioxidant protection. Spierenburg et al. (1984) report that pyruvate completely protects cultured lymhocytes against their death under Hepes phototoxicity. Moreover, Hornby (1982) has reported that adrenocortical cells, in the course of their propagation in culture, display a progressive increase in their dependence on glutamine rather than pyruvate (hence an increased involvement of transamination reactions for a partial operation of the TCA cycle), but would not do so in the presence of vitamin E and other antioxidants. Hornby's data and our own observations raise the intriguing speculation that free radicals may impose stepwise damage on the entire chain of energy-producing processes. Thus, glucose may be effectively utilized under protection by catalase, pyruvate becomes the required energy substrate in the absence of catalase for a period of time in culture, and transamination substrates such as glutamine take the place of pyruvate unless further damage is prevented.

It is attractive to include in these speculations a role for traditional *neuronotrophic factors* as defense agents against free radical damage. Inviting such a speculation, among several indirect considerations, is the direct observation by Tiffany-Castiglioni and Perez-Polo (1981) that *nerve growth factor* increases the resistance by centrain neuroblastoma cells to the cytotoxic damage imposed on them by 6-hydroxydopamine (6-OHDA). 6-Ohda is much more rapidly oxydized than dopamine, causing a generation of O_2^- and ·OH radicals that is too rapid for normal defenses (e.g., Heikkila and Cohen, 1972).

ACKNOWLEDGMENTS

This work was support by NSF grant BNS-8218366 and a grant from FIDIA Research Laboratories.

REFERENCES

Adler, R., and Varon, S. (1980) Cholinergic neuronotrophic factor. V. Segregation of survival- and neurite-promoting activities in heart conditioned media. Brain Res. 188:437–448.

Appel, S.H. (1981) A unifying hypothesis for the cause of amyotrophic lateral sclerosis, parkinsonism, and Alzheimer disease. Ann. Neurol. 10:499–505.

Bannon, M.J., Goedert, M., and Williams, B. (1984) The possible relation of glutathione, melanin and *l*-methyl-4-phenyl-1,2,5,6-tetrahydropyridine (MPTP) to Parkinson's disease. Biochem. Pharmacol. 33:2697–2698.

Barbin, G., Manthorpe, M., and Varon, S. (1984a) Purification of the chick eye ciliary neuronotrophic factor (CNTF) J. Neurochem. 43:1468–1478.

Barbin, G., Selak, I., Manthorpe, M., and Varon, S. (1984b) Use of central neuronal cultures for the detection of neuronotrophic agents. Neuroscience 12:33–43.

Barde, Y.-A., Edgar, D., and Thoenen, H. (1982) Purification of a new neurotrophic factor from mammalian brain. EMBO J. 1:549–553.

Björklund, A., and Stenevi, U. (1981) In vivo evidence for a hippocampal adrenergic neuronotrophic factor specifically released on septal deafferentation. Brain Res. 229:403–428.

Bors, W., Saran, M., and Tait, D. (eds.) (1984) Oxygen radicals in chemistry and biology. Berlin: De Gruyter.

Bottenstein, J.E., and Sato, G. (1979) Growth of a neuroblastoma cell line in serum-free supplemented medium. Proc. Natl. Acad. Sci. USA 76:514–517.

Bottenstein, J.E., Skaper, S.D., Varon, S., and Sato, G.H. (1980) Selective survival of neurons from chick embryo sensory ganglionic dissociates using defined, serum-free medium. Exp. Cell Res. 125:183–190.

Collins, F. (1978) Induction of neurite outgrowth by a conditioned-medium factor bound to the culture substratum. Proc. Natl. Acad. Sci. USA 75:5210–5213.

Cowan, W.M. (1973) Neuronal death as a regulative mechanism in the control of cell number in the nervous system. In: Rockstein, M. (ed.) Development and aging in the nervous system. New York: Academic Press, pp. 19–41.

Cowan, C.W., Fawcett, J.W., O'Leary, D.D.M., and Stanfield, B.B. (1984) Regressive events in neurogenesis. Science 225:1258–1265.

Crutcher, K.A., and Collins, F. (1982) In vitro evidence for two distinct hippocampal growth factors: basis of neuronal plasticity? Science 217:67–68.

Cunningham, T.J. (1982) Naturally occurring neuron death and its regulation by developing neural pathways. Int. Rev. Cytol. 74:163–186.

Cutler, R.G. (1985) Peroxide-producing potential of tissues: inverse correlation with longevity of mammalian species. Proc. Natl. Acad. Sci. USA 82:4798–4802.

Darfler, F.J., and Insel, P.A. (1983) Clonal growth of lymphoid cells in serum-free media requires elimination of H202 toxicity. J. Cell. Physiol. 115:31–36.

Davis, G.E., Manthorpe, M., and Varon, S. (1985a) Parameters of neuritic growth from ciliary ganglion neurons in vitro: influence of laminin, schwannoma polyornithine-binding neurite promoting factor and ciliary neuronotrophic factor. Dev. Brain Res. 17:75–84.

Davis, G.E., Manthorpe, M., Engvall, E., and Varon, S. (1985b) Isolation and characterization of rat schwannoma neurite-promoting factor: evidence that the factor contains laminin. J. Neurosci. 5:2662–2671.

Davis, G.E., Varon, S., Engvall, E., and Manthorpe, M. (1985c) Substratum-binding neurite promoting factors: relationships to laminin. Trends Neurosci. 8:528–532.

Edgar, D., and Thoenen, H. (1982) Modulation of NGF-induced survival of chick sympathetic neurons by contact with a conditioned-medium factor bound to the culture substratum. Dev. Brain Res. 5:89–92.

Facci, L., Skaper, S.D., and Varon, S. (1985) Specific replacements of pyruvate for trophic support of central and peripheral nervous system neurons. J. Neurochem. 45:926–934.

Facci, L., Skaper, S.D., and Varon, S. (1986) Transamination of glutamate to citric acid cycle intermediates in cultured neurons correlates with the ability of oxoacids to support neuronal survival in vitro. Biochem. J. 234:605–610.

Feringa, E.R., and Vahlsing, H.L. (1985) Labeled corticospinal neurons one year after spinal cord transection. Neurosci. Lett. 58:283–286.

Fridovich, I. (1983) Superoxide radical: an endogenous toxicant. Annu. Rev. Pharmacol. Toxicol. 23:239–257.

Gage, F.H., Björklund, A., and Stenevi, U. (1984) Denervation releases a neuronal survival factor in adult rat hippocampus. Nature (Lond.) 308:637–639.

Gage, F.H., Wictorin, K., Fischer, W., Williams, L.R., Varon, S., and Björklund, A. (1986) Retrograde cell changes in medial septum and diagonal band following fimbria–fornix transection:quantitative temporal analysis. Neuroscience 19:241–255.

Halliwell, G., and Gutteridge, J.M.C. (1985) Oxygen radicals and the nervous system. Trends Neurosci. 8:22–26.

Halks-Miller, M., Kane, J.P., Beckstead, J.H., and Smuckler, E.A. (1982) Vitamin E-enriched lipoproteins increase longevity of neurons in vitro. Dev. Brain Res. 2:439–447.

Hamburger, V., and Oppenheim, R.N. (1982) Naturally occurring neuronal death in vertebrates. Neurosci. Comment. 1:39–55.

Hefti, F. (1985) Nerve growth factor (NGF) promotes survival of septal cholinergic neurons after injury. Soc. Neurosci. Abstr. 11:660.

Heikkila, R.E., and Cohen, G. (1972) Further studies on the generation of hydrogen peroxide by 6-hydroxydopamine. Mol. Pharmacol. 8:241–248.

Hornby, P.J. (1982) The role of vitamin E in cellular energy metabolism in cultured adrenocortical cells. J. Cell. Physiol. 112:207–216.

Letourneau, P.C. (1975) Possible roles of cell-to-substratum adhesion in neuronal morphogenesis. Dev. Biol. 44:92–101.

Longo, F.M., Manthorpe, M., and Varon, S. (1982) Spinal cord neuronotrophic factors (SCNTFs). I. Bioassay of Schwannoma and other conditioned media. Dev. Brain Res. 3:277–294.

Longo, F.M., Manthorpe, M., Skaper, S.D., Lundborg, G., and Varon, S. (1983a) Neuronotrophic activities accumulate in vivo within silicone nerve regeneration chambers. Brain Res. 261:109–117.

Longo, F.M., Skaper, S.D., Manthorpe, M., Williams, L.R., Lundborg, G., and Varon, S. (1983b) Temporal changes in neuronotrophic activities accumulating in vivo within nerve regeneration chambers. Exp. Neurol. 81:756–769.

Longo, F.M., Selak, I., Zovickian, J., Manthorpe, M., Varon, S., and U, H.-S. (1984) Neuronotrophic activities in cerebrospinal fluid of head trauma patients. Exp. Neurol. 84:207–218.

Magistretti, P.J., Manthorpe, M., Bloom, F.E., and Varon, S. (1983) Functional receptors for vasoactive intestinal polypeptide in cultured astroglia from neonatal rat brain. Regul. Pept. 6:71–80.

Manthorpe, M., and Varon, S. (1985) Regulation of neuronal survival and neuritic growth in the avian ciliary ganglion. In: Guroff, G. (ed.) Growth and Maturation Factors, Vol 3. New York: John Wiley and Sons, pp. 77–177.

Manthorpe, M., Varon, S., and Adler, R. (1981) Neurite-promoting factor in conditioned medium from RN22 schwannoma cultures: bioassay, fractionation and properties. J. Neurochem. 37:759–767.

Manthorpe, M., Longo, F.M., and Varon, S. (1982) Comparative features of spinal neuronotrophic factors in fluids collected in vitro and in vivo. J. Neurosci. Res. 8:241–150.

Manthorpe, M., Nieto-Sampedro, M., Skaper, S.D., Lewis, E.R., Barbin, G., Longo, F.M., Cotman, C.W., and Varon, S. (1983a) Neuronotrophic activity in brain wounds of the developing rat: correlation with implant survival in the wound cavity. Brain Res. 267:47–56.

Manthorpe, M., Engvall, E., Ruoslahti, E., Longo, F.M., Davis, G.E., and Varon, S. (1983b) Laminin promotes neuritic regeneration from cultured peripheral and central neurons. J. Cell Biol. 97:1882:1890.

Manthorpe, M., Skaper, S.D., Williams, L.R., and Varon, S. (1986a) Purification of adult rat sciatic nerve ciliary neuronotrophic factor. Brain Res. 367:282–286.

Manthorpe, M., Rudge, J., and Varon, S. (1986b) Astroglial cell contributions to neuronal survival and neuritic growth. In: Fedoroff, S. and Vernadakis, A. (eds.) Astrocytes, Vol. 2. New York: Academic Press, pp. 315–376.

Nieto-Sampedro. M., Lewis, E.R., Cotman, C.W., Manthorpe, M., Skaper, S.D., Barbin, G., Longo, F.M., and Varon, S. (1982) Brain injury causes a time-dependent increase in neuronotrophic activity at the lesion site. Science 217:860–861.

Nieto-Sampedro, M., Manthorpe, M., Barbin, G., Varon, S., and Cotman, C.W. (1983) Injury-induced neuronotrophic activity in adult rat brain: correlation with survival of delayed implants in a wound cavity. J. Neurosci. 3:2219–2229.

Nieto-Sampedro, M., Whittemore, S.R., Needels, D.L., Larson, J., and Cotman, C.W. (1984) The survival of brain transplants is enhanced by extracts from injured brain. Proc. Natl. Acad. Sci. USA 81:6250–6254.

Pryor, W.A. (ed.) (1982) Free radicals in biology, vol. 5. New York: Academic Press.

Rudge, J.S., Manthorpe, M., and Varon, S. (1985) The output of neuronotrophic and neurite-promoting agents from rat brain astroglial cells: a microculture method for screening potential regulatory molecules. Dev. Brain Res. 19:161–172.

Sagone, A.L., Husney, R., Guter, H., and Clark, L. (1984) Effect of catalase on the proliferation of human lymphocytes to phorbol myristate acetate. J. Immunol. 133:1488–1494.

Schonfeld, A.R., Thal, L.J., Horowitz, S.Q., and Katzman, R. (1981) Heart conditioned medium promotes central cholinergic regeneration in vivo. Brain Res. 229:541–546.

Schonfeld, A.R., Heacock, A.M., and Katzman, R. (1985) Neuronotrophic factors: effects on central cholinergic regeneration in vivo. Brain Res. 336:297–301.

Selak, I., Skaper, S.D., and Varon, S. (1983) Age-dependent requirements of sympathetic neurons in serum-free culture. Dev. Brain Res. 7:171–179.

Selak, I., Skaper, S.D., and Varon, S. (1985) Pyruvate participation in the low molecular weight trophic activity for CNS neurons in glia-conditioned media. J. Neurosci. 5:23–28.

Skaper, S.D., and Varon, S. (1986) Age-dependent control of dorsal root ganglion neuron survival by macromolecular and low molecular weight trophic agents and substratum-bound laminins. Dev. Brain Res. 24:39–46.

Skaper, S.D., Adler, R., and Varon, S. (1979) A procedure for purifying neuron-like cells in cultures from central nervous tissues with a defined medium. Dev. Neurosci. 2:233–237.

Skaper, S.D., Selak, I., and Varon, S. (1982) Molecular requirements for survival of cultured avian and rodent dorsal root ganglionic neurons responding to different trophic factors. J. Neurosci. Res. 8:251–261.

Skaper, S.D., Selak, I., Manthorpe, M., and Varon, S. (1984) Chemically defined requirements for the survival of cultured 8-day chick embryo ciliary ganglion neurons. Brain Res. 302:281–290.

Spector, R., and Eells, J. (1984) Deoxynucleoside and vitamin transport into the central nervous system. Fed. Proc. 43:196–200.

Spierenburg, G.T., Oerlemans, F.T.J.J., Van Laarhoven, J.P.R.M., and De Bruyn, C.H.M.M. (1984) Phototoxicity of *N*-2-hydroxyethylpiperazine-*N'*-2-ethanesulfonic acid-buffered culture media for human leukemic cell lines. Cancer Res. 44:2253–2254.

Tiffany-Castiglioni, E., and Perez-Polo, J.R. (1981) Stimulation of resistance to 6-hydroxydopamine in a human neuroblastoma cell line by nerve growth factor. Neurosci. Lett. 26:157–161.

Varon, S. (1975) In vitro approaches to the study of neural tissue aging. In: Maletta, G. (ed.) Survey of the aging nervous system. DHEW Publ. (NIH) 74-296, pp. 59–76.

Varon, S. (1977) Neural growth and regeneration: a cellular perspective. Exp. Neurol. 54:1–6.

Varon, S. (1979) The culture of chick embryo dorsal root ganglion cells on polylysine-coated plastic. Neurochem. Res. 4:155–173.

Varon, S. (1985) Factors promoting the growth of the nervous system: In Discussions in neuroscience, vol. 2, no. 3. FESN (Foundation for the Study of the Nervous System), pp. 1–62.

Varon, S., and Adler, R. (1980) Nerve growth factors and control of nerve growth. Curr. Top. Dev. Biol. 16:207–252.

Varon, S., and Adler, R. (1981) Trophic and specifying factors directed to neuronal cells. adv. Cell. Neurobiol. 2:115–163.

Varon, S., and Somjen, G. (1979) Neuron–glia interactions. Neurosci. Res. Prog. Bull. 17:1–239.

Varon, S., Manthorpe, M., and Longo, F.M. (1982) Growth factors and motor neurons. Adv. Neurol. 36:453–472.

Varon, S., Manthorpe, M., Longo, F.M., and Williams, L.R. (1983) Growth factors in regeneration of neural tissues. In: Seil, F.J. (ed.) Nerve, organ, and tissue regeneration: research perspectives. New York: Academic Press, pp. 127–155.

Varon, S., Manthorpe, M., and Williams, L.R., (1984a) Neuronotrophic and neurite promoting factors and their clinical potentials. Dev. Neurosci. 6:73–100.

Varon, S., Skaper, S.D., Barbin, G., Selak, I., and Manthorpe, M. (1984b) Low molecular weight agents support survival of cultured neurons from the CNS. J. Neurosci. 4:654–658.

Varon, S., Williams, L.R., and Gage, F.H. (1986a) Exogenous administration of neuronotrophic factors in vivo protects CNS neuron against axotomy-induced degeneration. In Seip, F.J., Herbert, E. and Carlson, B.M. (eds.) 1985 PVA Regeneration Research Synip, Progress Brain Res. Vol. 71. New York: Elsevier, pp. 191–201.

Varon, S., Rudge, J., Davis, G.E., Manthorpe, M., and Skaper, S.D. (1986b) Astroglia production of neuronotrophic and neurite promoting agents. In: Grisar, T., and Franck, G. (eds.) Biochemistry of glial cells. (In press.)

Varon, S., Skaper, S.D., Facci, L., Rudge, J., and Manthorpe, M. (1986c) Trophic and metabolic couplings between astroglia and neurons. In: Althaus, H., and Seifert, W. (eds.) Glial–neuronal communications in development and regeneration. (In press.)

Vitorica, J., Machado, A., and Satrustegui, J. (1984) Age-dependent variations in peroxide-

utilizing enzymes from rat brain mitochondria and cytoplasm. J. Neurochem. 42:351–356.
Walicke, P., Varon, S., and Manthorpe, M. (1986) Purification of a human red blood cell protein supporting the survival of cultured CNS neurons, and its identification as catalase. J. Neurosci. 6:1114–1121.
Wang, R.J., and Nixon, B.T. (1978) Identification of hydrogen peroxide as photoproduct toxic to human cells in tissue-culture medium irradiated with "daylight" fluorescent light. In Vitro 14:715–5722.
Wewer, U., Albrechtsen, R., Manthorpe, M., Varon, S., Engvall, E., and Ruoslahti, E. (1983) Human laminin isolated in a nearly intact, biologically active form from placenta by limited proteolysis. J. Biol. Chem. 258:12654–12660.
Whitehouse, P.J., Price, D.L., Clark, A.W., Coyle, J.T., and De Long, M.R. (1982) Alzheimer's disease and senile dementia: loss of neurons in the basal forebrain. Science 215:1237–1239.
Williams, L.R., Selak, I., Skaper, S.D., Manthorpe, M., and Varon, S. (1985) Central nervous system-directed neuronotrophic activity present in red blood cells. Brain Res. 336:99–105.
Williams, L.R., Varon, S., Peterson, G., Wictorin, K., Fischer, W., Bjorklund, A., and Gage, F.H. (1986) Prevention of CNS neuronal death by continuous infusion of nerve growth factor. Proc. Nat. Acad. Sci. USA 83:9231–9235.
Yavin, E., and Yavin, Z. (1984) Attachment and culture of dissociated cells from rat embryonic cerebral hemispheres on poly-L-lysine-coated surfaces. J. Cell Biol. 62:540–546.

20. FACTORS INFLUENCING THE PROLIFERATION AND DIFFERENTIATION OF BRAIN NEURONS IN VITRO

YAIR AIZENMAN, DORIS K. WU AND JEAN DE VELLIS

1. INTRODUCTION

The development of an undifferentiated embryonic cell into a functioning neuron capable of participating in the working of the mind forms one of the most fascinating and central issues in neuroscience. Both genetically preprogrammed events and environmental factors shape the developing neuron. The environmental factors include hormones (Balazs, 1974; Graves, 1977; Arnold and Gorski, 1984), soluble growth factors released by neighboring cells such as glial cells (Patterson, 1978; Vernadakis et al., 1978; Barde et al., 1980), other neurons (Black, 1978), the choroid plexus (Dickson et al., 1985), and the meningeal cells (Barakat et al., 1981), and surface interactions among neurons (Goodman and Bastiani, 1984) or between neurons and glial cells (Rakic, 1971). It is difficult, and at times impossible, to identify the specific molecular factors involved in neuronal development, and to further study their modes of action in the intact brain because of the obscure milieu surrounding the neurons in this uncontrollable environment. It is here that the usefulness of cell culture which permits the study of viable neurons in a more defined and manipulatable environment becomes apparent. In this chapter, we focus on the hormonal and soluble growth factors that influence neuronal development in the central nervous system (CNS) as revealed by such cell culture studies.

The establishment of pure neuronal cultures from the peripheral nervous system (PNS), and the identification of new growth factors using these

Vernadakis, A. (ed.), *Model Systems of Development and Aging of the Nervous System*.

cultures, represent a major advance in developmental neurobiology. These cultures have stimulated a tremendous surge in research and have allowed the unraveling of some of the processes taking place during neuronal development. The ability of soluble growth factors to regulate neuronal development and influence neuronal destiny has been demonstrated. Various factors, many as yet uncharacterized, have been shown to affect different aspects of neuronal development such as neuronal survival, neurite extension, the choice of neurotransmitter-synthesizing enzyme, and electrical and synaptic functions (Varon and Adler, 1981; Puro and Egardh, 1984; Berg, 1984). Nerve growth factor (NGF), the prototype for neuronal growth factors, is capable of influencing all of the above (Varon and Adler, 1981). It appears that, while some factors act compatibly, like NGF and glial released factor(s) on the sensory ganglionic neuron survival (Barde et al., 1980), others seem to compete in determining the neuronal phenotype. Such is the case with NGF, which stimulates tyrosine hydroxylase in the precursor cells of adrenergic sympathetic neurons, while factors released by glial, muscle, and heart cells induce choline acetyltransferase, causing these neurons to become cholinergic (Patterson, 1978).

Ever since it was discovered that thyroid hormone deficiency results in mental retardation (Graves, 1977) and that androgens modulate sexual behavior (Balazs, 1974; Arnold and Gorski, 1984), hormones have been considered as one of the essential environmental influences on neuronal development in the brain. Critical developmental processes that shape the immature brain, such as the extent of dendritic growth, the formation of specific neuronal connections and the subsequent rearrangement of synaptic contacts, and the levels of neurotransmitter-synthesizing enzymes have all been shown to be affected by thyroid hormone and, in a more limited neuronal population, by androgens as well (Balazs, 1974; Graves, 1977). Both of these hormones have nuclear receptors and might be acting in part at the transcriptional level (Balazs, 1974; Oppenheimer and Samuels, 1983) influencing irreversibly the final character of the immature neuron and thus on a wider scale affecting brain function and behavior.

In the following sections, we describe the development of two new brain neuronal culture systems that are free of glial cells and are grown in a defined molecular environment free of serum (Aizenman and de Vellis, 1986a; Aizenman et al., 1986). These cultures have allowed us not only to observe brain neuronal growth, but also to determine some of the factors involved in brain neuronal proliferation and differentiation. The striking coincidence between the timing of growth factors action and the morphological and biochemical changes the neurons undergo is discussed as well.

2. THE NEURONAL CULTURE SYSTEM

The molecular study of environmental factors controlling neuronal development in the brain requires a pure neuronal population grown in a defined medium. However, previously reported CNS neuronal cultures use serum or a serum preincubation period, or they contain a large number of glial cells. Serum, with its ill-defined, complex, and variable composition (Rizzino et al., 1979; Barnes and Sato, 1980), interferes with various studies, including the identification of the factors critical for neuronal growth and differentiation. These same studies are also hampered by the presence of glial cells in neuronal cultures, since some growth factors might not be acting directly on the neurons in such a mixed glial–neuronal culture, but instead such factors might be supporting glial cell growth (Morrison and de Vellis, 1981; Aizenman and de Vellis, 1986b), which, in turn, influences neuronal development (Patterson, 1978; Barde et al., 1980; Berg, 1984). The mere presence of glial cells precludes identification, purification, and investigation of those very neuronal growth factors that they produce (Patterson, 1978; Barde et al., 1980). In addition, the cellular components being studied may exist in both neuronal and glial cells, making it difficult to determine in a mixed culture the exact effect of a substance on the neuronal system alone (Yong and Fellow, 1980).

The two serum-free culture systems we have developed are from 16- to 17-day-old rat embryo cerebral cortices, and from 6- to 9-day-old embryonic chick brains. The chick system was used to study proliferation, not only because a large quantity of neuroblasts can be obtained and can be easily dissected out of the surrounding tissue, but also because the embryonic date of the brains can be determined accurately—a quality that became important in studying the transition of neuronal cells from the neuroblast stage into postmitotic cells. The rat system was used to study differentiation because it is a mammalian system and pure astrocyte and oligodendrocyte cultures from the same species are already available (Morrison and de Vellis, 1981; Saneto and de Vellis, 1985).

Since historically the rat system was established first, its preparation and characterization shall be described here, but the same techniques and results apply to the chick cultures as well. To obtain a suspension of single cells, embryonic rat cortices were stripped of meninges, passed through a nylon mesh bag of 220-μm pore size and through a 130-μm filter, centrifuged for 6 minutes at 700 *g*, resuspended in fresh medium, and passed through a 33-μm mesh. Insulin (5 μg/ml; Sigma) and transferrin (100 μg/ml; Sigma) were added and the cells were plated in 24-well plates (2 cm^2/well; Falcon), which were coated with polylysine (mol. wt. 100,000, 2.5 μg/ml, 16 hours of incubation at 37°C; Sigma) at a density of 100,000 cells/cm^2 in 0.5 ml/well of the hormonally supplemented medium. The basic medium was made in pyrogen-free double-distilled water from a 1:1 mixture of Dulbecco–Vogt modification of Eagle's medium and Ham's F-12 medium (DMEM/F-12; Irvine Scientific) with 1.2 g of $NaHCO_3$/liter and 15 m*M* Hepes buffer. Cell cultures were incubated at 37°C in 100% humidity and 5% CO_2.

The neuronal purity of the culture, which is essential for the identification

Table 20-1. Cell Culture characteristics.

Cell markers	Treatment	Neuronal cultures	Glial cultures
Immunostaining			
Neurofilament		+	−
GFAP		−	+
Enzyme activity			
GS (nmol/min mg)	−HC	0.7 ± 0.2	14.0 ± 3.0
	+HC	0.8 ± 0.1	20.0 ± 1.0
GPDH (nmol U/mg)	−HC	4.0 ± 4.0	166.0 ± 2.0
	+HC	0.5 ± 0.9	304.0 ± 5.0

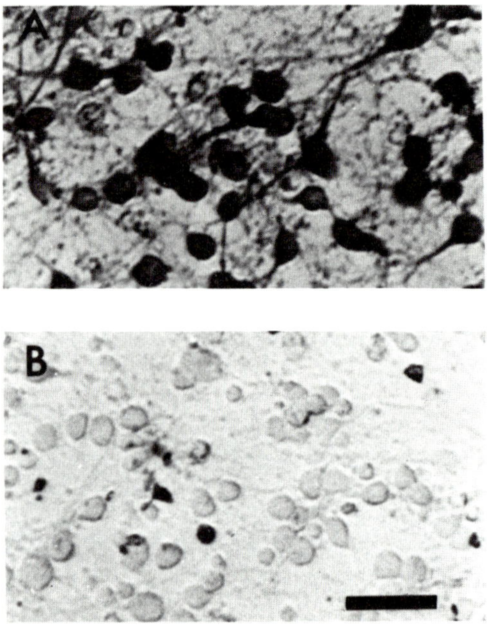

Figure 20-1. An embryonic rat culture immunostained with neurofilament antibody (1:10,000 dilution) (**A**) and a controlled culture immunostained with normal mouse immunoglobin G (**B**) using the avadin biotin complex immunocytochemical technique. Bar = 50 μm.

and investigation of factors influencing neuronal development, was established by a number of criteria (table 20-1). Immunocytochemically, the majority of the cells (>95%) stained with neurofilament antibody—a neuronal cell marker (Goldstein et al., 1982) (figure 20-1). Consistent with these findings, only a few cells (<0.01%) were found to stain with glial fibrillary acidic protein antibody—a glial cell marker (Morrison and de Vellis, 1981; Choi and Kim, 1984). Glial cell cultures used as controls showed the inverse of this staining pattern. Biochemically, glycerol phosphate dehydrogenase enzymatic activity, indicative of oligodendrocyte presence (Leveille et al., 1980), was negligible (table 20-1) and provided further evidence of glial absence from the culture. Glutamine synthetase (GS), once thought to be a definitive astrocyte marker, has more recently been found in much smaller quantities (10%–20%) in neurons (Patel et al., 1982) and synaptosomes (Dennis et al., 1980). The low levels of GS in our culture (only 5% of that found in control astrocytes) and its noninducibility by hydrocortisone (table 20-1), a characteristic of astroglial GS, are further evidence of the neuronal purity of the culture. The observed low GS level is consistent with the notion that a small pool of it exists within neurons. Several factors enabled us to obtain a pure neuronal population. Because glial cells only start rapid proliferation in the rat after embryonic day 17 (E17) (Abney et al., 1981) (reaching a peak in proliferation only after birth), we used E16–E17 brains; and 6- to 9-day-old embryos in the chick for the same reasons. Since serum favors glial cell survival and proliferation (McCarthy and de Vellis, 1980), its total absence minimized glial growth. In addition, other factors required for glial growth (Morrison and de Vellis, 1981; Saneto and de Vellis, 1985), such as fibroblast growth factor and prostaglandin $F_{2\alpha}$, were not added to the chemically defined medium.

3. NEURONAL PROLIFERATION

Cell division in the central nervous system is a tightly programmed phenomenon. In a given brain region, each subpopulation of neurons is generated in distinctive sequences, while glial cells are generated later during development. Usually, larger neurons are generated earlier than smaller neurons. For example, in the cerebellum, the Purkinje and deep nuclear neurons become postmitotic much earlier than granule cells and various other interneurons (Miale and Sidman, 1961). In addition, in a given segment of neuroaxis, e.g., brain stem or spinal cord, proliferation of cells at rostral levels precedes those at caudal levels. In the chick spinal cord, cells in the basal plate region precede those in alar plate, i.e., a ventral–dorsal gradient also exists (Hamburger, 1948). However, little is known about the regulation of such a delicately balanced phenomenon. The mechanisms involved for sequentially turning on and off the mitotic activity of neuroepithelial cells to generate neuronal subsets are virtually unknown. The possible role of hormones, such as steroids and thyroid hormones, in regulating this early developmental stage is also not known. Fetuses of guinea pigs that are rendered diabetic showed an increase in the cell number and DNA content of their fetal cerebrums (Saintonge and Cote, 1984). Progesterone (Ahmad and Zamenhof, 1979), serotonin (Ahmad and Zamenhof, 1978), or glycine (Zamenhof and Ahmad, 1979), injected onto the chorioallantoic membranes of E7 chick embryos (during neuronal proliferation), had a stimulating effect on weight, DNA, and protein content of cerebral hemispheres when examined at E10

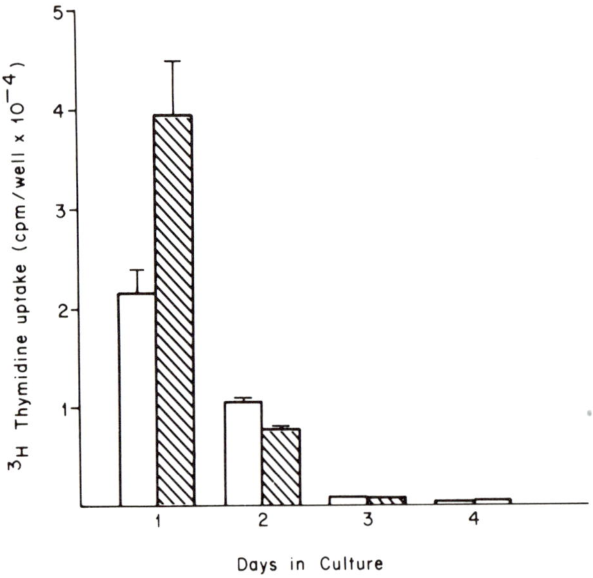

Figure 20-2. Cell cultures were prepared from 6-day-old embryonic chick brains and grown in DMEM/F-12 medium (□) or DMEM/F-12 medium plus 10 μg/ml of insulin (■). [^{3}H]-thymidine (1 μCi/ml) was added to the cultures 24 hours prior to harvesting. Cells were rinsed once with DMEM/F-12 medium, fixed with 5% cold trichloroacetic acid, and rinsed again with normal saline. Protein was solubilized in 0.5 *N* NaOH.

(end of neuronal proliferation). These data suggest that these substances may be acting as growth-promoting factors for embryonic brains. However, using these animal studies as models to study neuronal proliferation may be limited due to the complexity and heterogeneity of the nervous tissue. On the other hand, our neuronal cultures from E6 chick brains provide an excellent model to study neuronal proliferation. At this developmental age, the brain is largely composed of dividing neuroblasts. Cells from these E6 chick brains have been plated in a DMEM/F-12 medium and the effect of various factors on thymidine incorporation was measured. [^{3}H]-thymidine incorporation experiments indicate that these neuroblasts survive in vitro, continue to divide for up to 3–4 days in DMEM/F-12 medium alone, and later mature into differentiated neurons on schedule (figure 20-2). The addition of insulin during the first day of culture triggered a transient rise in the rate of DNA synthesis (figure 20-2) and resulted in a twofold to threefold increase in cell number at the end of the third day in culture. The presentation of insulin on the second day caused a transient increase in DNA synthesis for that day only (data not shown). Autoradiographic studies showed that 30%–40% of the cells incorporated the radioactive precursor into their

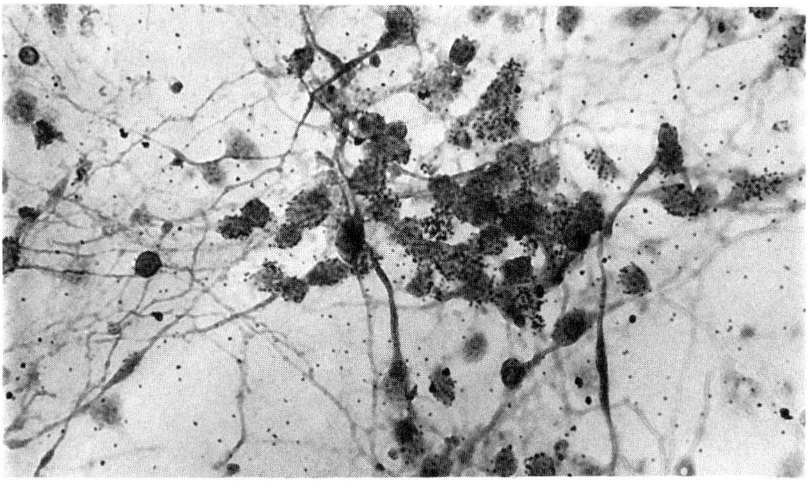

Figure 20-3. Brains from E6 chicks were dissected, dissociated, and cultured for 2 days in DMEM/F-12 medium + insulin + [^{3}H]-thymidine (1 μCi/ml). The cultures were stained with antineurofilament antibody diluted 1:500. Biotinylated goat antirabbit IgG was used as a secondary antibody. Then, cells were processed for autoradiography. Bar = 10 μm.

nuclei. Virtually all of the labeled cells also stained positively with antineurofilament antibodies (figure 20-3).

A variety of growth factors such as epidermal growth factor, insulin, insulin-like growth factor, and fibroblast growth factor are mitogenic for 3T3 cells (Gospodarowicz and Moran, 1976). More recently, many of these factors were also shown to be mitogenic for astrocytes (Morrison and de Vellis, 1981; Simpson et al., 1982) and oligodendrocytes (Saneto and de Vellis, 1985). We tested the effect of these growth factors on neuroblast proliferation. The factors that we have found to stimulate DNA synthesis in neuroblasts are endothelial cell growth factor (10–200 ng/ml), insulin-like growth factor I (1–100 ng/ml), and fibroblast growth factor (1–100 ng/ml). Interestingly, these factors are present in the brain (Maciag et al., 1979; Goodyer et al., 1984; Gospodarowicz et al., 1984), suggesting that they may play a role in regulating neuronal proliferation in vivo. These agents are unique, since many other known mitogens and neuroactive substances were found to be inactive on neuroblasts (table 20.2).

4. TRANSITION FROM NEUROBLAST TO NEURON

Neuroblasts become postmitotic on what appears to be a fixed schedule in vivo. Although the birthdates of various neuronal populations and their path of migration within the CNS have been thoroughly studied, little is known about the biochemical mechanisms underlying the transition from neuroblast

Table 20-2. Effect of factors on neuronal development.

Factors	Thymidine incorporation	Survival	Increased GAD
Insulin/IGF-1	+	+	
Transferrin		+	
T_3		−	+
FGF	+		
ECGF	+		
NGF 7S/βNGF	−	−	−
Hydrocortisone	−	−	−
EGF	−		
PDGF	−		
Interleukin II	−		
Forskolin/dbcAMP	−		−
$PGF_{d\alpha}$	−	−	
Progesterone	−	−	
Selenium	−	−	
KCl	−	−	
Putrescine		−	−
17-β Estradiol		−	−

ECGF, endothelial cell growth factor; FGF, fibroblast growth factor; PDGF, platelet-derived growth factor; $PGT_{2\alpha}$, prostaglandin $F_{2\alpha}$; and T_3, triiodothyronine.

to neuron. Nor do we know why, in mammals, once a neuroblast has become a neuron it can no longer proliferate. We describe here a system in which the transition from a neuroblast to a neuron can be studied at the molecular level. Specifically, we report on some changes in neuronal growth requirements and expression of antigens as the neuroblast terminally differentiates.

The 6- to 9-day-old embryonic chick brains provide a good system to study the transformation of a dividing neuroblast into a postmitotic neuron, because it is the period in which most neurons in these brains progress from a dividing into a postmitotic state (Tsai et al., 1981a). Dissociated cells from 6-, 7-, 8-, and 9-day-old chick embryos were plated in the basic medium (DMEM/F-12) and the effect on neuronal survival and growth of various factors added to the culture was quantitated by counting the number of cells possessing neurites after 5 days in vitro. Cells obtained from embryonic day 6 (E6) brains were found to grow in an insulin-supplemented medium (figure 20-4A). This supplementation alone, however, was insufficient for the growth of cells obtained from E9 brains (figure 20-4B). To survive and extend neurites in vitro, these E9 brain cells required, in addition to insulin, the presence of transferrin (figure 20-4C). In contrast to its effect on E9 brain cultures, transferrin had no effect on the number of growing cells in cultures prepared from E6 brains (figure 20-5). An intermediate stage existed in the

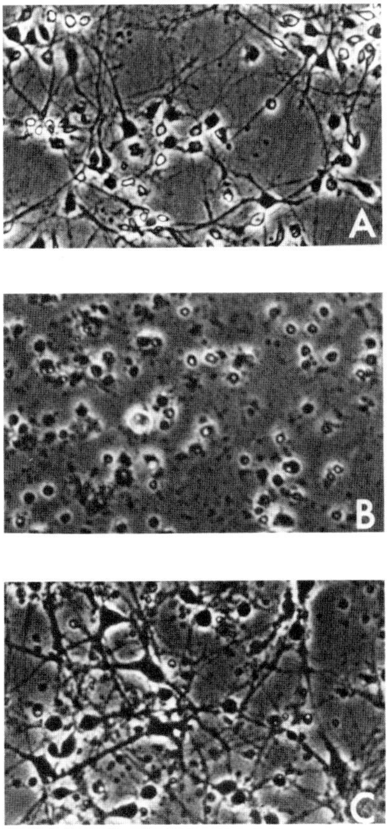

Figure 20-4. Neuronal cultures of E6 brain in a medium supplemented only with insulin (5 μg/ml) (**A**). Cultures of E9 brain in a medium supplemented only with insulin (5 μg/ml) (**B**), and in a medium supplemented with insulin (5 μg/ml) and transferrin (100 μg/ml) (**C**) Bar = 50 μm.

E7 brain cultures where 60% of the growing cells survived on insulin alone while 40% required transferrin as well (figure 20-5). Though all growing cells obtained from E8 brains required transferrin, one-half of these required only 25 μg/ml of transferrin—a level that could not sustain the growth of any of the cells obtained from E9 brains (figure 20-5). Thus, as the age of the embryos from which the neurons were obtained increased, there was an increase in both the proportion of neurons needing transferrin and the amount of transferrin required.

Most chick brain neurons in vivo proliferate and become postmitotic between E4 and E8, with a peak in proliferation at E6–E7 (Fujita, 1964; Tsai et al., 1981a). By the end of the E9, neuronal proliferation ceases. Similarly, in neuronal cell cultures, E3–E6 brain neurons divide in the presence of a

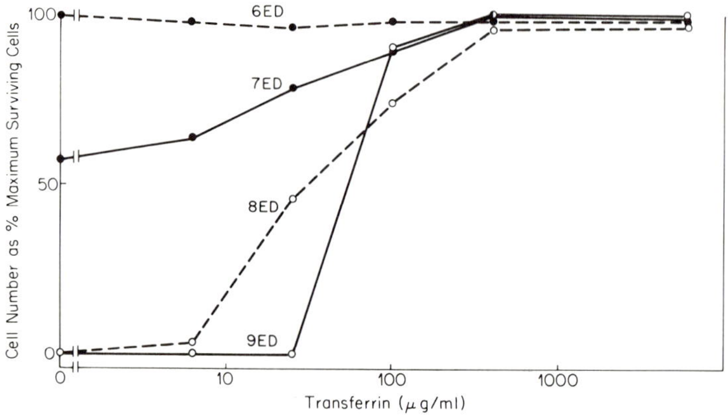

Figure 20-5. Growth response of neuronal cultures from E6, E7, E8, and E9 brain as a function of transferrin concentration. All cultures contained 5 μg/ml of insulin. Standard deviation was < 15%.

mitogenic factor from meningeal extracts, while E7 brain neurons do not (Barakat et al., 1981 and 1982). Thus, the change in the neuronal requirement for transferrin is associated with the transition of neuroblasts into postmitotic neurons, suggesting that transferrin is required by postmitotic neurons but not by neuroblasts. This stands in contrast to transferrin's well-known role in sustaining proliferation of a variety of cells in serum-free media, including 3T3, V79, hemopoietic cells, and the B104 neuroblastoma cell line (Bottenstein and Sato, 1979). However, astrocytes are an exception to this rule and do not require transferrin for division in a defined medium (Morrison and de Vellis, 1981).

Transferrin is a protein that carries iron in the blood. It is also found in the cerebral spinal fluid (Dziegielewska et al., 1981). It was proposed that transferrin might enter the CNS via an intracellular transport system (Mollgard et al., 1979). Alternatively, recent findings of high levels of transferrin mRNA in the brain, particularly in the choroid plexus, suggest that it might be synthesized there (Levin et al., 1984; Dickson et al., 1985). Further evidence for transferrin action on neurons is provided by the existence and active synthesis of transferrin receptors by CNS neurons (Markelonis et al., 1985) and transferrin detection within cortical neurons (Toran-Allerand, 1980). Thus, it seems that transferrin may be one of the few plasma proteins that are essential for brain development in vivo. This study shows, in contrast to astrocyte cultures (Morrison and de Vellis, 1981), the in vitro growth requirement for transferrin by cortical neurons. Given transferrin's classic role as an iron carrier, the change in neuronal requirement for it between E6 and E9 may signify an important developmental change in neuronal properties during this critical period of brain development.

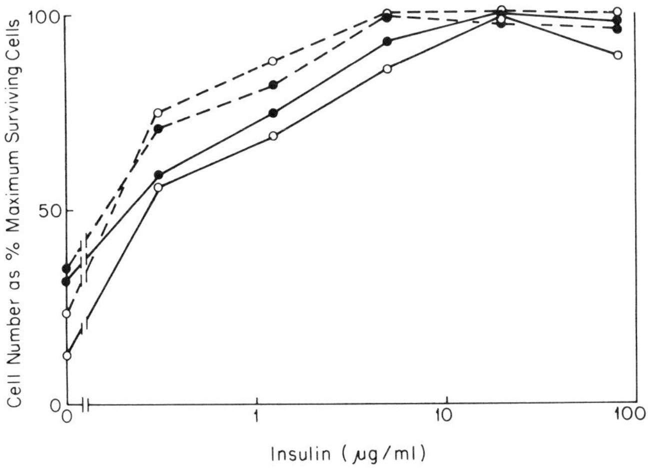

Figure 20-6. Growth response of neuronal cultures from E6 brain (○---○), E7 brain (●---●), E8 brain (●——●), and E9 brain (○——○) as a function of insulin concentration. Cultures of E7, E8, and E9 brain were supplemented with 100 μg/ml of transferrin. Standard deviation was < 14%.

Insulin, in contrast to transferrin, affected both neurons from E6 and E9 brains (figure 20-6). The difference between these two ages of brains not only manifests itself by the difference in transferrin requirement, but is also seen from the fact that one-third of the cells in the young E6 brain cultures supplemented with insulin could survive and extend neurites even without the addition of any growth factors to the culture (figure 20-6), which stands in contrast to cells obtained from E9 brains, all of which required factor supplementation to survive and grow (figure 20-4B). The finding that insulin affects both the young proliferating neurons from the E6 brain, and those cells from E9 brains that are postmitotic, demonstrates insulin action throughout neurogenesis both during proliferation and differentiation. The question remains, however, whether insulin's mode of action is the same or different in these different stages of neuronal development.

In addition to elucidating the requirements for neuronal survival at different developmental stages, we were also interested in neuronal markers that can delineate these developmental stages. The monoclonal antibody E/C8 was originally generated against chicken dorsal root ganglia, and was shown to decorate neurons during early development (Ciment and Weston, 1982). Thus, E/C8 was chosen as a candidate in our studies. When neuronal cultures from E6 chick brains were exposed to [^{3}H]-thymidine for the first day in culture and then processed for E/C8 staining and autoradiography, it was striking that nuclear radiolabeling and horseradish peroxidase (HRP) staining were mutually exclusive (figure 20-7). However, when [^{3}H]-thymidine was

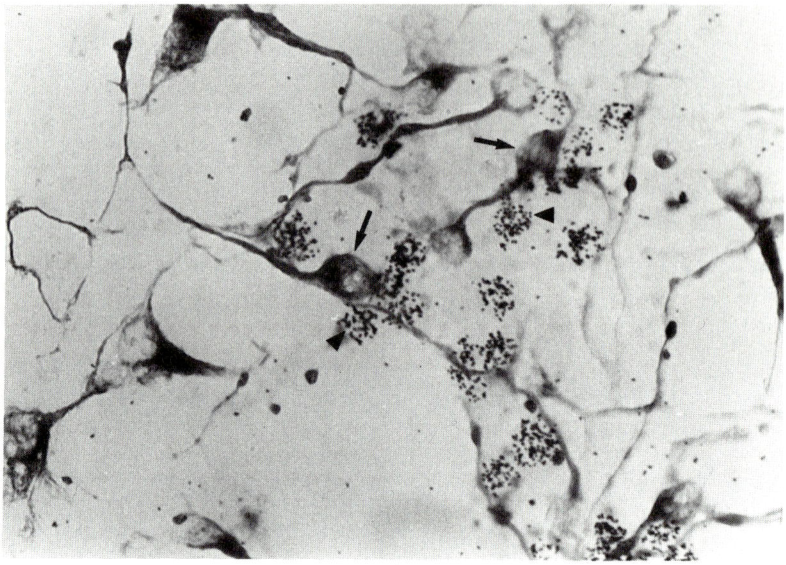

Figure 20-7. Brains from 6-day-old embryonic chicks were dissected, dissociated, and cultured on glass coverslips for 1 day in DMEM/F-12 medium + 10 μg/ml of insulin + [^{3}H]-thymidine (1 μCi/ml). At the time of harvest, cells were fixed with 4% paraformaldehyde, stained with E/C8 using the avidin–biotin complex technique, and processed for autoradiography. *Arrowheads* indicate cells that are radiolabeled only. Bar = 10 μm.

removed from the culture medium at the end of the first day and cultures were harvested at the end of five days, a time when cells were no longer synthesizing DNA (figure 20-2), an accumulation of double-labeled cells was observed (figure 20-8). These data suggest that E/C8 recognizes an epitope on postmitotic neurons.

5. NEURONAL DIFFERENTIATION

After a cell has become committed to a neuronal lineage and ceased to proliferate, it further differentiates into a highly specialized unit capable of receiving, transmitting, integrating, and storing information. To accomplish such processing of information, the neuron grows neurites (axons and dendrites) and forms synapses. Furthermore, it produces ionic channels to deal with the information electrically, and it synthesizes specific molecules such as neurotransmitters, neuromodulators, their receptors, and many other interneuronal components to process the information biochemically. Although the changes in these parameters during brain development have been reported, little is known about the molecular mechanism and the role environmental factors play in influencing the coordinated expression of these neuronal-specific molecules during development. For instance, whether hor-

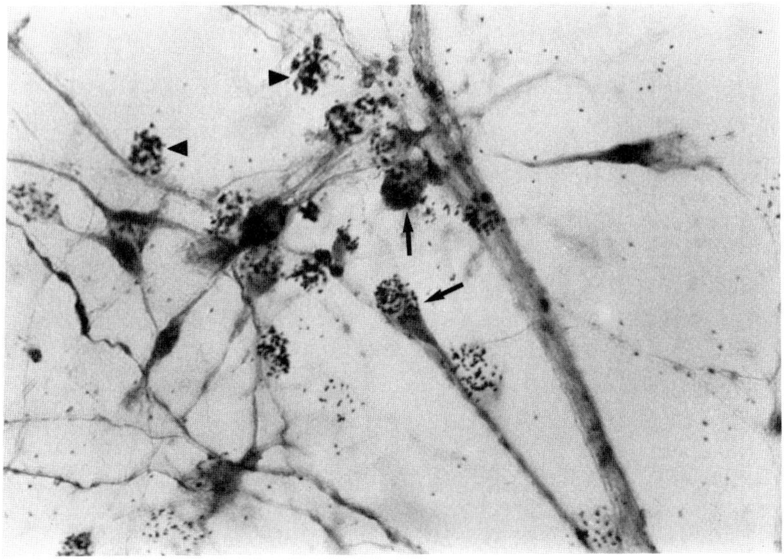

Figure 20-8. Brains from 6-day-old embryonic chicks were dissected, dissociated, and cultured for 5 days in DMEM/F-12 medium + 10 μg/ml of insulin on glass coverslips. [^{3}H]-thymidine (1 μCi/ml) was added at the beginning of culture and chased after 24 hours. At the time of harvest, cells were fixed with 4% paraformaldehyde, stained with E/C8 using the avidin–biotin complex technique, and processed for autoradiography. *Arrowheads* indicate cells that are radiolabeled only and *arrows* indicate cells that are double labeled. Bar = 10 μm.

mones such as hydrocortisone and insulin which have been shown to act on glial cells, directly act on the development of neurons in the brain is not known. The dissection of thyroid hormone and androgen direct effect on neurons from their indirect influences, as well as the understanding of their modes of action, remain obscure because of the difficulty of working with a complex and heterogeneous organ like the brain. In this section, we describe a simplified culture system, composed only of neurons, in which such studies can be done in a manipulatable environment. We report here on some of the morphological and biochemical changes the neurons undergo as they differentiate in culture, with emphasis on the extrinsic factors essential for this differentiation process.

Most of the cells obtained from the 16- to 17-day-old embryonic rat cerebral cortices possessed a spherical shape right after plating (figure 20-9A). In the next 3 days, the cells increased their body size and extended one or more major process(es) (figure 20-9C and D). The large cellular growth during this time period is also evidenced from the 100% increase in the culture total protein content (figure 20-10). Contacts between cells start forming on the second day (figure 20-9C), with the cells reaching an extensive neurite formation by the fifth day (figure 20-9F). Since γ-aminobutyric

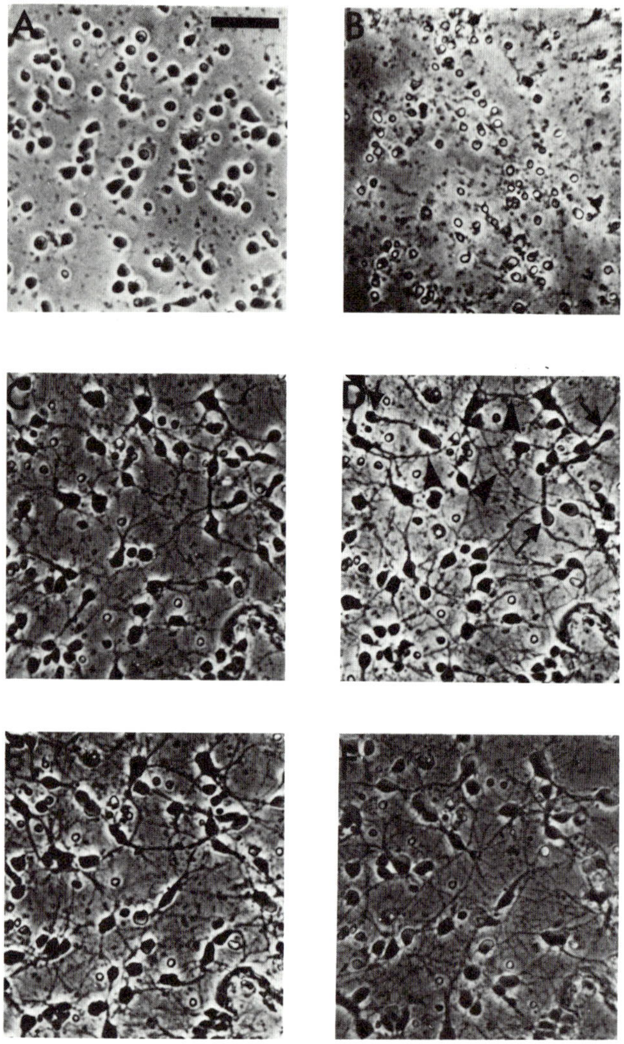

Figure 20-9. (**A**) A phase-contrast micrograph of neurons 4 hours after plating. (**B**) neuronal cultures 5 days later in the basic DMEM/F-12 medium without insulin and transferrin. (**C–F**) Time-lapse photography of the same area showing neuronal morphology after 2, 3, 4 and 5 days, respectively, in a medium supplemented with insulin and transferrin. *Arrowheads* in **D** (day 3) point to some of the neurites that have disappeared in the fourth (**E**) and Fifth (**F**) days, and the *arrows* point to some of the cells that move during this period. Bar = 50 μm.

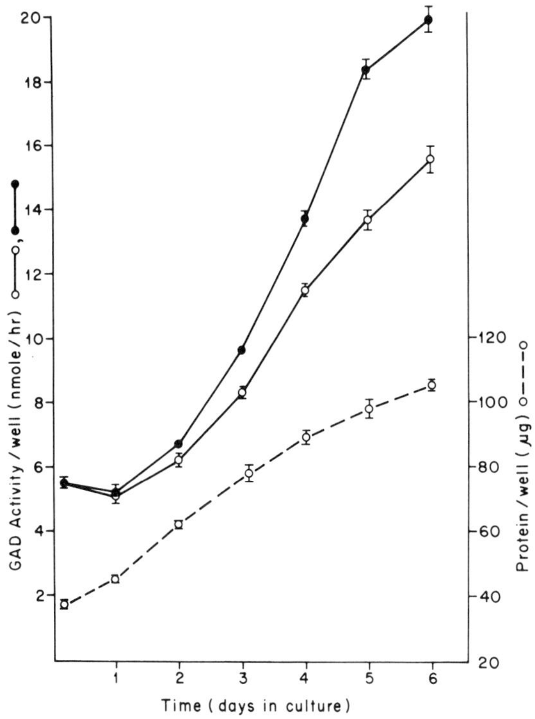

Figure 20-10. Time course of the protein (O---O) and GAD (O—O) increases as the neurons develop in a culture supplemented with insulin and transferrin. The effect of thyroid hormone addition to the culture medium on GAD activity is also shown (●—●). GAD was measured using the chromatographic method (Chude and Wu, 1976) and the proteins were measured using the Bradford (1976) procedure.

acid (GABA) is one of the major neurotransmitters in the cortex, glutamic acid decarboxylase (GAD)—its synthesizing enzyme—has been chosen as a differentiation parameter. GAD levels began to increase slightly on the second day and rose sharply on the third and following days—more than doubling by the sixth day (figure 20-10). Thus, GAD development in the culture lagged behind the large increase in protein, but correlated timewise with the increase in the number of neuronal processes and contacts that were forming in the culture.

In following the morphological development of neurons, some of the dynamics of the growth process were revealed. As can be seen from a sample of pictures taken during such a study (figure 20-9), the developmental progression involved more than just the mere growth of neurites. Not all the cells remained stationary—some were moving on the poly-D-lysine-coated plastic dish. Neither did all the neurites formed between cells remain stable—some lasted from the second day on, while others disappeared within a day of their formation. The small movements observed in the culture might also occur in the intact brain, allowing the neurons to establish their final position within a circuit and thus determining some of the connection they will eventually form. Furthermore, the fact that not all the connections between cells were stable suggests that, besides the initial projection of a neurite and the contacts it forms, there are as yet unidentified factor(s) that allow a permanent connection between cells to be established. Rearrangement of connections between neurons has also been reported to exist in vivo in the projections of the medial geniculate body that form the ocular dominance columns in the visual cortex (Hubel et al., 1977). However, in the brain, one cannot observe the rearrangement of connections while they take place as it can be seen in the culture; nor can one investigate the underlying molecular mechanisms.

Insulin and transferrin were the only two factors, out of a long list (table 20-2), that were necessary for the growth of the cortical neurons in culture. They allowed the survival (figure 20-9B), neurite extension (figure 20-9F), protein synthesis (figure 20-10), and an increase in the activity of a neurotransmitter-synthesizing enzyme (figure 20-10). Although insulin was shown to increase glucose uptake, and DNA, RNA, and protein synthesis in mixed brain cultures and brain slices, it has remained unclear whether these changes occur in the neuronal or the glial population (Yong and Fellows, 1980). Due to the neuronal purity of our culture, we were able to demonstrate the direct action of insulin on cortical neuron development. In many investigations, including our own, insulin was required at concentrations much higher than its physiological levels (Yong and Fellows, 1980; Ahmed et al., 1983). We found that insulin-like growth factor I (IGF) was able to replace insulin and did so at much lower concentration with half-maximal activity at 10 ng/ml. An even lower concentration was needed when IGF was added daily, indicating quick degradation of the factor in vitro. In vivo measurements also support the notion of IGF acting in the brain. IGF is found in the cerebral spinal fluid (Buckstrom et al., 1984) (at 5 ng/ml) and its receptors have been detected within the brain (Goodyer et al., 1984). Thus, it is suggested that IGF-I, and not insulin, is the essential factor for neuronal brain development.

After establishing the factors required for neuronal survival, we investigated the effect of hormones on neuronal differentiation. Each factor was added to the culture with insulin and transferrin at the time of plating and GAD levels were measured 6 days later. Out of the many factors examined (table 20-2), thyroid hormone (T_3) was the only one found to cause a 20% increase in GAD-specific activity. As its dose–response curve indicated, it was maximally active at a concentration of 50 n*M*. A time course of thyroid hormone effect on GAD levels is shown in figure 20-10. Thyroid hormone only began to affect GAD levels slightly on the second day in vitro and the effect became more pronounced on the third and following days (figure

20-10). Thus, thyroid hormone action started right after the initial neurite extension and the increase in its effect coincided with the increase in the number of neurites and contacts forming in the culture. Furthermore, thyroid hormone facilitated GAD development once GAD levels started to rise, but did not induce an earlier rise in GAD activity. This, combined with the fact that GAD levels sharply increased even in the absence of thyroid hormone (figure 20-10), suggests that GAD development is triggered by something else. For example, cell–cell contacts, which coincide with the rise in GAD, may be the inducing factor. Thus, thyroid hormone might be indirectly affecting GAD levels by enhancing such interneuronal connections. Alternatively, the increase in the number of contacts between neurons, the increase in GAD levels, and the neurons' ability to respond to thyroid hormone, all of which started on the second day and sharply increased on the following days, might be genetically preprogrammed to occur simultaneously, permitting the developmental process to proceed smoothly. Whatever the mechanism may be, the numerous changes that start on the second day indicate the beginning of a new and major stage in brain neuronal differentiation. This stage is preceded and somewhat overlapped by a large cellular growth as indicated by the increase in the total protein in the culture, possibly in preparation for the changes to come, and partly due to the extensive neurite outgrowth.

Due to the absence of glial cells in the neuronal culture, it became possible to examine their effect on brain neuronal development. Medium that has been conditioned over glial cells, in a defined serum-free environment, was collected and its effect on the neuronal cultures was studied. Glial conditioned medium (GCM) has been found to have no effect on GAD levels (tested in the same manner as thyroid hormone). However, GCM extended the longevity of the culture. In the insulin–transferrin-supplemented medium, neurons grew for 7–10 days even when the medium was replenished and thyroid hormone was added. This period could be extended to over 2 weeks by refeedng the cells with GCM. Since GCM did not affect GAD during the first 6 days in vitro, and since the neurons were able to survive and develop in the hormonally supplemented medium for this period without the presence of glial cells in the culture, early stages in neuronal differentiation are thus mainly dictated by genetic and hormonal factors, rather than by glial cells. Furthermore, since cultures plated at low densities did not survive, neuron–neuron interactions are likely to be another essential factor in early neuronal development. At later stages, after 1 week in culture, the neurons change their growth requirements and, from being influenced by hormones such as IGF-I and thyroid hormone, they become dependent on the more local factors produced by glial cells. This glial dependency seems to be preprogrammed in the neurons since it appears without the presence of glial cells in the culture. The effect of GCM on the longevity of cultured neurons supports the idea of glial participation in neuronal development of the cortex. For instance, glial cells in vivo might influence the selective neuronal cell death, a documented process during neurogenesis (Cowan, et al., 1984). GCM has also been shown to be important in the development of PNS neurons (Patterson, 1978; Barde et al., 1980; Arenander and de Vellis, 1983), but whether some of the same glial factor(s) active in the PNS are the same as those functioning in the CNS remains to be seen, and awaits purification of these factors.

The study of neuronal development in vitro provides some insight into the well-coordinated process of brain development. For example, the formation of neurites and consequently the initiation of contacts between neurons, the increase in neurotransmitter-synthesizing enzymes and the actions of thyroid hormone all occur within a narrow time period. In addition, the effect of thyroid hormone upon the cells begins on the second day in culture—corresponding to embryonic day 18 (E18), the point at which, in vivo, the thyroid gland (which started growing on E15) matures and becomes responsive to thyroid-stimulating hormone (Graves, 1977). Furthermore, factor(s) found in the GCM are required by the cells after a week in vitro corresponding to the second postnatal day in vivo—a time when glial proliferation is at its peak. By combining these in vitro and in vivo findings, the complex coordinate regulation that is observed during development within a cell, among cells, and between organs can be elucidated.

6. STAGES IN NEURONAL DEVELOPMENT

Changes in the morphology, biochemistry, electrophysiology, and responses to hormonal stimuli characterize the stages of neuronal development. It is important to interrelate these changes and to determine their time of onset if one is to unravel the developmental process. The model of brain neuronal development that we shall construct now (figure 20-11) is an attempt at providing a framework for the formulation of new hypotheses.

Neuronal development in the brain, as in other nervous tissues, progresses through a number of partially overlapping stages. During early embryogenesis, some ectodermal cells become committed to a neural cell lineage—forming the neuroepithelium (Patterson and Purves, 1982). These neuroepithelial cells give rise to neuroblasts and glioblasts. The neuroblasts proliferate. After they reach a postmitotic state, they migrate to their final positions in the nervous system (Rakic and Goldman-Rakic, 1982). The differentiation that follows is a continuous process whereby the neurons send out neurite-forming contacts with other neurons while concurrently the level of neurotransmitter-synthesizing enzymes rises. Subsequently, some of the connections stabilize and mature into functional synapses. Other neuronal characteristics such as neurotransmitter receptors and ionic channels appear as well (Black, 1983). In the mature brain, the neuron forms numerous connections with other neurons and with support from glial cells becomes fully capable of processing information both electrically and biochemically.

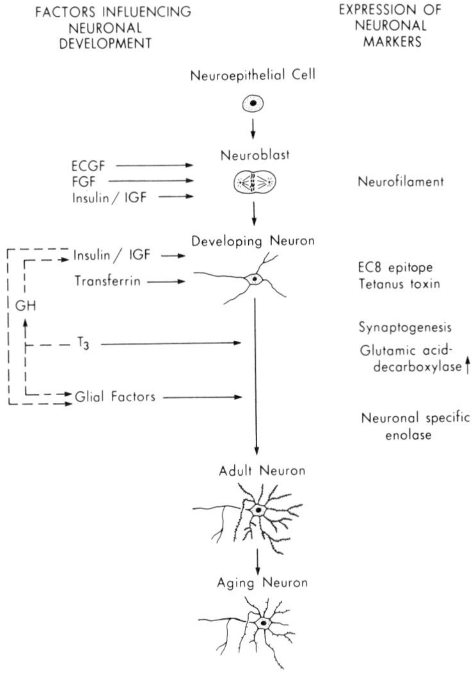

Figure 20-11. The time course of factors influencing brain neuronal development and the appearance of neuronal-specific antigens. *Dash lines* represent possible indirect action of some of the factors. *ECGF*, endothelial cell growth factor; *FGF*, fibroblast growth factor; *IGF*, insulin-like growth factor I; T_3, triiodothryonine (thyroid hormone); and *GH*, growth hormone.

Different environmental factors influence different stages in neuronal development. Endothelial cell growth factor (ECGF) and fibroblast growth factor (FGF) affect neuronal proliferation. Transferrin, on the other end, is only required after the cells have stopped dividing. Thyroid hormone specifically acts on the differentiation process, after the neurons have become postmitotic and during the onset of contact formation. Glial cells influence neuronal development at an even later stage. In contrast to the above factors, IGF, or insulin, is unique in its ability to influence both neuronal proliferation and differentiation. It works synergistically with ECGF in proliferation, with transferrin in neuronal survival, and possibly with thyroid hormone in neuronal differentiation. The changes in the growth requirements as the neurons develop is indicative of the changes in the developmental stages the cells go through. The timing of these changes can be used to detect defined and separate overlapping events in neuronal development. Understanding why the neurons start requiring these factors at the time they do and what changes these factors cause in the neurons will undoubtedly enlighten our knowledge of neuronal brain development.

The identification of the factors that directly act on neurons is also essential for revealing the other paths through which these and other factors can influence the developmental process indirectly. For example, thyroid hormone is known to increase growth hormone levels in pituitary cells (Oppenheimer and Samuels, 1983). In turn, growth hormone has been shown to increase IGF levels (Ceda et al., 1985). Since we have shown that IGF directly acts on neurons (table 20-2), this hormonal cascade may provide a complex and integrative endocrine regulation of neuronal development. Furthermore, within the brain, cellular interactions can be modified by hormones. For instance, thyroid hormone (Shanker et al., 1985; Aizenman and de Vellis, 1986b) and insulin (Morrison and de Vellis, 1981; Saneto and de Vellis, 1985; Aizenman and de Vellis, 1986b) influence glial growth and differentiation and thus can indirectly affect neuronal development via glial cells as discussed above.

In addition to establishing the role of environmental factors in neuronal development, it is crucial to delineate the sequence of events that a neuron undergoes as it matures. The use of cell markers has allowed investigators to approach this problem. The EC8 epitope marks the terminal differentiation of neurons (figure 20-7). Tetanus toxin binding to specific gangliosides (Dimpfel et al., 1977) also appears soon after CNS neurons stop dividing (Koulakoff et al., 1983). Recently the pp60$^{c\text{-}src}$ oncogene, a tyrosine kinase, has also been detected within the nervous system and it, too, appears to be expressed early in neuronal maturation, presumably in postmitotic neurons (Sorge et al., 1984). Neurotransmitter-synthesizing enzymes, like GAD, increase shortly after the initiation of neurite outgrowth and at the onset of contact formation between cells. Other enzymes, such as neuron-specific enolase, appear as well (Marangos et al., 1982). Undoubtedly, genetic engineering in concert with immunological and biochemical techniques will allow identification at the molecular level of many more changes taking place within the developing neuron.

The development of a pure neuronal culture system, free of glial cells and serum, permits the identification and investigation of environmental factors involved in brain neuronal development. Furthermore, the dynamic and complex events that occur as the neurons change their program from proliferation to differentiation can be more easily analyzed at the molecular level in a cell culture system. Future investigations of the mode of action of growth factors will increase the understanding of the sequence of events and their causality during the differentiation of an immature cell into a highly specialized neuron. This, in turn, will deepen our appreciation of the construction and function of the brain.

REFERENCES

Abney, E.R., Bartlett, P.P., and Raff, M.C. (1981) Astrocytes, ependymal cells, and oligodendrocytes develop on schedule in dissociated cell cultures of embryonic rat brain. Dev. Biol. 83:301–310.

Ahmad, G., and Zamenhof, S. (1978) Serotonin as a growth factor for chick embryo brain. Life Sci. 22:963–970.

Ahmad, G., and Zamenhof, S. (1979) The effect of progesterone on brain and body growth of chick embryos. Growth 43:58–61.

Ahmed, Z., Walker, P.S., and Fellows, R.E. (1983) Properties of neurons from dissociated fetal rat brain in serum-free culture. J. Neurosci. 3:2448–2462.

Aizenman, Y., and de Vellis, J. (1986a) Development of pure brain neurons in a defined environment: influences of insulin-like growth factor, thyroid hormone and transferrin. Brain Res. (in press).

Aizenman, Y., and de Vellis, J. (1986b) Insulin and thyroid hormone effect on astrocyte differentiation. Brain Res. (in press).

Aizenman, Y., Weichsel, M.E., and de Vellis, J. (1986) Changes in insulin and transferrin requirements of pure neuronal cultures during embryonic development. Proc. Natl. Acad. Sci. USA 83:2263–2266.

Arenander, A.T. and de Vellis, J. (1983) Frontiers of glial physiology. In: Rosenberg, R. (ed.) The clinical neurosciences. Section 5: Willis, W.D., Jr. (assoc. ed.) Neurobiology, chap. 4. New York: Churchill Livingstone, pp. 53–91.

Arnold, A.P., and Gorski, R.A. (1984) Gonadal steroid induction of structural sex differences in the CNS. Annu. Rev. Neurosci. 7:413–442.

Balazs, R. (1974) Influences of metabolic factors on brain development. Brain Med. Bull. 30:126–134.

Barakat, I., Wittendorp-Rechenman, E., Rechenman, R.V., and Sensenbrenner, M. (1981) Influence of meningeal cells on the proliferation of neuroblasts in culture. Dev. Neurosci. 4:363–372.

Barakat, I., Sensenbrenner, M., and Labourdette, G. (1982) Stimulation of chick neuroblast proliferation in culture by brain extracts. J. Neurosci. Res. 8:303–314.

Barde, A., Edgar, D., and Thoene, H. (1980) Sensory neurons in culture: changing requirements for survival factors during embryonic development. Proc. Natl. Acad. Sci. USA 77:1199–1203.

Barnes, D., and Sato, G. (1980) Serum-free cell culture: a unifying approach. Cell 22:649–655.

Berg, D.K. (1984) New neuronal growth factors. Annu. Rev. Neurosci. 7:149–170.

Black, I.B. (1978) Sympathetic neurone development. Trends Neurosci. 1:101–104.

Black, I.B. (1983) Cellular and molecular biology of neuronal development. New York: Plenum.

Bottenstein, J.E., and Sato, G.H. (1979) Growth of a rat neuroblastoma cell line in serum-free supplemental medium. Proc. Natl. Acad. Sci. USA 76:514–517.

Bradford, M. (1976) A rapid and sensitive method for the quantitation of microgram quantities of protein utilizing the principle of protein dye binding. Anal. Biochem. 72:248–254.

Buckstrom, M., Hall, K., and Sara, V. (1984) Somatomedian levels in cerebrospinal fluid from adults with pituitary disorders. Acta Endocrinol. 107:171–178.

Ceda, G.P., Hoffman, A.R., Silverberg, S.D., Wilson, D.M., and Rosenfeld, R.G. (1985) Regulation of growth hormone release from cultured human pituitary adenomos by somatomedians and insulin. J. Clin. Endocrinol. Metab. 60:1204–1209.

Choi, B.H., and Kim, R.C. (1984) Expression of glial fibrillary acidic protein in immature oligodendroglia. Science 223:407–408.

Chude, O., and Wu, N.J. (1976) A rapid method for assaying enzymes whose substrates and products differ by charge; application to brain L-glutamate decarboxylase. J. Neurochem. 27:83–86.

Ciment, G., and Weston, J.A. (1982) Early appearance in neural crest and crest-derived cells of an antigenic determinant present in avian neurons. Dev. Biol. 93:355–369.

Cowan, W.M., Fawcett, J.W., O'Leary, D.D.M., and Stanfield, B.B. (1984) Regressive events in neurogenesis. Science 225:1258–1265.

Dennis, S.C., Lai, J.C.K., and Clark, J.B. (1980) The distribution of glutamine synthetase in subcellular fractions of rat brain. Brain Res. 197:463–475.

Dickson, P.W., Aldred, A.R., Morley, P.D., Guo-Fen, T., Howlett, G.J., and Schreiber, G. (1985) High prealbumin and transferrin mRNA levels in the choroid plexus of rat brain. Biochem. Biophys. Res. Commun. 127:830–835.

Dimpfel, W., Huang, R.T.C., and Haberman, E. (1977) Gangliosides in nervous tissue cultures and binding of ^{125}I-labelled tetanus toxin, a neuronal marker. J. Neurochem. 29:329–334.

Dzisgielewska, K.M., Evans, C.A.N., Lai, P.C.W., Lorscheider, F.L., Malinowska, D.H., Mollgard, K., and Saunders, N.R. (1981) Protein in cerebrospinal fluid and plasma of fetal rats during development. Dev. Biol. 83:193–200.

Fujita, S. (1964) Analysis of neuron differentiation in the central nervous system by tritiated thymidine autoradiography. J. Comp. Neurol. 122:311–327.

Goldstein, M.E., Sternberger, N.H., and Sternberger, L.A. (1982) Developmental expression of neurotype revealed by immunocytochemistry with monoclonal antibodies. J. Neuroimmunol. 3:203–217.

Goodman, C.S., and Bastiani, M.J. (1984) How embryonic nerve cells recognize one another. Sci. Am. 251:58–66.

Goodyer, C.G., Stephano, L.D., Lai, W.H., Guyda, H.J., and Posner, B.I. (1984) Characterization of insulin-like growth factor receptors in rat anterior pituitary, hypothalamus, and brain. Endocrinology 114:1187–1195.

Gospodarowicz, D., Cheng, J., Lui, G.M., Baird, A., and Bohlent, P. (1984) Isolation of brain fibroblast growth factor by heparin–Sepharose affinity chromatography: identify with pituitary fibroblast growth factor. Proc. Natl. Acad. Sci. USA 81:6963–6967.

Gospodarowicz, D., and Moran, J. (1976) Growth factors in mammalian cell culture. Annu. Rev. Biochem. 45:531–558.

Graves, G.D. (1977) Thyroid hormones and brain development. New York: Raven.

Hamburger, V. (1948) The mitotic patterns in the spinal cord of the chick embryo and their relation to histgenic processes. J. Comp. Neurol. 88:221–283.

Hubel, D.H., Wiesel, T.N., and Le Vay, S. (1977) Plasticity of ocular dominance columns in monkey striate cortex. Philos. Trans. R. Soc. Lond. [B] 278:377–409.

Koulakoff, A., Bizzini, B., and Berwald-Netter, Y. (1983) Neuronal acquisition of tetanus toxin binding sites: relationship with the last mitotic cycle. Dev. Biol. 100:350–357.

Leveille, P.J., McGinnis, J.F., Maxwell, P.S., and de Vellis, J. (1980) Immunocytochemical localization of glycerol-3-phosphate dehydrogenase in rat oligodendrocytes. Brain Res. 196:287–305.

Levin, M.J., Tuil, D., Uzan, G., Dreyfus, J.C., and Kahn, A. (1984) Expression of the transferrin gene during development of non-hepatic tissues: high level of transferrin mRNA in fetal muscle and adult brain. Biochem. Biophys. Res. Commun. 122:212–217.

Maciag, T., Cerundolo, J., Ilsley, S., Kelley, P.R., and Forand, R. (1979) An endothelial cell growth factor from bovine hypothalamus: identification and partial characterization. Proc. Natl. Acad. Sci. USA 76:5674–5678.

Marangos, P.J., Polak, J.M., and Pearse, A.G.E. (1982). Neuron specific enolase: a probe for neurons and neuroendocrine cells. Trends Neurosci. 5:193–196.

Markelonis, G.J., Oh, T.E., Park, L.P., Cha, C.Y., Sofia, C.A., Kim, J.W., and Azari, P. (1985) Synthesis of the transferrin receptor by cultures of embryonic chicken spinal neurons. J. Cell Biol. 100:8–17.

McCarthy, K.S., and de Vellis, J. (1980) Preparation of separate astroglial and oligodendroglial cell cultures from rat cerebral tissue. J. Cell Biol. 85:890–902.

Miale, I.L., and Sidman, R.L. (1961) An autoradiographic analysis of histogenesis in the mouse cerebellum. Exp. Neurol. 4:277–296.

Mollgard, K., Jacobsen, M., Jacobsen, G.K., Clausen, P.P., and Saunders, N.R. (1979), Immunohistochemical evidence for an intracellular localization of plasma proteins in human foetal choroid plexus and brain. Neurosci. Lett. 14:85–90.

Morrison, R.S., and de Vellis, J. (1981) Growth of purified astrocytes in a chemically defined medium. Proc. Natl. Acad. Sci. USA 73:7208–7209.

Oppenheimer, J.H., and Samuels, H.H. (1983) Molecular basis of thyroid hormone action. New York: Academic Press.

Patel, A.J., Hunt, A., Gordon, R.D., and Balazs, R. (1982) The activities in different neuronal cell types of certain enzymes associated with the metabolic compartmentation glutamate. Rev. Brain Res. 4:3–11.

Patterson, P. (1978) Environmental determination of autonomic neurotransmitter functions. Am. Rev. Neurosci. 1:1–17.

Patterson, P.H., and Purves, D. (1982) Readings in developmental neurobiology. Cold Spring Harbor, NY: Cold Spring Harbor Laboratory.

Puro, D.G., and Egardh, E. (1984) Insulin-mediated regulation of neuronal maturation. Science 225:1170–1172.

Rakic, P. (1971) Neuron–glial relationship during granule cell migration in developing cerebellar cortex: a golgi and electron microscopic study in *Macacus rhesus*. J. Comp. Neurol. 141:283–312.

Rakic, P., and Goldman-Rakic, P.S. (1982) Development and modifiability of the cerebral cortex. Neurosci. Res. Prog. Bull. 20:4.

Rizzino, A., Rizzino, H., and Sato, G. (1979) Defined media and the determination of nutritional and hormonal requirements of mammalian cells in culture. Nutr. Rev. 37:369–378.

Saintonge, J., and Cote, R. (1984) Fetal brain development in diabetic guinea pigs. Pediatr. Res. 18:650–653.

Saneto, R., and de Vellis, J. (1985) Characterization of cultured rat oligodendrocytes proliferating in a serum-free, chemically defined medium. Proc. Natl. Acad. Sci. USA 82:3509–3513.

Shanker, G., Amur. S.G., and Pieringer, R.A. (1985) Investigation of myelinogenesis in vitro: a study of the critical period at which thyroid hormone exerts its maximum regulatory effect on the developmental expression of two myelin associated markers in cultured brain cells from embryonic mice. Neurochem. Res. 10:617–626.

Simpson, D., Morrison, R., de Vellis, J., and Herschman, H. (1982) Epidermal growth factor binding and mitogenic activity in purified populations of cells from the central nervous system. J. Neurosci. Res. 8:453–462.

Sorge, L.K., Levy, B.T., and Maness, P.F. (1984) pp60^{c-src} is developmentally regulated in the neuronal retina. Cell 36:249–257.

Toran-Allerand, C.D. (1980) Coexistence of α-fetoprotein, albumin and transferrin immunoreactivity in neurons of the developing mouse brain. Nature 286:733–735.

Tsai, H.M., Garber, B.B., and Larramendi, L.M.H. (1981a) ^{3}H-thymidine autoradiographic analysis of telencephalic histogenesis in the chick embryo. I. Neuronal birthdates of telencephalic compartments in situ. J. Comp. Neurol. 198:275–292.

Tsai, H.M., Garber, B.B., and Larramendi, L.M.H. (1981b) ^{3}H-thymidine autoradiographic analysis of telencephalic histogenesis in the chick embryo. II. Dynamics of neuronal migration, displacement and aggregation. J. Comp. Neurol. 198:293–306.

Varon, S., and Adler, R. (1981) Trophic and specifying factors directed to neuronal cells. Adv. Cell. Neurobiol. 2:115–163.

Vernadakis, A., Culver, B., and Nidess, R. (1978) Actions of steroid hormones on neural growth in culture: role of glial cells. Psychoneuroendocrinology 3:47–64.

Yong, J.W., and Fellows, R.E. (1980) Characterization of insulin stimulation of the incorporation of radioactive precursors into macromolecules in cultured rat brain cells. Endocrinology 107:1717–1724.

Zamenhof, S., and Ahmad, G. (1979) The effect of exogenous nutrients in growth of chick embryo brain. Growth 43:160–166.

21. ROLE OF STEROID HORMONES IN NEURAL GROWTH AND DIFFERENTIATION: IN VIVO AND IN VITRO COMPARISONS

FULVIA GREMO AND ANTONIA VERNADAKIS

1. INTRODUCTION

Although it is well known that hormones like thyroxine have a profound effect in regulating brain development and growth, the role played by steroid hormones such as glucocorticoids is still not well understood. Studies in vivo have shown that the effects of steroid hormones depend on the stage of maturation of neural tissue at the time of hormone treatment (Heim and Timiras, 1963; Vernadakis and Woodbury, 1963 and 1971). Cortisol and estradiol, when administered to developing rats, accelerate central nervous system (CNS) myelinogenesis (Casper et al., 1967). Other studies have shown that steroid hormones influence neurogenesis and brain cell proliferation (Bohn, 1980; Bohn and Lauder, 1978 and 1980; Cotterell et al., 1972; Patel and Balazs, 1980). More recently, neural culture studies have provided some basic information on mechanisms of actions of hormones at the cellular and molecular levels (for recent review, see Vernadakis, 1982). However, the question that frequently arises is whether the in vitro findings accurately and reliably reflect biological processes occurring in vivo. In this chapter, we compare specific data from in vivo and in vitro studies that allow us to evaluate the reproducibility and reliability of neural tissue cultures in predicting in vivo effects of exogenous and endogenous hormones.

2. EFFECTS OF HORMONES ON GLIAL CELLS

As will be discussed in more detail in other chapters of this book, the brain is a complex organ composed of neuronal "wiring cells" and the more supportive glial and connective tissue cells. It is the interaction of these cells directly or via the microenvironment that brings about nervous system activity and function. Glial cells play a key role in the modulation of the microenvironment and thus have a regulatory influence on neuronal growth and differentiation. Thus changes in glial growth and function would be ultimately reflected in neuronal function.

Extensive evidence derived primarily from culture studies shows that glucocorticoid hormones exert specific actions on glial cells (for reviews, see de Vellis et al., 1977; Kumar et al., 1985; Saneto and de Vellis, 1985; Vernadakis, 1982). Early work by Moscona and coworkers has established that glucocorticoid hormones induce the synthesis of glutamine synthetase (GS) in the chick embryonic retina (Moscona and Piddington, 1966). Glutamine synthetase is an enzyme of major importance in a variety of biosynthetic pathways (Meister, 1974) and is a key enzyme in the "small glutamate compartment" (Hertz, 1977, 1979; Van der Berg, 1970) that functions in salvaging neuronally released neurotransmitter metabolites. During embryonic development of the chick retina, the specific activity of GS begins to rise sharply at about day 16 of incubation, increases 100-fold with 5–6 days thereafter, and plateaus at this high level (Moscona, 1972; Moscona et al., 1980a). This growth pattern can be mimicked by treating the embryo in vivo or by adding cortisol to organ cultures of isolated retina tissue in a synthetic medium (Piddington and Moscona, 1967). Studies on organ cultures of retinas from 10-day-old embryos have demonstrated that cortisol enters into cells and induces accumulation of mRNA, which leads to a rapid increase of GS enzyme synthesis and levels in a few hours (Moscona and Wiens, 1975; Moscona et al., 1972; Sarkar and Moscona, 1973; Soh and Sarkar, 1978). It is of importance that GS can be induced precociously in the retina when corticoid receptors are present (Koeler and Moscona, 1966). Moreover, only cortisol and closely related 11-β-dihydrocorticosteroids can induce GS in the retina (Moscona and Piddington, 1966). Further studies on organ cultures have shown that precocious induction of GS synthesis takes place exclusively in the Muller cells (Linser and Moscona, 1979; Moscona et al., 1980b). Thus, the retinal target of glucocorticoids appears to be Muller cells, and competition for GS induction by specific corticosteroids and accumulation of GS represent regulatory features of these glial elements.

Although the effects of cortisol on GS synthesis were observed both in vivo and in vitro, some differences can be detected between organotypic and monolayer cultures. When cells remained dispersed and in random contacts, there was no measurable induction of GS at any time (Moscona, 1974). However, when trypsin-dissociated cells were reaggregated by rotation, treatment of such cell aggregates with cortisol resulted in GS induction.

Inductibility increased in parallel with the progressive histological organization of the cells and reconstruction of retinotypic tissue architecture (Morris and Moscona, 1971). Thus, specific cell–cell contacts, most likely between Muller cells and neurons, appear to be an essential condition for Muller cell responsiveness to the induction of GS by the steroid hormone.

In conclusion, these results show that many, but not all of the features of the intact retina are shared by in vitro cultures, unless the original architecture and histology ae maintained.

3. EFFECT OF THE HORMONE IN MODULATING NEUROTRANSMITTER EXPRESSION

Glucocorticoids influence both development and function of the nervous system (see review by Vernadakis, 1982). Moreover, maternal stress or treatments that elevate maternal glucocorticoid hormones may also result in neurological deficits (Moyer et al., 1978; Barlow et al., 1978; Ward and Weisz, 1980). Although the molecular mechanisms for these developmental defects have yet to be defined, there is increasing evidence that glucocorticoid hormones alter neurotransmitter phenotypic expression both in vivo and in vitro.

3.1 Effect of Glucocorticoid Hormones on Adrenergic Phenotype

Phenylethanolamine *N*-methyltransferase (PNMT) is an epinephrine-forming enzyme that in the adrenal is regulated by glucocorticoids (see review by Bohn, 1983). In adult rats, a decrease in glucocorticoid levels following hypophysectomy results in a fall in PNMT activity which can be reversed by glucocorticoids or adrenocorticotropic hormone (Wurtman and Axelrod, 1966). Glucocorticoids also regulate PNMT during development of the rat adrenal, as first suggested by Margolis et al. (1966). However, PNMT expression is not limited to the adrenal, but also occurs in sympathetic ganglia at the same embryologic age as in the adrenal (Bohn et al., 1982). In the rat, ganglionic PNMT activity increases from embryonic age 17.5 to term, but, in contrast to the adrenal, decreases to undetectable levels by 2 months after birth. If glucocorticoid levels are experimentally increased, PNMT-containing cells can be visualized by immunofluorescence in superior cervical ganglia (SCG), in other sympathetic ganglia, and in extra-adrenal chromaffin tissue (Bohn et al., 1982). This effect is observed only during a critical developmental period from embryonic age 18 to postnatal day, and it is confined to SIF-like (small intensely fluorescent) cells. Glucocorticoid treatment of neonatal rats during the first postnatal week increases the number of SIF cells in SCG, as visualized by formaldehyde-induced fluorescence (Ciaranello et al., 1973; Costa et al., 1974). When SCG are cultured in the presence of glucocorticoids, they contain elevated levels of PNMT activity and epinephrine as well as PNMT-immunofluorescent cells (Black et al., 1980; Bohn et al., 1984; Moore and Phillipson, 1975). A brief 2-hour pulse of dexamethasone (10^{-6} M) is sufficient to increase PNMT activity 11-fold after 2 days in culture (Bohn, 1983). This effect is mimicked by corticosterone, but not by other nonglucocorticoid steroids. Moreover, PNMT activity is significantly increased at 18 hours in vitro, but not earlier, which is interpreted to mean that the effect of the hormone requires some time for expression. However, brief exposure to a 2- to 4-hour pulse of dexamethasone results in the same increase in PNMT activity after 2 days in culture as with continuous exposure for the entire 2-day period. Observations in vivo and in vitro indicate that the number of PNMT-immunoreactive cells in the neonatal SCG increase with duration of dexamethasone treatment during the first 4 days. Thus, the effects observed in vivo are closely mimicked by glucocorticoid treatment in vitro. However, since these studies have been done in organ cultures, it is not possible to discriminate whether the effect is due to hormone action on neurons or on nonneuronal cells or both. Moreover, the role played by intercellular contacts cannot be evaluated.

3.2 Effects of the hormones on cholinergic phenotype

Muscarinic and nicotinic receptors have been described in the developing brain (Enna et al., 1976; Kuhar et al., 1980; Wade and Timiras, 1980), retina (Sugiyama et al., 1977), and tectum (Francis et al, 1980). We have used the chick embryo retinotectal system as a model to study the effects of glucocorticoids on the maturation of muscarinic receptors both in vivo and in vitro.

We first examined the distribution of corticosterone administered intracerebrally or intraocularly to the chick embryo in an attempt to determine whether localization of the hormone can be correlated with known actions of the hormone (Gremo and Vernadakis, 1981). We injected 1 μCi of ^{3}H-corticosterone into the brain of chick embryos at different stages of development and, at different times after the injection, retinas, cerebral hemispheres, optic tecta, and midbrain were dissected out and the percentage of total accumulated hormone was measured. We found that, at 8 days of incubation, there was a preferential accumulation into the retina, whereas in the other regions of the brain less than 10% of the hormone was detected (figure 21-1). This was in accord with the observations reported by Koeler and Moscona (1966), who showed that at this stage a peak of receptors was present, in contrast to the rest of the brain. In contrast to older stages of development, the hormone preferentially accumulated in the midbrain, suggesting that the hormone may play a role at different stages of development. The preferential accumulation into the retina was confirmed by the data obtained with injection of the hormone into the right eye. At 8 days of incubation, the hormone diffused and accumulated into the controlateral eye, whereas at later stages this phenomenon was no longer debectable. These data show that at 8 days of incubation the hormone can freely diffuse in and out of the retina, i.e., barriers to its accessability are not yet present.

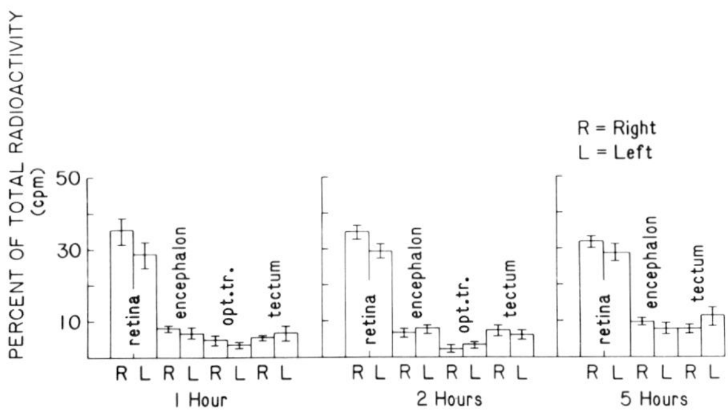

Figure 21-1. Percent of total accumulation of [^{3}H]-corticosterone recovered 1, 2, and 5 hours after injection into the brain of 8-day chick embryos. Total radioactivity recovered at *1 hour* = 23,331 ± 6290; *2 hours* = 22,185 ± 3092; and *5 hours* = 6231 ± 1974 (*opt. tr.*, optic tract; radioactivity was determined after 1 and 2 hours). Points represent means ± SE of 3–4 embryos. From Gremo and Vernadakis (1981).

In subsequent studies, we have found that the preferential accumulation of corticosterone correlated with the presence of cholinergic muscarinic receptors (Gremo and Vernadakis, 1983; Gremo et al., 1984 and 1985). Binding studies were performed using ^{3}H-quinuclidinyl benzilate (QNB) as the ligand. A single dose of 0.02 μg of corticosterone was injected intracerebrally into embryos at 8 days of incubation, and at 11 days the retinas and optic tecta were dissected and assayed for ^{3}H-QNB binding. We found that hormone treatment significantly affected the affinity of ^{3}H-QNB binding sites in retinas, whereas its effect in the optic lobes, where no preferential accumulation was detected, was very slight. Scatchard plot analysis of the saturation curves showed the presence of two ^{3}H-QNB binding sites (high and low affinity) in the treated retinas, whereas only the high-affinity one was present in the controls (figure 21-2 and 21-3). Thus, the hormone has a primary effect on retinal cells during early growth in the chick embryo. Agonist and antagonist binding showed also a difference between treated retinas and controls. Agonist binding showed that two binding sites (high and low affinity) were present in both treated and control retinas, already 24 hours after treatment. However, in the hormone-treated retinas, the two subpopulations significantly differed from the controls in their affinity and their relative percentage among the total receptor population. Moreover, using pirenzepine, an antagonist known to have the capacity to distinguish between muscarinic subclasses, two receptor subpopulations were found in the hormone-treated animals, whereas only a single one was present in the controls (Gremo et al., 1985; Marchisio et al., 1985). Thus, hormone treatment significantly affects muscarinic cholinergic binding sites, either by inducing the appearance of a new subpopulation of muscarinic cholinergic receptors or the maturation of a population of retinal cells carrying these

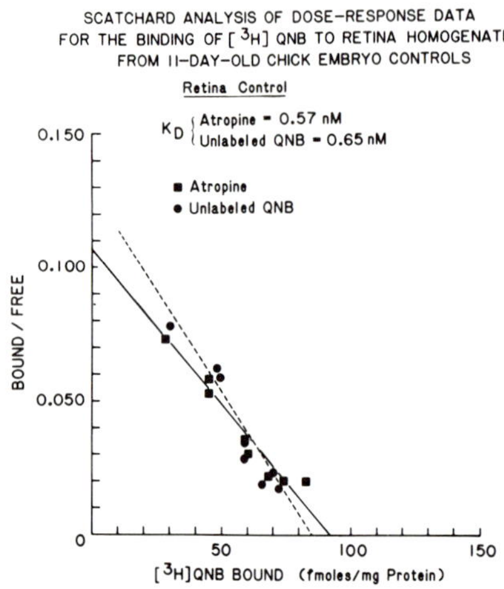

Figure 21-2. Scatchard analysis of dose-related data for the binding of ^{3}H-QNB to retina homogenates from 11-day-old chick embryo controls. Specific binding was determined with either 1 μM atropine (■-----■) or 1.01 μM unlabeled QNB (●-----●). Each point represents the mean of 3–5 experiments. From Gremo and Vernadakis (1983).

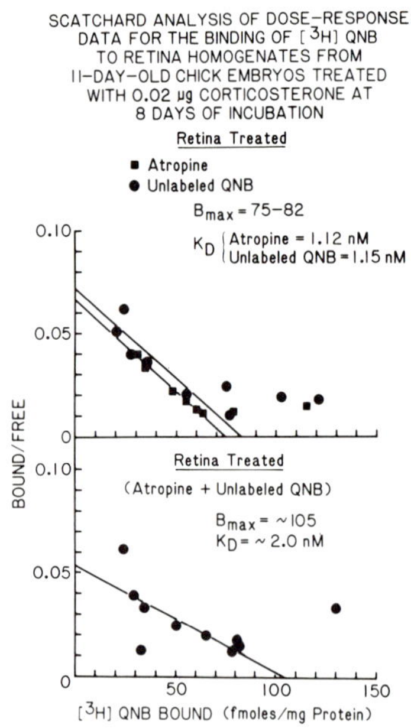

Figure 21-3. Scatchard analysis of dose–response data for the binding of ^{3}H-QNB to retina homogenates from 11-day-old chick embryos treated with 0.02 corticosterone at 8 days of incubation. (*Top half*) Specific binding was determined with either 1 μM atropine (■-----■) or 0.01 μM unlabeled QNB (●-----●). Points represent means of 3–5 experiments. (*Bottom half*) The data obtained using either atropine, 1 μM, or unlabeled QNB, 0.01 M, were combined and plotted; the points are mean values. From Gremo and Vernadakis (1983).

receptors. The possibility that corticosterone induces the expression of a new subpopulation of cholinergic binding sites recognized by pirenzepine is intriguing and warrants further investigation. Based on Moscona's findings discussed earlier, it can be speculated that the hormone is capable of interacting with the cellular genome when the neuron is at an early differentiating state.

We have speculated that corticosterone may affect retinal cells at very early stages of development by a mechanism involving cell surface modification and protein synthesis. However, the present studies have not provided any evidence concerning the hormone's alleged target cells in the retina (the Muller cells), or whether the hormone affects the cell itself or influences its behavior by modifying the microenvironment. In an attempt to further define the specific cellular effects of corticosterone, we have used retinal cultures. Retinas were dissected from 8-day-old chick embryos, dissociated by trypsinization, and cells were plated on salt-precipitated collagen. At day 5 in culture, cells were treated with corticosterone (from $10^{-9} M$ to $10^{-6} M$) for 24 hours (Gremo et al., 1984). The most striking observation was that corticosterone affects both neuronal and nonneuronal cells in culture (figure 21-4 through 21-6). However, the major effect of the hormone on neural cells was on the development of processes, and we have suggested that corticosterone may be a major modulatory factor in neuronal sprouting. Our results in neuronal sprouting by corticosterone agree with those reported by Unsicker et al. (1978), who reported that corticosterone impaired the outgrowth of processes in cultured sympathetic neurons and adrenal chromattin cells. Moreover, Cotman and associates (Scheff et al., 1980) found that corticoid hormones retarded postlesion sprouting of rat-dentate gyrus and have suggested that this effect may be due to astrocytic hypertrophy. In our study, however, we observed an inhibition of proliferation and confluency flat cells, presumptively glial cells, in hormone-treated cultures. Based on these findings, we have proposed that the lack of flat cell confluency in hormone-treated cultures can be a synergistic factor in the direct effects that the hormone may have on neuronal sprouting. The findings of Moscona and associates discussed earlier support this notion since they have shown that the Muller cells are the target cells of corticoids in the retina. Therefore, the effect of the hormone on cholinergic neuronal expression could be mediated via glial cells. Further supportive evidence is also provided by the studies of Fukada (1980) and McLennan et al. (1980), who found that corticosterone treatment greatly inhibited the cholinergic development in superior cervical ganglia in culture and suggested that the effect may be mediated via nonneuronal cells (McLennan et al., 1980). Moreover, Fukada (1980) has shown that glucocorticoids and epidermal growth factor exert antagonistic effects on the adrenergic cholinergic transmitter choice, presumably indirectly, by controlling and ability of heart cells to produce cholinergic-conditioned medium.

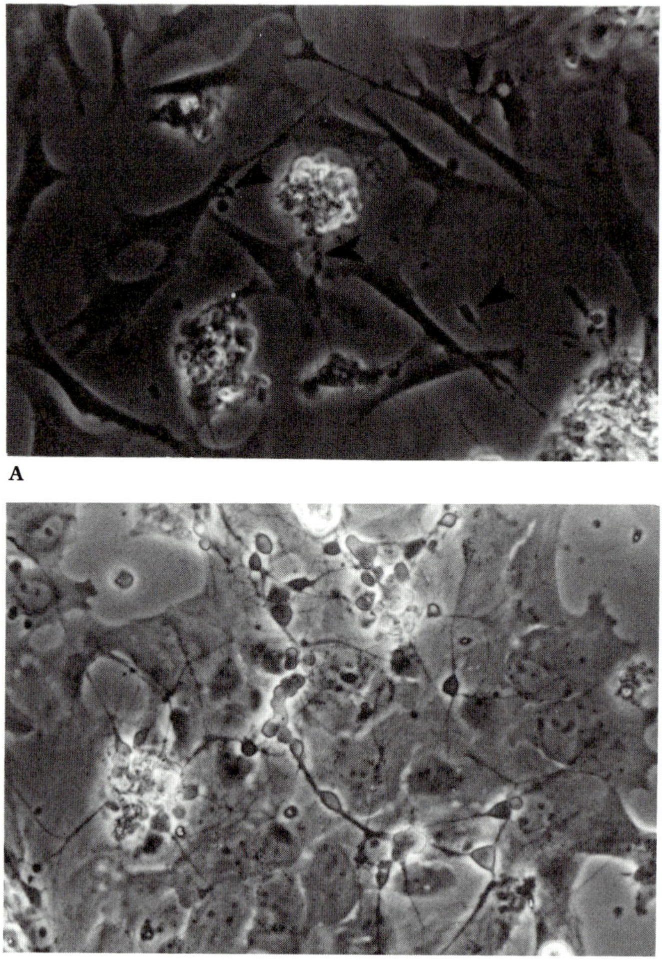

A

B

Figure 21-4. Retinal cultures at day 6. (**a**) Cortisone-treated cultures ($1 \times 10^{-7} M$). The neuronal network is poorly developed; the few isolated neurons (see *arrows*) are undifferentiated and aggregates are mostly unconnected. Flat cells are fewer and not confluent. (**b**) Control cultures. Both isolated neurons and aggregates show well-developed processes that form a network. A carpet of confluent flat cells underlies aggregates and cells. × 730. From Gremo et al. (1984).

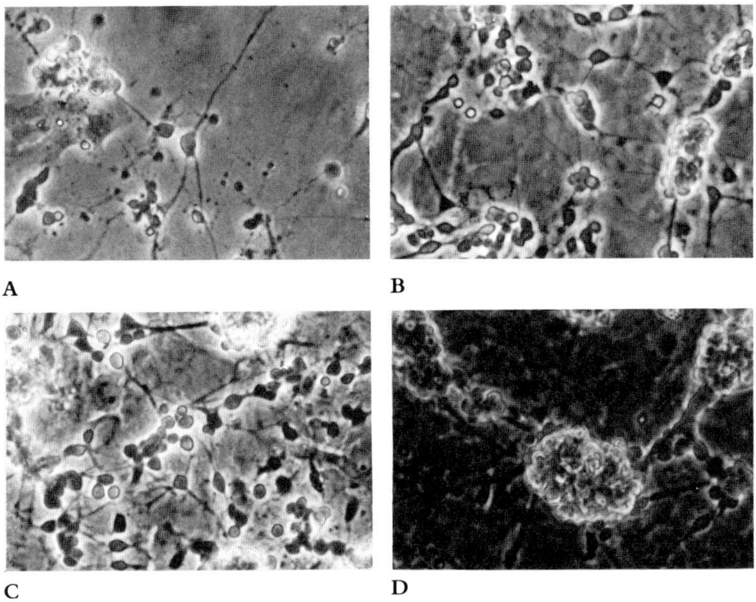

Figure 21-5. Retinal cultures at day 7. The effect of the hormone is evident, even at lower doses (1×10^{-9} M). (**a** and **b**) Corticosterone-treated cultures show fewer processes and flat cells; aggregates are smaller. (**c** and **d**) Control cultures show well-differentiated neurons that connect with each other and with the aggregates. × 730. From Gremo et al. (1984).

4. CONCLUSIONS

As will be discussed throughout this book, in vitro systems have provided basic information on regulatory mechanisms involved in growth and differentiation of nervous systems. The role of endogenous substances such as hormones as regulatory factors of growth and differentiation of the nervous system has long been advocated based primarily on in vivo observations. However, in order to investigate the basic cellular interrelationships and the molecular mechanisms involved, in vitro models such as neural cultures have been utilized. Comparative data of in vivo and in vitro hormone studies presented in this chapter provide evidence that several processes are closely mimicked by in vitro models and closely reflect in vivo phenomena.

ACKNOWLEDGMENTS

The research work by the authors included in this chapter was partially supported by a NATO grant 148.80 and by a Developmental Psychobiology Research Group Endowment fund.

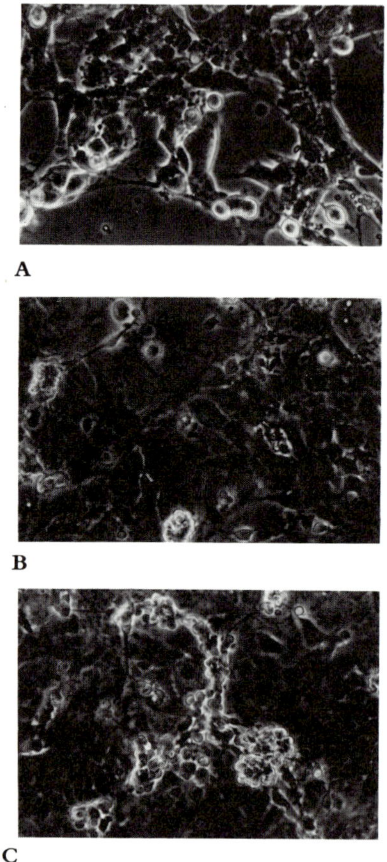

Figure 21-6. Retinal cultures at day 11. (**a**) The effect of the hormone ($1 \times 10^{-9} M$) continues to be evident at 11 days in vitro. (**b** and **c**) In contrast, interconnecting aggregates on top of a carpet of flat cells continue to be present in the control cultures.

REFERENCES

Barlow, S.M., Knight, A.F., and Succivam, F.M. (1978) Delay in postnatal growth and development of offspring produced by maternal restraint stress during pregnancy in the rat. Teratology 18:211.

Black, I.B., Bohn, M.C., and Bloom, E. (1980) Glucocorticoids induce expression of the adrenergic phenotype in a rat sympathetic ganglia. Soc. Neurosci. Abstr. 6:608.

Bohn, M.C (1980) Granule cell genesis in the hippocampus of rats treated neonatally with hydrocortisone. Neuroscience 5:2003–2012.

Bohn, M.C. (1983) Role of glucocorticoids in expression and development of phenylethanolamine *N*-methyltransferase (PNMT) in cells derived from the neural crest: a review. Psychoneuroendocrinology 8:381–390.

Bohn, M.C., and Lauder, J.H. (1978) The effects of neonatal hydrocortisone on rat cerebellum development: an autoradiographic and light microsope study. Dev. Neurosci. 1:250–266.

Bohn, M.C., and Lauder, J.M. (1980) Cerebellar granule cell genesis in the hydro cortisone-treated rat. Dev. Neurosci. 3:81–89.
Bohn, M.C., Goldstein, M., and Black, I. (1982) Expression of phenylethanolamine N-methyltransferase in rat sympathetic ganglia and extra-adrenal chromaffin tissue. Dev. Biol.89:299–308.
Bohn, M.C., Bloom, E., Goldstein, M., and Black, I. (1984) Glucocorticoid regulation of phenylethanolamine N-methyltransferase (PNMT) in organ culture of superior cervical ganglia. Dev. Biol. 105:130–136.
Casper, R., Vernadakis, A., and Timiras, P.S. (1967) Influence of estradiol and cortisol on lipids and cerebrosides in the developing brain and spinal cord of the rat. Brain Res. 5:524–526.
Ciaranello, R.D., Jacobowitz, D., and Axelrod, J. (1973) Effect of dexamethasone on phenylethanolamine N-methyltransferase in chromaffin tissue of the neonatal rat. J. Neurochem. 20:799–805.
Costa, M., Eranko, O., and Eranko, L. (1974) Hydrocortisone-induced increase in the histochemically demonstrable catacholamine content of sympathetic neurons of the newborn rat. Brain Res. 67:457–466.
Cotterell, M., Balazs, R., and Johnson, A.L. (1972) Effects of corticosteroids on the biochemical maturation of rat brain: postnatal cell formation. J. Microchem. 19:2151–2167.
de Vellis, J., McGinnis, J.F., Breen, G.A.M., Leveille, P., Bennett, K., and McCarthy, K. (1977) Hormonal effects on differentiation in neural cultures. In: Fedoroff, S., and Hertz, L. (eds.) Cell, tissue and organ cultures in neurobiology. New York: Academic Press, pp. 485–511.
Enna, S., Yamamura, H., and Snyder, S. (1976) Development of muscarinic cholinergic and GABA receptor binding in chick embryonic brain. Brain Res. 101:177–183.
Francis, A., Tagmmath, A., and Schecter, N. (1980) Stability of muscarinic cholinergic receptor activity in the deafferented retino-tectal pathway. Brain Res. 185:161–183.
Fukada, K. (1980) Hormonal control of neurotransmitter choice in sympathetic neural cultures. Nature 286:553–555.
Gremo, F., and Vernadakis, A. (1981) Preferential accumulation of ^{3}H-corticosterone in chick brain during embryonic development. Neurochem. Res. 6:343–351.
Gremo, F., and Vernadakis, A. (1983) Muscarinic cholinergic binding in chick embryo retino-tectal system: effect of corticosterone. J. Neurosci. Res. 9:27–36.
Gremo, F., Porru, S., and Vernadakis, A. (1984) Effects of corticosterone in chick embryonic retinal cells in culture. Dev. Brain Res. 15:45–52.
Gremo, F., Marchisio, A.M., and Vernadakis, A. (1985) Muscarinic receptor subclasses in the chick embryo retina: influence of corticosterone treatment. J. Neurochem. 45:345–351.
Heim, L.M., and Timiras, P.S. (1963) Gonad–brain relationship: precocious brain maturation after estradiol in rats. Endocrinology 72:598–606.
Hertz, L. (1977) Biochemistry of glial cells. In: Fedoroff, S., and Hertz, L. (eds.) Cell, tissue, and organ cultures in neurobiology. New York: Academic Press, pp. 39–97.
Hertz, L. (1979) Functional interactions between neurons and astrocytes. I. Turnover and metabolism of putative amino acid transmitters. Prog. Neurobiol. 13:277–323.
Koeler, D.E., and Moscona, A.A. (1966) Corticosteroid receptors in the neural retina and other tissues of the chick embryo. Arch. Biochem. Biophys. 170:102–103.
Kuhar, M.J., Birdsall, N.J.M., Burgen, A.S.V., and Hulme, E.C. (1980) Ontongeny of muscarinic receptors in rat brain. Brain Res. 184:375–383.
Kumar, S., Sachar, K., Huber, J., Weingarter, D.P., and de Vellis, J. (1985) Glucocorticoids regulate the transcription of glycerol phosphate dehydrogenase in cultured glial cells. J. Biol. Chem. 260:14743–14747.
Linser, P., and Moscona, A.A. (1979) Induction of glutamine synthetase in embryonic neural retina: localization in Muller fibers and dependence on cell interactions. Proc. Natl. Acad. Sci. USA 76:6476–6480.
Marchisio, A.M., Palomba, N., Mucas, M.L., and Gremo, F. (1985) Heterogeneity of muscarinic cholinergic receptors in the developing chick embryo retina. Brain Res. 325:381–389.
Margolis, F.L., Roff, J., and Jost, A. (1966) Norepinephrine methylation in fetal rat adrenals. Science 159:275–276.
McLennan, I.S., Hill, C.E., and Hendry, I.A. (1980) Glucocorticoids modulate transmitter choice in developing superior cervical ganglion. Nature 283:206–207.

Meister, A. (1974) Glutamine synthetase in mammals. In: Boyer, P.D. (ed.) The enzymes, vol. 10, 3rd edn. New York: Academic Press, pp. 699–754.

Moore, K.E., and Phillipson, O.T. (1975) Effects of dexamethasone on phenylethanolamine *N*-methyltransferase and adrenalin in the lens and superior cervical ganglia of adult and neonatal rats. J. Neurochem. 25:289–294.

Morris, J.E., and Moscona, A.A. (1971) The induction of glutamine synthetase in aggregates of embryonic neural retina cells: correlations with differentiation and multicellular organization. Dev. Biol. 25:420–444.

Moscona, A.A. (1972) Induction of glutamine synthetase in embryonic neural retina: a model for the regulation of specific gene expression in embryonic cells. In: Monroy, A., and Tsaner, R. (eds.) Biochemistry of cell differentiation, vol. 24. New York: Academic Press, pp. 1–23.

Moscona, A.A. (1974) Surface specification of embryonic cells: lectin receptors, cell recognition and specific cell ligands. In: Moscona, A.A. (ed.) The cell surface in development. New York: John Wiley and Sons, pp. 67–99.

Moscona, A.A., and Piddington, R. (1966) Stimulation by hydrocortisone of premature changes in the developmental pattern of glutamine synthetase in embryonic retina. Biochem. Biophys. Acta 121:409–411.

Moscona, A.A., and Wiens, A.W. (1975) Proflavine as a differential probe of gene expression: inhibition of glutamine synthetase induction in embryonic retina. Dev. Biol. 44:33–45.

Moscona, M., Frenkel, N., and Moscona, A.A. (1972) Regulatory mechanisms in the induction of glutamine synthetase in the embryonic retina: immunochemical studies. Dev. Biol. 28:229–241.

Moscona, A.A., Linser, P., Mayerson, P., and Moscona, M. (1980a) Regulatory aspects of the induction of glutamine synthetase in embryonic neural retina. In: Imora, J., and Palacios, R. (eds.) Glutamine; metabolism, enzymology and reproduction. New York: Academic Press, pp. 299–313.

Moscona, A.A., Mayerson, P., Linser, R., and Moscona, M. (1980b) Induction of glutamine synthetase in the neural retina of the chick embryo: localization of the enzyme in Muller fibers and effects of BrdU and cell separation. In: Giacobini, E., Vernadakis, A., and Shahar, A.A. (eds.) Tissue culture in neurobiology. New York: Raven, pp. 111–127.

Moyer, J.A., Herrenkohl, L.R., and Jacobowitz, D.M. (1978) Stress during pregnancy; effect on catecholamines in discrete brain regions of offspring as adult. Brain Res. 144:173–178.

Patel, A.J., and Balazs, R. (1980) Hormones and cell proliferation in the rat brain. In: Brambilla, F., Racagmi, R., and De Wied, D. (Eds.) Progress in psychoneuroendocrinology. Amsterdam: Elsevier, pp. 621–632.

Piddington, R., and Moscona, A.A. (1967) Precocious induction of retina glutamine synthetase by hydrocortisone in the embryo and in culture: age-dependent differences in tissue response. Biochim. Biophys. Acta 141:429–432.

Saneto, R.P., and de Vellis, J. (1985) Hormonal regulation of the proliferation and differentiation of astrocytes and oligodendrocytes in primary culture. In: Bottenstein, J., and Sato, G. (eds.) Cell culture in the neurosciences. New York: Plenum, pp. 125–167.

Sarkar, P.K., and Moscona, A.A. (1973) Glutamine synthetase induction in embryonic neural retina: immunochemical identification of polyamines involved in enzyme synthesis. Proc. Natl. Acad. Sci. USA 70:1667–1671.

Scheff, S.W., Bernardo, L.S., and Cotman, C.W. (1980) Hydrodortisone administration retards axon sprouting in rat dentate gyrus. Exp. Neurol. 68:195–201.

Soh, B.M., and Sarkar, P.K. (1978) Control of glutamine synthetase mRNA by hydrocortisone in the embryonic chick retina. Dev. Biol. 64:316–328.

Sugiyama, H., Daniels, M.P., and Nirenberg, M. (1977) Muscarinic acetylcholine receptors of the developing retina. Proc. Natl. Acad. Sci. USA 74:5524–5528.

Unsicker, K., Krisch, B., Otten, U., and Thoenen, H. (1978) Nerve growth factor-induced fiber outgrowth from isolated rat adrenal chromattin cells. Proc. Natl. Acad. Sci. USA 75:3498–3502.

Van der Berg. C.J. (1970) Glutamate and glutamine. In: Lajtha, A. (ed.) Handbook of neurochemistry, vol. 3. New York: Plenum, pp. 355–379.

Vernadakis, A. (1982) Endocrine regulation of neural development: physiological and biochemical aspects. In: Vernadakis, A., and Timiras, P. (eds.) Hormones in development and aging. New York: S.P. Medical and Scientific Books, pp. 207–271.

Vernadakis, A., and Woodbury, D.M. (1963) Effect of cortisol on the electroshock seizure thresholds in developing rats. J. Pharmacol. Exp. Ther. 139:110–113.

Vernadakis, A., and Woodbury, D.M. (1971) Effects of cortisol on maturation of the CNS. In: Ford, D.H. (ed.) Influence of hormones on the nervous system. Basel: S. Karger, pp. 85–97.

Wade, P.D., and Timiras, P.S., (1980) Whole brain regional [^{125}I]-α-bungarotoxin binding in developing rat. Brain Res. 181:381–389.

Ward, I.L., and Weisz, J. (1980) Maternal stress alters plasma testosterone in fetal males. Science 207:328–329.

Wurtman, R.J., and Axelrod, J. (1966) Control of enzymatic synthesis of adrenalin in the adrenal medulla by adrenal corticol steroids. J. Biol. Chem. 241:2301–2304.

22. NEURAL FACTOR-INDUCED FORMATION OF ACETYLCHOLINE RECEPTOR AGGREGATES ON MYOTUBES: AN IN VITRO MODEL FOR DEVELOPMENT OF THE POSTSYNAPTIC CELL SURFACE COMPLEX

MATHEW P. DANIELS, JACQUELINE G. KRIKORIAN,
ALICE LING, AND ANTHONY J. OLEK

1. INTRODUCTION

Signaling by neurons of the vertebrate central nervous system (CNS) is mainly the function of discrete chemical synapses. As generally understood, the presynaptic terminal releases neurotransmitter molecules that cross the synaptic cleft and bind to receptors in the postsynaptic membrane. As a result, ion channels open and the postsynaptic cell membrane potential is changed. The neurotransmitter receptors appear to be concentrated and anchored in place (Kelly et al., 1976) by a transmembrane complex of extracellular matrix and cytoskeletal structures, which we will refer to as the postsynaptic cell surface complex (see also Gulley and Reese, 1981). It is possible that this structural complex is also involved in the adhesion between presynaptic and postsynaptic membranes (see, for example, Bloom, 1970; Steinbach and Bloch, 1986), and in the maintenance or modulation of synaptic form (see Crick, 1982).

The complexity of the CNS has led many investigators who are interested in synaptic development and function to study the peripheral synapses formed by central neurons, in particular the skeletal neuromuscular junction (NMJ). The study of this synapse has been facilitated by its anatomic accessibility, and by the discovery and development of useful probes for the nicotinic acetylcholine receptor (AChR) of muscle, most notably the α-neurotoxins from snake venoms (see, for example, Lee et al., 1967; Changeux et al., 1970). These peptide toxins bind specifically and virtually

Vernadakis, A. (ed.), *Model Systems of Development and Aging of the Nervous System*.

irreversibly to the AChR and can be labeled to study the number and distribution of the receptors.

The level of experimental control and accessibility possible in the analysis of NMJ development is enhanced by the ability of some classes of central and peripheral cholinergic neurons to form synapses with skeletal myotubes in cell and tissue cultures (see Shimada and Kano, 1971; Fischbach, 1972; Steinbach et al., 1973; Nurse and O'Lague, 1975; Cohen, 1980). Correlated analysis of cholinergic synapse formation and AChR distribution in culture has shown that, as in vivo (Bevan and Steinbach, 1977; Betz et al., 1980; Chow and Cohen, 1983; Smith and Slate, 1983), the onset of functional innervation is rapidly followed by the aggregation of AChRs at the synaptic site (Anderson et al., 1979; Frank and Fischbach, 1979; Kidokoro et al., 1980). This receptor aggregation in vitro is probably the result of both redistribution of existing surface receptors (Anderson and Cohen, 1977; Anderson et al., 1977; Ziskind-Conhaim et al., 1984) and insertion of newly synthesized receptors from the internal pool (Role et al., 1985). The aggregation of AChRs at sites of innervation in vitro has been directly correlated with the accumulation of extracellular matrix (Frank and Fischbach, 1979; Anderson and Fambrough, 1983; Anderson et al., 1984), acetylcholinesterase (Frank and Fischbach, 1979; Rubin et al., 1979), and electron-dense specializations of the submembrane cytoskeleton (Anderson et al., 1984), as in vivo (reviewed in Steinbach and Bloch, 1986).

It is not known whether this localized effect of cholinergic neurons on muscle cells is mediated by the release of soluble factors from the neuron, or by direct contact between components of the nerve and muscle cell surfaces. However, it has been shown that the aggregation of AChRs and associated specialization of the muscle cell surface can be induced by exposing myotubes in culture to soluble factors from nervous tissues and neurons (Christian et al., 1978; Podleski et al., 1978; Jessell et al., 1979; Salpeter et al., 1982; Schaffner and Daniels, 1982; Sanes et al., 1984), or a detergent-insoluble preparation from the electric organ of *Torpedo* (Rubin et al., 1980; Godfrey et al., 1984), as well as by establishing contact between the muscle cell surface and various artificial surfaces (Bloch and Geiger, 1980; Peng et al., 1981; Peng and Cheng, 1982).

The discovery that "exogenous" factors such as these can induce muscle cell surface changes similar to those induced by innervation has provided a new experimental approach to studying the development of the postsynaptic cell surface complex in muscle. In this chapter, we summarize our recent work in devising a system that uses this approach, and describe some of our experimental observations. Except where otherwise stated, these methods and observations have appeared in published form or are in press (Olek et al., 1983 and 1986a and b; Krikorian and Daniels, 1985). We relate these findings to those from other laboratories, especially with reference to the mechanisms

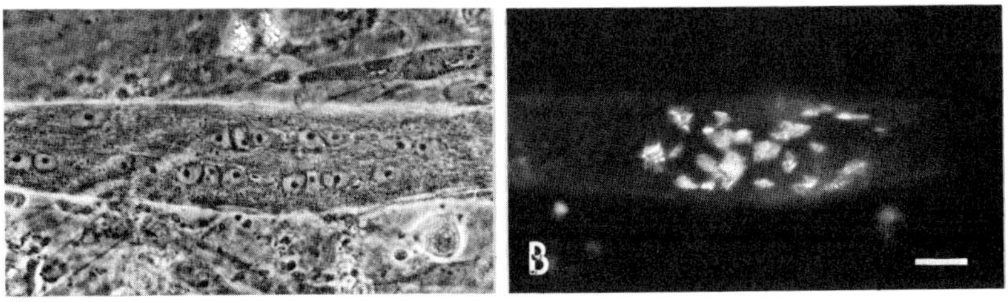

A B

Figure 22-1. EBX-induced AChR aggregates. Phase contrast (**A**) and fluorescence (**B**) photomicrographs of a myotube that was exposed to EBX for 4 hours, and stained with TMR-BTX. A group of brightly stained AChR aggregates is seen on the upper surface of the myotube (**B**). Bar = 20 μm.

involved in AChR organization and, finally, we point out some of the similarities between the phenomena that have been observed using exogenous factors and those that occur during synapse development in vivo and in vitro.

2. THE EXPERIMENTAL SYSTEM

Our initial goal was to develop an experimental system that would allow us first to observe sequential changes in ACh receptor distribution on individual living myotubes and then to examine the same myotubes by ultrastructural and immunocytochemical techniques. We believed that this approach could reveal early and intermediate steps in the formation of receptor aggregates and provide evidence for the involvement of specific structures and macromolecules. A rapid, vigorous AChR aggregation response was obtained in primary cultures of rat myotubes when they were treated with a saline extract of embryonic pig brain (EBX). In a typical experiment, about 60% of the myotubes were highly responsive, approximately 10% of their upper surfaces being covered with distinct receptor aggregates, 2–10 μm in longest dimension, after 4 hours (figure 22-1). Autoradiographic studies showed that these newly formed aggregates had receptor site densities 5–9 times higher than adjacent "diffuse" regions. EBX-induced aggregates could be observed with the fluorescence microscope in cultures that had been labeled with tetramethylrhodamine-conjugated α-bungarotoxin (TMR-BTX, which specifically labels the ACh receptors) before exposing the cells to EBX. There was little or no difference apparent in the 4-hour aggregation response between such prelabeled cultures and cultures labeled at the end of the 4 hours, indicating that the aggregation in this system was largely the result of rearrangement of receptors already present on the myotube surface before exposure to EBX. However, the results of autoradiographic experiments in which newly inserted receptors were labeled with ^{125}I-α-bungarotoxin 1 hour

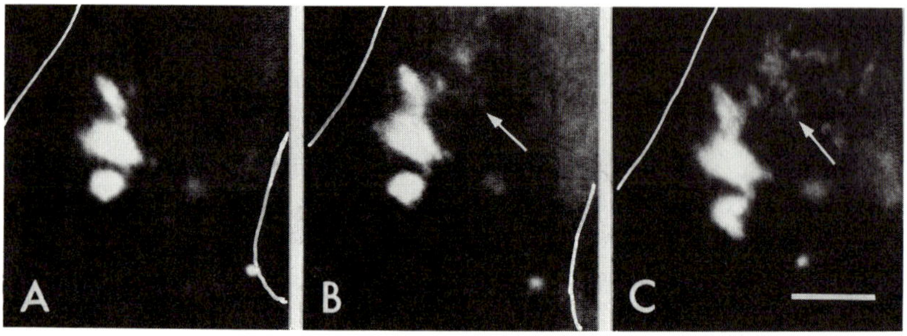

Figure 22-2. Formation of AChR microaggregates. Sequential VIF photomicrographs of the upper surface of a myotube that was stained with TMR-BTX and then exposed to EBX for 0 (**A**), 45 (**B**), and 90 (**C**) minutes. A cloud of AChR microaggregates (*arrows*) is seen to form near a preexisting AChR aggregate. *White lines* indicate the edges of the myotubes. Bar = 20 μm.

after prelabeling and blocking surface receptors with TMR-BTX indicated that newly synthesized or newly inserted receptors were also present in the EBX-induced aggregates (A.J. Olek, unpublished).

In our initial attempts to make sequential observations of AChR aggregation on TMR-BTX-labeled living cells, we found that exposure of the labeled myotubes to incident (excitation) illumination for the 10 seconds needed for conventional fluorescence photomicrography prevented EBX-induced changes in receptor distribution. This effect probably resulted from the excitation of the fluorochrome, since cells exposed to the excitation illumination without fluorochrome present showed a vigorous aggregation response. We found that sequential observations showing changes in the distribution of fluorescently labeled AChRs could be made by reducing the excitation intensity with neutral-density filters and reducing exposure time. Under these conditions, the aggregation response in sequentially observed cells was quantitatively comparable to that in cells observed only at the termination of EBX exposure. This reduction of incident illumination necessitated the use of image intensification, for which we used a video image intensification camera (see Willingham and Pastan, 1978).

With this system, it was possible to observe several myotubes repeatedly in a given culture, thus following the changes that occurred in AChR distribution. The culture could be fixed at a given time point and the directly observed myotubes could be relocated after processing for electron microscopy.

3. EARLY STAGES OF ACHR AGGREGATION

In our sequential observations of AChR aggregation, the earliest and most subtle change that we could detect after exposure of the myotubes to EBX was the appearance of groups of punctate fluorescent structures, less than 1 μm in diameter. We have celled these structures "microaggregates," and the groups of them "microaggregate clouds" (figure 22-2). By inspection,

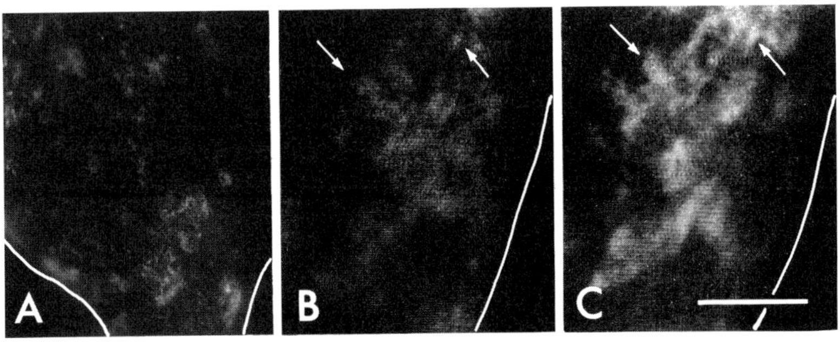

Figure 22-3. Formation of AChR aggregates from microaggregate clouds. Conventional (**A**) and VIF (**B** and **C**) photomicrographs of myotubes that were stained with TMR-BTX and exposed to EBX. (**A**) A myotube exposed to EBX for 4 hours at 23°C. The upper surface of the myotube shows several clouds of microaggregates. (**B**) Another myotube treated as in **A**. The microaggregate clouds can be seen, but the individual microaggregates are not well resolved under these photographic conditions. (**C**) The same field as in **B**, photographed 90 minutes after raising the temperature to 36°C in the continued presence of EBX. AChR aggregates have formed at the former sites of microaggregate clouds. (The transition is most clearly seen at the *arrows*.) Bar = 20 μm. Panels **B** and **C** are modified from Olek et al. (1986b).

fluorescence intensity within the microaggregate clouds was higher than that of surrounding (diffuse) cell surface regions, but less than that in the receptor aggregates found after longer exposure to EBX,. Results obtained by electron mcroscopy (see the following section) suggested that microaggregates were indeed small regions of elevated AChR density forming in the plane of the plasma membrane. Microaggregate clouds were observed within 30 minutes after exposing the myotubes to EBX and continued to appear for at least 2 hours more. However, the microaggregate clouds were transient structures, usually evolving into receptor aggregates within 2 hours.

The concept that microaggregate cloud formation is a discrete step in the early formation of AChR aggregates was supported by the results of experiments in which the culture incubation temperature was varied during exposure of the myotubes to EBX. The standard incubation temperature used in our experiments was 36° ± 0.5°C. At temperatures between 18° and 24°C, clouds of microaggregates were formed and microaggregates accumulated for several hours, but did not evolve into aggregates. The formation of microaggregate clouds at 23°C was reversibly inhibited by sodium azide, suggesting that the process requires energy in the form of ATP. When microaggregate clouds were induced to form at 23°C and the temperature was subsequently raised to 36°C in the continued presence of EBX, the microaggregate clouds evolved into aggregates within 90 minutes (figure 22-3). The overall pattern of the aggregates formed in this way closely resembled that of the precursor microaggregate clouds. Therefore, the microaggregate cloud appears to be the direct precursor of the aggregate, and the transition to aggregate may involve an increase in number and receptor site density of microaggregates within a given region of the myotube surface.

This temperature-sensitive transition should facilitate more detailed studies of microaggregate structure and composition, and the microaggregate-aggregate transition, since microaggregates can be accumulated at low temperatures, and the transition can be synchronized.

As already mentioned, under standard experimental conditions, microaggregate clouds appear to be transient structures. Our results also suggest that they are more labile than aggregates. Cells were exposed to EBX at 23°C for 4 hours to induce formation of microaggregate clouds. When EBX was removed and the cultures were warmed to 36°C, about 50% of the microaggregate clouds disappeared within 90 minutes, while the other 50% became aggregates. Aggregates formed after 4 hours of exposure to EBX (4-hour aggregates) are apparently more stable than microaggregate clouds in that the average half-life of an aggregate in the absence of EBX at 36°C was about 6 hours.

4. ULTRASTRUCTURAL CHANGES IN THE EARLY STAGES OF ACHR AGGREGATION

Studies from a number of laboratories have shown that AChR aggregates on cultured muscle (Burrage and Lentz, 1981; Peng and Cheng, 1982; Salpeter et al., 1982; Sanes et al., 1984; Anderson et al., 1984) and in muscle developing in vivo (Jacob and Lentz, 1979; Burrage and Lentz, 1981; Matthews-Bellinger and Salpeter, 1983) are associated with ultrastructural specializations of the extracellular matrix (i.e., basal lamina) and submembrane cytoskeleton (i.e., submembrane density or filaments). Thus, it has been suggested that these structrues might be involved in the formation or maintenance of the receptor aggregates (see the same references). One of our goals was to determine at which stage in the initial formation of AChR aggregates these specializations appeared, and whether steps in the assembly of this cell surface complex could be detected. To do this, we followed the EBX-induced formation of receptor aggregates by direct observation with video-intensified fluorescence (VIF). The observed cultures were fixed at the appropriate time points, embedded and sectioned for transmission electron microscopy. Since we were able to transpose a fluorescence map of the location of aggregates on the myotube surface onto electron micrographs of a series of thin sections with an accuracy of $\pm$ 1 μm, we could readily compare the cell surface ultrastructure within newly formed AChR aggregates to that in other regions. Although the location of microaggregate clouds could similarly be mapped, the individual microaggregates were too small to map accurately. For this reason, most experiments designed to study the structures associated with microaggregates utilized a mixture of TMR-BTX and horseradish peroxidase-conjugated α-bungarotoxin (HRP-BTX) to allow electron-microscopic visualization of individual microaggregates within the clouds.

These studies revealed a relatively subtle specialization of the ultrastructure

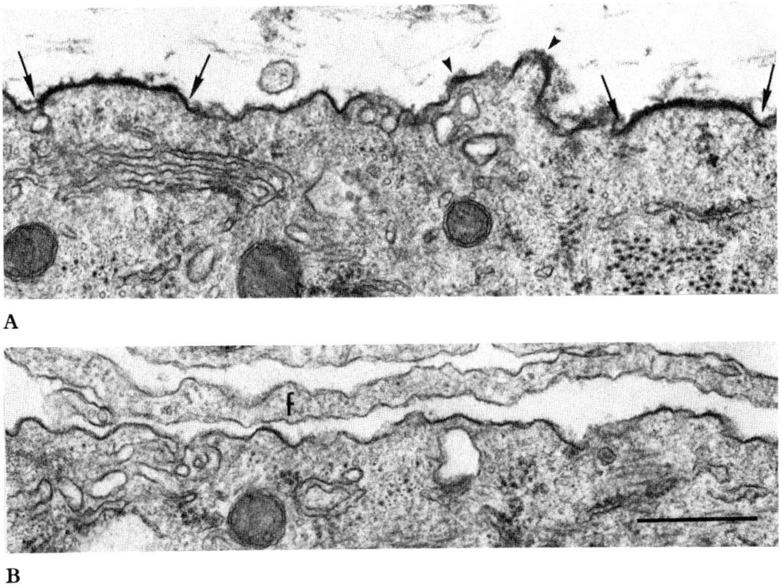

Figure 22-4. Ultrastructure of AChR microaggregates. Electron micrographs of transverse sections through a myotube that was stained with a mixture of TMR-BTX and HRP-BTX and then exposed to EBX for 90 minutes. The formation of microaggregate clouds was observed by VIF and the cell was fixed, reacted to reveal bound HRP-BTX, and processed for electron microscopy. The sections passed through (**A**) a newly formed microaggregate cloud and (**B**) a nearby region with diffuse AChR distribution. Two distinct HRP-positive mounds in the cell surface (between *arrows*) are presumed to be microaggregates. Smaller HRP-positive protrusions can also be seen (*arrowheads*). In **B**, a fibroblast process (*f*) overlies the muscle cell surface. Bar = 0.5 μm.

of the myotube surface associated with microaggregate clouds that had formed within 90 minutes of EBX treatment at 36°C. We found many small (usually 0.2–1.0 μm wide in profile) discrete mounds in the myotube surface with elevated AChR site density (figure 22-4), as seen by HRP-BTX labeling. The cytoplasm immediately below the mounds contained some fine filamentous material, and usually appeared to exclude organelles that closely approached the cell surface in other regions. Most of the mounds were not associated with a basal lamina or an electron-dense (by uranyl and lead staining) specialization of the submembrane cytoskeleton. This result is in agreement with the obervations reported by Peng and Phelan (1984), who found simply a meshwork of fine filaments that excluded most other cytoplasmic organelles, at charged bead contact sites that would later become AChR aggregates on *Xenopus* muscle cells.

In aggregates, formed within 4–6 hours of exposing the myotubes to EBX, a more elaborate structure was observed (figrues 22-5 and 22-8). The plasma membrane mounds were more numerous and closely packed than in

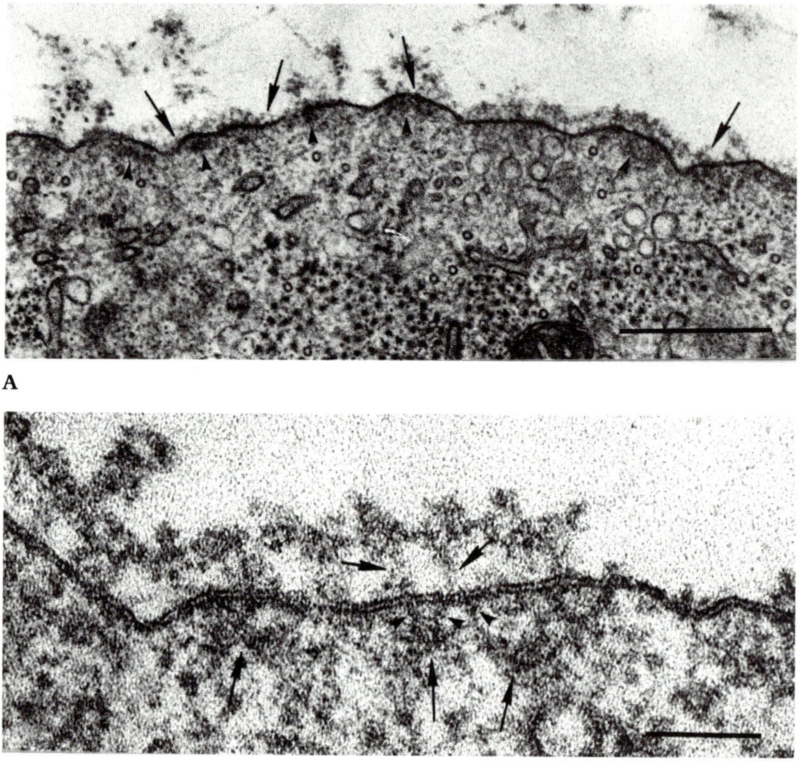

A

B

Figure 22-5. Ultrastructure of AChR aggregates. Electron micrographs of transverse sections through a myotube that was stained with TMR-BTX and then exposed to EBX for 4 hours. The formation of AChR aggregates was observed by VIF and the cell was fixed and processed for electron microscopy. The sections passed through newly formed AChR aggregates. At intermediate magnification (**A**), cell surface mounds with overlying basal lamina (seen most clearly between *arrows*) and an underlying dense cytoskeletal specialization (examples at *arrowheads*) are visible. At high magnification in a very thin section (**B**), it is possible to see structures that traverse the lamina lucida and apparently contact the plasma membrane (*short arrows*) where they appear to be aligned with short cytoplasmic filaments (*arrowheads*). These filaments are connected in turn with a dense structrue deeper in the cytoplasm (*long arrows*). Bars = 0.5μm (**A**) and 0.1 μm (**B**).

the microaggregate clouds and were clearly associated with a basal lamina. This is in agreement with earlier observations that immunoreactivity for laminin, a major glycoprotein of the basal lamina, was preferentially associated with both endogenous (Daniels et al., 1984; Bayne et al., 1984) and brain extract-induced AChR aggregates (Daniels et al., 1984). The submembrane cytoskeleton in aggregates was more obviously differentiated than in microaggregates. We observed an electron-dense layer, consisting of filaments and granules, located between 20 and 100 nm from the plasma membrane, as well as a less dense region immediately adjacent to the membrane, containing short segments of thin filaments. These filaments often appeared to contact both the dense cytoplasmic layer and the plasma membrane and were aligned with wispy columns of extracellular matrix connecting the lamina densa of the basal lamina to the plasma membrane. The detailed structure of the intracellular and extracellular specializations found at the newly formed AChR aggregates thus suggests the presence of an interconnected transmembrane complex, the structure and molecular composition of which merit further intensive study (see Bloch and Geiger, 1980; Anderson and Fambrough, 1983; Bloch and Resneck, 1983; Peng, 1984; Bayne et al., 1984; Daniels et al., 1984; Sanes et al., 1984; and the final section of this chapter).

It is probable that the aggregate-associated transmembrane complex was largely assembled de novo. Quantitation of electron micrographs showed that the extent of these cell surface specializations increased about threefold over controls in myotubes treated for 4 hours with EBX. In addition, as mentioned above, the microaggregate clouds, which would later become aggregates, did not have extensive specializations. Thus, it appears unlikely that receptor aggregates are formed in the presence of EBX as a result of the trapping of receptors in an elaborate, preassembled surface complex such as can be seen in the fully formed aggregates. More likely, relatively subtle changes in cell surface organization may lead to the initiation of receptor aggregation, which is followed by more extensive surface changes. The elaborate cell surface complex may be involved in the transition from microaggregate cloud to aggregate (by increasing the capacity to immobilize receptors) and in the stabilization of aggregates, once formed.

In some regions of the newly formed AChR aggregates, the surface contour was modified to a greater extent than the shallow "mounding" seen throughout the aggregates. Deep folds or invaginations of the plasma membrane and overlying basal lamina were found; the dense submembrane specialization was present virtually throughout the folds, and the cytoplasm within the folds was rich in microtubules. This deep folding of the myotube surface suggests that there is extensive remodeling of the myotube cytoskeleton in the vicinity of receptor aggregates.

5. REORGANIZATION AND STABILIZATION OF AChR AGGREGATES

Although 4-hour receptor aggregates appear to be relatively stable, compared to microaggregate clouds, they are nevertheless dynamic structrues. As mentioned above, 4-hour aggregates disappear slowly in the absence of EBX at 36°C. The rate of this disappearance of aggregates could be slowed down to about one-third by lowering the incubation temperature to 24° or 30°C. To our surprise, the rate of disappearance of 4-hour aggregates in the absence of EBX was increased about threefold by elevating the temperature from 36° to only 38°C (figure 22-6). Under these conditions, the half-life of 4-hour aggregates was approximately 2 hours, as opposed to 6 hours at 36°C. If the

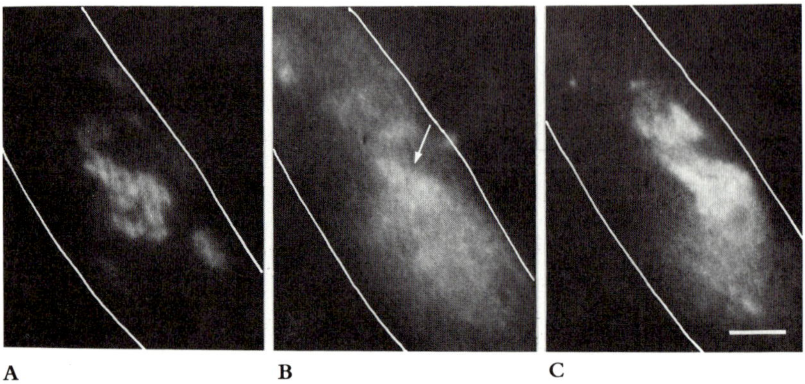

Figure 22-6. Reversible dispersal of AChR aggregates. Sequential VIF micrographs of the upper surface of a myotube. The myotube was exposed to EBX for 4 hours at 36°C and stained with TMR-BTX during the last hour, to reveal AChR aggregates (**A**). EBX was then removed and the temperature was raised to 38°C, causing the aggregates to disperse within 4 hours, leaving only a remnant of the original structure (*arrow* in **B**). When the temperature was returned to 36°C and EBX re-added, an aggregate of similar shape and location reformed within 4 hours (**C**). Bar = 20 μm. From Olek et al. (1986b).

temperature was lowered to 36°C and EBX added again, aggregates reformed within 4 hours, usually in close proximity to the site of the original aggregates and with a similar topographic configuration. This indicated that the disappearance of aggregates at 38°C was reversible, and suggested that some residual surface structure might predispose a given region to reformation of aggregates. Since the reformed aggregate could be visualized without relabeling surface AChRs, the disappearance of the aggregates at 38°C was probably due predominantly to dispersal of the aggregated receptors, which were subsequently reaggregated, and did not simply result from internalization of the receptors. Rapid dispersal of aggregates also occurred in the presence of 7 m*M* sodium azide and in culture medium with the Ca^{2+} concentration reduced to 0.1 m*M*, suggesting that the stability of 4-hour EBX-induced aggregates is dependent on the formation of ATP and extracellular Ca^{2+} ions. These last-mentioned results agree with the observations that contact-induced aggregates on the bottom of rat myotubes in culture are destabilized by exposure to energy metabolism inhibitors (Bloch, 1979) or low Ca^{2+} medium (Bloch, 1983; Bursztajn et al., 1984).

As mentioned above, 4-hour receptor aggregates tend to disappear slowly at 36°C in the absence of EBX. When exposure of the myotubes to EBX was continued beyond 4 hours, a gradual reorganization of surface AChRs, including the newly inserted receptors, occurred such that a group of small, irregularly shaped 4-hour aggregates (32 μm^2 mean area per aggregate) was replaced over a period of 20 hours by one or a few large, oval-shaped aggregates (483 μm^2 mean area per aggregate). The 24-hour aggregates were distinct from the 4-hour aggregates not only in size and shape, but also in stability (figure 22-7). They persisted approximately twice as long as 4-hour aggregates in the absence of EBX at 36° or 38°C, in the presence of 7 m*M* sodium azide, or in medium with 0.1 m*M* Ca^{2+}. In agreement with our results, the stability of nerve-induced AChR aggregates increased with days of nerve–muscle coculture (Kuromi and Kidokoro, 1984).

6. MECHANISMS OF ACh RECEPTOR AGGREGATION INDUCED BY EXOGENOUS FACTORS

Studies from several laboratories on the formation and dispersal of AChR aggregates induced by exogenous factors on myotubes in culture, and studies on the structure, molecular composition, and other physical properties of these aggregates, have led to progress (briefly assessed here) in delineating the kinds of mechanisms that may be involved in aggregate formation and stabilization in vivo as well as in vitro (for a detailed recent review, see Steinbach and Bloch, 1986).

Among the observations that should be taken into account are the following: 1) The formation and stability of AChR aggregates are sensitive to inhibitors of energy metabolism (Bloch, 1979; Christian et al., 1980; Olek et al., 1986b), are dependent on the concentration of extracellular Ca^{2+} (Bloch, 1979 and 1983; Christian et al., 1980; Bursztajn et al., 1984; Peng, 1984), and are sensitive to colchicine and other inhibitors of microtubule assembly (Bloch, 1979 and 1983; Christian et al., 1980; Connolly, 1984). 2) Aggregation of AChRs appears to involve their immobilization (Axelrod et al., 1976; Stya and Axelrod, 1983b) and increased binding to the detergent insoluble "cytoskeleton" (Prives et al., 1982; Stya and Axelrod, 1983a). These observations are complemented by ultrastructural, immunocytochemical, and some biochemical evidence for association of aggregated receptors with components of the cytoskeleton (Bloch and Geiger, 1980; Burrage and Lentz, 1981; Peng and Cheng, 1982; Salpeter et al., 1982; Bloch and Resneck, 1983; Peng and Phelan, 1984; Peng and Froehner, 1985; Olek et al., 1986a). 3) Components of the extracellular matrix also tend to be preferentially associated with the aggregated AChRs (Burrage and Lentz, 1981; Peng and Cheng, 1982; Salpeter et al., 1982; Silberstein et al., 1982; Anderson and Fambrough, 1983; Bayne et al., 1984; Daniels et al., 1984; Sanes et al., 1984; Olek et al., 1986a), and there is experimental evidence for activity or involvement of extracellular matrix components in the aggregation process (Rubin et al., 1980; Kalcheim et al., 1982a–c; Vogel et al., 1983; Godfrey et al., 1984).

There is some evidence for the role of increased synthesis and preferential insertion of AChRs into developing aggregates (Podleski et al., 1978; Jessell et al., 1979; Buc-Caron et al., 1983; Bursztajn and Fischbach, 1984; Bursztajn et al., 1984; Neugebauer et al., 1985), but it is clear that receptors already in the myotube surface can be mobilized to form aggregates (Christian et al., 1978; Schaffner and Daniels, 1982; Stya and Axelrod, 1983a; Godfrey et al.,

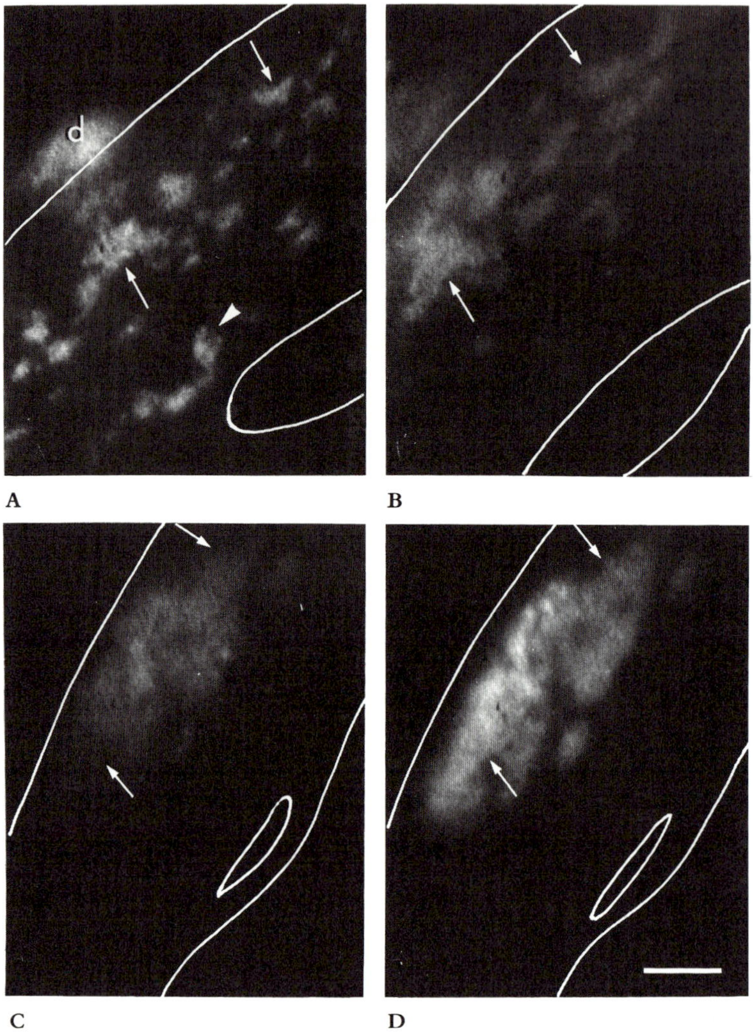

Figure 22-7. Reorganization of 4-hour AChR aggregates in the presence of EBX. Sequential VIF micrographs of the upper surface of a myotube. The myotube was exposed to EBX for 4 hours, stained with TMR-BTX during the last hour, and photographed (**A**). Exposure to EBX was continued to 10.5 (**B**) and 23 (**C**) hours, and then the myotube was restained to label newly inserted AChRs and photographed at 25 hours (**D**). *Arrows* indicate examples of the enlargement of 4-hour aggregates to form parts of the 24-hour aggregate. Some 4-hour aggregates dispersed (example at *arrowhead*). The decrease in aggregate brightness in **B** and **C** reflects the turnover of AChRs and the spreading of labeled AChRs over a larger area as the aggregates grow. The fluorescent object near the upper left edge of the myotube (*d*) is cellular debris. Bar = 20 μm.

1984). It is possible that these two mechanisms play proportionately different roles in different culture systems (see, for example, Neugebauer et al., 1985; and compare Christian et al., 1978, with Jessell et al., 1979).

In view of these observations, three kinds of mechanisms presently appear to provide viable (and not mutually exclusive) explanations of AChR aggregation: 1) active translocation of receptors into a given location, where they are immobilized, presumably by association with other molecules or macromolecular structures; 2) passive diffusion of receptors into a region in which they are similarly bound (see Edwards and Frisch, 1976); and 3) preferential insertion of newly synthesized receptors into a given region.

Although receptor aggregation and stabilization of aggregates are active processes, it is not known whether metabolic activity is required to provide energy for active translocation of the receptors, for events involved in the formation of a domain in which the receptors are immobilized, or in the intracellular traffic of receptors leading to their insertion into the cell surface.

7. COMPARISON OF EBX-INDUCED AChR AGGREGATION WITH NEUROMUSCULAR JUNCTION DEVELOPMENT IN VIVO AND IN VITRO

The EBX-induced changes in AChR organization and cell surface architecture, as observed with our experimental system and supported by descriptions from other laboratories, can be summarized as follows (figures 22-8 and 22-9): By some energy-dependent mechanism, AChRs are first organized into clouds of labile microaggregates that are associated with subtle changes in plasma membrane contour and cytoskeletal organization. In a temperature-sensitive step, these microaggregate clouds evolve into aggregates that have a higher receptor site density and are more stable than microaggregates. This transition is associated with the appearance of a distinctive transmembrane complex that includes a basal lamina and specializations of the submembrane cytoskeleton and with the appearance of more extreme changes in cell surface contour in some parts of the aggregate. With continued exposure of the myotubes to EBX, receptors in groups of small, irregularly shaped aggregates are gradually reorganized, along with newly inserted receptors, to form larger, oval-shaped aggregates that are again more stable.

The developmental events occurring in the course of NMJ formation that may be compared to these EBX-induced changes include the initial formation of junctional AChR aggregates and their subsequent reorganization, stabilization, and remodeling (see figure 22-9). In their reports on the early stages of NMJ formation in vivo and in vitro, other investigators have described the presence of loose clusters of "speckles" ($\leq 1\ \mu m$ in diameter) or small aggregates of AChR in the region of innervation (Anderson et al., 1979; Kidokoro et al., 1980; Steinbach, 1981; Ishikawa and Shimada, 1982; Role et al., 1985). Some of these speckles may correspond to the microaggregates that we have described as the first stage of aggregate formation on EBX-treated rat

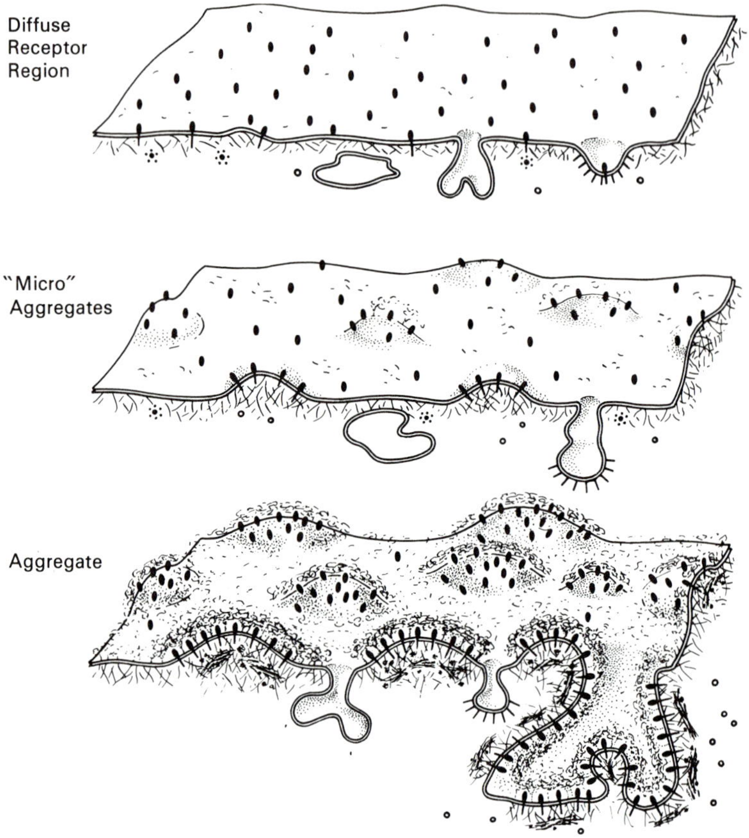

Figure 22-8. Diagrammatic representation of the changes in AChR distribution and cell surface ultrastructure that occur during EBX-induced formation of an AChR aggregate (see text). The *dark ovoids* represent AChRs. From Olek et al. (1986a).

myotubes. This similarity should be further examined by direct observation of the earliest stages of synaptic aggregate formation in vitro (see Kuromi and Kidokoro, 1982), closely correlated electron-microscopic observations, and quantitation of the receptor site density in both microaggregates and speckles.

The loosely organized junctional AChR cluster that forms initially is remodeled after 2–3 days to form a more compact, oval-shaped aggregate. The structural reorganization of EBX-induced aggregates that occurs with prolonged exposure of myotubes to EBX, as described here, may be comparable to this in vivo reorganization. In the rat, the stability of junctional aggregates (as seen by resistance to dispersal in medium with low Ca^{2+} or high K^{+} concentrations) increases between embryonic day 16 and birth (Bloch and Steinbach, 1981). Kuromi and Kidokoro (1984) have reported that the resis-

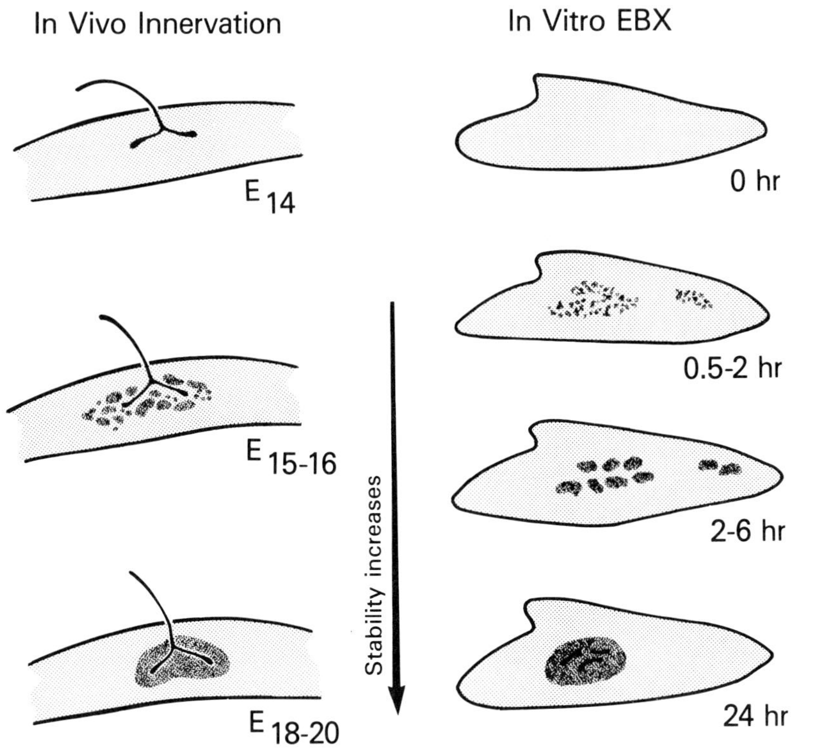

Figure 22-9. Diagrammatic representation of the changes in AChR distribution and the stability of aggregates in the rat NMJ developing in vivo (based on Steinbach 1981; Bloch and Steinbach, 1981) and analogous changes in rat myotubes in culture, exposed to EBX (see text).

tance of junctional aggregates to dispersal after denervation in *Xenopus* nerve–muscle cocultures increases with the number of days of coculture. Similar mechanisms may be involved in the stabilization of aggregates that we have observed with prolonged exposure of rat myotubes in cultures to EBX. It is possible that this stabilization results from the progressive accumulation of components associated with the AChR aggregate such as reported for certain basal lamina components (Weinberg et al., 1981; Sanes et al., 1984) for actin (Hall et al, 1981) and for an M_r 43,000 protein associated with the AChRs on the cytoplasmic face of the postsynaptic membrane (Peng and Froehner, 1985).

The precise relationship between the accumulation of AChRs and the formation of basal lamina and electron-dense cytoskeletal specializations at the developing NMJ remains unclear. Association of such specializations with newly formed junctions and junctional AChR aggregates has been reported both in vivo (Teravainen, 1968; Kelly and Zacks, 1969; Lentz, 1969; Jacob and Lentz, 1979; Burrage and Lentz, 1981) and in vitro (Weldon and

Cohen, 1979; Nakajima et al., 1980; Burrage and Lentz, 1981; Anderson et al., 1984) and is essentially in agreement with the findings made for exogenous factor-induced aggregates in vitro (see above). However, results of electron-microscopic autoradiographic studies on the rat NMJ in vivo (Matthews-Bellinger and Salpeter, 1983) suggest that the extensive formation of these structures may lag considerably behind the accumulation of AChRs. This apparent discrepancy may be explained by the extensive remodeling of the junctional receptor aggregate that occurs throughout prenatal development of the junction and for some weeks after birth (Steinbach, 1981; Slater, 1982). This remodeling could be reflected in the presence of many cell surface regions in a state of transition. It is also possible that some of the regions with elevated receptor density, but little obvious specialization, are microaggregate regions.

The strength of the analogies drawn between soluble factor-induced development of a postsynaptic-like complex and similar developmental events at the NMJ will depend in part on the similarity (or identity) of the neuronal macromolecules that provide the signals. Studies to identify and isolate the active molecules present in embryonic brain extracts or junctional basal lamina have been in progress in several laboratories for a number of years (Rubin et al., 1980; Bauer et al., 1981; Buc-Caron et al., 1983; Godfrey et al., 1984; Fallon et al., 1985; Knaack and Podleski, 1985; Neugebauer et al., 1985). The results of these studies suggest that there may be multiple factors of widely different molecular weights involved in this nerve–muscle interaction.

For the present, the identity of the molecules that mediate the nerve-induced development of the postsynaptic cell surface copmplex in muscle remains largely uncertain. However, there is extensive similarity between the developing postsynaptic complex at the neuromuscular junction and the complex associated with AChR aggregates induced by exogenous factors (as described above), including the overall pattern of receptor aggregation and stabilization of aggregates, the structure, and the composition of the complex. It seems likely, in fact, that the formation of a postsynaptic complex is largely under the control of the muscle cell, and need only be triggered by one of several types of signals. Thus, in vitro model systems should continue to generate useful information about the mechanisms of synaptic development.

REFERENCES

Anderson, M.J., and Cohen, M. (1977) Nerve-induced and spontaneous redistribution of acetylcholine receptors on cultured muscle cells. J. Physiol. 268:757–773.

Anderson, M.J., and Fambrough, D.M., (1983) Aggregates of acetylcholine receptors are associated with plaques of a basal lamina heparan sulfate proteoglycan on the surface of skeletal muscle fibers. J. Cell Biol. 97:1396–1411.

Anderson, M.J., Cohen, M., and Zorychta E. (1977) Effects of innervation on the distribution of acetylcholine receptors on cultured muscle cells. J. Physiol. 268:731–756.

Anderson, M.J., Kidokoro, Y., and Gruener, R. (1979) Correlation between acetylcholine receptor localization and spontaneous synaptic potentials in cultures of nerve and muscle. Brain Res. 166:185–190.

Anderson, M.J., Klier, F.G., and Tanguay, K.E. (1984) Acetylcholine receptor aggregation parallels the deposition of a basal lamina proteoglycan during development of the neuromuscular junction. J. Cell. Biol. 99:1769–1784.

Axelrod, D., Ravdin, P., Koppel, D.E., Schlessinger, J., Webb, W.W., Elson, E.L., and Podleski, T.R. (1976) Lateral mobility of fluorescently labeled acetylcholine receptors in membranes of developing muscle fibers. Proc. Natl. Acad. Sci. USA 73:4594–4598.

Bauer, H.C., Daniels, M.P., Pudimat, P.A., Jacques, L., Sugiyama, H., and Christian, C.N. (1981) Characterization and partial purification of a neuronal factor which increases acetylcholine receptor aggregation on cultured muscle cells. Brain Res. 209:395–404.

Bayne, E.K., Anderson, M.J., and Fambrough, D.M. (1984) Extracellular matrix organization in developing muscle: correlation with acetylcholine receptor aggregates. J. Cell Biol. 99:1486–1501.

Bevan, S., and Steinbach, J.H. (1977) The distribution of α-bungarotoxin binding sites on mammalian skeletal muscle developing in vivo. J. Physiol. 267:195–213.

Betz, H., Bourgeois, J.P., and Changeux, J.P. (1980) Evolution of cholinergic proteins in developing slow and fast skeletal muscles in chick embryo. J. Physiol. 302:197–218.

Bloch, R.J. (1979) Dispersal and reformation of acetylcholine receptor clusters of cultured rat myotubes treated with inhibitors of energy metabolism. J. Cell Biol. 82:626–643.

Bloch, R.J. (1983) Acetylcholine receptor clustering in rat myotubes: requirement for Ca^{2+} and effects of drugs which depolymerize microtubules. J. Neurosci. 3:2670–2680.

Bloch, R.J., and Geiger, B. (1980) The localization of acetylcholine receptor clusters in areas of cell substrate contact in cultures of rat myotubes. Cell 21:25–35.

Bloch, R.J., and Hall, Z.W. (1983) Cytoskeletal components of the vertebrate neuromuscular junction: vinculin, α-actinin and filamin. J. Cell Biol. 97:217–223.

Bloch, R.J., and Resneck, W.G. (1983) A molecule resembling actin is involved in clustering of acetylcholine receptors in cultured rat myotubes. Soc. Neurosci. Abstr. 9:757.

Bloch, R.J., and Steinbach, J.H. (1981) Reversible loss of acetylcholine receptor clusters at the developing rat neuromuscular junction. Dev. Biol. 81:386–391.

Bloom, F.E. (1970) Correlating structure and function of synaptic ultrastructure. In: Schmitt, F.O. (ed.) The neurosciences: second study program. New York: Rockefeller University Press, pp. 729–747.

Buc-Caron, M.H., Nystrom, P., and Fischbach, G.F. (1983) Induction of acetylcholine receptor synthesis and aggregation: partial purification of a low molecular weight activity. Dev. Biol. 95:378–386.

Burrage, T.G., and Lentz, T.L. (1981) Ultrastructural characterization of surface specializations containing high-density acetylcholine receptors on embryonic chick myotubes in vivo and in vitro. Dev. Biol. 85:267–286.

Bursztajn, S., and Fischbach, G. (1984) Evidence that coated vesicles transport acetylcholine receptors to the surface membrane of chick myotubes. J. Cell Biol. 98:498–506.

Bursztajn, S., McManaman, J.L., and Appel, S.H. (1984) Organization of acetylcholine receptor clusters in cultured rat myotubes is calcium dependent. J. Cell Biol. 98:507–517.

Bursztajn, S., Berman, S.A., McManaman, J.L., and Watson, M.L. (1985) Insertion and internalization of acetylcholine receptors at clustered and diffuse domains on cultured myotubes. J. Cell Biol. 101:104–111.

Changeux, J.P., Kasai, M., and Lee, C.Y. (1970) Use of a snake venom toxin to characterize the cholinergic receptor protein. Proc. Natl. Acad. Sci. USA 67:1241–1247.

Chow, I., and Cohen, M.W. (1983) Developmental changes in the distribution of acetylcholine receptors in the myotubes of *Xenopus laevis*. J. Physiol. 339:553–571.

Christian, C.N., Daniels, M.P., Sugiyama, H., Vogel, Z., Jacques, L., and Nelson, P.G., (1978) A factor from neurons increases the number of acetylcholine receptor aggregates on cultured muscle cells. Proc. Natl. Acad. Sci. USA 75:4011–4015.

Christian, C.N., Bergey, G.K., Daniels, M.P., and Nelson, P.G. (1980) Cell interactions in nerve and muscle cell cultures. J. Exp. biol. 89:85–101.

Cohen, M.W. (1980) Development of an amphibian neuromuscular junction in vivo and in culture. J. Exp. Biol. 89:43–56.

Connolly, J.A. (1984) Role of the cytoskeleton in formation, stabilization, and removal of acetylcholine receptor clusters in cultured muscle cells. J. Cell Biol. 99:148–154.

Crick, F. (1982) Do dendritic spines twitch? Trends Neurosci. 5:44–46.

Daniels, M.P., Vigny, M., Sonderegger, P., Bauer, H.C., and Vogel, Z. (1984) Association of Laminin and other basement membrane components with regions of high acetylcholine receptor density on cultured myotubes. Int. J. Dev. Neurosci. 2:87–99.

Edwards, C., and Frisch, H.L. (1976) A model for the localization of acetylcholine receptors at the muscle endplate. J. Neurobiol. 4:377–381.

Fallon, J.R., Nitkin, R.M., Reist, N.E., Wallace, B.G., and McMahan, U.J., (1985) Acetylcholine receptor-aggregating factor is similar to molecules concentrated at neuromuscular junctions. Nature 315:571–574.

Fischbach, G.D. (1972) Synapse formation between dissociated nerve and muscle cells in low density cell cultures. Dev. Biol. 28:407–429.

Frank, E., and Fischbach, G.D. (1979) Early events in neuromuscular junction formation in vitro: induction of acetylcholine receptor clusters in the postsynaptic membrane and morphology of newly formed synapses. J. Cell Biol. 83:143–158.

Godfrey, E.W., Nitkin, R.M., Wallace, B.G., Rubin, L.L., and McMahan, U.J. (1984) Components of *Torpedo* electric organ and muscle that cause aggregation of acetylcholine receptors on cultured muscle cells. J. Cell Biol. 99:615–627.

Gulley, R.L., and Reese, T.S. (1981) Cytoskeletal organization at the postsynaptic complex. J. Cell Biol. 91:298–302.

Hall, Z.W., Lubit, B.W., and Schwartz, J.H. (1981) Cytoplasmic actin in postsynaptic structures at the neuromuscular junction. J. Cell Biol. 90:789–792.

Ishikawa, Y., and Shimada, Y. (1982) Acetylcholine receptors and cholinesterase in developing chick skeletal muscle fibers. Dev. Brain Res. 5:187–197.

Jacob, M., and Lentz, T.L. (1979) Localization of acetylcholine receptors by means of horseradish peroxidase-α-bungarotoxin during formation and development of the neuromuscular junction in chick embryo. J. Cell. Biol. 82:195–211.

Jessell, T.M., Siegel, R.E., and Fischbach, G.D. (1979) Induction of acetylcholine receptors on cultured skeletal muscle by a factor extracted from brain and spinal cord. Proc. Natl. Acad. Sci. USA 76:5397–5401.

Kalcheim, C., Duksin, D., and Vogel, Z. (1982a) Involvement of collagen in the aggregation of acetylcholine receptors on cultured muscle cells. J. Biol. Chem. 257:12722–12727.

Kalcheim, C., Duksin, D., and Vogel, Z. (1982b) Aggregation of acetylcholine receptors in nerve–muscle cocultures is decreased by inhibitors of collagen production. Neurosci. Lett. 31:265–270.

Kalcheim, C., Vogel, Z., and Duksin, D. (1982c) Embryonic brain extract induces collagen biosynthesis in cultured muscle cells: involvement in acetylcholine receptor aggregation. Proc. Natl. Acad. Sci. USA 79:3077–3081.

Kelly, A.M., and Zachs, S.I. (1969) The fine structure of motor endplate morphogenesis. J. Cell Biol. 42:154–169.

Kelly, P., Cotman, C.W., Gentry, C., and Nicolson, G.L. (1976) Distribution and mobility of lectin receptors on synaptic membranes of identified neurons in the central nervous system. J. Cell Biol. 71:487–496.

Kidokoro, Y., Anderson, M.J., and Gruener, R. (1980) Changes in synaptic potential properties during acetylcholine receptor accumulation and neurospecific interactions in *Xenopus* nerve–muscle cell culture. Dev. Biol. 78:464–483.

Knaack, D., and Podleski, T. (1985) Ascorbic acid mediates acetylcholine receptor increase induced by brain extract in L_5 myogenic cells. Proc. Natl. Acad. Sci. USA 82:575–579.

Krikorian, J.G., and Daniels, M.P. (1985) Reorganization and stabilization of acetylcholine receptor (AChR) aggregates in the presence of embryonic brain extract (EBX). J. Cell Biol. 101:131a.

Kuromi, H., and Kidokoro, Y. (1982) Development of acetylcholine receptor clusters at nerve–muscle contacts in *Xenopus* cultures. Soc. Neurosci. Absr. 8:863.

Kuromi, H., and Kidokoro, Y. (1984) Denervation disperses acetylcholine receptor clusters at the neuromuscular junction in *Xenopus* cultures. Dev. Biol. 104:421–427.

Lee, C.Y., and Tseng, L.F. (1966) Distribution of *Bungarus multicinctus* venom following envenomation. Toxicon 3:281–290.

Lentz, T.L. (1969) Development of the neuromuscular junction. I. Cytologic and cytochemical studies on the neuromuscular junction of differentiating muscle in the regenerating limb of the newt, *Triturus*. J. Cell Biol. 42:431–443.
Matthews-Bellinger, J.A., and Salpeter, M.M. (1983) Fine structural distribution of acetylcholine receptors at developing mouse neuromuscular junctions. J. Neurosci. 3:644–657.
Nakajima, Y., Kidokoro, Y., and Klier, F.G. (1980) The development of functional neuromuscular junctions in vitro: an ultrastructural and physiological study. Dev. Biol. 77:52–72.
Neugebauer, K., Salpeter, M.M., and Podleski, T.R. (1985) Differential responses of L_5 and rat primary muscle cells to factors in rat brain extract. Brain Res. 346:58–69.
Nurse, C.A., and O'Lague, P.H. (1975) Formation of cholinergic synapses between dissociated sympathetic neurons and skeletal myotubes of the rat in cell culture. Proc. Natl. Acad. Sci. USA 72:1955–1959.
Olek, A.J., Pudimat, P.A., and Daniels, M.P. (1983) Direct observation of the rapid aggregation of acetylcholine receptors on identified cultured myotubes after exposure to embryonic brain extract. Cell 34:255–264.
Olek, A.J., Ling, A., and Daniels, M.P. (1986a) Development of ultrastructural specializations during the formation of acetylcholine receptor aggregates on cultured myotubes. J. Neurosci. 6:487–497.
Olek, A.J., Krikorian, J.G., and Daniels, M.P. (1986b) Early stages in the formation and stabilization of acetylcholine receptor aggregates on cultured myotubes: sensitivity to temperature and azide. Dev. Bio. 117:24–34.
Peng, H.B. (1984) Participation of calcium and calmodulin in the formation of acetylcholine receptor clusters. J. Cell Biol. 98:550–557.
Peng, H.B., and Cheng, P.C. (1982) Formation of postsynaptic specializations induced by latex beads in cultured muscle cells. J. Neurosci. 2:1760–1774.
Peng, H.B., and Froehner, S.C. (1985) Association of the postsynaptic 43K protein with newly formed acetylcholine receptor clusters in cultured muscle cells. J. Cell Biol. 100:1698–1705.
Peng, H.B., and Phelan, K.A. (1984) Early cytoplasmic specialization at the presumptive acetylcholine receptor cluster: a meshwork of thin filaments. J. Cell Biol. 99:344–349.
Peng, H.B., Cheng, P.C., and Luther, P.W. (1981) Formation of acetylcholine receptor clusters induced by positively charged latex beads. Nature 292:831–834.
Podleski, T.R., Axelrod, D., Ravdin, P., Greenberg, I., Johnson, M.M., and Salpeter, M.M. (1978) Nerve extract induces increase and redistribution of acetylcholine receptors on cloned muscle cells. Proc. Natl. Acad. Sci. USA 75:2035–2039.
Prives, J., Fulton, A.B., Penman, S., Daniels, M.P., and Christian, C.N. (1982) Interaction of the cytoskeletal framework with acetylcholine receptor on the surface of embryonic muscle cells in culture. J. Cell Biol. 92:231–236.
Role, L.W., Matossian, V.R., O'Brien, J., and Fischbach, G.D. (1985) On the mechanism of acetylcholine receptor accumulation at newly formed synapses on chick myotubes. J. Neurosci. 5:2197–2204.
Salpeter, M.M., Spanton, S., Holley, K., and Podleski, T.R. (1982) Brain extract causes acetylcholine receptor redistribution which mimics some early events at developing neuromuscular junctions. J. Cell Biol. 93:417–425.
Sanes, J.R., Feldman, D.H., Cheney, J.M., and Lawrence, J.C., Jr. (1984) Brain extract induces synaptic characteristics in the basal lamina of cultured myotubes. J. Neurosci. 4:464–473.
Schaffner, A.E., and Daniels, M.P. (1982) Conditioned medium from cultures of embryonic neurons contains a high molecular weight factor which induces acetylcholine receptor aggregation on cultured myotubes. J. Neurosci. 2:623–632.
Shimada, Y., and Kano, M. (1971) Formation of neuromuscular junctions in embryonic cultures. Arch. Histol. Jpn. 33:95–114.
Silberstein, L., Inestrosa, N.C., and Hall, Z.W. (1982) Aneural muscle cell cultures make synaptic basal lamina components. Nature 295:143–145.
Slater, C.R. (1982) Postnatal maturation of nerve–muscle junctions in hindlimb muscles of the mouse. Dev. Biol. 94:11–22.
Smith, M.A., and Slater, C.R. (1983) Spatial distribution of acetylcholine receptors at developing chick neuromuscular junctions. J. Neurocytol. 12:993–1005.
Steinbach, J.H. (1981) Developmental changes in acetylcholine receptor aggregates at rat skeletal neuromuscular junctions. Dev. Biol. 84:267–276.

Steinbach, J.H., and Bloch, R.J. (1986) The distribution of acetylcholine receptors on vertebrate skeletal muscle cells. In: Price, G.B., and Gorczynski, R. (eds.) Receptors in cell recognition and differentiation. New York: Academic Press.

Steinbach, J.H., Harris, A.J., Patrick, J., Schubert, D., and Heinemann, S.F., (1973) Nerve–muscle interaction in vitro: the role of acetylcholine. J. Gen. Physiol. 62:255–270.

Stya, M., and Axelrod, D. (1983a) Diffusely distributed acetylcholine receptors can participate in cluster formation on cultured rat myotubes. Proc. Natl. Acad. Sci. USA 80:449–453.

Stya, M., and Axelrod, D. (1983b) Mobility and detergent extractability of acetylcholine receptors on rat myotubes: a correlation. J. Cell Biol. 97:48–51.

Teravainen, H. (1968) Development of the myoneural junction in the rat. Z. Zellforsch. Mikrosk. Anat. 87:249–265.

Vogel, Z., Christian, C.N., Vigny, M., Bauer, H.C., Sonderegger, P., and Daniels, M.P. (1983) Laminin induces acetylcholine receptor aggregation on cultured myotubes and enhances the receptor aggregation activity of a neuronal factor. J. Neurosci. 3:1058–1068.

Weinberg, C.B., Sanes, J.R., and Hall, Z.W. (1981) Formation of neuromuscular junctions in adult rats: accumulation of acetylcholine receptors, acetylcholinesterase and components of synaptic basal lamina. Dev. Biol. 84:255–266.

Weldon, P.R., and Cohen, M.W. (1979) Development of synaptic ultrastructure at neuromuscular contacts in an amphibian cell culture system. J. Neurocytol. 8:239–259.

Willingham, M.C., and Pastan, I. (1978) The visualization of fluorescent proteins in living cells by video intensification microscopy (VIM). Cell 13:501–507.

Ziskind-Conhaim, L., Geffen, I., and Hall, Z.W. (1984) Redistribution of acetylcholine receptors on developing rat myotubes. J. Neurosci. 4:2346–2349.

23. CURRENT FINDINGS ON THE MORPHOLOGICAL INDUCTION OF SYNAPTOGENESIS IN NEURONAL CELLS IN VITRO

POLYXENIE E. SPOERRI

1. INTRODUCTION

Formation of a synapse requires growth of the presynaptic axon, contact with, and recognition of, the appropriate postsynaptic neuron, replacement of growth cone organelles by synaptic typical organelles, and assembly at the active zone of presynaptic and postsynaptic dense material (Rees, 1978). After initial formation, maturational changes occur and certain conditions are required for the synapse to be maintained (Changeux and Danchin, 1976). In this chapter we describe current in vitro findings on the morphological induction of synpathogenesis in neuronal cell cultures with special emphasis on the order of events and the origin of organelles involved. Neuritic growth cones are being described growing over the surface of neuronal somata (Spoerri et al., 1980b; Spoerri and Dresp, 1981; Spoerri and Wolff, 1981 and 1982). The traffic of pinocytotic and exocytotic coated vesicles and their known and hypothesized functions are indicated as well as bilateral membrane specializations and their possible function in recognition are demonstrated. Early synapses in which only postsynaptic dense material and a few synaptic vesicles are present, as well as mature synapses with full complement of presynaptic and postsynaptic dense material are illustrated (Spoerri and Wolff, 1981 and 1982; Spoerri, 1983 and 1986).

Current research focusing on the neuritogenic and neuronotrophic properties of gangliosides (Spoerri, 1983 and 1986), as exemplified by time-lapse cinematography (Spoerri, 1984), in their ability to potentiate growth factors

Vernadakis, A. (ed.), *Model Systems of Development and Aging of the Nervous System*.

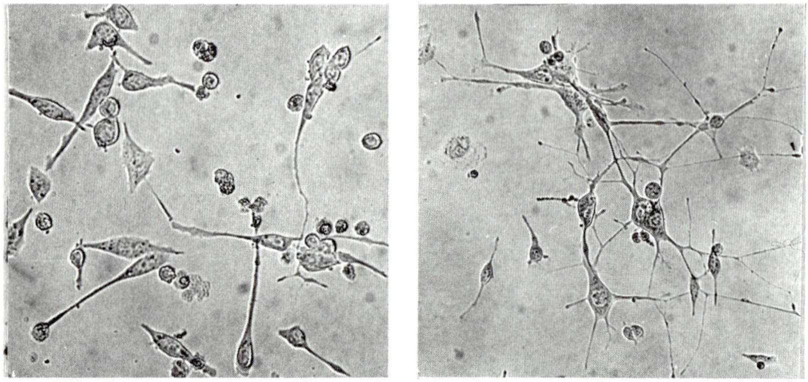

A B

Figure 23-1a and b. Light micrographs of N-2A cells fixed for TEM (**a**) Proliferating cells grown in the presence of 10% FCS. (**b**) Differentiated cells with long processes grown in medium containing only 2% FCS. Contacts can be seem between the cells and other neurites. × 113.

and to induce differentiation and early maturation in monolayers of a neuronal cell line, emphasizes the importance of the underlying molecular mechanisms involved in neuronal recognition and of the assembly and maintenance of a functional synaptic active zone.

2. MORPHOLOGICAL INDUCTION OF SYNAPTOGENESIS

2.1 Effect of serum concentration

Neuroblastoma cells, clone N-2A, when grown in the presence of 10% FCS, proliferate rapidly and differentiate slowly (figure 23-1a). When serum is deleted or low concentrations are being used, the reverse occurs (figure 23-1b). The cells extend neurites and form numerous contacts. When such fixed cells are sliced for transmission electron microscopy (TEM) and selected areas are ultrastructurally examined, several observations are made: the neurites of the neurons form only apparent contacts (figure 23-2a and b); bilateral membrane specializations that represent early stages of synapse formation preceding the appearance of synaptic contacts are numerous (figure 23-3); synaptic vesicles are initially found within the somata (figure 23-4) of neurons forming apparent contacts; and, after slicing numerous selected areas of N-2A-fixed differentiated cells, the axon-like process reveals terminal swellings with a few, or filled with, synaptic vesicles. The latter appear to originate from the smooth endoplasmic reticulum (SER; figures 23-5a and b, and 23-6a–d).

2.2 Effect of GABA

After 5 days in culture, the N-2A cells were fed with normal medium or with the same medium containing 1%–2% fetal calf serum instead of 10% to

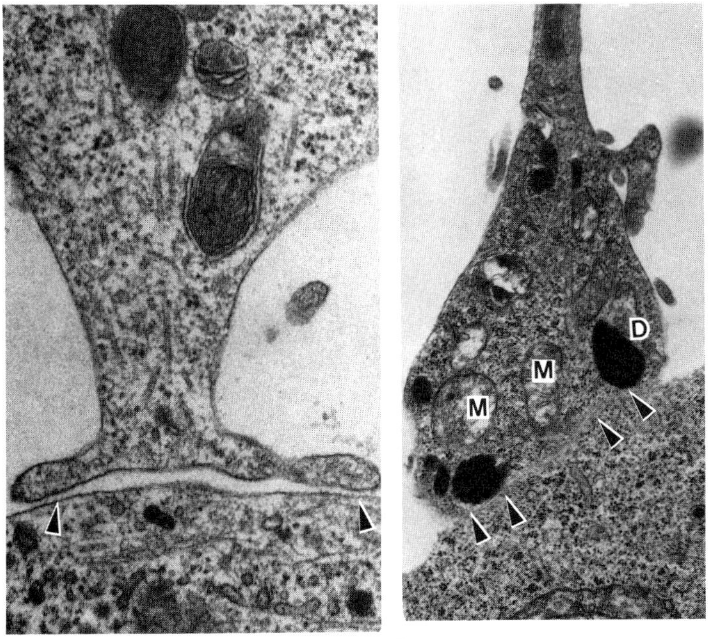

A B

Figure 23-2a and b. Cell bodies and neurite terminals in contact. (**a**) A short neurite with tentacles making an apparent contact. × 31,365. (**b**) The neuritic terminal in contact contains no vesicles but mitochondria (*M*), dense bodies (*D*), and ribosomes. Note a slight membrane thickening at the area of contact (*arrowheads*). × 22,950.

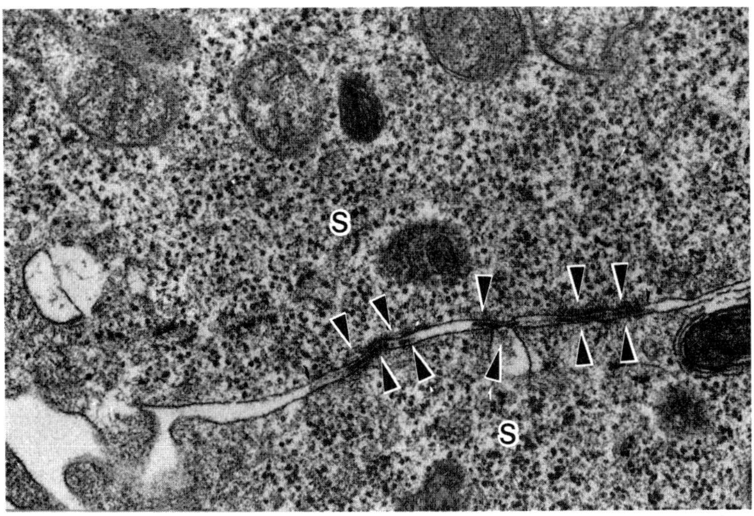

Figure 23-3. Two somata (*S*) in close apposition. The limits of contact are shown by *arrowheads*. The bilateral membrane specializations are characterized by a very dense material in some areas of the intercellular space and the fuzzy, slightly electron-dense substance that lies on either side of the apposition. × 31,365.

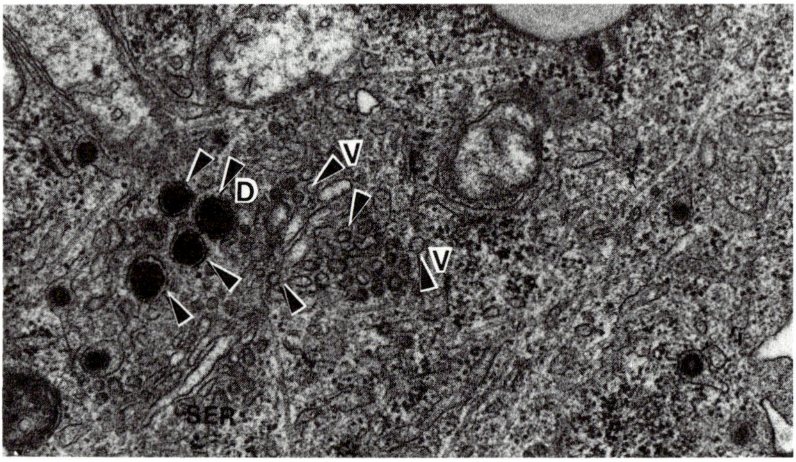

Figure 23-4. A profile of a soma depicting clear vesicles (*V*) 20–60 nm in diameter (*arrowheads*) and dense-core vesicles (*D*) 100–200 nm in diameter. × 54,315.

which γ-aminobutyric acid (GABA) was added (Eins et al., 1983; Spoerri and Wolff, 1981 and 1982). These substances were dissolved in double-distilled water and added to the normal medium, producing concentrations of 10^{-4} to 10^{-5} *M*. GABA was applied to the cultures for a maximum of 2 days. The effect of a 2-day application of 10^{-4} to 10^{-5} *M* GABA to the cultures advanced the stage of neuronal maturation. There is a proliferation of the pino-exocytotic (100 nm in diameter) coated vesicle (figure 23-7a–e). These vesicles appear to originate from the Golgi cisternae and the rough endoplasmic reticulum (RER; figures 23-8 and 23-9). The coated vesicle has a lattice-like coat with fine projections (figure 23-7e). These vesicles are found in continuity with unilateral and bilateral membrane specializations contributing undercoating to the postsynaptic density (figures 23-7a, b, and d, and 23-10a and b). Following the appearance of the postsynaptic density, a small number of synaptic vesicles are seen in the growth-cone cytoplasm in some stage of its reorganization into a synaptic bouton. However, the synaptic vesicles do not cluster at the active zone. The growth-cone cytoplasm still contains a few mound areas and primitive contacts or symmetrical densities (figures 23-11a–c). As clear synaptic vesicles accumulate at the forming presynaptic site, counts of large dense-cone vesicles gather at the periphery of somata or neurites, suggesting that they may play a role in the process of synapse formation (figures 23-12a–c).

2.3 Effect of Gangliosides

Purified bovine brain gangliosides were purchased singly and were then added together to make the mixture consisting of 27% GM1, 40% GD1a,

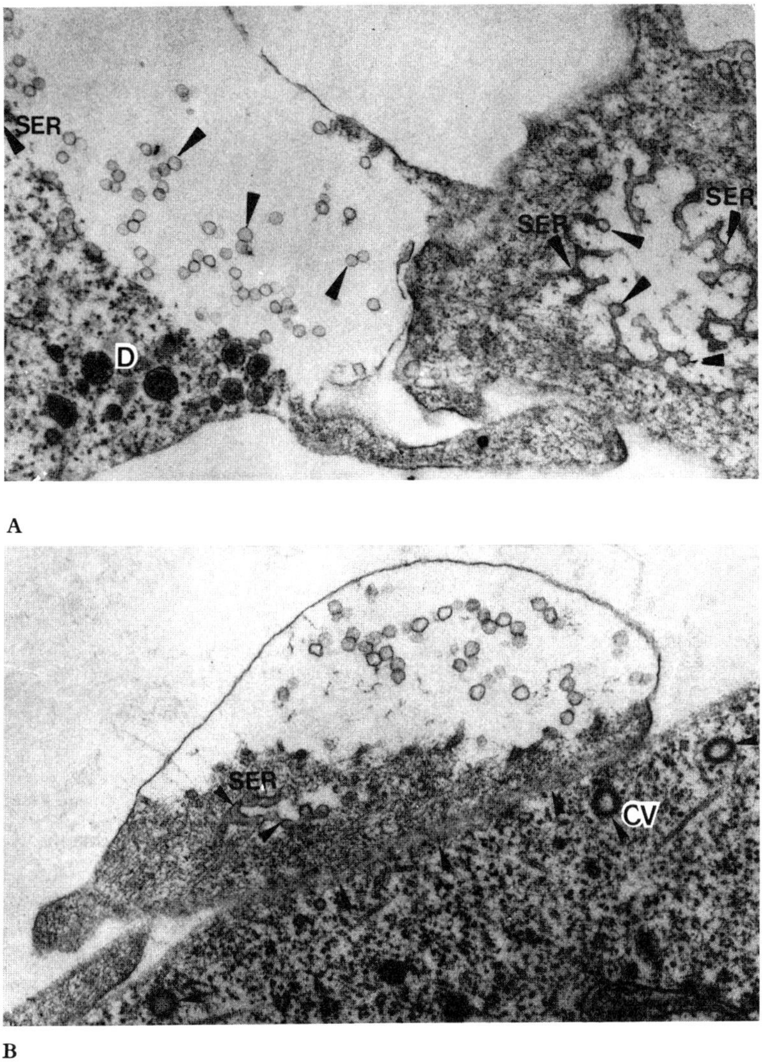

Figure 23-5a and b. Neurite terminals containing round to polygonal small vesicles appearing to bud from the smooth endoplasmic reticulum (*SER*). Dense-core vesicles (*D*) and coated vesicles (*CV*) are also seen within the somata. Note the membrane thickening at the site of the contact in **b** (*arrowheads*). × 26,775.

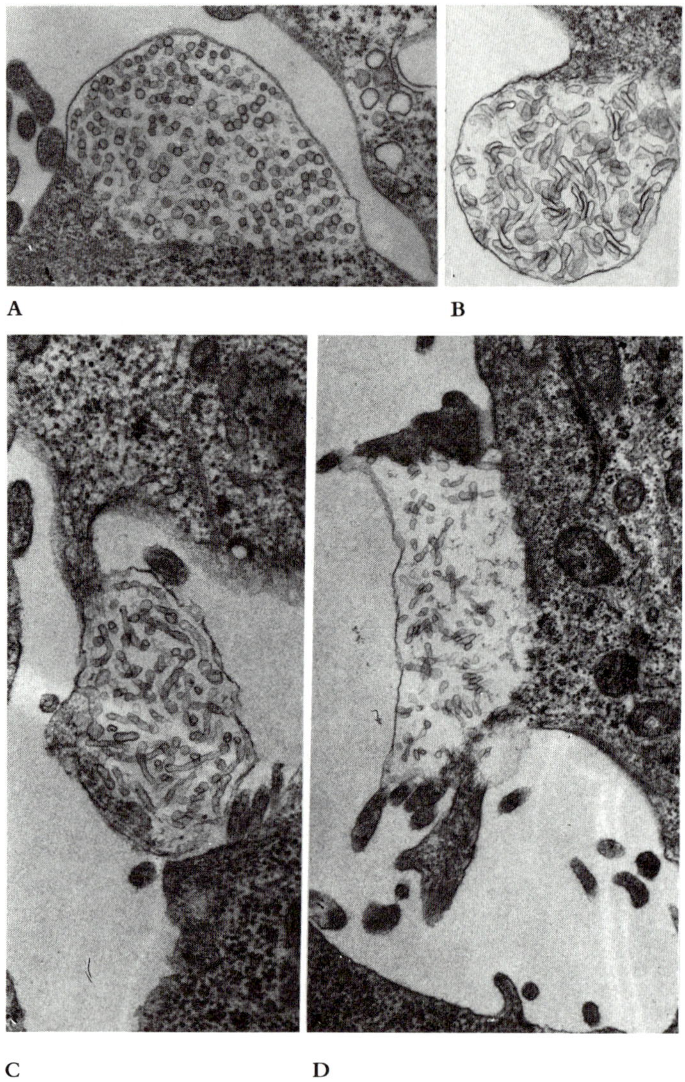

Figure 23-6a–d. Small polygonal vesicles 20–40 nm in diameter and flat vesicles having a length of 80–200 nm are also seen in large numbers at the periphery of the neurons or in terminal swellings of neurites. The number of vesicles present is age dependent: (**a–c**) from 30-day-old cultures, and (**d**) from 15-day-old cultures. (**a** and **b**) × 260,000, (**c**) × 31,365, and (**d**) × 30,600.

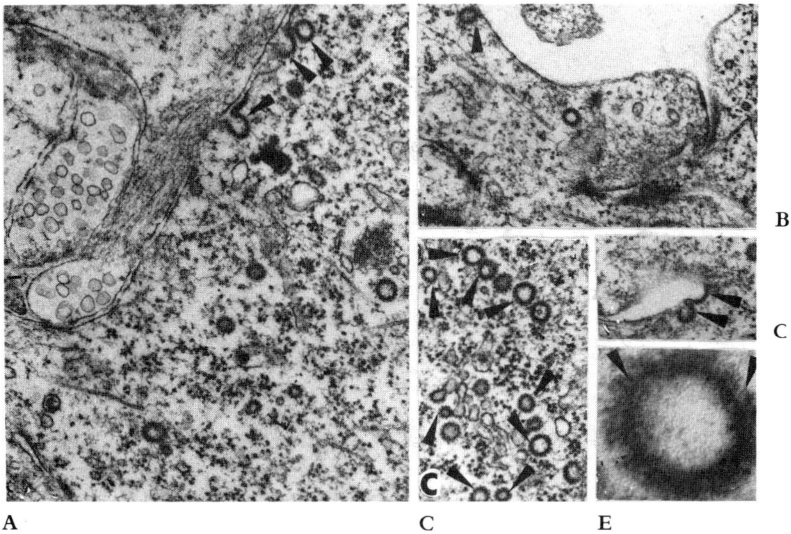

Figure 23-7a–e. Coated vesicles (*CV*) displaying clear centers 80–120nm in diameter a filamentous dense substance, and "striae" or thin cytoplasmic protrusions extending radially from the outer surface of their limiting membrane. Such vesicles are seen within the cytoplasm and at the periphery of the cell (*arrowheads*). From 18-day-old structures treated with GABA. (**a–d**) × 25,500 and (**e**) × 100,000.

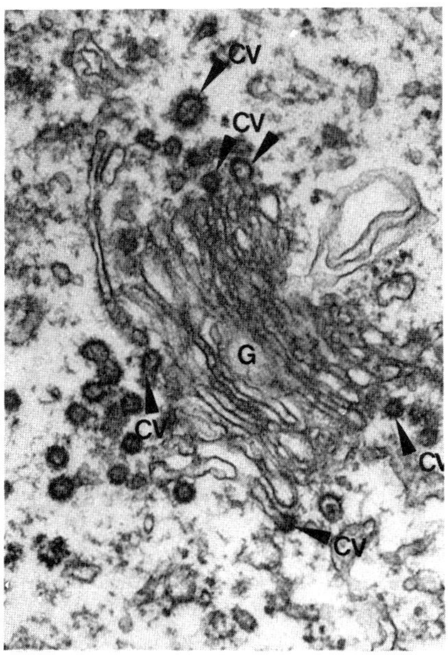

Figure 23-8. Coated vesicles located within the Golgi (**G**) apparatus of N-2A cells and being confluent with the Golgi cisternae. × 38,250.

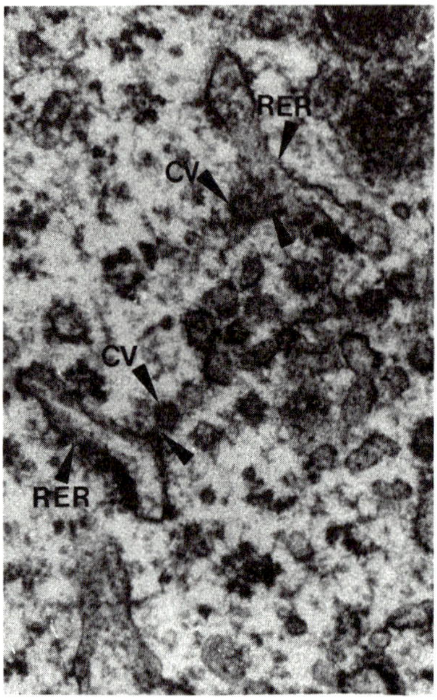

Figure 23-9. Coated vesicles associated with the rough endoplasmic reticulum (*RER*) and apparently budding off from the cisternae. × 56,950.

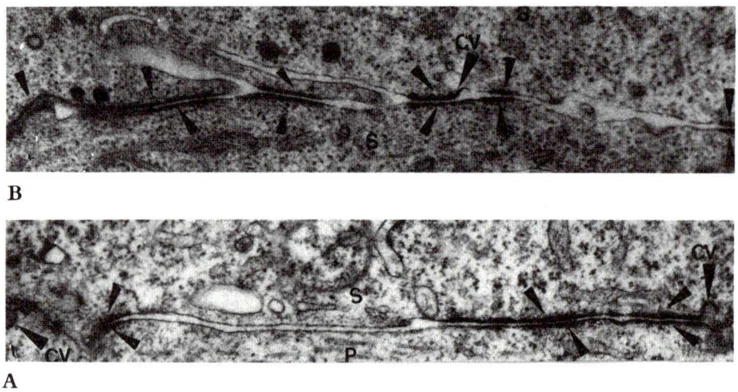

Figure 23-10a and b. Bilateral membrane specializations, neuronal somata (*S*), and neuronal processes (*P*) in close apposition. Note the dense membrane invaginations and the coated vesicles (*CV*) in continuity with the electron-dense membranes. From 38-day-old treated cultures. × 25,500.

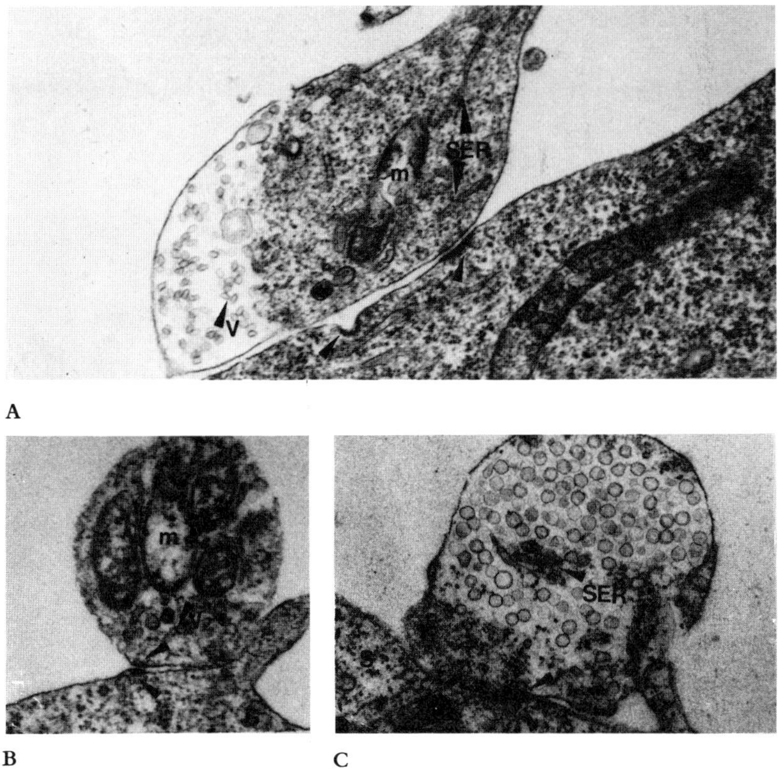

Figure 23-11a–c. Varicosities displaying mitochondria, large and small synaptic vesicles (*V*), and single cisternae of the smooth endoplasmic reticulum (*SER*). (**a**) Vesicles, 20–40 nm, a mitochondrion (*m*), and single cisternae of the SER (*arrowheads*). Opposite to the varicosity at the periphery of the neuronal soma (*S*), a membrane specialization and an electron-dense membrane invagination (*arrowheads*) are shown, both remote from the vesicles (*V*). (**b**) Membrane thickenings or bilateral membrane specializations, one on the varicosity and the other on the apposed soma (*arrowheads*). Note also the presence of vesicles (*V*) and mitochondria (*m*). (**c**) Within the terminal swelling, numerous large, empty-appearing, smooth, round vesicles of varying diameter (up to 20 nm), *SER*, and membrane specializations, the latter on both sides. (**a**) 10-day-old culture, (**b**) 18-day-old culture, and (**c**) 15-day-old culture. All treated with 10^{-4} *M* GABA. × 25,500.

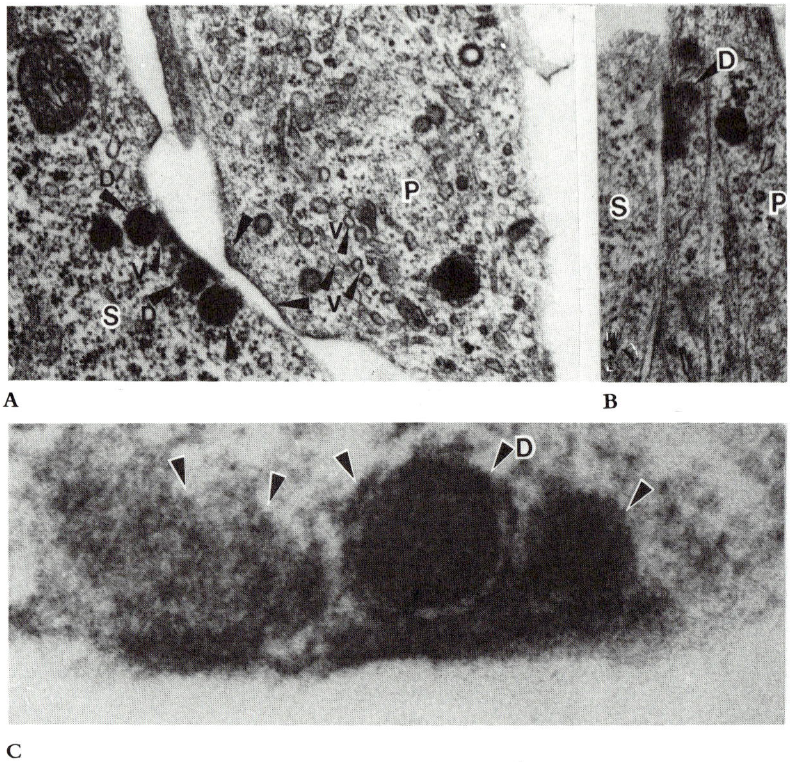

Figure 23-12a–e. Dense-core (*D*) vesicles (*arrowheads*) gather near the plasmalemma. Linear arrangement. In all cases, a dense substance is present at the outermost edge and occasionally masks these vesicles. In **b** and **c**, similar, more electron-lucent vesicles are seen. (**a** and **b**) × 25,500 and (**c**) × 58,650.

16% GD1b, and 19% GT. The gangliosides (200–250 μg/ml) were added to 6- to 7-day-old GABA-pretreated N-2A cultures (Spoerri, 1984). The effect of added ganglioside mixture on preconfluent cells 24–48 hours after seeding was determined as described (Spoerri, 1986). Addition of gangliosides enhances the rate and degree of neurite formation. The elaborate long neurites with side branches form networks (figures 23-13a and b). Early membrane-relationship studies in this system, using time-lapse cinematography (Spoerri, 1984), reveal that growth-cone filopodia initially establish an extensive contact area and are observed as a flattened sheet lying on the surface of the neuronal soma (figures 23-14a–f). A period of some 8–12 hours elapses before postsynaptic dense material and synaptic vesicles are seen by TEM. Gangliosides induced mature synapses among differentiated cells.

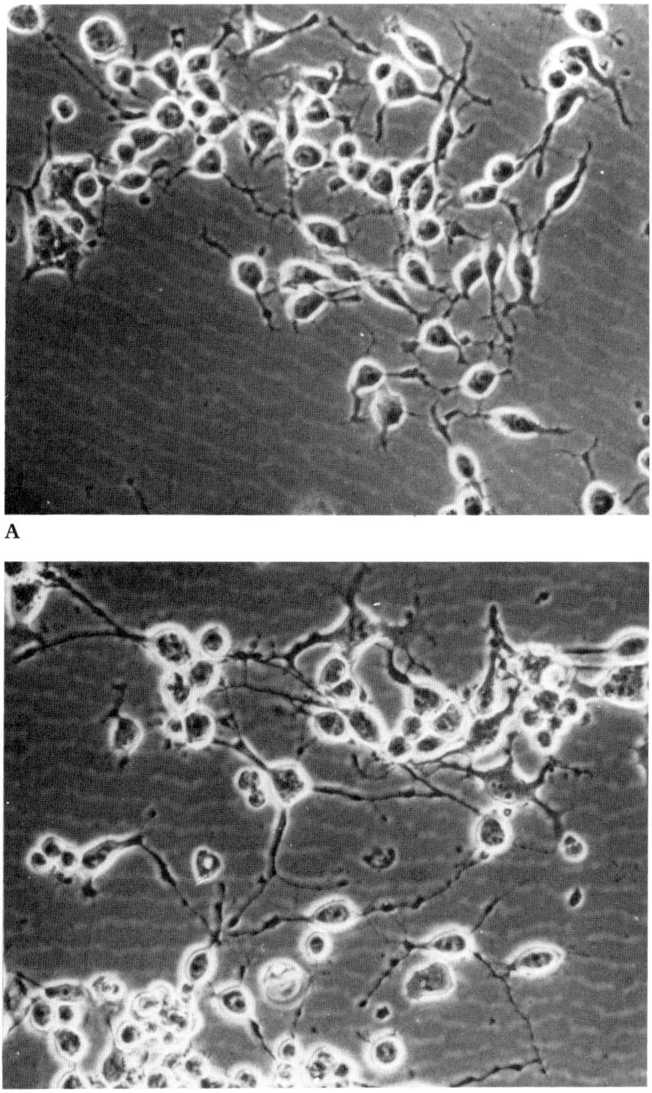

A

B

Figure 23-13a and b. N-2A cells living preparation. (**a**) Some cells after 24 hours in medium containing 1% FCS and 200 μg/ml of gangliosides (GM1, GD1a, GD1b, and GT) depict neurite formation with contacts. (**b**) After 48 hours, network formation due to intercommunicating neurites with fine side branches. Phase contrast, × 480.

3. NATURE OF THE SYNAPSE

The mature synaptic contact (figure 23-15a) has the presynaptic element opposite the postsynaptic element. With the appearance of a recognizable synaptic active zone, the growth-cone organelles, endoplasmic reticulum, lysosomes, and mound areas diminish and disappear. Synaptic vesicles increase markedly in number and are present in dense clusters near the presynaptic membrane. They are 40–200 nm in diameter, spherical to ovoid in shape. The typical mature bouton is filled with synaptic vesicles, some mitochondria, and an occasional dense-core vesicle (not shown). The presynaptic dense projection is narrow while the postsynaptic density is high, having a relatively dark profile. The synapse is asymmetrical and similar to those found in primary cultures resembling Gray's type 1, which have a somewhat wider cleft and more pronounced synaptic density than those of type 2. An immature synaptic bouton filled with synaptic vesicles and a dense-core vesicle are depicted on the same electron micrograph (figure 23-15). There is a beginning postsynaptic dense projection or a unilateral membrane specialization; a presynaptic dense projection with vesicle aggregation has still to be formed.

4. ANALYSIS OF SYNAPTOGENESIS IN CULTURE

4.1 Increased membrane dynamics

In the in vitro system described above, several opposed membranes displayed extracellular coat material within the small electron-dense areas of bilateral membrane specialization. At these areas, opposed membranes were parallel and the coat material covered the 10-nm large intercellular gap. The underlying dense material was present on the cytoplasmic side of the inner membrane in both presynaptic and postsynaptic neurons. Such bilateral membrane specializations have been described previously during synaptogenesis (Rees et al., 1976; Vaughn et al., 1976). These authors postulated them as possible sites of low-affinity recognition requiring further confirmation for synapse formation. As reviewed by Rees (1978), interneuronal recognition is based on a chemospecific surface glycoprotein. Such a glycoprotein can be combined with the present and past concept, that the membrane and its coat or glycocalyx may act as a transducer of external

Figure 23-14a–f. Membrane-relationship studies using time-lapse cinematography. Note the extensive neurite elongation with bulges and microspikes or filopodia. The growth-cone filopodia initially establish an extensive contact area with the ruffling membrane, observed as a flattened sheet lying on the surface of the neuronal somata, or in the case of the small neuron, making contact with a neurite. When a synaptic contact is being made, the filopodia and the undulating membrane disappear. The cells round up when serum is added to the medium (**e**). 1 hour 20 minutes/24 frames/second. × 900.

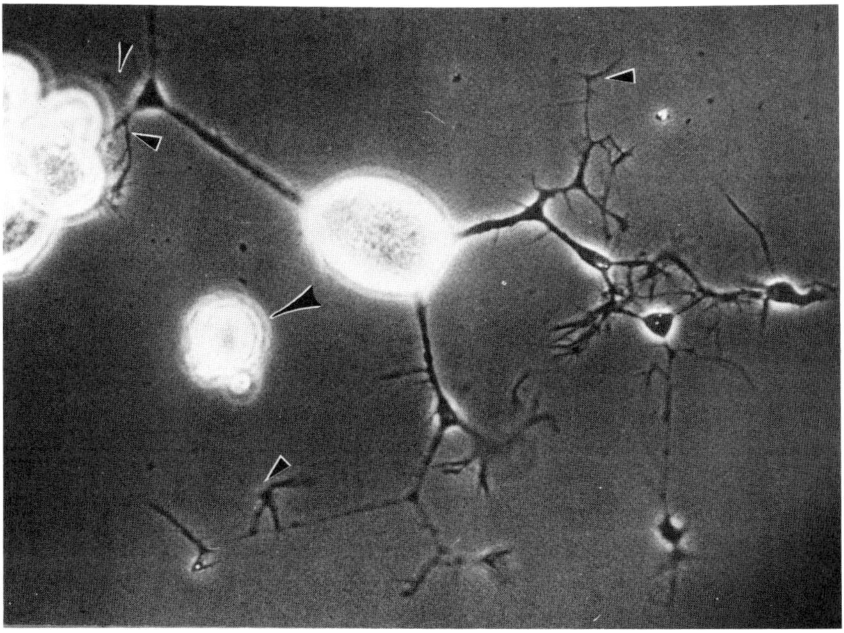

A

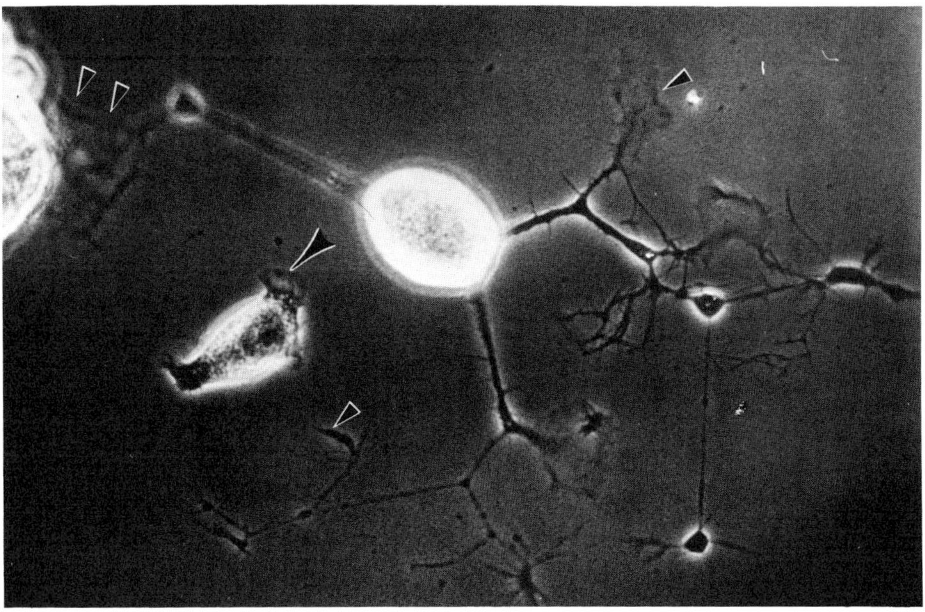

B

Figure 23-14 (continued)

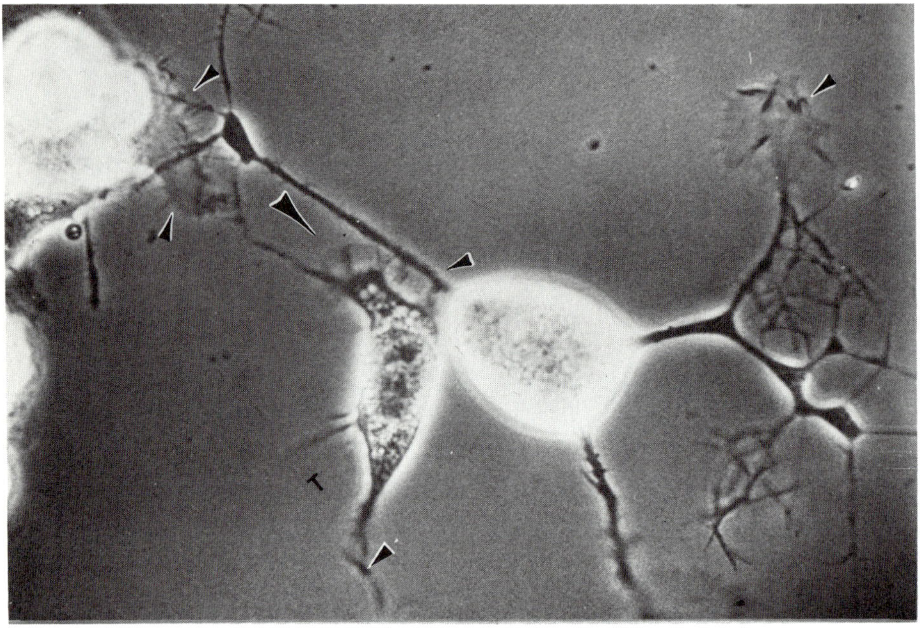

C

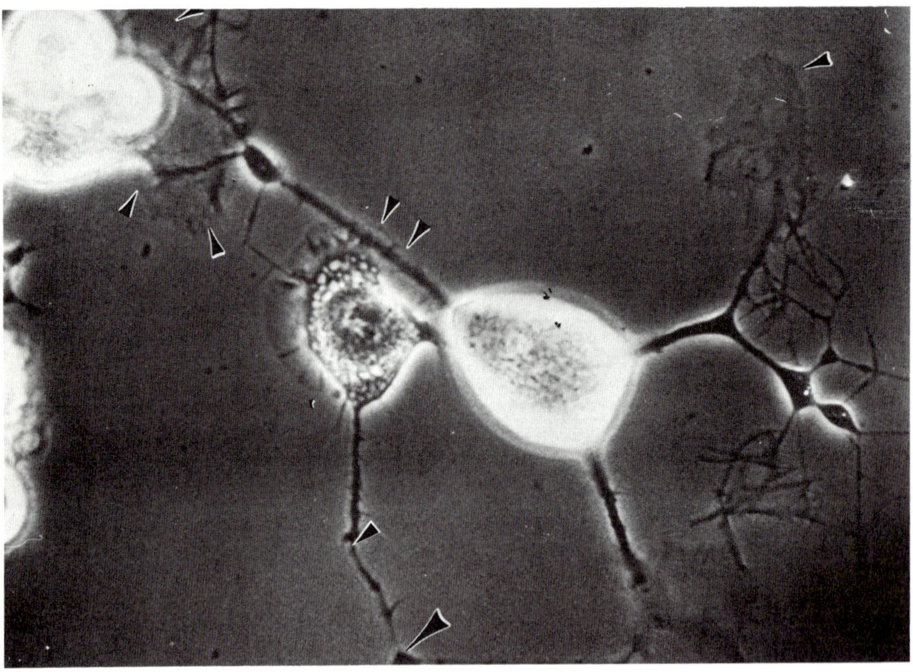

D

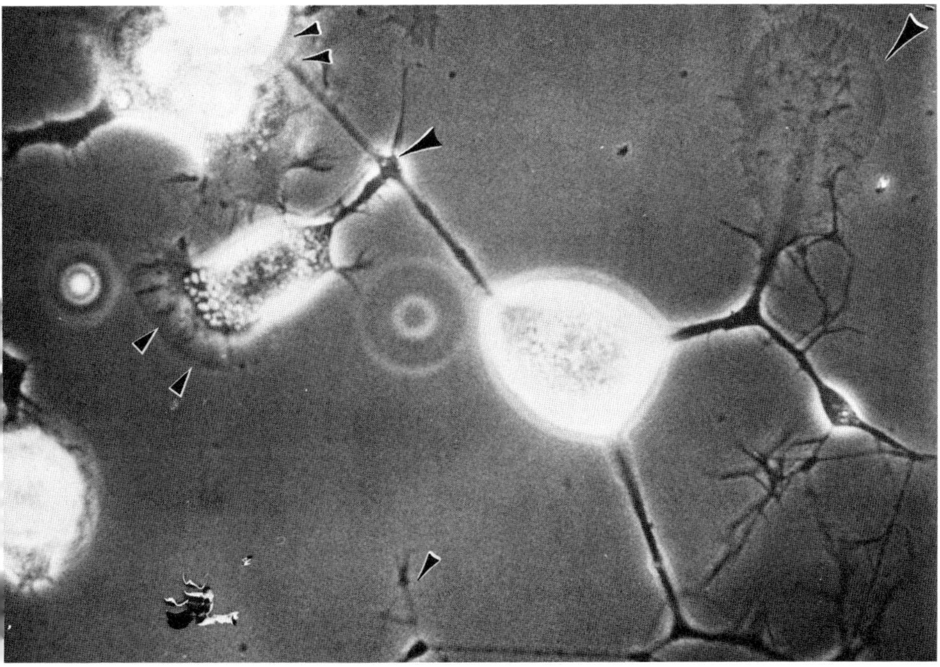

E

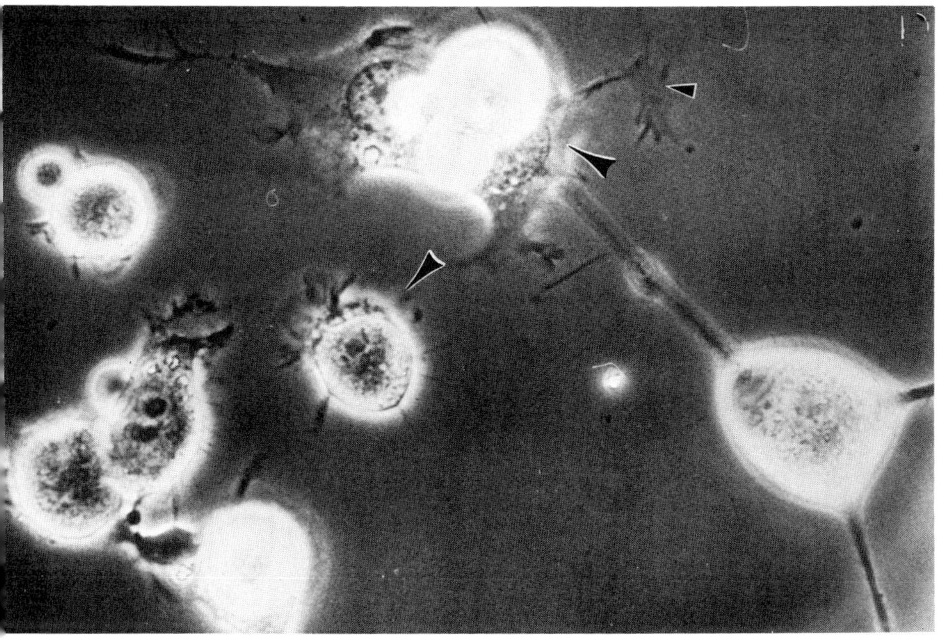

F

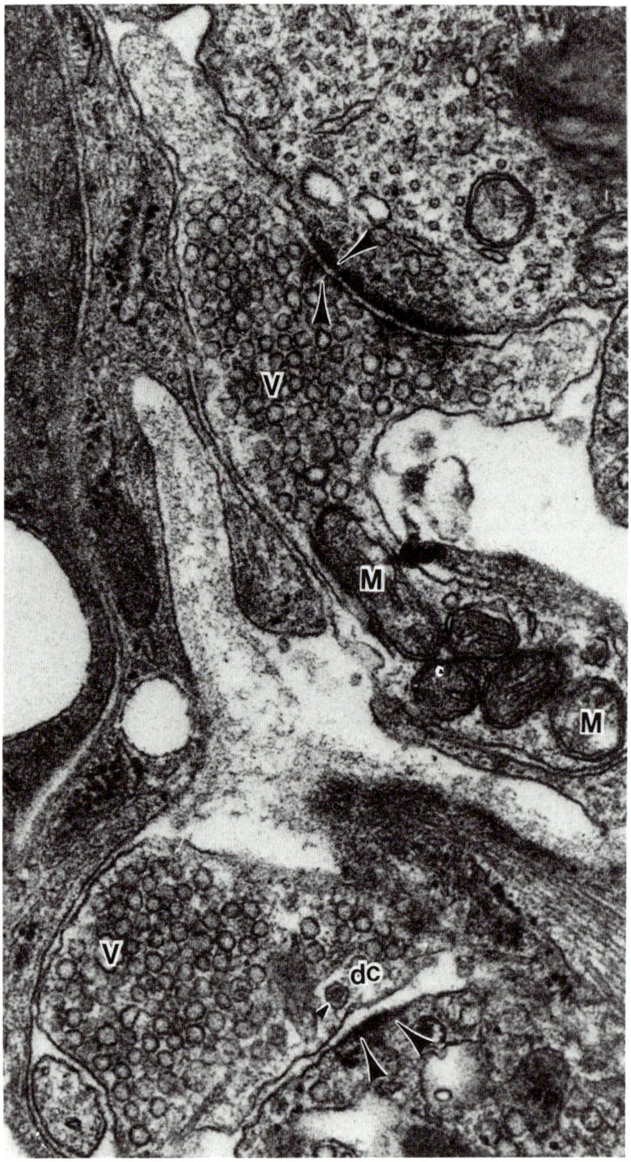

Figure 23-15. Electron micrograph displaying a mature synapse. Note the vesicle (*V*) aggregation on the presynaptic side and the presence of mitochondria (*M*) within the terminal swelling. An immature bouton (*B*) without membrane thickenings with the exception of a unilateral membrane specialization on the postsynatpic side (*arrowheads*). An occasional dense-core (*dc*) vesicle is also seen. × 9000.

stimuli, each of which can be translated by a different membrane messenger and transferred to the cytoplasm or cell organelles.

Coated vesicles were prominent in the neuronal cytoplasm at the time of synaptogenesis. They were present in considerable numbers and in continuity with surface membrane, especially the postsynaptic membrane, or near the area of the presynaptic contact. Similar vesicles were seen in continuity with Golgi cisternae (Rees et al., 1976) and the RER (Spoerri and Wolff, 1981 and 1982), and there was an increase in number during the synaptogenic period. Indeed it is not established whether the coated vesicles are exocytotic or pinocytotic in nature (Altman, 1971; Hinds and Hinds, 1976; Rees et al., 1976; Vaughn et al., 1976). Tracer studies confirm both viewpoints (Rees et al., 1976; Waxmann and Pappas, 1969). Many supporters consider them as exocytotic and seem to be passing from the Golgi apparatus to the surface, contributing areas of specialized surface membrane, complete with glycoprotein (Rees, 1978). Such areas are then postulated to function in possible recognition and in the formation of the initial bilateral membrane specialization described. Such vesicles were seen to contribute undercoating to the postsynaptic density and to be building up the material in the synaptic cleft. The biological role of coated vesicle may depend on the function of the cell in which it occurs. Undoubtedly, pinocytotic coated vesicles were also present in the neuron studies. Neuronotrophic factors may bind to receptors on the surface membrane of responsive neurons that can be pinocitized and transported retrogradely to the neuronal soma (Frazier et al., 1973). Coated vesicles may bud from either Golgi or RER to contribute membrane to the neuronal surface, at a time when significant molecular changes occur—such as assembly of a synaptic active zone.

Following the initial contact and the formation of bilateral membrane specialization described above, additional unilateral areas of membrane with undercoating were observed on the postsynaptic neuron. Such areas were described as being structurally similar to the postsynaptic active zones and appeared to be the first definitive sign of synapse formation (Rees, 1978). It has been suggested that the postsynaptic dense material may derive wholly or in part from the exocytic 100-nm coated vesicle presently and previously described (Altman, 1971; Rees et al., 1976; Spoerri and Wolff, 1981 and 1982; Stelzner et al., 1973). Following the appearance of the postsynaptic density, growth-cone filopodia diminish as observed by time-lapse cinematography (Spoerri, 1984) presently shown, and a small number of synaptic vesicles were seen in the growth-cone cytoplasm. The early synaptic vesicles appeared to arise by budding from the SER of the growth cone (Spoerri et al., 1980b; Spoerri and Dresp, 1981).

4.2 Presynaptic organelles

The present studies also indicated that the presynaptic dense projection became evident after the postsynaptic dense material. As clear synaptic vesicles aggregated at forming synaptic sites, counts of large dense-core vesicles gathered, suggesting that they may play some specific but yet unknown role in the processes of synapse formation (Oppenheim and Foelix, 1972; Pfenninger and Rees, 1976; Spoerri and Wolff, 1981 and 1982; Spoerri et al., 1982). With the appearance of a recognizable synaptic active zone, the growth cone, endoplasmic reticulum, and lysosomal and mound areas diminished. Synaptic vesicles increased markedly in number and were present in dense clusters near the forming presynaptic membrane. There were consequently increases in the height and width of the presynaptic dense projection. The typical synaptic bouton was filled with vesicles, some mitochondria, and an occasional dense-core vesicle.

4.3 Synapse formation

The results strongly indicate that GABA, in addition to being an important neurotransmitter, serves as a trophic factor in the development of at least certain types of neurons (Hansen et al., 1984). The possible role played by gangliosides, sialic-containing glycospingolipids, in promoting trophic agents leading to the establishment of the synapse has been approached by several investigators differently (Ledeen, 1984). It has been pointed out that the negative charges of the sialic acid terminals, present in gangliosides, may have an effect on the polarization of the neuronal membrane, which can promote the establishment of contacts (Massarelli et al., 1984). Others have suggested a correlation between sialocompounds and Ca^{2+} ions, the release of neurotransmitter being Ca^{2+} dependent (Probst and Rahmann, 1980; Svennerholm, 1980). In vitro studies, morphological and biochemical, attribute a role of gangliosides in cellular recognition ensuring the linking between the presynaptic and postsynaptic membrane (Dreyfus et al., 1980 and 1984). Moreover, these substances, especially the GM1, seem to function as mediator molecules, as shown in neuronotrphic interactions regulating neurite outgrowth (Katoh-Semba, 1980; Leon et al., 1984), but most likely the state of the postsynaptic neuron, which in turn maintains the presynaptic element and synaptic adhesiveness (Changeux and Danchin, 1976). The cells finally converted from an unstable intermediate state into a stably differentiated state, which ultrastructurally depicted synaptic contacts (Morgan and Seifert, 1984).

Though much has been described about the order of events and the origin of organelles involved in synaptogenesis, the molecular basis underlying these events remains unanswered.

ACKNOWLEDGMENTS

The work presented was supported by the German Science Foundation. The author expresses her appreciation to Springer-Verlag, Heidelberg, for copyright permission to include some of the published micrographs.

REFERENCES

Altman, J. (1971) Coated vesicles and synaptogenesis: a developmental study in the cerebellar cortex of the rat. Brain Res. 30:311–322.

Changeux, J.P., and Danchin, A. (1976) Selective stabilization of developing synapses as a mechanism for the specification of neuronal networks. Nature (Lond.) 264:705–712.

Dreyfus, H., Louis, J.C., Harth, S., and Mandel, P. (1980) Gangliosides in cultured neurons. Neuroscience 5:1847–1855.

Dreyfus, H., Ferret, B., Harth S., Gorio, A., Durand, M., Freysz, L., and Massarelli, R. (1984) Metabolism and function of gangliosides in developing neurons. J. Neurosci. Res. 12:311–322.

Eins, S., Spoerri, P.E., and Heyder, E. (1983) GABA or sodium bromide induced plasticity of neurites of mouse neuroblastoma cells: a quantitative study. Cell Tissue Res. 229:457–460.

Frazier, W.A., Boyd, L.F., and Bradshaw, R.A. (1973) Interaction of nerve growth factor with surface membranes: biological competence of insolubilized nerve growth factor. Proc. Natl. Acad. Sci. USA 70:2931–2935.

Hansen, G.H., Meier, E., and Schousboe, A. (1984) GABA influences the ultrastructure composition of cerebellar granule cells during development in culture. Int. J. Dev. Neurosci. 2:247–257.

Hinds, J.W., and Hinds, P.L. (1976) Synapse formation in the mouse olfactory bulb. I. Quantitative studies. J. Comp. Neurol. 169:15–40.

Katoh-Semba, R., Skaper, S.D., and Varon, S. (1984) Interaction of GM1 ganglioside with PC12 pheochromacytoma cells: serum- and NSG-dependent effects on neuritic growth and proliferation. J. Neurosci. Res. 12:299–310.

Ledeen, R.W. (1984) Biology of gangliosides: neuritogenic and neuronotrophic properties. J. Neurosci. Res. 12:147–159.

Leon, A., Benvegnu, D., Dal Toso, R., Presti, D., Facci, L., Giorgi, O., and Toffano, G. (1984) Dorsal root ganglia and nerve growth factor: a model for understanding the mechanism of GM1—effects on neuronal repair. J. Neurosci. Res. 12:277–287.

Massarelli, R., Wong, T.H., Harth, S., Louis, J.C., Freysz, L., and Dreyfus, H. (1984) Possible role of sialocompounds in the uptake of choline into synaptosomes and nerve cell cultures. Neurochem. Res. 7:301–316.

Morgan, J.I., and Seifert, W. (1984) Growth factors and gangliosides: a possible new perspective in neuronal growth control. J. Supramol. Struct. 10:111–124.

Oppenheim, R.W., and Foelix, R.F. (1972) Synaptogenesis in the chick embryo spinal cord. Nature (Lond.) 235:126–128.

Pfenninger, K.H., and Rees, R.P. (1976) From the growth cone to the synapse: properties of membranes involved in synapse formation. In: Barondes, S.H. (ed.) Neuronal recognition. New York: Plenum, pp. 131–178.

Pfenninger, K.H., Sandri, C., Akert, K., and Engster, C.H. (1969) Contribution to the problem of structural organization of the presynaptic area. Brain Res. 12:10–18.

Probst, W., and Rahmann, H. (1980) Influence of temperature changes on the ability of gangliosides to complex with Ca^{++}. J. Therm. Biol. 5:243–247.

Rees, R.P. (1978) The morphology of interneuronal synaptogenesis: a review. Fed. Proc. 37:2000–2009.

Rees, R.P., Bunge, M.B., and Bunge, R.P. (1976) Morphological changes in the neuritic growth cone and target neuron during synaptic function and development in culture. J. Cell Biol. 68:240–263.

Spoerri, P.E. (1983) Effects of gangliosides on the in vitro development of neuroblastoma cells: an ultrastructural study. Int. J. Dev. Neurosci. 1:383–391.

Spoerri, P.E. (1984) Mouse neuroblastoma cells in vitro: substances regulating synapse-like formations. Film B 1544 sect. Med. ser. 6, no. 15. Göttingen: Institut für den Wissenschaftlichen Film.

Spoerri, P.E. (1987) Facilitated establishment of contacts and synapses in neuronal cultures: ganglioside-mediated neurite sprouting and outgrowth. In: Tettamanti, G., Ledeen, R.W., Nagal, V., Sandhoff, K., Toffano, G. (eds.) Neuronal plasticity and gangliosides. Fidia research series. Vol. 6 Padova: Liviana, pp. 309–325.

Spoerri, P.E., and Dresp, W. (1981) Smooth endoplasmatic reticulum and synaptic vesicles in cultures murine neuroblastoma cell. Z. Mikrosk. Anat. Forsch. 95:521–530.

Spoerri, P.E., and Wolff, J.R. (1981) Effect of GABA administration on murine neuroblastoma cells in culture. I. Increased membrane dynamics and formation of specialized contacts. Cell Tissue Res. 218:567–579.

Spoerri, P.E., and Wolff, J.R. (1982) Morphological changes induced by sodium bromide in murine neuroblastoma cells in vitro. Cell Tissue Res. 222:379–388.

Spoerri, P.E., Dresp. W., and Heyder, E. (1980a) A simple embedding technique for monolayer neuronal cultures grown in plastic flasks. Acta Anat. 107:221–223.

Spoerri, P.E., Glees, P., and Dresp. W. (1980b) The time course of synapse formation of mouse neuroblastoma cells in monolayer cultures. Cell Tissue Res. 205:421–441.

Spoerri, P.E., Dresp, W., and Heyder, E. (1982) Modified growth cones in mouse neuroblastoma cultures: an electron microscopic study. J. Hirnforsch. 23:287–293.

Stelzner, D.J., Martin, A.H., and Scott, G.I. (1973) Early stages of synaptogenesis in the cervical spinal cord of the chick embryo. Z. Zellforsch. Mikrosk. Anat. 136:475–488.

Svennerholm, L. (1980) Gangliosides and synaptic transmission. In: Svennerholm, L., Mandel, P., Dreyfus, H., Urban, P.F. (eds.) Structure and function of gangliosides. New York: Plenum, pp. 533–544.

Vaughn, J.E., Hendrikson, C.K., and Wood, J.D. (1976) Surface specialization of neurites in embryonic mouse spinal cord. Brain Res. 110:431–445.

Waxmann, S.C., and Pappas, G.D. (1969) Pinocytosis at postsynaptic membranes: electron microscopic evidence. Brain Res. 14:240–244.

24. ARE AGE-RELATED CHANGES IN RECEPTOR ACTIVITY AN EXPRESSION OF ALTERED MEMBRANE FLUIDITY?

KERN VON HUNGEN AND CLAUDE F. BAXTER

Among the various theories of aging, the membrane hypothesis of aging is particularly appealing (Sun and Sun, 1979; Grinna, 1977). Cellular membranes function in mediating and regulating active transport of substances across the cell boundary as well as serving as an important permeability barrier. Enzyme activities and receptor functions are also modulated by the lipid composition of membranes. The nature of membranes may change during aging either as a result of oxidative damage (Tappel, 1973) or from a deterioration of homeostatic mechanisms that maintain proper lipid composition (Kates and Kuksis, 1980). In vitro and in vivo manipulation of membrane lipids may thus provide a model of aging.

Impairment of synaptic neurotransmission mechanisms is one of the most prominent concomitants of neuronal aging (Bonetti et al., 1983). While neurotransmission is a multistep process that could be impaired at several different stages, the proper interaction of neurotransmitters with their respective receptors is crucial. This process has come under active study during the last decade with the development and proliferation of receptor-binding assay procedures. Analysis of binding data permits the calculation of both receptor affinity and receptor number. On the whole, neurotransmitter-receptor density has been found to decrease with age, although there are also many reports of no change, and results differ widely depending on the receptor investigated, animal species, tissue, and, in the case of brain, brain region.

Neurotransmitter receptors are currently viewed as proteins floating within

Vernadakis, A. (ed.), *Model Systems of Development and Aging of the Nervous System*.

membranes, their transmitter-binding sites exposed on the external surface of the membrane, but with much of the protein submerged within the hydrophobic bilayer. Receptors are in a dynamic state. Their functional regulation may involve de novo synthesis, internalization, and processing as well as mobility within the membrane. The nature of lipids within the membrane can influence receptor function either by specific interaction with the receptors or by altering the fluidity of the membrane. Membrane fluidity depends on the molecular packing of lipids within the membrane and is reflected by the rotational restriction of fluorescent probes. Restriction of the probe can be indicated by maintenance of polarization of emitted light and measured as fluorescence polarization (P). Membrane fluidity is inversely related to microviscosity and has been shown to affect vertical displacement as well as lateral mobility of proteins within membranes (Borochov and Shinitzky, 1976). Membrane fluidity may thus modulate the accessibility of surface receptors for neurotransmitter substances or modulate signal transduction by influencing the interaction between receptors and effector molecules such as adenylate cyclase (Shinitzky, 1984).

There is now considerable evidence for a gradual decrease in membrane fluidity in various tissues, particularly brain, during aging. Evidence is also available that the binding of various ligands to transmitter receptors can be modulated by altering the fluidity of the membranes in which the receptors reside. Fluorescence polarization studies show that the fluidity of brain membranes from old rats (Bonetti et al., 1983; Calderini et al., 1983) is significantly reduced. This effect is most prominent in the purified synaptic membrane fraction and not apparent in the myelin or microsomal fractions. The use of selective electron-spin resonance probes has permitted the demonstration that the increased membrane viscosity is confined to the hydrophobic core of the membrane (Nagy et al., 1983; Cimino et al., 1984). The fluidity of membranes is determined primarily by two factors: the size and degree of unsaturation of the fatty acids, and the amount of cholesterol in the membranes. It is an increase in membrane cholesterol that appears to be responsible for the increased viscosity in membranes from old animals (Cooper, 1978). The brain cholesterol-to-phospholipid ratio in rodents appears to increase during development (Hitzemann and Johnson, 1983; Wood et al., 1984) and aging (Bonetti et al., 1983; Horrocks et al., 1975; Norton and Poduslo, 1973), although one group failed to see an increase in this ratio in rat brain synaptic membranes (Calderini et al., 1983). It should be noted that, in contrast to rodents, the cholesterol-to-phospholipid ratio may not increase in human brain during aging (Horrocks et al., 1975).

There have been numerous studies on changes in brain neurotransmitter receptors during aging, but the information is by no means complete or clear-cut. In many cases there appears to be a decrease in receptors with aging. Both alpha-adrenergic (Misra et al., 1980) and beta-adrenergic (Misra et al., 1980; Maggi et al., 1979; Greenberg and Weiss, 1978) receptors decline with age in rats, an effect that is more prominent in cerebellum and brain stem than cerebral cortex. In man, beta receptors have been shown to decrease in cerebellar but not cortical membranes (Maggi et al., 1979). Rat brain cholinergic receptors measured by the binding of quinuclidyl benzylate (Morin and Wasterlain, 1980) or γ-aminobutyric acid (GABA)ergic receptors measured with muscimol (Maggi et al., 1979) appear to decrease in cerebellar but not cortical membranes. Age-related alterations in dopaminergic systems have recently been reviewed (Joseph and Roth, 1983). Striatal dopamine receptors decline with age in both rats (Misra et al., 1980; Severson and Finch, 1980) and man (Severson et al., 1983). There is direct evidence for declines in D2 and D3 receptors and indirect evidence for declines in D1 receptors (linked to adenylate cyclase) as well. A loss of high-affinity serotonin receptors in cortical membranes from aged humans has been reported (Shih and Young, 1978). More recent studies on human brain observe only a single high-affinity binding site at all ages (Marcusson et al., 1984). A small decline with age was found in membranes from frontal cortex and hippocampus, while membranes from putamen showed an increase with age. Basal binding to serotonin receptors from age mice was found to be elevated approximately 50%, although young mice had more reserve receptors, which could be revealed by rigidifying the membranes with cholesterol (Shinitzky et al., 1983). These investigators also reported a loss of high-affinity receptors in the old animals.

While information concerning the role of lipids in membrane function is increasing rapidly (Sandermann, 1978; Hawthorne and Pickard, 1979; Loh and Law, 1980; Stubbs and Smith, 1984; Yeagle, 1985), there are as yet few studies on the influence of lipids on specific brain receptors. The first indication of the importance of phospholipids for receptor binding was the demonstration that beta-adrenergic binding in frog erythrocytes was destroyed by phospholipase treatment (Limbird and Lefkowitz, 1976). Solubilized beta-adrenergic or dopaminergic receptors must be reconstituted with phospholipid vesicles to obtain appropriate ligand binding (Thang et al., 1980; Moroi and Hsu, 1983). Using dietary lipids to change the unsaturation level of heart membrane lipids, it was shown that increasing the level of unsaturation, which would increase fluidity, lowered the number of beta-adrenergic receptors measured by the binding of dihydroalprenolol (DHA) (Wince and Rutledge, 1981). Treatment of rat lung membranes with cholesteryl hemisuccinate, which would decrease membrane fluidity, increased DHA binding (O'Connor et al., 1984). Similarly, cholesteryl hemisuccinate strongly increased DHA binding in brain membranes (Hershkowitz et al., 1982). These studies demonstrate an inverse correlation between beta-adrenergic receptor binding and membrane fluidity, or a direct correlation of the binding with membrane microviscosity. The data are thus consistent with the model which says that certain membrane proteins, in this case receptors, will be vertically displaced as the membrane microviscosity increases (Borochov

Table 24-1. Binding data analysis.

	P	B_{max}	Kd
Control	.290 ± .004	37.2 ± 1.4	5.3
DOPC	.270 ± .008	17.9 ± 0.6	7.5
Active lipid	.150 ± .008	22.2 ± 1.7	4.3

B_{max} is in units of femtomoles per milligram of protein. Kd is in nanomolar units.

and Shinitzky, 1976). Receptors would thereby become more exposed and display enhanced ligand binding as the microviscosity of the membrane increases.

There is little evidence that cholinergic or dopaminergic receptors are much affected by membrane fluidity, but serotonin receptors show a strong correlation between membrane viscosity and binding (Heron et al., 1980; Hershkowitz et al., 1982). This effect is restricted to serotonin-1 receptors, thought to be linked to adenylate cyclase; serotonin-2 receptors, labeled with spiperone are unaffected (Shih and Ohsawa, 1983). Serotonin binding can be increased several fold by increasing membrane microviscosity with cholesteryl hemisuccinate. When the microviscosity is increased about 70%, the binding decreases (Heron et al., 1980). That this decrease is due to solubilization and shedding of the receptors has been clearly demonstrated by reconstitution experiments (Shih and Ohsawa, 1983).

We have examined the effect of dioleoylphosphatidylcholine (DOPC) and active lipid (AL) on serotonin binding to cortical membranes from mouse brain (Von Hungen et al., 1987). DOPC contains two oleic acid moieties that have one element of unsaturation and can thus be expected to increase the fluidity of membranes. Active Lipid is a special lipid mixture containing phosphatidylcholine, phosphatidylethanolamine, and neutral lipids in a specific ratio that has a very strong fluidizing effect on membranes (Lyte and Shinitzky, 1985). Membranes were incubated with DOPC or AL and then washed by centrifuging through a sucrose cushion. The membranes were then tested for serotonin binding and membrane fluidity using the fluorescence polarization technique (Shinitzky and Berenholz, 1978) and diphenylhexatriene (DPH) as a probe. The binding data was analyzed by Scatchard analysis, and the results in table 24-1 show that both DOPC and AL caused a profound decrease in serotonin binding (Bmax) with little effect of binding affinity (Kd). DOPC caused a modest drop in fluorescence polarization (P), while AL caused a very large decrease. Fluorescence polarization depends on the mobility of the fluorescent probe and thus is a measure of membrane fluidity (Shinitzky and Berenholz, 1978).

If changes in receptors that develop during aging are due to changes in membrane fluidity, it might be possible to compensate for these membrane changes by external means. Indeed, recent reports indicate that age-related increases in membrane viscosity can be counteracted. One approach to increasing membrane fluidity would be by the stepwise methylation of phosphatidylethanolamine (PE) to phosphatidylcholine (PC) using S-adenosylmethionine as a methyl donor. Chronic treatment of old rats for 3 months with S-adenosylmethionine lowered microviscosity to normal (young) values for membranes from corpus striatum (Cimino et al., 1984). There was a restoration of depressed DHA binding to beta-adrenergic receptors, but no restoration of spiperone binding to dopaminergic receptors. Other studies discussed above led to the expectation that membrane fluidization would decrease DHA binding. Dramatic restoration of serotonin receptor binding when old mice were treated with AL has been reported (Shinitzki et al., 1983). Not only was membrane fluidity restored to young levels, but chronic treatment appeared to prevent the loss of reserve receptors. Untreated animals had fewer reserve receptors, possibly because, as viscosity increased during aging, more receptors were exposed to degradative processes. Recently, an additional lipid mixture termed membrane-modulating lipid (MML) has been shown to reduce cholesterol and restore brain membrane fluidity in old rats to young levels (Kessler et al., 1985).

Are age-related changes in receptor activity an expression of altered membrane fluidity? At this time, only two brain neurotransmitter receptors have been shown to be significantly affected by membrane fluidity. There is a tendency for beta-adrenergic receptors to decline with age, but this occurs primarily in the cerebellum and brain stem and little if at all in the cortex. Such regional differences suggest that the decrease in beta receptors is more likely due to cellular changes other than age-related membrane properties. In vitro experiments indicate that beta receptor binding is inversely correlated with membrane fluidity, yet one in vivo treatment that increased membrane fluidity actually increased beta-adrenergic binding (Cimino et al., 1984). The case is somewhat stronger for serotonin receptors. The relationship between high- and low-affinity binding sites and how these change with age is not entirely clear. However, the restoration of (young animal) serotonin-binding characteristics in old mice treated with AL is impressive, particularly the restoration (or preservation) of reserve receptors (Shinitzky et al., 1983). On the other hand, some data suggest that lipid properties other than fluidity may be important in modulating serotonin receptors (table 24-1; Von Hungen et al., 1987). The conclusion must be that membrane lipid composition is an important modulator of certain receptors and this field demands additional study, but, on the whole, evidence that age-related changes in membrane fluidity per se are important determinants of neurotransmitter-receptor function remains weak.

ACKNOWLEDGMENTS

This research was conducted with the encouragement of the Sepulveda Geriatric, Research, Education and Clinical Center and the support of the Medical Research Service of the Veterans Administration.

REFERENCES

Bonetti, A.C., Battistella, A., Calderini, G., Teolato, S., Crews, F.T., Gaiti, A., Algeri, S., and Toffano, G. (1983) Biochemical alterations in the mechanisms of synaptic transmission in aging brain. In: Samuel, D., et al. (eds.) Aging of the brain. New York: Raven, pp. 171–181.

Borochov, H., and Shinitzky, M. (1976) Vertical displacement of membrane proteins mediated by changes in microviscosity. Proc. Natl. Acad. Sci. USA 73:4526–4530.

Calderini, G., Bonetti, A.C., Batistella, A., Crews F.T., and Toffano, G. (1983) Biochemical changes of rat brain membranes with aging. Neurochem. Res. 8:483–492.

Cimino, M., Valtini, G., Algeri, S., Curatola, G., Pezzoli, G., and Stramentinoli, G. (1984) Age-related modification of dopaminergic and beta-adrenergic receptor system: restoration to normal activity by modifying membrane fluidity with S-adenosylmethionine. Life Sci. 34:2029–2039.

Cooper, R.A. (1978) Influence of increased membrane cholesterol on membrane fluidity and cell function in human red blood cells. J. Supramol. Struct. 8:413–430.

Greenberg, L.H., and Weiss, B. (1978) Beta-adrenergic receptors in aged rat brain: reduced number and capacity of pineal gland to develop supersensitivity. Science 201:61–63.

Grinna, L.S. (1977) Changes in cell membranes during aging. Gerontology 23:452–464.

Hawthorne, J.N., and Pickard, M.R. (1979) Phospholipids in synaptic function. J. Neurochem. 32:5–14.

Heron, D.S., Shinitzky, M., Hershkowitz, M., and Samuel, D. (1980) Lipid fluidity markedly modulates the binding of serotonin to mouse brain membranes. Proc. Natl. Acad. Sci. USA 77:7463–7467.

Hershkowitz, M., Heron, D., Samuel, D., and Shinitzky, M. (1982) The modulation of protein phosphorylation and receptor binding in synaptic membranes by changes in lipid fluidity: implications for ageing. In: Gispen, W.H., and Routtenberg, A. (eds.) Progress in brain research, vol. 56. Amsterdam: Elsevier.

Hitzemann, R.J., and Johnson, D.A. (1983) Developmental changes in synaptic membrane lipid composition and fluidity. Neurochem. Res. 8:121–131.

Horrocks, L.A., Sun, G.Y., and D'Amato, R.A. (1975) Changes in brain lipids during aging. In: Ordy, J.M., and Brizzee, K.R., (eds.) Neurobiology of aging. New York: Plenum.

Joseph, J.A., and Roth, G.S. (1983) Age-related alterations in dopaminergic mechanisms. In: Samuel, D., et al., (eds.) Aging of the brain. New York: Raven.

Kates, M., and Kuksis, A. (1980) Membrane fluidity. In: Biophysical techniques and cellular regulation. Clifton, NJ: Humana.

Kessler, A.R., Kessler, B., and Yehuda, S. (1985) Changes in the cholesterol level, cholesterol-to-phospholipid mole ratio, and membrane lipid microviscosity in rat brain induced by age and a plant oil mixture. Biochem. Pharmacol. 34:1120–1121.

Limbird, L.E., and Lefkowitz, R.J. (1976) Adenylate cyclase-coupled beta adrenergic receptors: effect of membrane lipid-perturbing agents on receptor binding the enzyme stimulation by catecholamines. Mol. Pharmacol. 12:559–567.

Loh, H.H., and Law, P.Y. (1980) The role of membrane lipids in receptor mechanisms. Annu. Rev. Pharmacol. Toxicol. 20:201–234.

Lyte, M., and Shinitzky, M. (1985) A special lipid mixture for membrane fluidization. Biochim. Biophys. Acta 812:133–138.

Maggi, A., Schmidt, M.J., Ghetti, B., and Enna, S.J. (1979) Effect of aging on neurotransmitter receptor binding in rat and human brain. Life Sci. 24:367–374.

Marcusson, J., Oreland, L., and Winblad, B. (1984) Effect of age on human brain serotonin (S-1) binding sites.

Misra, C.H., Shelat, H.S., and Smith, R.C. (1980) Effect of age on adrenergic and dopaminergic receptor binding in rat brain. Life Sci. 27:521–526.

Morin, A.M., and Wasterlain, C.G. (1980) Aging and rat brain muscarinic receptors as measured by quinuclidinyl benzylate binding. Neurochem. Res. 5:301–308.

Moroi, K., and Hsu, L.L. (1983) Dopamine receptors: effects of phosphatidyl choline. Trans. Am. Soc. Neurochem. 14:158.

Nagy, K., Simon, P., and Zs.-Nagy, I. (1983) Spin label studies on synaptosomal membranes of rat brain cortex during aging. Biochem. Biophys. Res. Commun. 117:688–694.

Norton, W.T., and Poduslo, S.E. (1973) Myelination in rat brain: changes in myelin composition during brain maturation. J. Neurochem. 21:759–773.
O'Connor, S.W., Scarpace, P.J., and Abrass, I.B. (1984) The effect of age and cholesterol on the rat lung beta-adrenergic system. Biochim. Biophys. Acta 778:497–502.
Sandermann, H. (1978) Regulation of membrane enzymes by lipids. Biochim. Biophys. Acta 515:209–237.
Severson, J.A., and Finch, C.E. (1980) Reduced dopaminergic binding during aging in rodent striatum. Brain Res. 192:147–162.
Severson, J.A., Marcusson, J., Winblad, B., and Finch, C.E. (1983) Age-correlated loss of dopaminergic binding sites in human basal ganglia. J. Neurochem. 39:1623–1631.
Shih, J.C., and Ohsawa, R. (1983) Differential effect of cholesterol on two types of 5-hydroxytryptamine binding sites. Neurochem. Res. 8:701–710.
Shih, J.C., and Young, H. (1978) The alteration of serotonin binding sites in aged human brain. Life Sci. 23:1441–1448.
Shinitzky, M. (1984) Membrane fluidity and cellular functions. In: Shinitzky, M. (ed.) Physiology of membrane fluidity, vol. 1. Boco Raton, FL: CRC Press, pp. 1–51.
Shinitzky, M., and Berenholz, Y. (1978) Fluidity parameters of lipid regions determined by fluorescence polarization. Biochim. Biophys. Acta 515:367–394.
Shinitzky, M., Heron, D.S., and Samuel, D. (1983) Restoration of membrane fluidity and serotonin receptors in the aged mouse brain. In: Samuel, D., et al., (eds.) Aging of the brain, New York: Raven.
Stubbs, C.D., and Smith, A.D. (1984) The modification of mammalian membrane polyunsaturated fatty acid composition in relation to membrane fluidity and function. Biochim. Biophys. Acta 779:89–137.
Sun, A.Y., and Sun G.Y. (1979) Neurochemical aspects of the membrane hypothesis of aging. Interdiscipl. Top. Gerontol. 15:34–53.
Tappel, A.L., Fletcher, B., and Deamer, D. (1973) Effect of antioxidants and nutrients on lipid peroxidation fluorescent products and aging parameters in the mouse. J. Gerontol 28:415–424.
Thang, N.X., Borsodi, A., and Wolleman, M. (1980) Effects of phospholipids on the binding of ^{3}H-dihydroalprenolol to the beta adrenergic receptor of rabbit heat membranes. Biochem. Pharmacol. 29:2791–2797.
Von Hungen, K., Derby, P., and Baxter, C.F. (1987) Modulation of serotonin receptors by specific phosphatidylcholines. Neurochem. Internat. (in press).
Wince, L.C., and Rutledge, C.O. (1981) The effect of dietary lipid on the ^{3}H-dihydroalprenolol and adenylate cyclase activity in rat atria. J. Pharmacol. Exp. Ther. 219:625–631.
Wood, W.G., Strong, R., Williamson, L.S., and Wise, R.W. (1984) Changes in lipid composition of cortical synaptosomes from different age groups of mice. Life Sci. 35:1947–1952.
Yeagle, P.L. (1985) Cholesterol and the cell membrane. Biochim. Biophys. Acta 822:267–287.

25. METABOLIC REGULATION IN BRAIN CELLS

J. TYSON TILDON AND LOIS M. ROEDER

1. INTRODUCTION

Recent studies of metabolic regulation in brain have focused on the complexity and fine-tuning involved in the control of metabolic processes. Gaps in our knowledge are rapidly being filled as previously unidentified "factors" are purified and characterized, new compounds are identified, and new functions for familiar compounds are discovered. Often new information is obtained initially from tests of substances that support survival and growth of neural tissue in vitro, and the mechanisms involved are not fully investigated until much later. Such studies also have revealed the importance of cell–cell interactions (Mandel et al., 1977; Seeds and Hawkins, 1985), and several workers have identified specific metabolic processes involved in these exchanges (Cummins et al., 1979; Hertz, 1979; Pentreath and Kai-Kai, 1982; Sykora, 1983; Hertz and Richardson, 1984). Different factors seem to be involved in central nervous system (CNS) and peripheral nervous system (PNS) (Seifert and Muller, 1984). However, nerve growth factor (NGF), which is essential for the survival, both in vivo and in vitro, of sympathetic and sensory neurons (Levi-Montalcini and Angeletti, 1968) may also have a role in brain (Honegger and Lenoir, 1982). The proposed mechanisms of the neurotrophic action of NGF (and of some new putative growth factors) were recently reviewed by Thoenen and Edgar (1985). Kauffman and coworkers have begun to identify some of the specific metabolic alterations that accompany the growth effect of NGF on the superior cervical ganglion (Dumbrowski et al., 1983).

Vernadakis, A. (ed.), *Model Systems of Development and Aging of the Nervous System*.

Ions and ion fluxes have been assigned important new roles in cascades of metabolic events associated with activities of receptors and transporters at various membrane sites (Kuffler and Nicholls, 1966; Pentreath and Kai-Kai, 1982; Walz and Hertz, 1983), and this has prompted a host of investigations of potassium-induced alterations in specific metabolic events (Hertz, 1978; Brookes and Yarowsky, 1985). Protein phosphorylations have key roles in the control of cell metabolism, especially in ion channel modification and enzyme activation (Reichardt and Kelly, 1983); the enzymes responsible for the phosphorylations (protein kinases) are themselves regulated by one or more of a growing list of effectors including cyclic AMP, calcium–calmodulin, and calcium–phospholipid. The number of peptides that have been proposed to have a neurotransmitter or neuromodulator function is now extensive (Iverson, 1984). The list includes most of the peptide hormones, the opioid peptides, and several other compounds. The coexistence and cosecretion of the neuropeptides with one of the monoamines suggest the possibility of a very large number of specific chemical signals associated with combinations of two, three, or more of these substances.

It is beyond the scope of this chapter to review comprehensively all proposed regulatory factors and the entire range of metabolic activities associated with brain development and function. We have chosen to focus on some recent findings that open possibilities for obtaining information about new sites of metabolic regulation, new roles for familiar compounds, and/or new types of cell–cell interactions. Emphasis is placed on the utilization of substrates for production of energy and synthesis of complex lipids and includes effects of various factors on the transport and metabolic fates of substrates in these pathways. Carrier-mediated transport systems, recently shown to be present on brain cell plasma membranes, provide sites for regulation of substrate utilization beyond the well-recognized barrier operating in brain capillary endothelial cells (which is not reviewed here). Although established cell lines of either glial (Pfeiffer et al., 1977) or neuronal (Nelson, 1977) origin have specific advantages in neurobiology, this chapter focuses only on factors that regulate metabolic processes in freshly prepared or primary cultures of neural cells.

2. REGULATION BY SUBSTRATE INTERACTIONS

Studies in vivo (Hawkins et al., 1971; Webber and Edmond, 1977) and in vitro (Itoh and Quastel, 1970; Yeh et al., 1977) and using isolated cells (Pleasure et al., 1979; Roeder et al., 1982; Tildon et al., 1983) have shown that the brain uses multiple substrates for both anabolic and catabolic reactions. It is also well established that these metabolic processes during early neonatal development are different from those found in adults (Robinson and Williamson, 1980). Among the major substrates that the brain is capable of oxidizing and using as fuel for energy are glucose, ketone bodies, pyruvate, lactate (Cremer, 1981), and glutamine (Yu and Hertz, 1983; Yu et al., 1982; Tildon, 1983; Tildon and Roeder, 1984).

The regulation of substrate utilization is controlled by a variety of mechanisms including the levels of specific enzyme activities and the concentrations of effectors as well as the uptake and transport of substrates into specific cells and subcellular organelles. In addition, the vast heterogeneity of the brain provides separate compartments for metabolic activities; this localization and the different sizes of individual metabolic pools also contribute to the regulation of metabolism. Within a specific metabolic pool, the interaction among two or more alternatives has been shown to have profound effects on the overall homeostasis within the cell.

In a series of studies conducted more than a decade ago, we demonstrated that hyperketonemia (elevated levels of 3-hydroxybutyrate and acetoacetate) had striking effects on both glucose and glutamine metabolism in the brain (Ozand et al., 1975 a and b; Tildon et al., 1975). At least seven metabolic sites were shown to be affected. These included glycogen degradation, the conversion of fructose-6-phosphate to fructose 1,6,-diphosphate, and the conversion of pyruvate to lactate (the results revealed changes in both the cytosolic and mitochondrial redox potential). There were changes also in enzyme activities including thiolase, coenzyme-A (CoA) transferase, glutamate dehydrogenase, and glutaminase. It was apparent from these reports that no single mechanism was responsible for these multiple changes. Because of the apparent physiological relevance of these studies and the obvious intracellular changes that resulted from the presence of elevated levels in ketone bodies, we designed a series of experiments to determine the effect of alternative substrates on the metabolism of a specific substance.

In recent studies, we evaluated the competition among substrates as a regulatory factor in substrate oxidation. Using both whole homogenates and dissociated cells obtained from the contralateral hemispheres of brains from the same animal, we were able to demonstrate interactive effects of several different substrates (Roeder et al., 1984a and b; Tildon et al., 1985a). The results are summarized in table 25-1; the data revealed that the oxidation of ^{14}C-glucose was inhibited by about 40% when it was metabolized in the presence of unlabeled glutamine. This effect was observed using either homogenates or dissociated cells and with preparations from young (4 to 6 days old) and adult animals. However, in the converse situation (the oxidation of labeled glutamine in the presence of unlabeled glucose), the production of $^{14}CO_2$ was not affected when homogenates were used, but there was a 30%–35% decrease in the rate in dissociated cells. If two substrates were oxidized via the tricarboxylic acid cycle in the same compartment, one would expect a reciprocal dilution of the $^{14}CO_2$ produced when one unlabeled substrate was added to a reaction system with another labeled compound. However, in the present results, unlabeled glutamine was effective in competing with glucose, but the addition of unlabeled glucose, even at 5 m*M*,

Table 25-1. Summary of results of competition among alternative substrates in brain.

Substrates	Homogenates (%)		Dissociated cells (%)	
	Young	Adult	Young	Adult
[6-^{14}C]-glucose + glutamine	−40	−40	−40	−60
[U-^{14}C]-glutamine + glucose	NS	NS	−30	−35
[U-^{14}C]-glutamine + 3-hydroxybutyrate	+40	NS	−25	NS
[U-^{14}C]-glutamine + acetoacetate	+75	NS	NS	NS
[3-^{14}C]-hydroxybutyrate + glutamine	−28	NS	−17	−33
[3-^{14}C]-acetoacetate + glutamine	−19	NS	NS	NS
[6-^{14}C]-glucose + 3-hydroxybutyrate	−37	−18	−59	−21
[6-^{14}C]-glucose + acetoacetate	−32	NS	−42	NS
[3-^{14}C]-3-hydroxybutyrate + glucose	NS	+40	NS	+60
[3-^{14}C]-acetoacetate + glucose	−10	+60	−15	−10

Hemispheres of brains from young (4- to 6-day old) and adult (90-day or older) rats were used to prepare whole homogenates (in 0.25 *M* sucrose) or suspensions of dissociated cells (in 0.9% NaCl), and the rates of $^{14}CO_2$ production from ^{14}C-labeled substrates were determined as previously described (Roeder et al., 1984a and b). The concentrations of labeled substrates were 2 m*M* 6-^{14}C-glucose, 2 m*M* U-^{14}C-glutamine, 1 m*M* 3-^{14}C-3-hydroxybutyrate, and 1 m*M* 3-^{14}C-acetoacetate. The effects of additions of each unlabeled alternative substrate (5 m*M*) are expressed as the percentage changes from the rate with labeled substrate alone (control). NS indicates no significant change. The numbers of rats per group were 3–13 for controls and 3–7 for groups with added unlabeled substrate.

had no effect on glutamine oxidation in whole brain homogenates. These results suggest that glutamine or one of its metabolites may inhibit a step in the pathway for glucose utilization in the brain. In liver, glutamine inhibits pyruvate dehydrogenase activity (Haussinger et al., 1982); other amino acids have been shown to regulate glycolytic enzymes, including the inhibitory action of phenylalanine on pyruvate kinase and hexokinase of brain (Weber, 1969) and the inhibition of hepatic pyruvate kinase by alanine (Weber et al., 1968).

The results with other substrate pairs were also revealing (table 25-1). When labeled glutamine oxidation was measured in the presence of ketone bodies (either 3-hydroxybutyrate or acetoacetate), there was an enhancement of the rate of oxidation with brain homogenates from young animals, but not with adults. All other preparations (i.e., homogenates from adult animals and dissociated brain cells from either young or adult animals) showed little or no change. The mechanism accounting for this enhancement may be related to a release of inhibition of the activity of α-oxoglutarate dehydrogenase. It has been shown (Garland, 1964; Smith et al., 1974) that succinyl-CoA is an effective inhibitor of this enzyme, and the addition of either ketone body would decrease the concentration of succinyl-CoA by using it for the synthesis of acetoacetyl-CoA (yielding free succinate). As a consequence, the oxidation of glutamine via the tricarboxylic acid cycle would be less inhibited. It was also found that the addition of unlabeled glutamine to labeled 3-hydroxybutyrate caused a reduction in the rate of oxidation by dissociated brain cells from both young and adult animals and that unlabeled glutamine caused a reduction in the oxidation of 3-hydroxybutyrate by homogenates from young animals. The finding that glutamine enhanced the oxidation of ketone bodies in young rats, but not in adults, is consistent with the observation that the levels of ketone-body enzymes are higher in the brains of young animals than in adults (Page et al., 1971; Tildon et al., 1971) and consistent with the higher rates of oxidation of both 3-hydroxybutyrate and acetoacetate in 10- to 15-day-old animals compared to adults (Tildon et al., 1983). These results suggest an unusual synergistic action between these particular substrates. The interactions between glutamine and the two ketone bodies in brain tissues of young animals support the hypothesis that glutamine is used as an energy source in the brain and that this utilization is greater in younger animals.

The results with glucose and either of the ketone bodies were also unusual. When labeled glucose oxidation was measured in the presence of either unlabeled 3-hydroxybutyrate or acetoacetate, the rate of $^{14}CO_2$ production was decreased significantly using either homogenates or dissociated cells from young animals. However, when preparations from adults were used, the decrease was much less with 3-hydroxybutyrate and not significant with acetoacetate. In the converse situation, when adult animals were used, unlabeled glucose caused an enhancement in the oxidation of either 3-hydroxybutyrate or acetoacetate by homogenate preparations and a similar effect was observed for 3-hydroxybutyrate oxidation by dissociated cells. Essentially no effect was observed with either homogenates or dissociated cells from young animals. In summary, out results revealed a complex set of interactions between many substrates. They suggest several possible mechanisms by which certain substrates may regulate the utilization of other compounds, and emphasize that developmental changes in brain play a role in these processes.

Studies by Miller et al., (1982) also revealed patterns of interactions between glucose and ketone bodies that were similar to our studies. In addition to the observation that increased levels of ketone bodies caused a decrease in the rate of glucose utilization, they reported that the decrease was additive to the effects caused by hypercapnia, including changes in brain metabolites such as glucose, glucose 6-phosphate, pyruvate, α-ketoglutarate, and glutamate. The authors conclude that hyperketonemia diminishes flux through hexokinase, phosphofructokinase, and pyruvate dehydrogenase. These findings are consistent with our present studies and our ealier reports (Ozand et al., 1975a and b; Tildon et al., 1975) showing complex interactions between ketone bodies, glucose, and glutamine metabolism. The results by Miller et al. appear to be somewhat different from those of Wapnir et al. (1973), Dierks-Ventling (1971), and Dierks-Ventling and Cone (1971), who described an adaptive increase in use of ketone bodies by the brain in hyperketonemic animals.

Pyruvate appears to be another substrate that acts as an important regulator of metabolic processes in the brain. Several lines of evidence support this conclusion, including studies by Manthorpe et al. (1982) and Varon et al. (1984), who demonstrated that dissociated cultures of neurons can be maintained without the addition of exogenous neurotrophic factors if a low molecular weight fraction ($< 1000\,M_R$) from glial-conditioned media is added to the culture. Other studies provide evidence that exogenous pyruvate is required for dissociated neurons in addition to a "hormonal complement" (Skaper et al., 1984). Studies by Patel (1974) and by Shank et al. (1985) have shown that pyruvate carboxylase is the predominant anaplerotic enzyme in CNS tissue; using immunocytochemical techniques, they also demonstrated that this enzyme is found mainly in astrocytes. Although the low molecular weight ($< 1000\,M_R$) factor has not been identified, the collective results suggest a role for pyruvate as a "trophic" factor, which may be mediated through its contribution as an anaplerotic agent. This suggests that pyruvate would be the major substrate to provide citric acid cycle intermediates such as α-ketoglutarate and malate, which could be used as precursors for glutamate, aspartate, γ-aminobutyric acid (GABA), and other components of the CNS.

Collectively, these results provide cogent evidence for the intracellular regulation of metabolic events by substrate interactions. Although this would not appear to be unusual, the broader implication of this aspect of regulation has received only sparse attention. The statements above should help to focus additional investigations to identify the specific sites where these regulatory processes occur. Underlying these considerations is the obvious complexity of brain because of the heterogeneity of the cells. To understand the overall aspects of metabolic regulation requires the identification of specific substrate interactions in each cell type.

3. KNOWN FACTORS

Among the more interesting recent reports on the effects of known or identified factors on brain metabolism are those of some hormones, especially thyroid hormone, ACTH, and insulin. Some low molecular weight peptides can enter the brain interstitial space via those regions that lack a blood–brain barrier, i.e., the circumventricular organs (Pardridge, 1983). Receptors for some of these, including ACTH and insulin, have been found on the plasma membranes of nerve endings (Van Houten et al., 1979 and 1981) and, for insulin, on the endothelia of the luminal side of brain capillaries (Van Houten and Posner, 1979; Frank and Pardridge, 1981) and in all brain regions (Havrankova et al., 1978). A transport system for thyroid hormone (both T_3 and T_4) has been described (Pardridge, 1981) and the biologically active form of the hormone, T_3, is generated in situ in brain cells (Leonard et al., 1981). These observations have stimulated renewed interest in identifying specific metabolic events modulated by these compounds.

Table 25-2. Effects of thyroid hormone (T_3) on sulfolipid synthesis and activities of lipid sulfotransferases in cultures of dissociated brain cells from embryonic mice.

Culture medium	Sulfolipid synthesis (%)	Lipid sulfotransferase activities (%)
+ Normal calf serum (control)	100	100
+ Hypothyroid calf serum	60	70
+ Hypothyroid calf serum + T_3	114	103
Serum-free	44	—
Serum-free + T_3	86	—

Dissociated brain cells were obtained from 15-day-old embryos and maintained in culture in the presence of 20% calf serum for 10 days, and then transferred to one of the above media; 3 days later, the cells were either exposed to $^{35}SO_4$ for 16 hours for determination of incorporation into lipids or harvested, homogenized, freeze-dried, and assayed for the activities of cerebroside (CER) and of monogalactosyl diacylglycerol (MGD) sulfotransferases. The results were expressed as cpm of ^{35}S from ^{35}S SO_4 incorporated into lipid per milligram of protein (lipid synthesis) or as pmoles of ^{35}S from ^{35}S adenosine-3′-phosphate, 5′-phosphosulfate transferred to lipid (CER or MGD) per milligram protein per hour (enzyme activities). The values from three tables summarizing results of 2–4 experiments each are expressed as the average percentages of corresponding control values set equal to 100%. From Bhat et al. (1981).

The ability of thyroid hormone to affect the course of myelination in neonatal brain has been reassessed in cultured cells by Pieringer and associates (Bhat et al., 1981) with the aim of identifying the underlying mechanisms involved. In a series of experiments summarized in table 25-2, they showed that, in cells maintained in a thyroid hormone (T_3)-deficient environment (medium containing serum from a hypothyroid calf), there was a reduced synthesis of myelin-associated lipids and this was accompanied by decreased activities of two sulfotransferases. Both lipid synthesis and the enzyme activities could be restored if T_3 were added to the culture medium. The assigned role of T_3 in these effects was strengthened by the demonstration that addition of T_3 to serum-free culture medium also reversed the reduction of lipid synthesis in T_3-deprived cells. Later studies by this group (Shanker et al., 1985) showed an age-dependent loss of sensitivity to the myelinogenic effects of T_3 in these cultured cells. Hoeben et al. (1985) also observed that T_3 enhanced the synthesis of sulfolipids without altering the general lipogenic activity of cultured oligodendrocytes.

Atterwill et al. (1985), using several cell culture systems, found some very selective effects of thyroid hormone on the activity of Na^+-K^+-ATPase, a plasma membrane-associated enzyme that controls ion fluxes in the CNS. In a mixed population of brain cells, these workers observed that T_3 increased the activity of this enzyme (but not that of MG^{2+}-ATPase). However, in cultures enriched either in granule cell cerebellar neurons or in cerebellar astrocytes, T_3 had no effect on either enzyme. The authors felt that these results supported their proposal (Balazs et al., 1975) that the primary targets of thyroid hormone action in brain are the Purkinje cells. In serum-free cultures of either neonatal brain cells, most of which were judged by microscopic examination to be neurons (Akuzawa and Wakabayashi, 1985) or

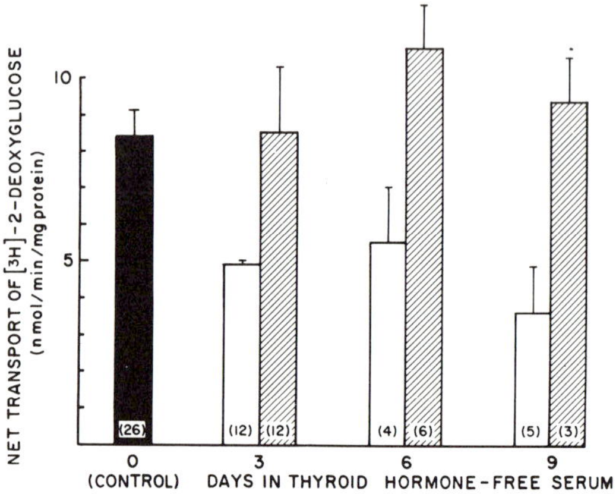

Figure 25-1. The acute effect of thyroid hormone (T_3) on transport of ^{3}H-2-deoxyglucose in astrocytes. Cells from newborn rat brain were prepared for culture as previously described (Roeder et al., 1985a). After growth in media with normal serum, cells were transferred to media containing thyroid hormone-free serum for the number of days indicated. The acute effect of the hormone on the rate of uptake of ^{3}H-2-deoxyglucose was determined by addition of T_3(50 μM) to the reaction mixture. Hypothyroid cells (□) had significantly ($p < 0.05$) reduced rates of transport of ^{3}H-2-deoxyglucose than controls (■) after 3 and 9 days of exposure to thyroid hormone-free serum. Hypothyroid cells incubated with T_3(▨) during the 60-second transport assay transported the hexose at significantly ($p < 0.05$) higher rates than cells without added T_3 (□) at 6 and 9 days.

aggregates of fetal rat brain cells (Honegger and Lenoir, 1980), supplementation with thyroid hormone improved cell survival and enhanced the activities of choline acetyltransferase and acetylcholinesterase.

Recent findings on effects of thyroid hormone in primary cultures of astrocytes suggest a new key role in brain for this substance (Roeder et al., 1985a). As shown in figure 25-1, cells that had been maintained in medium containing thyroid hormone-free serum transported ^{3}H-2-deoxyglucose at rates significantly less than control cultures (cells maintained in complete serum). Addition of exogenous T_3 to the transport incubation medium brought about a rapid (within 60 seconds) restoration in the rate of hexose transport to control levels. Koenig et al. (1984), working with cerebral cortex slices from hypothyroid rats, found very rapid effects of T_3 on hexose and amino acid transport and on the synthesis of polyamines and flux of intracellular calcium. They proposed a coupling of these events in a cascade of phenomena required for membrane transport function. This action of thyroid hormone in brain may be analogous to that of insulin in other tissues. Indeed the rapidity of the effect of T_3 suggests a mechanism such as translocation of a transporter as proposed for insulin in adipose cells (Cushman et al., 1984; Kono, 1984) rather than a nuclear transcription-mediated event (Oppenheimer et al., 1976; Samuels and Tsai, 1973).

Interestingly, this response of astrocytes to thyroid hormone was observed

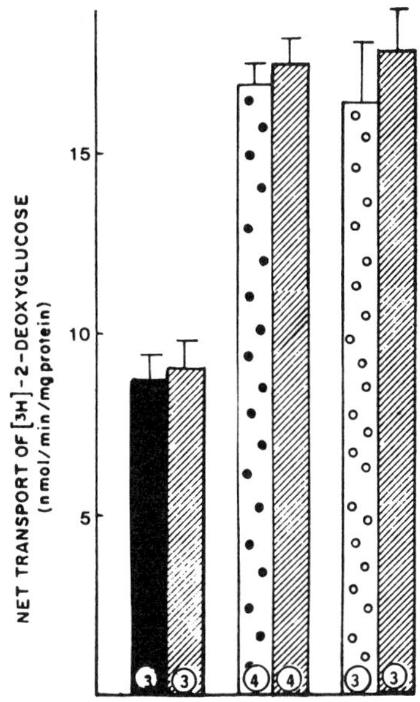

Figure 25-2. Cells from newborn rat brain were prepared for culture as previously described (Roeder et al., 1985a). After growth in media with normal serum for 14 days, samples (number shown in *circles*) were exposed to 0.25 mM dBcAMP for 10 days either with (°°) or without (∴) transfer 3 days later to media containing thyroid hormone-free serum. Controls (■) remained in regular media. *Hatched bars* (▨) represent rates of uptake in the presence of 50 μM T_3 added to the incubation medium.

only in "immature" cells, i.e., flat glial cells that had not been stimulated to "differentiate" by addition of dibutyryl cAMP (dBcAMP) to the culture medium as described by others (Sensenbrenner, 1977; Hertz et al., 1982) or by treatment with norepinephrine or brain maturation factor or deprived of serum (Kimelberg, 1983). When astrocytes were treated with dBcAMP, the cytoplasm retracted so that a network of thin, elongated processes resulted (Trimmer et al., 1982). Such treatment was accompanied both by a marked increase in the rate of 2-deoxyglucose transport and elimination of the response to thyroid hormone (figure 25-2). This lack of response of dBcAMP-treated cells to T_3 may reflect the morphological characteristics of the cells rather than a direct effect of the cyclic nucleotide on the transport system. Indeed, Cummins et al. (1979) reported that addition of dBcAMP to the incubation mixture containing immature astrocytes did not alter uptake of 2-deoxyglucose. Thus, the glucose transporter in the "differentiated" cells may only reside at the membrane surface and may not require translocation from a storage site stimulated by thyroid hormone.

Timiras and colleagues have reported several aspects of age-related differences in the effects of thyroid hormone on the brain (Valcana and Timiras,

1978; Draves and Timiras, 1980). Using their in vitro models, they observed a striking change in the response of neural cells to T_3 that was associated with cell differentiation (Safaei and Timiras, 1985). Their study revealed that 1) the activity of tyrosine hydroxylase was low in neuroblastoma cells maintained in a hypothyroid condition, 2) the enzyme activity increased with exposure of immature, undifferentiated cells to T_3 for 10 days, and 3) that in differentiated cells the enzyme activity was elevated and no longer responsive to T_3 status. Although these results were obtained using neuroblastoma cell lines, they may have important clinical implications since it has been shown that tyrosine hydroxylase activity is elevated in several regions of the brains of infants that die of sudden infant death syndrome (SIDS) (Ozand and Tildon, 1983) and other studies revealed that SIDS victims have elevated T_3 levels (Chacon and Tildon, 1981). These findings also suggest that there are multiple mechanisms of T_3 action in brain. In the studies by Timiras and coworkers, the dose response of tyrosine hydroxylase activity to T_3 was correlated with an increase in the number of nuclear T_3 receptors, and they proposed that transcription-mediated events may account for the T_3 regulation of this enzyme activity. In the studies of T_3 regulation of glucose transport in normal astrocytes, the rapidity of the response to T_3 led us to conclude that the mechanism was not likely to involve a transcription process, but rather a translocation of existing transporter molecules. Thus, there appear to be at least two different mechanisms present in brain whereby thyroid hormone regulates metabolic activities.

A variety of metabolic effects of the behaviorally active peptide, $ACTH_{1-24}$, have been reported, including elevation of intracellular cyclic AMP levels (Van Calker et al., 1983) and inhibition of protein phosphorylation (Zwiers et al., 1978) with a concomitant increase in lipid phosphorylation, i.e., formation of phosphatidylinositol biphosphate (Jolles et al., 1981). The latter finding suggests an important role of ACTH in regulating membrane permeability, and a later report by Daval et al. (1983) is relevant in this regard; they found a marked stimulation of the uptake of 2-deoxyglucose in cultured neurons by this peptide.

Reports in the literature regarding the effects of insulin on glucose transport in the CNS have been conflicting. Some investigators found insulin stimulated glucose transport across the blood–brain barrier (Chapman et al., 1977; Hertz et al., 1981) and in cultured glial cells (Clark et al., 1984). Others observed no influence of insulin (Diamond and Fishman, 1973; Betz et al., 1973; Roeder et al., 1985a). However, insulin did enhance glycogen synthesis in brain (Daniel et al., 1975; Nelson et al., 1968; Strang and Bachelard, 1971) and stimulated lipid synthesis in cultured oligodendrocytes (Hoeben et al., 1985) and in intact rats (Reed, personal communication). Other functions for insulin are suggested by the reports by Roth and Bornstein (1985), who observed that in organotypic cultures insulin stimulated neuritic outgrowth, myelination, and the activity of the myelin-associated enzyme, 2′-3′-cyclic

Table 25-3. Effect of treatment with dBcAMP on rates of substrate oxidation by cultured astrocytes from rat brain.

Substrate (m*M*)	*n*	Untreated	After dBcAMP	Ratio
6-^{14}C-glucose (2 m*M*)	22	5.66 ± 0.64*	16.54 ± 2.90**	2.69 ± 0.37**
U-^{14}C-glutamine (2 m*M*)	10	52.80 ± 5.59	66.23 ± 7.15	1.27 ± 0.10**
3-^{14}C-3-hydroxybutyrate (1 m*M*)	19	17.18 ± 0.91	16.09 ± 1.23	1.02 ± 0.06
3-^{14}C-3-acetoacetate (1 m*M*)	8	18.85 ± 1.15	19.63 ± 1.48	1.05 ± 0.08

*Mean nmol/hour/mg protein ± SEM. **$p < 0.05$.

Astrocytes were obtained from brains of newborn rats and cultured in 25-cm^2 flasks as previously described (Roeder et al., 1985a) in the presence of 5% fetal bovine serum. After 2 weeks in vitro, half of the flasks in each experiment were then maintained in the presence of 0.25 m*M* dibutyryl cyclic AMP (dBcAMP). Rates of oxidation of the substrates were measured from 1 to 3 weeks later (average 2 weeks) in treated and untreated cells by incubation in the presence of one of the above ^{14}C-labeled substrates in the same culture flasks. Buffered medium and the procedures for collecting $^{14}CO_2$ on hyamine-hydroxide-saturated filter paper in hanging center wells suspended from rubber septa capping the flasks were essentially as described for measurement of substrate oxidation of homogenates and dissociated brain cells incubated in test tubes (Tildon et al., 1984). From three to six separate brain cell cultures were used for each substrate.

nucleotide 3′-phosphohydrolase, and by Roger and Fellows (1980), who found that insulin stimulated ornithine decarboxylase in brain. Some investigators (Van Houten and Posner, 1981; Havrankova et al., 1981) have even proposed completely new roles for insulin in brain as a neurotransmitter, neuromodulator, or a signal to CNS structures involved in metabolic regulation.

The enhanced rate of glucose uptake in the dBcAMP-treated cells (figure 25-2) may be responsible for the increased rate of glucose oxidation in these cells. As shown in table 25-3, there was nearly a threefold elevation in the rate of conversion of 6-^{14}C-glucose to $^{14}CO_2$ following treatment of cultured rat brain astrocytes with dBcAMP. In contrast, little or no change in rate of oxidation was seen when either U-^{14}C-glutamine, 3-^{14}C-3-hydroxybutyrate, or 3-^{14}C-3-acetoacetate was used as substrate.

4. UNIDENTIFIED "FACTORS"

Although there are many factors whose chemical and biological nature have been established, an increasing number of observations suggest that there are various agents in serum and biological fluids whose regulatory actions have not been characterized fully. These putative "factors" have been shown to have significant effects on several aspects of neural cell function. In some instances, the effects are manifested by the growth or propagation of cells. In other instances, these effects are revealed by changes in the uptake and transport properties of specific substrates, and some studies have provided information about the effect of the putative "factor" on the metabolic fate of specific substrates.

In addition to the classic factors that regulate brain metabolism, there are now several new putative effectors of brain metabolism. Much of the evidence for these putative factors is the outgrowth of studies to define the most appropriate culture conditions for isolated brain cells. In many instances, some of the most potent evidence for the presence of uncharacterized effectors has been obtained from studies using defined media and from studies using "conditioned" media. These media were usually derived from glial cells that secreted critical substances that supported the survival of neurons or that were essential for differentiation as indicated by neurite outgrowth. These approaches have been helpful in the identification of several hormones and agents including insulin, progesterone, putrescine, transferrin, and thyroid hormone as required factors for cellular maintenance.

Several laboratories have identified specific protein fractions of serum that act as trophic agents or effectors, some of which modulate neuronal survival in vitro (Varon and Adler, 1981). Neurons could be maintained as dissociated cell cultures if they were supplied with low molecular weight fractions from conditioned media (Varon et al., 1984). More recent studies have shown that pyruvate is the main trophic ingredient in glia-conditioned media; however, several other low molecular weight substances (not yet defined) can reduce the requirement for pyruvate by CNS neurons (Selak et al., 1985).

Using neurite outgrowth and neuronal survival, Kaufman and Barrett (1983) described an active serum fraction that supported neuronal survival, but did not support either glia or fibroblast proliferation. The factor has an apparent molecular weight of 55,000 with an isoelectric point of 5.6, and it is distinct from other well-defined growth factors. These workers also reported the presence of serum factor(s) that were toxic to cultured neurons and several that caused cell death. Several of these had molecular weights around 35,000 and others were greater than 100,000 M_R. However, the specific biochemical and biological characteristics of these factors are yet to be defined.

On several occasions, the evidence for putative effectors has been obtained through serendipitous experiences. For example, in our attempts to evaluate the optimum system for substrate oxidation, we observed inhibitory effects of serum on glucose metabolism. At a serum protein concentration of approximately 0.55 mg/ml, oxidation of 6-^{14}C-glucose to $^{14}CO_2$ was decreased more than 50%. In contrast to the inhibitory effects on glucose, the rates of oxidation of 3-^{14}C-3-hydroxybutyrate and U-^{14}C-glutamine were decreased much less (figure 25-3). Serum from cows, rats, horses, and humans produced similar effects, as did serum from young and old animals and from both sexes. The effect on 6-^{14}C-glucose oxidation was proportional to serum protein concentration, and significant inhibitory activity was obtained with dialyzed serum. Heating (80°C for 10 minutes) significantly reduced the inhibitory activity.

Additional experiments with dissociated cells revealed that the inhibitory effects of fetal bovine serum were only slightly changed over more than a 50-fold range in initial glucose concentration. Fetal bovine serum had no

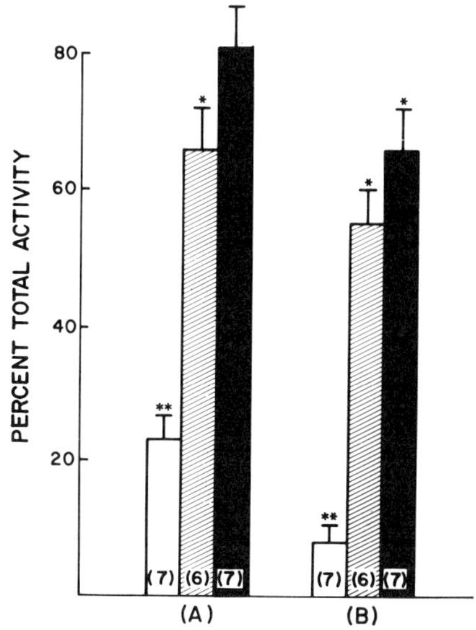

Figure 25-3. The effect of serum on the oxidation of 6-^{14}C-glucose (*open bars*), U-^{14}C-glutamine (*cross-hatched bars*), and 3-^{14}C-3-hydroxybutyrate (*closed bars*) by dissociated brain cells. Values are percentages (± standard errors) of the control values (rates obtained in the absence of serum). The average rates for controls were 4.24, 3.69, and 8.95 nmol/hour/mg for glucose, 3-hydroxybutyrate, and glutamine, respectively. The numbers of individual animals are shown in *parentheses*. Concentrations of serum protein added were (*A*) 2.56 and (*B*) 10.2 mg/ml. Specific activities of the substrates were 69,000, 195,000, and 32,700 dpm/μmol for 6-^{14}C-glucose, 3-^{14}C-3-hydroxybutyrate, and U-^{14}C-glutamine, respectively. Statistical significance was evaluated with the Scheffe's test for multiple comparisons. From Tildon and Stevenson (1984).

effect on the rate of glucose transport as measured by the uptake of 1,3-^{3}H-2-deoxyglucose into dissociated cells (Tildon et al., 1985b). The inhibitory effect of serum was greater on $^{14}CO_2$ production from 6-^{14}C-glucose than 1-^{14}C-glucose (80% vs 55%). Other studies revealed that $^{14}CO_2$ production from 1-^{14}C-pyruvate was more than five times the rate obtained using 3-^{14}C-pyruvate; however, the inhibitory effect of serum was much greater on the latter (20% vs 60% inhibition, respectively) at 2 m*M* pyruvate and in the presence of 1% fetal bovine serum. These results lead us to postulate that the apparent decrease in $^{14}CO_2$ production reflects the regulation of pyruvate utilization as a precursor of oxalacetate versus its use as a source of acetyl-CoA. When pyruvate is labeled on the first carbon, the $^{14}CO_2$ produced is liberated at either the pyruvate dehydrogenase step or at the isocitrate dehydrogenase step following is carboxylation to form oxalacetic acid. Therefore, little or no label could accumulate in glutamate, glutamine, or at other anaplerotic reactions. The only opportunity for intermediate dilution would be via oxaloacetic acid conversion to malate or aspartate. In contrast, when pyruvate is labeled on the third carbon, $^{14}CO_2$ production can occur only

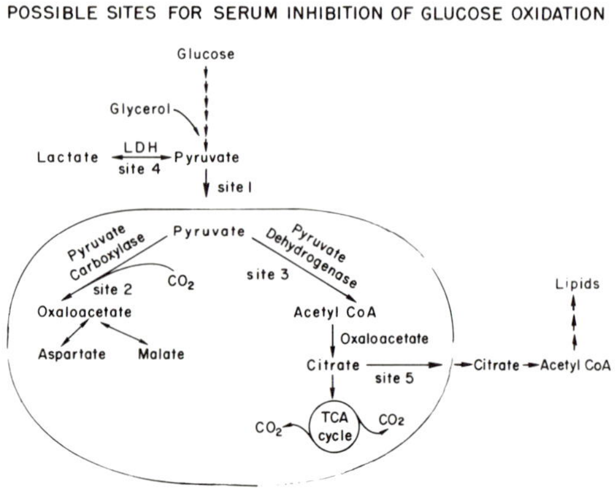

Figure 25-4. Possible sites for serum inhibition of glucose oxidation. These include: 1) pyruvate uptake by mitochondria or synaptosomes, 2) pyruvate carboxylase (oxaloacetate formation), 3) pyruvate dehydrogeanse (acetyl CoA formation), 4) lactate dehydrogenase (interconversion of pyruvate and lactate), and 5) citrate transport.

after the second turn of the tricarboxylic acid (TCA) cycle, regardless of the fraction of pyruvate that enters the TCA cycle via the carboxylase reaction or the pyruvate dehydrogenase reaction. Therefore, when pyruvate is labeled on the third carbon, a greater decrease in $^{14}CO_2$ production can occur by diverting the label into other intermediates, and a relatively larger amount of label can be removed from the TCA cycle as glutamate, aspartate, or by utilization as succinate. The possible sites of regulation are shown in figure 25-4. We postulate that the effect occurs at the step catalyzed by pyruvate carboxylase (site 2) and experiments by Sykes and Lopes-Cardozo (1985) as well as from our laboratory (Tildon and Stevenson, unpublished results) suggest that serum may alter pyruvate carboxylase activity. Attempts to characterize the factor using Amicon filtration showed the highest inhibitory activity in a 10,000-M_R fraction, although some inhibitory activity was found in commercial preparations of bovine serum albumin. Delipidation of serum had no effect.

Our observations may have relevance to some recent studies by Shank and his coworkers (Shank and Campbell, 1984; Shank et al., 1985). In their attempts to determine possible regulators of substrate transport, Shank and Campbell (1984) demonstrated that glutamine altered the uptake of α-ketoglutarate by decreasing the substrate-carrier affinity and by increasing the Vmax. These and other results led the authors to conclude that α-ketoglutarate and malate uptakes by synaptic terminals are modulated by glutamine and other compounds. They further suggested that the regulation of this flux contributes to the metabolism and synthesis of several neurotransmitters. Their results and conclusions support several statements put forth by our laboratories suggesting a role for alternative substrates in the regulation of metabolic processes in the brain (see above).

As indicated above, the transport and uptake of substrates by individual cells and factors affecting events on the cellular membrane also play major roles in the regulation of metabolic processes in the cell. A study by Wu et al. (1985) suggested the presence of a substance in serum that makes astrocytes grown in culture responsive to beta-adrenergic agonists. The substance appears to be larger than 3500 M_R, stable to heat inactivation, and extractable by methanol/chloroform. In their studies, they examined several possible mechanisms as being responsible for this activity including: 1) a decrease in adenylate cyclase activity, 2) increased phosphodiesterase activity, or 3) decreased receptor numbers or affinity. Their results indicate that the effect appears to be the result of a decrease in receptor number and an increase in the rate of cAMP degradation. Although the nature of the substance and the exact mechanism is yet to be identified, this study is indicative of a growing number of investigations that demonstrate the usefulness of tissue culture systems as a probe for determining the presence of uncharacterized factors that regulate cell metabolism.

5. CONCLUSIONS

Presented above is a broad spectrum of results suggesting the presence of a variety of agents in serum and tissue extracts that regulate many different kinds of metabolic activity in brain cells. Most of the conclusions are the outgrowth of experiments that have measured gross changes in the biochemical or morphological properties of the cell. These usually include cell proliferation, neurite outgrowth, and substrate transport and utilization. Although many of these findings are mostly phenomenological, they provide valuable clues for understanding metabolic regulation in isolated cell systems. In view of the evidence that the brain uses a variety of substrates for energy and anabolic processes, especially during development, the number of putative factors will surely increase. Therefore it becomes important to define the underlying mechanisms responsible for the observed biochemical effects.

ACKNOWLEDGMENTS

The authors gratefully acknowledge Joseph Stevenson and Irene Williams for their able technical assistance, Dr. Stephen R. Max for his helpful comments on the manuscript, and Bernadette Pasko and Janet Coleman for their expertise in preparing the typescript.

REFERENCES

Akuzawa, K., and Wakabayashi, K. (1985) A serum-free culture of the neurons in the septal, preoptic, and hypothalamic region: effects of triiodothyronine and estradiol. Endocrinology (Jpn.) 32:163–173.

Atterwill, C.K., Atkinson, D.J., Bermudez, I., and Balazs, R., (1985) Effect of thyroid hormone and serum on the development of Na^+, K^+-adenosine triphosphatase and associated ion fluxes in cultures from rat brain. Neuroscience 14:361–373.

Balazs, R., Lewis, P.D., and Patel, A.J. (1975) Effects of metabolic factors on brain development. In: Brazier, M.A.B. (ed.) Growth and development of the brain. New York: Raven, pp. 83–115.

Betz, A.L., Gilboe, D.D., Yudilevich, D.L., and Drewes, L.R. (1973) Kinetics of undirectional glucose transport into the isolated dog brain. Am. J. Physiol. 225:586–592.

Bhat, N.R., Subba, Rao, G., and Pieringer, R.A. (1981) Investigations on myelination in vitro: regulation of sulfolipid synthesis by thyroid hormone in cultures of dissociated brain cells from embryonic mice. J. Biol. Chem. 256:1167–1171.

Brookes, N., and Yarowsky, P.J. (1985) Determinants of deoxyglucose uptake in cultured astrocytes: the role of the sodium pump. J. Neurochem. 44:473–479.

Chacon, M., and Tildon, J.T. (1981) Elevated levels of triiodothyronine (T3) in victims of SIDS. J. Pediatr. 99:758–761.

Chapman, A.G., Meldrum, B.B., and Siesjo, B.K. (1977) Cerebral metabolic changes during prolonged epileptic seizures in rats. J. Neurochem. 28:1025–1035.

Clark, D.W., Boyd, F.T., Jr., Kappy, M.S., and Raizoda, M.K. (1984) Insulin binds to specific receptors and stimulates 2-deoxy-D-glucose uptake in cultured glial cells from rat brain. J. Biol. Chem. 259:11672–11675.

Cremer, J.E. (1981) Nutrients for the brain: problems in supply. Early Hum. Dev. 5:117–132.

Cummins, C.J., Glover, R.A., and Sellinger, O.Z. (1979) Neuronal cues regulate uptake in cultured astrocytes. Brain Res. 170:190–193.

Cushman, S.W., Wardzala, L.J., Simpson, I.A., Karnieli, E., Hissin, P.J., Wheeler, T.J., Hinkle, P.C., and Salans, L.B. (1984) Insulin-induced translocation of intracellular glucose transporters in the isolated rat adipose cell. Fed. Proc. 443:2251–2255.

Daniel, P.M., Love, E.F., and Pratt, O.E. (1975) Insulin and the way the brain handles glucose. J. Neurochem. 25:471–476.

Daval, J.L., Louis, J.C., Gerard, M.J., and Vincendon, G. (1983) Influence of adrenocorticotropic hormone on the growth of isolated neurons in culture. Neurosci. Lett. 36:99–304.

de Vellis, J., McGinnis, J.F., Breen, G.A.M., Leveille, R., Bennett, K., and McCarthy, K. (1977) Hormonal effects on differentiation in neural cultures. In: Federoff, S., and Hertz, L. (eds.) Cell, tissue and organ cultures in neurobiology. New York: Academic Press, pp. 485–511.

Diamond, I., and Fishman, R.A. (1973) High affinity transport of 2-deoxyglucose in isolated synaptic nerve endings. Nature (Lond.) 242:122–123.

Dierks-Ventling, C. (1971) Prenatal induction of ketone body enzymes in rat. Biol. Neonate 19:426–433.

Dierks-Ventling, C., and Cone, A.L. (1971) Acetoacetyl-coenzyme A thiolase in brain, liver, and kidney during maturation of rat. Science 172:380–383.

Draves, D.J., and Timiras, P.S. (1980) Thyroid hormone effects in neural (tumor) cell culture: differential effects on triiodothyronine nuclear receptors Na^+-K^+-ATPase activity and intracellular electrolyte levels. In: Ciacobini, E., Vernadakis, A., and Shahar, A. (eds.) Tissue culture in neurobiology. New York: Raven, pp. 291–301.

Dumbrowski, A.M., Jerkins, A., and Kauffman, F.C. (1983) Muscarinic receptor binding and oxidative enzyme activities in the adult rat superior cervical ganglion: effects of 6-hydroxydopamine and nerve growth factor. J. Neurochem. 3:1963–1970.

Frank, H.J., and Pardridge, W.M. (1981) A direct in vitro demonstration of insulin binding to isolated brain microvessels. Diabetes 30:557–561.

Garland, P.B. (1964) Some kinetic properties of pig-heart oxoglutarate dehydrogenase that provide a basis for metabolic control of the enzyme activity and also a stoichiometric assay for coenzyme A in tissue extracts. Biochem. J. 92:10c–12c.

Gjedde, A., and Crone, C. (1975) Induction processes in blood brain transfer of ketone bodies during starvation. Am. J. Physiol. 229:1165–1169.

Haussinger, D., Gerok, W., and Sies, H. (1982) Inhibition of pyruvate dehydrogenase during the metabolism of glutamine and proline in hemoglobin-free perfused rat liver. Eur. J. Biochem. 126:69–76.

Havrankova, J., Roth, J., and Brownstein, M. (1978) Insulin receptors are widely distributed in the central nervous system of the rat. Nature 272:827–829.

Havrankova, J., Brownstein, M., and Roth, J. (1981) Insulin and insulin receptors in rodent brain. Diabetologia 20:268–273.

Hawkins, R.A., and Biebuyck, J.F. (1979) Ketone bodies are selectively used by individual brain regions. Science 205:325–327.

Hawkins, R.A., Williamson, D.H., and Krebs, H.A. (1971) Ketone body utilization by adult and suckling rat brain in vivo. Biochem. J. 122:113–118.

Hernandez, M.J., Vanmucci, R.C., Salcedo, A., and Brennan, R.W. (1980) Cerebral blood flow and metabolism during hypoglycemia in newborn dogs. J. Neurochem. 35:622–628.

Hertz, L. (1978) Energy metabolism in glial cells. In: Schoffeniels, E., Franck, G., Hertz, L., and Tower, D.B. (eds.) Dynamic properties of glial cells. New York: Pergamon, pp. 121–132.

Hertz, L. (1979) Functional interactions between neurons and astrocytes. I. Turnover and metabolism of putative amino acid transmitter. Prog. Neurobiol. 13:277–323.

Hertz, L., and Richardson, J.S. (1984) Is neuropharmacology merely the pharmacology of neurons—or are astrocytes important too? Trends Pharmacol. 5:272–276.

Hertz, L., Juurlink, B.H.J., Fosmark, H., and Schousbot, A. (1982) Astrocytres in primary culture. In: Pfeiffer, S.E. (ed.) Neuroscience approached through cell culture.

Hertz, M.M., Paulson, O.B., Barry, D.I., Christiansen, J.S., and Svendsen, P.A. (1981) Insulin increases glucose transfer across the blood–brain barrier in man. J. Clin. Invest. 67:597–604.

Hoeben, R.C., Koper, J.W., and Lopes-Cardozo, M. (1985) Hormonal regulation of oligodendrocyte differentiation and sulfolipid synthesis in glial cell cultures. J. Neurochem. 44:S57D.

Honegger, P., and Lenoir, D. (1980) Triiodothyronine enhancement of neuronal differentiation in aggregating fetal rat brain cells in a chemically defined medium. Brain Res. 199:425–434.

Honegger, P., and Lenoir, D. (1982) Nerve growth factor (NGF) stimulation of cholinergic telencephalic neurons in aggregating cell cultures. Dev. Brain Res. 3:229–238.

Itoh, T., and Quastel, J.H. (1970) Acetoacetate metabolism in infant and adult rat brain in vitro. Biochem. J. 116:641–655.

Iverson, L.L. (1984) Amino acids and peptides: fast and slow chemical signals in the nervous system? Proc. R. Soc. Lond. [B]221:245–260.

Jolles, J., Zwiers, H., Dekker, A., Wirtz, K.W.A., and Gispen, W.H. (1981) Corticotropic-(1-24)-tetracosapeptide affects protein phosphorylation and polyphosphoinositide metabolism in rat brain. Biochem. J. 194:283–291.

Kaufman, L.M., and Barrett, J.N. (1983) Serum factor supporting long-term survival of rat central neurons in culture. Science 220:1394–1396.

Kimelberg, H.K. (1983) Primary astrocyte cultures: a key to astrocyte function. Cell. Mol. Neurobiol. 3:1–16.

Koenig, H., Fan, C.-C., and lqbal, Z. (1984) Polyamines mediate T_3 stimulation of Ca^{2+} influx and membrane transport. Trans. Am. Soc. Neurochem. 15:220.

Kono, T. (1984) Translocation hypothesis of insulin action on glucose transport. Fed. Proc. 43:2256–2257.

Kuffler, S.W., and Nichols, J.G. (1966) The physiology of neuroglial cells. Ergebn. Physiol. Biol. Chem. Exp. Pharmakol. 57:1–90.

Leonard, J.L., Kaplan, M.M., Visser, T.J., Silva, J.E., and Larsen, P.R. (1981) Cerebral cortex responds rapidly to thyroid hormones. Science 214:572–573.

Levi-Montalcini, R., and Angeletti, P. (1968) Nerve growth factor. Physiol. Rev. 48:534–569.

Mandel, P., Ciesielski-Treska, J., and Stefanovic, V. (1977) Neuroblast–glioblast interactions: ecto-enzymes. In: Federoff, S., and Hertz, L. (eds.) Cell, tissue and organ cultures in neurobiology. New York: Academic Press, pp. 593–615.

Manthorpe, M., Longo, F.M., and Varon, S. (1982) Comparative features of spinal neuronotrophic factors in fluids collected in vitro and in vivo. J. Neurosci. Res. 8:241–250.

Miller, A.L., Kiney, C.A., Corddry, D.H., and Staton, D.M. (1982) Interactions between glucose and ketone body use by developing brain. Dev. Brain Res. 4:443–450.

Monard, D., Solomon, F., Rentsch, M., and Gysin, R. (1973) Glial-induced morphological differentiation in neuroblastoma cells. Proc. Natl. Acad. Sci. USA 70:1894–1897.

Morrison, R.S., and de Vellis, J. (1984) Preparation of a chemically defined medium for purified astrocytes. In: Barnes, D.W., Sirbasku, D.A. and Sato, T.H. (eds.) Methods for serum-free culture of neuronal and lymphoid cells. New York: Alan R. Liss, pp. 15–22.

Nelson, P.G. (1977) Neuronal cell lines. In: Federoff, S., and Hertz, L. (eds.) Cell, tissue and

organ cultures in neurobiology, New York: Academic Press, pp. 347–365.
Nelson, S.R., Schulz, D.W., Passonneau, J.V., and Lowry, O.H. (1968) Control of glycogen levels in brain. J. Neurochem. 15:1271–1279.
Oppenheimer, J.H. Schwartz, H.L., Surks, M.I., Koerner, D.H., and Dillman, W.H. (1976) Nuclear receptors and initiation of thyroid hormone action. Recent Prog. Horm. Res. 32:529–565.
Ozand, P.T., Stevenson, J.H., Tildon, J.T., and Cornblath, M. (1975a) Effects of hyperketonemia on glycolytic intermediates in developing rat brain. J. Neurochem. 25:61–65.
Ozand, P.T., Stevenson, J.H., Tildon, J.T., and Cornblath, M. (1975b) Effects of hyperketonemia on glutamate and glutamine metabolism in developing rat brain. J. Neurochem. 25:67–71.
Ozand, P.T., and Tildon, J.T. (1983) Alterations of catecholamine enzymes in several brain regions of victims of sudden infant death syndrome. Life Sci. 32:1765–1770.
Page, M.A., Krebs, H.A., and Williamson, D.H. (1971) Activities of enzymes of ketone body utilization in brain and other tissues of suckling rats. Biochem. J. 121:49–53.
Pardridge, W.M. (1981) Transport of protein-bound hormones into tissues in vivo. Endocr. Rev. 2:103–123.
Pardridge, W.M. (1983) Neuropeptides and the blood brain barrier. Annu. Rev. Physiol. 45:73–82.
Patel, M.S. (1974) The relative significance of CO_2-fixing enzymes in the metabolism of rat brain. J. Neurochem. 22:717–724.
Pentreath, V.W., and Kai-Kai, M.A. (1982) Significance of the potassium signal from neurones to glial cells. Nature 295:59–61.
Pfeiffer, S.E., Betschart, B., Cook, J., Mancini, P., and Morris, R. (1977) Glial cell lines. In: Federoff, S., and Hertz, L. (eds.) Cell, tissue and organ cultures in neurobiology. New York: Academic Press, pp. 287–346.
Pleasure, D., Lichtman, C., Eastman, S., Lieb, M., Abramsky, O., and Silberberg, D. (1979) Acetoacetate and D(−)-beta-hydroxybutyrate as precursors for sterol synthesis by calf oligodendrocytes in suspension culture: extramitochondrial pathway for acetoacetate metabolism. J. Neurochem. 32:1447–1450.
Reichardt, L.F., and Kelly, R.B. (1983) A molecular description of nerve terminal function. Annu. Rev. Biochem. 52:871–926.
Robinson, A.M., and Williamson, D.H. (1980) Physiological roles of ketone bodies as substrates and signals in mammalian tissues. Physiol. Rev. 60:143–189.
Roeder, L.M., Poduslo, S.E., and Tildon, J.T. (1982) Utilization of ketone bodies by established neural cell lines. J. Neurosci. Res. 8:671–682.
Roeder, L.M., Tildon, J.T., and Holman, D.C. (1984a) Competition among oxidizable substrates in brains of young and adult rats: dissociated cells. Biochem. J. 219:131–135.
Roeder, L.M., Tildon, J.T., and Stevenson, J.H., Jr (1984b) Competition among oxidizable substrates in brains of young and adult rats: whole homogenates. Biochem. J. 219:125–130.
Roeder, L.M., Tildon, J.T., and Williams, I.B. (1985a) Transport of 2-deoxyglucose by dissociated brain cells. Brain Res. 345:298–305.
Roeder, L.M., Williams, I.B., and Tildon, J.T. (1985b) Glucose transport in astrocytes: regulation by thyroid hormone, J. Neurochem. 45:1653–1657.
Roger, L.J., and Fellows, R.E. (1980) Stimulation of ornithine decarboxylase activity by insulin in developing rat brain. Endocrinology 106:619–625.
Roth, G.A., and Bornstein, M.B. (1985) Effect of insulin and insulin-related substances on myelination and remyelination of organotypic nerve tissue culture. J. Neurochem. [suppl] 44:S136C.
Safaei, R., and Timiras, P.S. (1985) Thyroid hormone binding and regulation of adrenergic enzymes in two neuroblastoma cell lines. J. Neurochem. 45:1405–1410.
Samuels, H.H., and Tsai, J.S. (1973) Thyroid hormone action in cell culture: demonstration of nuclear receptors in intact cells and isolated nuclei. Proc. Natl. Acad. Sci. USA 70:3488–3492.
Seeds, N.W., and Hawkins, R.L. (1985) Extracellular proteases and their inhibitors influence neural development. J. Neurochem. 44:S37C.
Seifert, W., and Muller, H.W. (1984) Neuron–glia interaction in mammalian brain: preparation and quantitative bioassay of a neurotrophic factor (NTF) from primary astrocytes. In: Barnes, D.W., Sirbasku, D.A., and Sato, G.H. (eds.) Methods for serum-free culture of neuronal and

lymphoid cells. Cell culture methods for molecular and cell biology, vol. 4. New York: Alan R. Liss, pp. 66–77.
Selak, I., Skaper, S.D., and Varon, S. (1985) Pyruvate participation in low molecular weight trophic activity for central nervous system neurons in glia-conditioned media. J. Neurosci. 5:23–28.
Sensenbrenner, M. (1977) Dissociated brain cells in primary cultures. In: Federoff, S., and Hertz, L. (eds.) Cell, tissue and organ cultures in neurobiology. New York: Academic Press, pp. 191–213.
Shank, R.P., and Campbell, G. Le M. (1984) Glutamine, glutamate and other possible regulators of α-ketoglutarate and malate uptake by synaptic terminals. J. Neurochem. 42:1162–1169.
Shank, R.P., Bennett, G.S., Freytag, S.O., and Campbell, G.Le M. (1985) Pyruvate carboxylase: an astrocyte specific enzyme implicated in the replenishment of amino acid neurotransmitter pools. Brain Res. 329:364–367.
Shanker, G., Amur, S.G., and Pieringer, R.A. (1985) Investigations on myelinogenesis in vitro: a study of the critical period at which thyroid hormone exerts its maximum regulatory effect on the developmental expression of two myelin associated markers in cultured brain cells from embryonic mice. Neurochem. Res. 10:617–625.
Skaper, S.D., Selak, I., Manthorpe, M., and Varon, S. (1984) Chemically defined requirements for the survival of cultured 8-day chick embryo ciliary ganglionic neurons. Brain Res. 302:281–290.
Smith, C.M., Bryla, J., and Williamson, J.R. (1974) Regulation of mitochondrial α-ketoglutarate metabolism by product inhibition at α-ketoglutarate dehydrogenase. J. Biol. Chem. 249:1497–1505.
Strang, R.H.C., and Bachelard, H.S. (1971) Effect of insulin on levels and turnover of intermediates of brain carbohydrate metabolism in vivo. J. Neurochem. 181:1799–1807.
Sykes, J.E.C., and Lopes-Cardozo, M. (1985) Effects of serum on energy and lipid metabolism of oligodendrocyte-enriched glial cell cultures. J. Neurochem. 44:S47B.
Synkova, E. (1983) Extracellular K^+ accumulation in the central nervous system. Prog. Biophys. Mol. Biol. 42:135–189.
Thoenen, H., and Edgar, D. (1985) Neurotrophic factors. Science 279:238–242.
Tildon, J.T. (1983) Glutamine: a possible energy source for the brain. In: Hertz, L., Kvamme, E., McGeer, E., and Schousboe, A. (eds.) Metabolic relationship between glutamine, glutamate and GABA in the CNS. New York: Alan R. Liss, pp. 415–429.
Tildon, J.T., and Cornblath, M. (1972) Succinyl-CoA-3 ketoacid CoA-transferase deficiency: cause for ketoacidosis in infancy. J. Clin. Invest. 51:493–498.
Tildon, J.T., and Roeder, L.M. (1984) Glutamine oxidation of dissociated cells and homogenates of rat brain: kinetic and inhibitor studies. J. Neurochem. 42:1069–1076.
Tildon, J.T., and Roeder, L.M. (1985) Developmental patterns of glucose and 3-hydroxybutyrate transport by dissociated brain cells. J. Neurochem. 44:S131C.
Tildon, J.T., Cone, A.L., and Cornblath, M. (1971) Coenzyme A transferase activity in rat brain. Biochem. Biophys. Res. Commun. 43:225–231.
Tildon, J.T., Merrill, S., and Roeder, L.M. (1983) Differential substrate oxidation by dissociated brain cells and homogenates during development. Biochem. J. 216:21–25.
Tildon, J.T., Ozand, P.T., and Cornblath, M. (1975) The effects of hyperketonemia on neonatal brain and metabolism. In: Hommes, F.A. and Van den Berg, C.J. (eds.) Normal and pathological development of energy metabolism. New York: Academic Press, pp. 143–154.
Tildon, J.T., Roeder, L.M., and Stevenson, J.H., Jr. (1985a) Substrate oxidation by isolated rat brain mitochondria and synaptosomes. J. Neurosci. Res. 14:207–215.
Tildon, J.T., Stevenson, J.H., Jr., and Roeder, L.M. (1985b) Factors regulating $^{14}CO_2$ production from labelled substrates by brain cells. Trans. Am. Soc. Neurochem. 16:148.
Trimmer, P.A., Reier, P.J., Oh, T.H., and Eng, L.F. (1982) An ultrastructural and immunocytochemical study of astrocytic differentiation in vitro. J. Neuroimmunol. 2:235–260.
Valcana, T., and Timiras, P.S. (1978) Nuclear triiodothyronine receptors in the developing rat brain. Mol. Cell. Endocrinol. 11:31–41.
Van Calker, D., Loffler, F. and Hamprecht, B. (1983) Corticotropin peptides and melanotropins elevate the level of adenosine 3′-5′-cyclic monophosphate in cultured brain cells. J. Neurochem. 40:418–427.

Van Houten, M., and Posner, B.I. (1979) Insulin binds to brain blood vessels in vivo. Nature 282:623–625.

Van Houten, M., and Posner, B.I., (1981) Cellular basis of direct insulin action in the central nervous system. Diabetologia 20:255–267.

Van Houten, M., Khan, M.N., Khan, R.J., and Posner, B.I. (1981) Blood-borne adrenocorticotropin binds specifically to the median eminence-arcuate region of the rat hypothalamus. Endocrinology 108:2385–2387.

Van Houten, M., Posner, B.I., Kopriwa, B.M., and Brawer, J.F., (1979) Insulin-binding sites in the rat brain: in vivo localization to the circumventricular organs by quantitative radioautography. Endocrinology 105:666–673.

Varon, S., and Adler, R. (1981) Trophic and specifying factors directed to neuronal cells. Adv. Cell. Neurobiol. 2:115–163.

Varon, S., and Bunge, R.P. (1978) Trophic mechanisms in the peripheral nervous system. Annu. Rev. Neurosci. 1:327–361.

Varon, S., Skaper, S.D., Barbon, G., Selak, I., and Manthorpe, M. (1984) Low molecular weight agents appear to support survival of cultured neurons from central nervous system. J. Neurosci. 4:654–658.

Walz, W., and Hertz, L. (1983) Functional interactions between neurons and astrocytes. II. Potassium homeostasis at the cellular level. Prog. Neurobiol. 20:133–183.

Wapnir, R.A., Tildon, J.T., and Cornblath, M. (1973) Metabolic differences in offspring of rats fed high-fat and control diets. Am. J. Physiol. 224:596–599.

Weber, G. (1969) Inhibition of human brain pyruvate kinase and hexokinase by phenylalanine and phenylpyruvate: possible relevance to phenylketonuric brain damage. Proc. Natl. Acad. Sci. USA 63:1365–1369.

Weber, G., Lea, M.A., and Stamm, N.B. (1968) Sequential feedback, inhibition and regulation of liver carbohydrate metabolism through control of enzyme activity. Adv. Enzyme Regul. 6:101–123.

Weber, R.J., and Edmond, J. (1977) Utilization of L(+)-3-hydroxybutyrate, D(−)-3-hydroxybutyrate, acetoacetate and glucose for respiration and lipid synthesis in the 18-day-old rat. J. Biol. Chem. 252:5222–5226.

Wu, D.K., Morrison, R.S., and de Vellis, J. (1985) Modulation of beta-adrenergic response in rat astrocytes by serum and hormones. J. Cell. Physiol. 122:73–80.

Yeh, Y.Y., Streuli, V.L., and Zee, P. (1977) Ketone bodies serve as important precursors of brain lipids in the developing rat brain. Lipids 12:957–964.

Yu, A.C., and Hertz, L. (1982a) Uptake of glutamate, GABA and glutamine into predominantly GABA-ergic and predominantly glutamatergic nerve cell populations in culture. J. Neurosci. Res. 7:23–35.

Yu, A.C., and Hertz, L. (1982b) Metabolic fate of labeled glutamate in astrocytes in primary cultures. J. Neurochem. 39:954–968.

Yu, A.C.H., and Hertz, L. (1983) Metabolic sources of energy in astrocytes. In: Hertz, L., Kvamme, E., Greer, E.G., and Schousboe, A. (eds.) Glutamine, glutamate and GABA in the central nervous system. New York: Alan R. Liss, pp. 205–217.

Yu, A.C., Schousboe, A., and Hertz, L. (1982) Metabolic fate of ^{14}C-labeled glutamate in astrocytes in primary cultures. J. Neurochem. 39:954–960.

Zwiers, H., Wiegant, V.M., Schatman, P., and Gispen, W.H. (1978) ACTH-induced inhibition of endogenous rat brain protein phosphorylation in vitro: structure activity. Neurochem. Res. 3:455–463.

III. AGING AND REGENERATION

26. MONOAMINERGIC AND CHOLINERGIC INTRACEREBRAL GRAFTS: USE IN MODEL SYSTEMS OF AGING

FRED H. GAGE

1. INTRODUCTION

Intracerebral grafting of monoaminergic and cholinergic neurons has particular relevance in the context of model systems. In two neurodegenerative disease syndromes—Parkinson's and Alzheimer's diseases—in particular, the progressive and substantial loss of such types of neurons is likely to underlie some of the major clinical symptoms of the disease. Thus, in Parkinson's disease the severe motor disturbances can be related to a loss of some 80%–90% of the neurons in the nigrostriatal dopamine (DA) system. And in both Alzheimer's and Parkinson's diseases there is a strong correlation between the degree of dementia and the loss (or possibly atrophy) of cholinergic neurons in the basal forebrain projection system, which is the source for the cholinergic afferents to the hippocampal formation and the neocortical mantle. There is experimental evidence in rats to support these relationships. Thus, the interference with DA transmission (e.g., by administration of the DA-depleting drug reserpine) or selective lesioning of the nigrostriatal DA neurons [by the toxins 6-hydroxydopamine (6-OHDA), or 1-methyl-4-phenyl-tetrahydropyridine (MPTP)] induces sensorimotor impairments in rats and monkeys that can be said to be analogous to the dysfunctions seen in parkinsonian patients (Burns et al., 1983; Carlsson, 1959; Marshall and Teitelbaum, 1977; Ungerstedt, 1974). Interference with cholinergic transmission, on the other hand (e.g., by cholinergic receptor blockade), or surgical lesions to the septohippocampal or basal forebrain–neocortical pathways

Vernadakis, A. (ed.), *Model Systems of Development and Aging of the Nervous System*. Copyright © 1987 by Martinus Nijhoff Publishing.

carrying the cholinergic projection systems, causes learning and memory impairments in rats (Bartus et al., 1982; Coyle et al., 1983). The analogies between such experimentally induced conditions and the clinical disease states with respect to neuropathological, neurochemical, and behavioral changes have justified the use of animals with 6-OHDA lesions of the nigrostriatal DA pathway, on one hand, and surgical lesions of the septohippocampal or basal forebrain–neocortical cholinergic pathways, on the other, as experimental models for at least some aspects of the pathology involved in Parkinson's and Alzheimer's diseases.

In the first part of the chapter, I review briefly the results obtained by implantation of dopaminergic and cholinergic neurons in animals with lesions of the nigrostriatal or septohippocampal systems, i.e., in conditions that can be said to represent *analogous* models of Parkinson's and Alzheimer's diseases. Parallel to this work we have used the aged rat as an experimental model for age-dependent impairments in brain function and behavior. To the extent that aging affects the brain similarly in rodents and in man, the aged rat can be viewed as a *homologous* model of neurological impairments associated with aging in man. Neural grafting in aged rats offers new opportunities to study both the factors underlying the age-related decrements in motoric and cognitive function, and the possibilities to improve the performance of selected brain systems in the aged rat.

2. YOUNG RATS WITH SELECTIVE BRAIN DAMAGE

2.1 Graft survival and outgrowth

The survival and growth of DA-rich suspension grafts have been studied in different brain sites (Björklund et al., 1983c and 1984; Dunnett et al., 1984; Gage and Björklund, 1984). The implanted DA neurons survive in clusters of aggregates at the injection site. Some cells have apparently migrated up to about 0.5 mm into the seemingly intact host tissue. Although the DA neurons have been seen to survive quite well in several different brain sites, their fiber outgrowth into the surrounding host tissue varies markedly from one area to another (Björklund et al., 1983a). Rich fiber outgrowth is thus seen only into areas normally richly innervated by DA fibers, such as the caudate putamen and nucleus accumbens, whereas DA cells implanted into normally poorly innervated areas, such as parietal cortex, globus pallidus, lateral hypothalamus, and substantia nigra, demonstrate very little fiber growth into the surrounding host tissue. In the striatum, the DA fibers radiate out in a halo with a radius of about 1.5–2.0 mm around the implant. Each DA cell deposit reaches only a subportion of the striatum, and extensive reinnervation of the striatum will require several implants in different sites. The regionally restricted fiber outgrowth from each implant will, on the other hand, allow observations of regionally confined effects on behavior.

Reinnervation of the denervated target is only efficient if the DA neurons are placed into or immediately adjacent to the target. DA implants placed along the course of the nigrostriatal pathway (i.e., into the lateral hypothalamus, internal capsule, or globus pallidus) have not been seen to grow any axons into the striatum (Björklund et al., 1983c). However, this does not necessarily mean that grafted mesencephalic DA neurons are unable to extend their axons over longer distances. In fact, in a recent study we grafted fetal mesencephalic tissue over the superior colliculus within the caudal end of a 2- to 3-cm-long heterologous sciatic nerve segment laid longitudinally on the skull. It was shown that the use of a peripheral nerve segment as a bridge between a distant neuronal transplant and a selected region of the adult host brain resulted in a growth of monoaminergic axons a distance of 2–3 cm into the denervated striatum from the extracerebrally located grafted neurons (Gage et al., 1985c).

Experiments with septal grafts in the hippocampal formation have provided evidence that the ingrowth from the grafted neurons into the surrounding host tissue is markedly modified by the presence or absence of the homologous intrinsic innervation. Thus, the extension of acetylcholine esterase (AChE)-positive, presumably cholinergic, fibers into the hippocampal formation is considerably greater in animals denervated of their hippocampal cholinergic afferents by a fimbria–fornix lesion than in animals with intact cholinergic inputs (Gage and Björklund, 1986). With intrahippocampal septal suspension grafts, the choline acetyltransferase (ChAT) level in the host hippocampus reached about 75% of normal in animals with lesioned fimbria–fornix at the time of transplantation, as compared to about 25% in rats whose cholinergic inputs were left intact until 1 week before sacrific (see figure 26-1). The parallel AChE histochemistry suggested that the denervating fimbrial lesion, made at the time of transplantation, made the cholinergic axons from the graft grow over greater distances within the host hippocampus and ramify more extensively in some of the normal terminal zones of the intrinsic septohippocampal cholinergic projection. In an attempt to further examine the importance of denervation in the graft survival and the potential mechanisms involved in the response to denervation, we implanted neonatal superior cervical ganglia [SCG, which are dependent on nerve growth factor (NGF) at this time in their development for survival] to the hippocampus in the presence or absence of fimbria–fornix transection, and concluded from our results that the denervation resulted in the release of a trophic factor that was sufficiently similar to NGF to support the survival of the SCG while the SCG did not survive grafting to the hippocampus in the absence of the fimbria-fornix transection (Gage et al., 1984a) (see figure 26-2).

These various observations suggest that the ingrowth and patterning of axons from implanted neurons are greatly influenced by the target tissue, and that this regulatory influence is relatively (fairly) specific for different types of neurons within the graft. The effects of denervating lesions suggest, more-

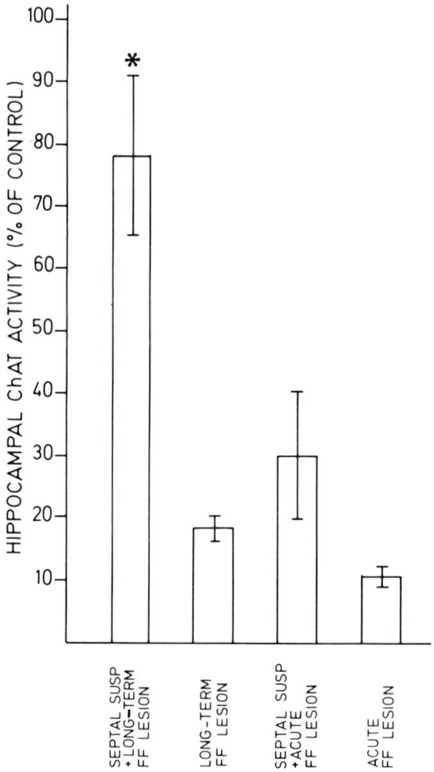

Figure 26-1. Choline acetyltransferase activity in the hippocampal formation following fetal septal suspension grafts, with acute or chronic fimbria–fornix transection.

over, that the proliferation and terminal patterning of the reinnervating axons may in some way be related to the filling of vacated terminal space in the denervated target, analogous to what has been described for lesion-induced collateral sprouting in the septohippocampal system. The monoaminergic and cholinergic neurons in the implants can also reinnervate the adult host hippocampus or striatum in a reproducible and orderly manner even if the fibers are made to enter from abnromal directions and along routes that are inconspicuous in normal animals. The extension, proliferation, and patterning of the ingrowing fibers can be drastically modified by the removal of intrinsic homologous and heterologous afferents (of both homologous and heterologous kinds) (Björklund et al., 1976 and 1986; Björklund and Stenevi, 1977).

2.1 Biochemical Measures of Graft Function

The activity of the cholinergic innervation of the denervated hippocampus, derived from solid or suspension grafts of the septal–diagonal band area, has

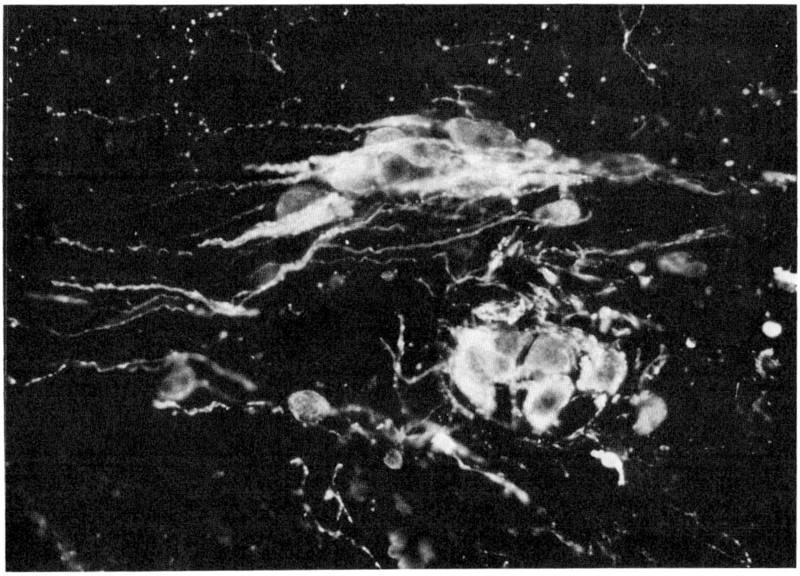

Figure 26-2. A photomicrograph of catecholamine histofluorescent superior cervical ganglia cells grafted as neonatal cells to the hippocampus of a rat with a fimbria–fornix transection.

been monitored biochemically by measurements of the acetylcholine-synthesizing enzyme, ChAT, and of acetylcholine synthesis rates in vitro. Although AChE enzyme is not a specific marker for cholinergic neurons, the synthetic enzyme ChAT (by contrast) is an enzyme specifically localized in cholinergic neurons and is therefore a better marker of cholinergic neurotransmission. Together the anatomical marker AChE and the biochemical marker ChAT have been used to measure the time course and magnitude of fiber outgrowth from both solid and suspended septal grafts (Björklund et al., 1983a and b; Gage et al., 1985a). Graft-derived ChAT activity was barely detectable 10 days after cell suspension grafts, but sharply increased between 10 days and 1 month in the region of the host hippocampus close to the graft. By 6 months, ChAT activity was restored to near normal levels in all segments of the previously denervated hippocampal formation. When comparing the total amount of ChAT activity derived from the solid grafts and the cell suspension grafts, the cell suspension grafts appeared to be about twice as effective as the solid grafts although the amount of tissue grafted was about the same in each case.

The functional activity of the septal grafts was further assessed by measurements of [^{14}C]acetylcholine synthesis from [^{14}C]glucose in vitro in fimbria–fornix-lesioned rats with septal suspensions implanted into the depth of the denervated hippocampus (Björklund et al., 1983a). The overall hippocampal [^{14}C]acetylcholine synthesis was restored to normal levels in the grafted animals, and estimates of acetylcholine turnover rate suggested that the transmitter machinery of the newly established septohippocampal connections operated at a rate similar to that of the intrinsic septohippocampal pathway. Thus, these septal cell suspensions seem capable of maintaining function at a relatively "physiological" level despite their abnormal position.

In a more recent study, Kelly et al. (1984) investigated the magnitude of lesion-induced alterations in the hippocampal formation as reflected in the local rates of 2-deoxy[^{14}C]glucose (2-DG) utilization, and the degree to which this index of functional activity could be normalized following reinnervation by solid septal grafts. By 6 months after transection of the septohippocampal pathway by a unilateral fimbria–fornix lesion, the 2-DG utilization was reduced by 30%–50% throughout the ipsilateral hippocampal formation of the nongrafted animals. Fimbria–fornix-lesioned rats that had received septal grafts displayed a significant recovery in hippocampal 2-DG utilization, to within the normal range, as compared to the rats with lesion alone. In addition, the changes in 2-DG utilization were significantly correlated with the density of AChE staining in adjacent sections from the same brains ($r = 0.75$; p 0.01), thus suggesting a relationship between the cholinergic reinnervation from the septal grafts and the restoration of functional glucose utilization. These results strongly suggest that the cholinergic component of the grafts is functioning at the biochemical level and influencing, or normalizing, the function of the deafferented hippocampal formation.

Also for grafted DA neurons, measurements of DA turnover and synthesis rates in vivo indicate that the implanted DA neurons are capable of restoring DA neurotransmission in the initially denervated target (Schmidt et al., 1983). Nigral suspension implants restored neostriatal DA levels to an average of 13%–18%, with the highest individual values reaching about 50% of control. Dioleoylphosphatidylcholine (DOPAC) was restored from about 5% of control (in the 6-OHDA-denervated rats) to an average of about 20% of normal in the grafted animals. The DOPAC–DA ratio and the DOPA accumulation rates indicated that DA turnover and synthesis in the grafted neurons were, on the average, some 50%–100% higher than in the intrinsic nigrostriatal path (Schmidt et al., 1983).

A recent set of experiments was designed to test the possibilities for neuronal replacement in the striatum previously depleted of its constituent neurons by excitotoxic lesions or aging (Isacson et al., 1984). We have developed a large ibotenic acid lesion ($4 \times 10\ \mu g$ in caudate putamen on each side) that gives reliably about 85% degeneration of the neurons in the head of the caudate putamen. Grafts of embryonic striatal neurons, injected into the ibotenate-lesioned striatum in the form of a cell suspension, survive excellently. The normal striatal neuron types are found in these grafts and glutamic acid decarboxylase (GAD) is restored to normal levels. In our first functional analyses, we have seen that the grafts 1) normalize glucose metabolism in several striatal target areas (such as globus pallidus, subthalamic nucleus, and deep superior colliculus), as measured by the autoradiographic 2-DG technique; and 2) compensate completely for the nocturnal locomotor hyperactivity seen in the ibotenic acid-lesioned animals. This model system is now applicable to the aged rat model.

3. AGED RATS

3.1 Characterization of functional deficits

In an attempt to characterize the behavioral deficits that result in our rats as a function of age, we found early on that not all of the aged rats showed functional deficits that distinguished them significantly from young control rats on all behavioral measures that we were interested in measuring (Gage et al., 1984d). We subsequently tried to determine whether there were physiological correlates that were predicted by the presence or absence of behavioral deficits. To this end we found, using the 2-DG technique, that impairments in water-maze performance (a measure of spatial memory) were highly correlated with a decrease in 2-DG in the prefrontal cortex, hippocampal formation, and septal area (Gage et al., 1984e). Subsequent attempts to reverse functional deficits in cognitive performance were made (see section 2.3) based on this characterization.

3.2 Graft survival and outgrowth

The first stage in our study of neuronal implants in the aged rat brain was to demonstrate that such transplants could survive well in the aged brain and that they were able to provide appropriate reinnervation of the host (Gage et al., 1983a). To this end, neuronal cell suspensions prepared from the ventral mesencephalon and septal–diagonal band of rat embryos were implanted into the depth of the intact neostriatum or hippocampus of 21- to 23-month-old female rats of the same strain. Graft survival assessed 3–4 months after grafting was comparable to that seen in our previous studies of young adult recipients. Fiber outgrowth into the host brain was evaluated in animals that were subjected to lesions of the intrinsic nigrostriatal or septohippocampal system 6–10 days before killing. Dense dopaminergic fiber outgrowth was seen within a zone up to about 1-mm radius around the nigral implants, and dense outgrowth of AChE-positive fibers occurred up to about 2 mm from the septal implants. The overall magnitude of fiber outgrowth was less than that generally seen in previously denervated targets in young adult recipients. However, the outgrowth seen in the aged rats with intact afferents appeared to be as extensive as in young recipients when the grafts are placed in nondenervated targets (Gage and Björklund, 1986). The distribution of the AChE-positive fibers from the septal implants in the host hippocampus suggested that the pattern found in the nondenervated target of the aged recipients was more diffuse, and partly different, from normal. It is an interesting possibility, therefore, that synapse loss in intrinsic connections (Geinisman et al., 1977) may influence the patterning of the graft-derived innervation, and, in fact, improve the implant's ability to terminate in the otherwise intact target of the aged host brain.

3.3 Behavioral effects

The next step in the study was to determine whether and to what extent the grafts could ameliorate any of the motor coordination (Gage et al., 1983b and 1984b) or learning deficits seen in the aged animals prior to grafting (Gage et al., 1984c and 1985b).

Motor coordination skills were assessed on four measures adapted from the battery described by Campbell et al. (1980): 1) the ability to maintain balance on, and successfully reach, a safety platform at either end of a narrow bridge of square cross section; 2) the ability to maintain balance on a similar bridge of round cross section; 3) the period of time that the animal could sustain its own weight clinging suspended from a taut wire; and 4) the ability to descend in a coordinated manner on a vertical pole covered with wire mesh. Prior to transplantation, the aged rats were significantly impaired with respect to the young controls on the four measures. On the square and round bridges, young rats had no difficulty walking and exploring along the rod, and they generally reached the platform within 30–60 seconds. The aged rats before transplantation, by contrast, had greater difficulties maintaining balance. Most of these animals fell off the bridges or alternatively lay on the bridges without attempting to walk, and clung tightly within all four paws or with the forepaws while the hind paws hung freely. By 12 weeks after transplantation, the aged rats with nigral grafts, but not the aged control or the septal-grafted rats, showed marked and significant improvement in their balance and limb coordination on both the square and round bridges. Typically they could walk along the bridge without falling, displayed gait posture similar to the young rats, and fell less frequently than the other aged rats. On the wire-mesh-covered pole, the aged rats descended more rapidly before transplantation than did the young controls, frequently falling, slipping, or sliding down backward. Although the aged rats with nigral transplants had a tendency to descend in a more controlled and coordinated manner, with head-down orientation as seen in the young controls, the differences from the other aged rats did not reach significance on this measure. The aged rats showed no difference between groups either in body weight or in latency to fall from the taut wire; this suggests that enhanced motor coordination in the aged rats with nigral grafts was not attributable to nonspecific differences in weight or strength of these animals.

Spatial learning was assessed in the Morris water-maze task 1 week prior to transplantation and 2.5–3.0 months after transplantation (Morris, 1981). This test requires that the rat use spatial cues in the environment to find a platform hidden below the surface of a pool of opaque water. Normal young rats have no trouble learning this task with speed and accuracy. Since our previous studies using this task (Isacson et al., 1984; Gage et al., 1984e) showed that only a portion (one-fourth to one-third) of our rats was markedly impaired, a pretransplant test served to identify those impaired individuals in the aged rat group. Based on the performance of the young controls, we set the criterion for impaired performance in the aged rats such that the mean escape latency (i.e., swim time to find the submerged platform) would be above an upper 99% confidence limit of the escape latencies recorded in the young control group (Gage et al., 1984c). A subgroup of old rats showed mean swim times greater than the criterion and were thus allocated to the "old impaired" group, which was used for subsequent transplantation. The remaining subgroup of aged rats constituted the "old nonimpaired" group. This latter group, together with a young control group, served as reference groups. A portion of the old impaired group received bilateral suspension grafts prepared from the septal–diagonal band area obtained from 14- to 16-day-old embryos of the same rat strain. Three implant deposits were made stereotaxically into the hippocampal formation on each side. The remaining old impaired rats were left unoperated upon and served as the "old impaired" control group. On the posttransplantation test, 2.5–3.0 months after grafting, the nongrafted groups remained impaired, while the grafted animals, as a group, showed a significant improvement in performance as indicated by escape latency. This improvement of the grafted group was demonstrated by comparisons to its pretransplantation performance as well as to the performance of the nongrafted old controls in the second test.

The ability of the rats to use spatial cues for the location of the platform in the pool was assessed by analyzing their search behavior after removal of the platform on the fifth day of testing. While the young rats and rats in the old impaired groups focused their search on the fourth quadrant, where the platform had previously been placed, the old impaired rats failed to do so in the pretransplant test. In the posttransplantation test, the grafted rats, but not the nongrafted old impaired group, showed significantly improved performance. Swim distance in the fourth quadrant was increased 83% and they swam significantly more in the fourth quadrant than in other quadrants of the pool. By contrast, the nongrafted controls showed no significant change over their pretransplant performance.

These results (Gage et al., 1983b and 1984c) demonstrate, for the first time, the ability of intracerebral grafts to ameliorate age-related impairments in motoric and complex cognitive behaviors. To what extent this effect is dependent on the dopaminergic and cholinergic neurons in the nigral and septal grafts, respectively, is not at present clear. It is interesting to note, however, that intrahippocampal septal grafts, identical to the ones used here, had no effect on the motor coordination disabilities described above, while the fetal substantia nigra implanted into the striatum was effective in reducing these motor-related disabilities. This suggests that the behavioral effects of the grafts are regionally specific and may be mediated through a direct action on elements in the surrounding region of the host brain.

4. MECHANISMS OF ACTION OF INTRACEREBRAL IMPLANTS

The combined morphological, biochemical, electrophysiological, and behavioral data available show that implanted embryonic nerve cells in some cases can substitute quite well for a lost intrinsic neuronal system in mammals. The intracerebral implants probably exert their effects in several ways. The functional effects seen with grafts placed into one of the cerebral ventricles, such as in the studies by Perlow et al. (1979), Freed et al. (1980 and 1981), and Gash et al. (1980), are thus probably explained on the basis of a diffuse release of an active amine or peptide into the host central nervous system (CNS) and adjacent brain tissue. In other instances, such as in animals with DA-rich grafts reinnervating the neostriatum or acetylcholine-rich grafts reinnervating the hippocampus, we believe that the available data provide quite substantial evidence that the behavioral recovery is directly related to the ability of the grafted neurons to reinnervate relevant parts of the host brain. This is illustrated by the studies showing that the degree of functional recovery in 6-OHDA-lesioned rats with nigral transplants is directly correlated with the extent of striatal DA reinnervation, and that the "profile" of functional recovery is dependent on which area of the striatal complex is reinnervated by the graft (Björklund et al., 1980 and 1981; Dunnett et al., 1983).

To what extent the intracerebral implants can be functionally integrated with the host brain is, however, still poorly known and remains, therefore, an important question for further investigation. The chances for extensive integration may be greatest for neuronal suspension grafts implanted as deposits directly into the depth of the brain, but even solid grafts inserted as whole pieces into the brain have in several cases been seen to become reinnervated from the host brain both in adult and in developing recipients.

Available data, derived primarily from studies on monoaminergic and cholinergic systems, indicate that implanted embryonic central neurons can substitute to some degree for a lost set of afferents to a denervated brain region in adult rats, and also replace a lost intrinsic neuronal system in normalizing the rat's behavior. This indicates a remarkable plasticity of the mature rat CNS in incorporating new neuronal elements into its already established circuitries. There is now abundant evidence that the adult CNS can reorganize and rebuild itself in response to damage and, in some cases, continued neurogenesis and incorporation of newly formed neurons are likely to play a normal physiological role.

Neuronal replacement by neuronal implants is a striking further example of how the brain can allow new elements to be inserted and linked into its own functional subsystems. Obviously there must be definite limitations as to which types of neurons or functional subsystems can successfully be manipulated in this way. Neuronal implants seem, in fact, most likely to have behaviorally meaningful functional effects only with types of neurons that normally do not convey, or link, specific or patterned messages, i.e., not in sensoric or motoric input and output systems.

4.1 Clinical aspects

The demonstration of the functional capacity of DA-rich grafts to ameliorate behavioral impairments following experimental lesions of brain DA systems has excited speculation that the transplantation technique may have clinical application to Parkinson's disease and Alzheimer's disease. This has already led to the first clinical trials of intracerebral grafts in parkinsonian patients. The effectiveness of such an approach will depend on the resolution of several issues relating to the etiology of the disease itself, to practical techniques to be used in human patients, and to ethical considerations about sources of graft tissue.

4.2 Practical aspects of transplantation

Virtually all successful experimental work with neuronal grafts has so far been conducted in rodents, although the first primate work is now under way. Apart from the issue of appropriate source of donor tissue (see below), practical questions remain as to appropriate surgical techiques, placement of graft, and immunological factors, which require more extensive study to be fully answered. Nevertheless, immunological factors may be less severe than might be supposed. The brain has been described as an "immunologically privileged site" and, indeed, functional brain grafts can survive, though not as well, across major histocompatibility barriers (Björklund et al., 1982 and 1983c; Low et al., 1985; Brundin et al., 1985). However, with DA-rich grafts in rats it is clear that the pattern of graft placement is a critical feature for functional recovery and, given the size of the forebrain DA terminal area in primates and humans, a large number of small grafts in different locations may be required to achieve an adequate functional compensation. A further caution needs to be raised from the observation that, in rodents, there remain functional impairments, such as adipsia, that are not amenable to graft-derived amelioration by any procedure yet adopted.

4.3 Sources of donor tissue

Perhaps the most contentious issue to be resolved in any clinical application of transplants is the identification of a suitable source of donor tissue. In rats, the most successful results have been achieved using fetal tissue from the brain of embryos of the same strain, yet the utilization of human embryonic tissue in analogous clinical application is likely to prove highly controversial. In order to circumvent this problem, three approaches have been considered: use of cross-species grafts, autografts of peripheral tissue taken from the patient himself, and use of cultured cells. Viable cross-species grafts have been obtained from mouse to rat (Low et al., 1985; Björklund et al., 1982; Brundin et al., 1985), but, although such grafts are also functional, the histological appearance or extent of functional recovery is not as successful as in comparable within-strain grafts. Recently, however, it has been demonstrated that, if the host animals are treated chronically with the immunosuppressant drug cyclosporin A, cross-species graft survival and function are greatly improved (Brundin et al., 1986). The alternate approach that has attracted most attention and that was adopted in the first clinical application to Parkinson's patients has been the use of adrenal medullary grafts, which can be taken from the patient himself. Adrenal medullary tissue survives grafting to the rat lateral ventricle (Freed et al., 1981), in which catecholamine-fluorescent cells survive, but only very limited development of fibers was seen growing into the host brain. The adrenal grafts were effective in compensating unilateral lesion-induced rotation to apomorphine, as are grafts of embryonic substantia nigra, but, in the absence of extensive ingrowth, this compensation is probably attributable to diffusion of catecholamines from the graft into the deafferented host straitum (Freed et al., 1981). A third approach is the grafting of previously cultured cells. Brundin et al. (1986) recently demonstrated that fetal rat DA neurons cultured in vitro for a 6-day period prior to being grafted to the neostriatum of adult rats revealed surviving catecholamine-fluorescent cells in the grafts 7 weeks after transplantation. In addition, both Jeager (1986) and Hefti et al. (1985) have demonstrated that the cell line PC12 grown in culture will survive and produce catecholamines when grafted to the adult brain.

4.4 Etiological problems

At least two aspects of the etiology of neurodegenerative diseases such as parkinsonism and Alzheimer's disease that are as yet unknown bear upon the viability of transplantation as a novel clinical treatment for patients. First, the causative agent is unidentified. If that agent is both selective for specific neurons and has become established in the patient's system, then any replacement graft tissue might be equally susceptible to assault. Alternatively, there may be some genetic or developmental deficiency in the system that does not affect normal neurons, in which case transplantation may provide a viable replacement therapy. The second etiological factor that remains unresolved is whether the more widespread neuronal degeneration and accompanying dementia seen in later stages of these disease processes is secondary to or independent of the primary neuronal loss. If these processes are indeed secondary, then transplantation-derived replacement of normal neurons may be most effective as a prophylactic treatment, but this requires transplantation at an early stage of the disease when the patient is still responsive to conventional chemotherapy.

Although there remain many outstanding questions before the general applicability of grafts can be fully assessed, each of the issues raised is potentially solvable, though several of these (such as the two aspects of the etiology of the diseases raised above) may only be resolved through actual clinical trials.

ACKNOWLEDGMENTS

This research was supported in part from an NIH grant AG03766. In addition, Anders Björklund and Stephen B. Dunnett are gratefully acknowledged for their role in the studies presented.

REFERENCES

Bartus, R.T., Dean, R.L., Beer, B., and Lippa, A.S. (1982) The cholinergic hypothesis of geriatric memory dysfunction. Science 217:408–417.

Björklund, A., and Stenevi, U. (1977) Reformation of the severed septohippocampal cholinergic pathway in the adult rat by transplanted septal neurons. Cell Tissue Res. 185:298–302.

Björklund, A., Stenevi, U., and Svendgaard, N.A. (1976) Growth of transplanted monoaminergic neurons into the adult hippocampus along the perforant path. Nature 262:787–790.

Björklund, A., Schmidt, R.H., and Stenevi, U. (1980) Functional reinnervation of the neostriatum in the adult rat by use of intraparenchymal grafting of dissociated cell suspensions from the substantia nigra. Cell Tissue Res. 212:39–45.

Björklund, A., Stenevi, U., Dunnet, S.B., and Iversen, S.D. (1981) Functional reactivation of the deafferented neostriatum by nigral transplants. Nature 289:497–499.

Björklund, A., Stenevi, U., Dunnett, S.B., and Gage, F.H. (1982) Cross-species neural grafting in the rat model of Parkinson's disease. Nature 298:652–654.

Björklund, A., Gage, F.H., Schmidt, R.H., Stenevi, U., and Dunnett, S.B. (1983a) Intracerebral grafting of neuronal cell suspensions. VII. Recovery of choline acetyltransferase activity and acetylcholine synthesis in the denervated hippocampus reinnervated by septal suspension implants. Acta Physiol. Scand. [Suppl.] 522:59–66.

Björklund, A., Gage, F.H., Stenevi, U., and Dunnett, S.B. (1983b) Intracerebral grafting of neuronal cell suspensions. VI. Survival and growth of intrahippocampal implants of septal cell suspensions. Acta Physiol. Scand. [Suppl.] 522:49–58.

Björklund, A., Stenevi, U., Schmidt, R.H., Dunnett, S.B., and Gage, F.H. (1983c) Intracerebral grafting of neuronal cell suspensions. II. Survival and growth of nigral cell suspensions implanted in different brain sites. Acta Physiol. Scand. [Suppl.] 522:9–18.

Björklund, A., Dunnett, S.B., Gage, F.H., and Stenevi, U. (1984) Implantation of suspensions of dissociated dopamine neurons in the rat brain. In: Usdin, E., and Dahlstrom, A. (eds.) Fifth catecholamine meeting; proceedings. New York: Alan R. Liss (in press).

Björklund, A., Nornes, H., and Gage, F.H. (1986) Cell suspension grafts of noradrenergic locus coeruleus neurons in rat hippocampus and spinal cord: reinnervation and transmitter turnover. Neuroscience 18(3):685–698.

Brundin, P., Nilsson, O.G., Gage, F.H., and Björklund, A. (1985) Cyclosporin A increases survival of cross-species intrastriatal grafts of embryonic dopamine-containing neurons. Exp. Brain Res. 60:204–208.

Brundin, P., Barbin, G., Isacson, O., Mallat, M., Prochiantz, A., Gage, F.H., and Björklund, A. (1985) Survival of intracerebrally grafted rat dopamine neurons previously cultured in vitro. Neurosci. Lett. 61:79–84.

Burns, R.S., Chiuch, C.C., Markey, S.P., Ebert, M.H., Jacobowitz, D.M., and Kopin, I.J. (1983) A primate model of parkinsonism: selective destruction of dopaminergic neurons in the pars compacta of the substantia nigra by *N*-methyl-4-phenyl-1,2,3,6-tetrahydropyridine. Proc. Natl. Acad. Sci. USA 80:4546–4550.

Campbell, B.A., Krauter, E.E., and Wallace, J.E. (1980) Animal models of aging: sensory-motor and cognitive function in the aging rat, In: Stein, D.G. (ed.) Psychology of aging: problems and prospectives. Amsterdam: Elsevier/North Holland, pp. 201–226.

Carlsson, A. (1959) The occurrence, distribution and physiological role of catecholamines in the nervous system. Pharmacol. Rev. 11:490–493.

Coyle, J., Price, D., and De Long, M. (1983) Alzheimer's disease: a disorder of cortical cholinergic innervation. Science 219:1184–1190.

Dunnett, S.B., Björklund, A., Schmidt, R.H., Stenevi, U., and Iversen, S.D. (1983) Intracerebral grafting of neuronal cell suspensions. V. Behavioral recovery in rats with bilateral 6-OHDA lesions following implantation of nigral cell suspensions. Acta Physiol. Scand. [Suppl.] 522:39–47.

Dunnett, S.B., Bunch, S.T., Gage, F.H., and Björklund, A. (1984) Transplantation of dopamine-rich tissue into rats with 6-OHDA lesions of the ventral tegmental area. I. Effects on spontaneous and drug-induced activity. Behav. Brain Res. 13:71–82.

Freed, W.J., Perlow, M.J., Karoum, F., Seiger, A., Olson, L., Hoffer, B.J., and Wyatt, R.J. (1980) Restoration of dopaminergic function by grafting of fetal rat substantia nigra to the caudate nucleus: long-term behavioural, biochemical and histochemical studies. Ann. Neurol. 8:510–519.

Freed, W.J., Morishisa, J.M., Spoor, E., Hoffer, B.J., Olson, L., Seiger, A., and Wyatt, R.J. (1981) Transplanted adrenal chromaffin cells in rat brain reduce lesion-induced rotational behaviour. Nature 292:351–352.

Gage, F.H., and Björklund, A. (1984) Intracerebral grafting of neuronal cell suspensions into the adult brain. Centr. Nerv. Syst. Trauma 1:47–56.

Gage, F.H., Björklund, A. (1986) Enhanced graft survival in the hippocampus following selective denervation. Neuroscience 17:89–98.

Gage, F.H., Björklund, A., Stenevi, U., and Dunnett, S.B. (1983a) Intracerebral grafting of neuronal cell suspensions. VII. Cell survival and axonal outgrowth of dopaminergic and cholinergic cells in the aged brain. Acta Physiol. Scand. [Suppl.] 522:67–75.

Gage, F.H., Dunnett, S.B., Björklund, A., and Stenevi, U. (1983b) Aged rats: recovery of motor impairments by intrastriatal nigral grafts. Science 221:966–969.

Gage, F.H., Björklund, A., and Stenevi, U. (1984a) Denervation releases a neuronal survival factor in adult rat hippocampus. Nature 308:637–639.

Gage, F.H., Björklund, A., Stenevi, U., and Dunnett, S.B. (1984b) Intracerebral grafting in the aged brain. In: Gispen, W.H., and Taber, J. (eds.) Aging of the brain. Amsterdam: Elsevier, pp. 125–137.

Gage, F.H., Björklund, A., Stenevi, U., Dunnett, S.B., and Kelly, P.A.T. (1984c) Intrahippocampal septal grafts ameliorate learning impairments in aged rats. Science 225:533–535.

Gage, F.H, Dunnett, S.B., and Björklund, A. (1984d) Spatial learning and motor deficits in aged rats. Neurobiol. Aging 5:43–48.

Gage, F.H., Kelly, P.A.T., and Björklund, A. (1984e) Regional changes in brain glucose metabolism reflect cognitive impairments in aged rats. J. Neurosci. 4:2856–2866.

Gage, F.H., Björklund, A., Stenevi, U., and Dunnett, S.B. (1985a) Grafting of embryonic CNS tissue to the damaged adult hippocampal formation. In: Björklund, A., and Stenevi, U. (eds.) Neural grafting in the mammalian CNS. Amsterdam: Elsevier, pp. 559–574.

Gage, F.H., Björklund, A., Stenevi, U., and Dunnett, S.B. (1985b) Intracerebral grafting in the aged rat brain: anatomical and functional characterization. In: Björklund, A., and Stenevi, U. (eds.) Neural grafting in the mammalian CNS. Amsterdam: Elsevier, pp. 585–594.

Gage, F.H., Stenevi, U., Carlstedt, T., Foster, G., Björklund, A., and Aguayo, A.J. (1985c) Anatomical and functional consequences of grafting mesencephalic neurons into a peripheral nerve "bridge" connected to the denervated striatum. Exp. Brain Res. 60:584–589.

Gash, D., Sladek, J.R., and Sladek, C.D. (1980) Functional development of grafted vasopressin neurons. Science 21:1367–1369.

Geinisman, U., Bondareff, W., and Dodge, J.T. (1977) Partial deafferentation of neurons in the dentate gyrus of the senescent rat. Brain Res. 134:541–545.

Hefti, F., Hartikka, J., and Schlumpf, M. (1985) Implantation of PC12 cells into the corpus striatum of rats with lesions of the dopaminergic nigrostriatal neurons. Brain Res. 348:283–288.

Hoff, S.F., Scheff, S.W., Bernardo, L.S., and Cotman, C.W. (1982) Lesion induced synaptogenesis in the dentate gyrus of aged rats. I. Loss and reacquisition of normal synaptic density. J. Comp. Neurol. 205:246–252.

Isacson, O., Brundin, B., Kelly, P.A.T., Gage, F.H., and Björklund, A. (1984) Functional neural replacement by grafted striated neurons in the ibotenic acid lesioned rat striatum. Nature 311:458–460.

Jaeger, C.B. (1986) Immunocytochemical study of PC12 cells grafted to the brain of Sprague Dawley rats. J. Comp. Neurol. (in press).

Kelly, P.A.T., Gage, F.H., Ingvar, M., Lindvall, O., Stenevi, U., and Björklund, A. (1984) Functional reactivation of the deafferented hippocampus by embryonic septal grafts as assessed by measurements of local glucose ultilization. Exp. Brain Res. 58:570–579.

Low, W.C., Daniloff, J.K., Bodony, R.P., and Wells, J. (1985) Cross-species transplants of cholinergic neurons and the recovery of function. In: Björklund, A., and Stenevi, U., (eds.) Neural grafting in the mammalian CNS. Amsterdam: Elsevier, pp. 575–584.

Marshall, J.F., and Teitelbaum, P. (1977) New considerations in the neuropsychology of motivated behaviors. In: Iversen, L.L., Iversen, S.D., and Snyder, S.H., (eds.) Handbook of psychopharmacology, vol. 7. New York: Plenum, pp. 201–229.

Morris, R.G.M. (1981) Spatial localization does not require the presence of local cues. Learn. Motiv. 12:239–260.

Perlow, M.J., Freed, W.J., Hoffer, B.J., Seiger, A., Olson, L., and Wyatt, R.J. (1979) Brain grafts reduce motor abnormalities produced by destruction of nigrostriatal dopamine system. Science 204:643–647.

Schmidt, R.H., Björklund, A., Stenevi, U., Dunnett, S.B., and Gage, F.H. (1983) Intracerebral grafting of neuronal cell suspensions. III. Activity of intrastriatal nigral suspension implants as assessed by measurement of dopamine synthesis and metabolism. Acta Physiol. Scand. [Suppl.] 522:19–28.

Ungerstedt, U. (1974) Brain dopamine neurons and behavior. In: Schmitt, F.O., and Worden, F.G. (eds.) The neurosciences: third study program. Cambridge: MIT Press, pp. 695–703.

27. THE IRIS AS A MODEL OF AGING

EZIO GIACOBINI AND ISABELLA MUSSINI

1. INTRODUCTION

Ideally, a model of aging of neurotransmission should be as close as possible to the physiological process, so that normal mechanisms of aging can be studied separately from pathological conditions such as vascular and metabolic changes, degenerative processes, and chronic infections. A second requirement is that the pattern of aging of the animal should be long enough to reflect the several steps of the complex process of aging of higher vertebrates, including primates. Finally, the model should be amenable to physiological, biochemical, and pharmacological manipulations at the in vivo and in vitro levels. We believe that the neuromuscular junction of aging animals, particularly the avian iris, provides an excellent model of aging to study both neurotransmission in the synapse and function at the effector organ level.

2. AGING OF THE NEUROMUSCULAR JUNCTION

2.1 Skeletal muscles of rodents

Studies of the motor endplates in aging animals and humans provide models for analyzing changes in cholinergic mechanisms affecting neuromuscular function during old age. Inactivity and muscle-mass decrease during senescence cannot alone explain the deficit in muscular function observed during old age. The pioneer work of Gutmann and Hanzlikova (1965 and 1966) and of Russian authors (Frolkis et al., 1976) has provided the basis for a quantitia-

Vernadakis, A. (ed.), *Model Systems of Development and Aging of the Nervous System*.

Table 27-1. Aging of the neuromuscular junction: changes in motoneurons, motor units, and motor axons.

Parameter	Species/age	Variation	Author
Nr. motoneurons	Human Over 60 years	Decreased	Rebeiz et al. (1972)
Nr. motor units	Human Over 60 years	Decreased	Campbell et al. (1973)
Ventral root axons	Rat 27 months	10% decrease	Duncan (1934)
Nr. motor axons	Rat 24 months	Unchanged	Gutmann and Hanzlikova (1966)
Nr. motor axons	Cat 10–38 years	Unchanged	Moyer and Kaliszewski (1958)
Preterminal nerves	Rat 24 months	Deformed	Frolkis et al. (1976)
Terminal innervation ratio	Human Over 60 years	Unchanged	Coers et al. (1973)
Preterminal axons small terminals/e.p.	Rat 18–25 months	Shrunken Multiple	Fujisawa (1976)
Terminal branches/ e.p. and nodal sprouts	Cat 6 years	Increased	Tuffery (1971)
Nr. axons entering the junctions	Mouse 29 months	Unchanged	Fahim et al. (1983)
Endplate size	Rat (EDL, S) 24 months	Unchanged or decreased	Gutmann and Hanzlikova (1965)

EDL, ext. dig. longus; S, soleus; and e.p., endplate.

tive analysis of such complex phenomena as reduction of neuromuscular transmission, modified neurotrophic relationship, and muscular atrophy in old age. The abundance of information provided by physiology and biochemistry of the cholinergic apparatus at the neuromuscular junction (NMJ) has made this structure a favorite object of detailed interdisciplinary observation. The changes in the basic components of the NMJ that are summarized in table 27-1 provide a first necessary background. The number of motor axons both in rat and in cat do not seem to change with age (Moyer and Kaliszewski, 1958; Gutmann and Hanzlikova, 1966) while in humans a decrease in number of motoneurons and motor units has been observed (Rebeiz et al., 1972; Campbell et al., 1973). In rodents, preterminal axons are shrunken (Fujisawa, 1976) and terminal branches and nodal sprouts are increased (Tuffery, 1971). The size of the endplate in the 24-month-old rat is either unchanged or decreased (Gutmann and Hanzlikova, 1965). These findings suggest a disturbance of neuromuscular relations during old age and a diminished control of the cholinergic apparatus over the muscle.

2.2 Avian iris neuromuscular junction (NMJ)

The NMJ of the iris of the aging chicken has been extensively studied biochemically and ultrastructurally in our laboratory (for review, see Marchi and Giacobini, 1981; Giacobini, 1982a and b and 1983). This structure is innervated by cholinergic (ciliary ganglion) and adrenergic (superior cervical) as well as peptidergic fibers (trigeminal) (table 27-2) and presents a unique possibility of experimental approaches in vitro and in vivo (Marchi and Giacobini, 1981; Giacobini, 1982a and b) (figure 27-1). Biochemical and structural changes in cholinergic synapses of ciliary ganglia and irises of the chicken during aging are summarized in table 27-3. The nerve terminal arborizations on the iris muscle and their synaptic boutons can be stained and visualized in light microscopy and quantitative data can be obtained (figure 27-2).

Based on the results summarized in table 27-3 (Giacobini, 1982a and b), a hypothesis of aging of the cholinergic synapse has been formulated (Giacobini, 1983). This hypothesis contemplates age-related changes in carrier-mediated mechanisms of uptake and release of the neurotransmitter and its precursor (choline), leading to a "chemical denervation" at the cholinergic synapse (figure 27-3). In order to establish this hypothesis firmly, several basic questions need to be answered.

In the cholinergic terminals of the chicken iris, signs of aging have been identified morphologically and biochemically as early as 18–24 months of age (Giacobini et al., 1984; Mussini and Giacobini, 1985). Polymorphic signs of damage were seen in the nerve ending as well as in the axoplasm of the nerve fiber (Giacobini et al., 1984). Kinetic changes (V_{max}) in the uptake process of choline have been described in the same period, leading to a decreased access of the precursor (Marchi et al., 1980b).

Qualitative ultrastructural analysis of nerve terminals and nerve fibers in iris muscles confirmed previous findings of morphological alterations common to both 3- and 5-year-old chickens (Giacobini et al., 1984; Mussini and Giacobini, 1985). These alterations included the presence of dense granular lysosomes in the adaxonal space of a large number of nerve fibers. Lysosomes and mitochondria accumulated in the axoplasm and, occasionally, enormously swollen, degenerating axons were observed. Irregularities at the level of the nerve terminal were also seen. Polymorphic synaptic vesicles with little tendency to cluster and form active zones were observed (figure 27-4). Lysosomes were also found in the terminals together with accumulations of neurofilaments. Of major interest was the observation that Schwann cell cytoplasmic processes progressively enveloped the terminal (figure 27-4), infiltrating and protruding into the bouton. This process leads to a progressively wider synaptic cleft, which can be filled by a multilayered basal lamina (figure 27-4).

Quantitative morphometric analysis suggested a possible deficit in the

Table 27-2. Neurotransmitter systems in the iris: a comparison of rodent versus avian.

Neurotransmitter	Species	Type of fibers	Function	Origin	Author
Acetylcholine	Rat	Motor	Constrictor pup. Dilator pup.	Ciliary g. Cerv. sup.	Ehinger and Falck (1966)
Norepinephrine	Rat	Motor	Dilator pup.	Cerv. sup.	Malmfors (1965)
Substance P	Rat	Sensory	Sensory	Trigeminal	Seiger et al. (1985)
Leu-enkephalins	Rat	Sensory	Sensory	Trigem. (??)	Bjorklund et al. (1984a)
Vasoactive intestinal polypeptide (VIP)	Rat	Sensory	cotransmitter (?)	Trigem. (??)	Schultzberg et al. (1980)
Other					
Neurofilament (NF) protein	Rat	Sensory	Transport (?)	Trigeminal	Seiger et al. (1985)
Glial fibrillary acid (GFA) protein	Rat	Schwann cells	??	Iris Nerves	Bjorklund et al. (1984b)
Acetylcholine	Chicken	Motor	Constrictor pup.	Ciliary g.	Zenker and Krammer (1967) Lucchi et al. (1974) Chiappinelli et al. (1976)
Norepinephrine	Chicken	Motor	Dilator pup.	Cerv. sup.	Yurkewicz et al. (1981) Ehinger (1967) Kirby et al. (1978)
Substance P	Chicken	Sensory	??	Trigem. (?)	Giacobini (unpubl.)
Enkephalins	Chicken	Sensory	??	Trigem. (?)	Giacobini (unpubl.)

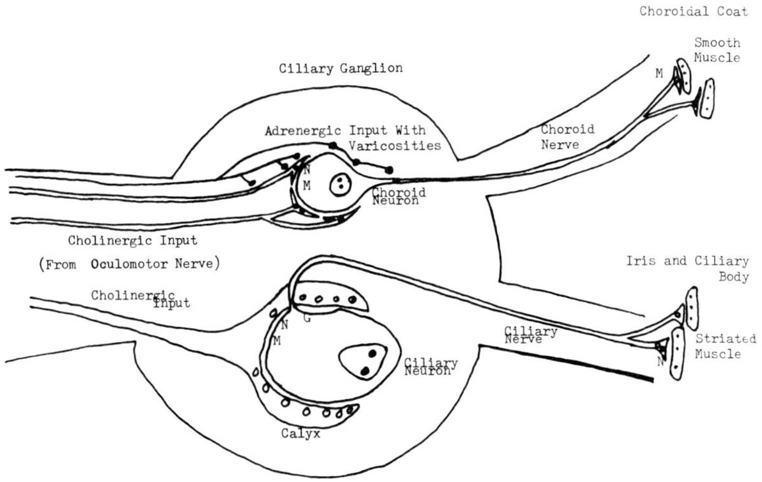

Figure 27-1. Diagram of the ciliary-ganglion–iris preparation of the chicken: *N*, nicotinic receptors; *M*, muscarinic receptors; and *G*, gap junctions. Modified from Giacobini (1982b).

Table 27-3. Acetylcholine metabolism and ultrastructural changes in the peripheral autonomic nervous system of aging animals.

Parameter	Species	Organ	Variation	Author
ChAc, AChE, BuChE sensitiv. to ACh	Rat	Auton. ganglia	Decrease	Frolkis et al. (1973)
ACh and Ch levels	Chick	Ciliary ganglion Iris	Decrease	Marchi et al. (1979 and 1981a) Marchi and Giacobini (1980)
AChE and ChAc	Chick	Ciliary ganglion Iris	Decrease	Marchi et al. (1980d)
Nicotinic binding	Chick	Ciliary ganglion Iris	Decrease	Marchi et al. (1981b)
ACh release	Chick	Ciliary ganglion Iris	Decrease	Giacobini et al. (1984 and 1986)
Ch uptake	Chick	Ciliary ganglion Iris	Decrease	Marchi et al. (1980b and c)
Nr. of boutons Nr. of vesicles Apposition. membr. Vesic. volume	Chick	Iris	Decrease	Giacobini et al. (1984 and 1986) Mussini and Giacobini (1985)
Neurofilaments	Chick	Iris	Increase	Mussini and Giacobini (1985)

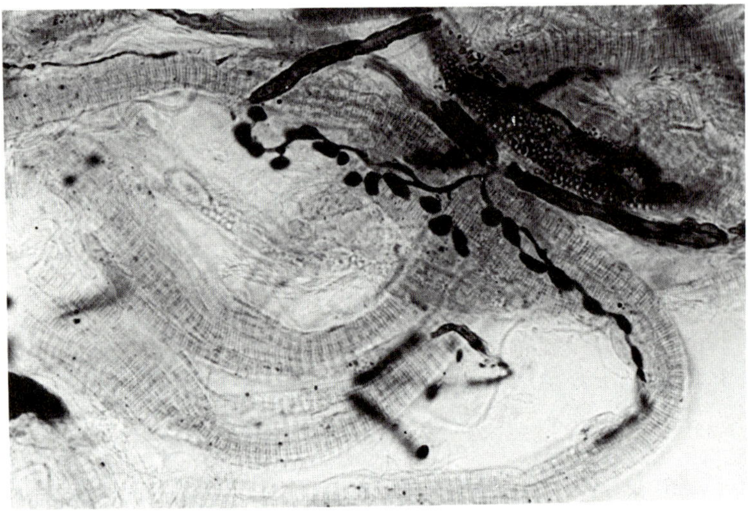

Figure 27-2. Light microscopy of a nerve terminal arborization on iris muscle fibers of the chick (4 months) dissociated and stained with the zinc iodide–osmic acid (ZIO) impregnation method. Preterminal axons and their synaptic boutons are intensely stained in *black*. The average number of boutons per arborization, as well as the average length of boutons, can be determined by this method.

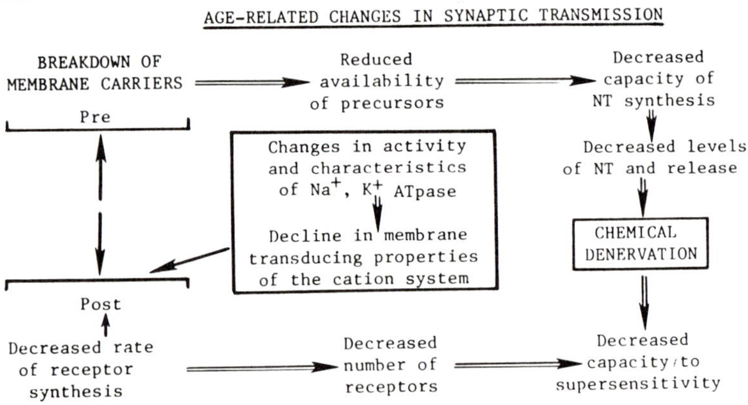

Figure 27-3. A new hypothesis of aging of the cholinergic synapse. Modified from Giacobini (1983).

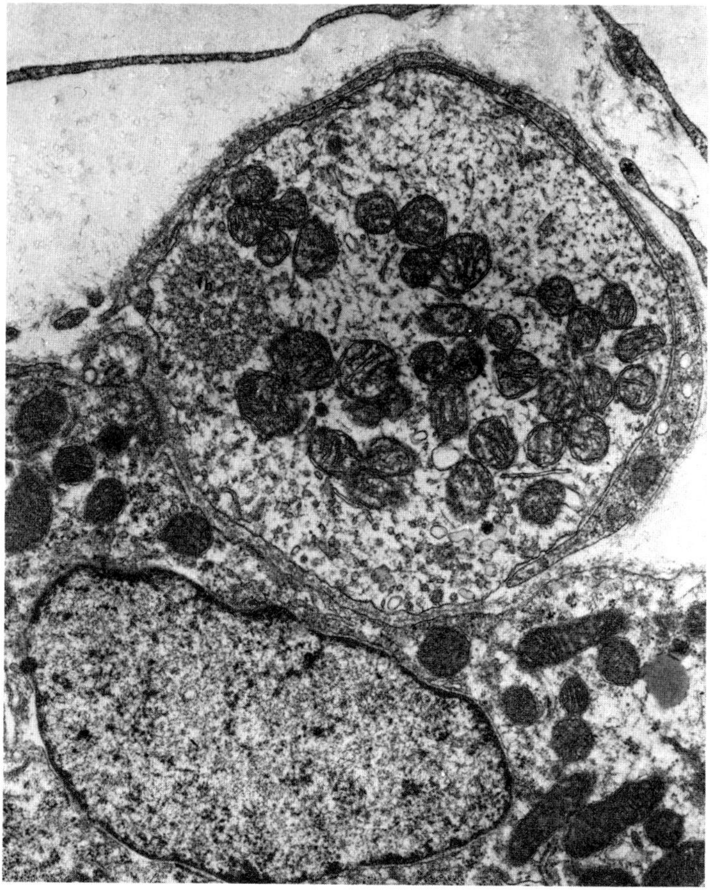

Figure 27-4. Iris muscle of a 3-year-old chicken. The nerve terminal contains a reduced number of vesicles that are polymorphic. Schwann cell cytoplasmic processes infiltrate the synaptic cleft.

ability of the 3- and 5-year-old animals to sustain neurotransmission actively. The decrease in vesicular volume and in synaptic membrane length confirmed the hypothesis that one of the consequences of aging at the neuronal terminal is a decreased ability for neurotransmitter release ("chemical denervation") (figure 27-3). Experiments were designed to study the release of acetylcholine (ACh) from 2- to 4-month and 3-year chicken irises (figure 27-5) under more (type II) or less (type I) severe conditions of K^+ stimulation (Giacobini et al., 1984).

The main effects for age showed that the 3-year tissue released significantly less ^{3}H-ACh than did the 4-month tissue as determined by the area under the release curve (peak area) (figure 27-6). Also, the 3-year tissue showed a lower

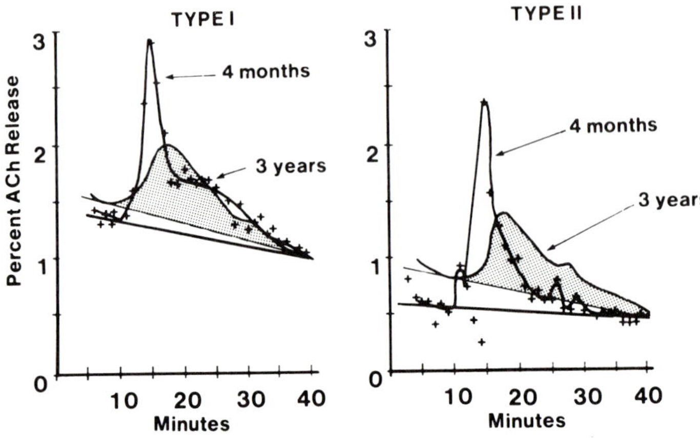

Figure 27-5. Typical release curves in type-I and type-II experiments. ACh release is represented as the percent ^{3}H-ACh released of the total ^{3}H-ACh present in the tissue at the time of release (see text). Modified from Giacobini et al. (1984).

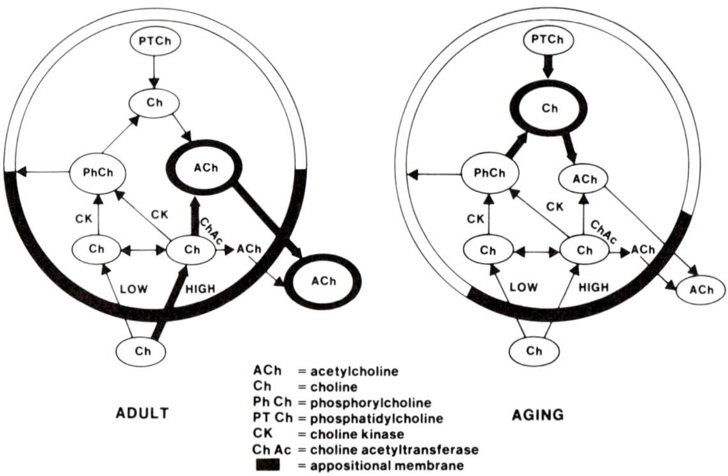

Figure 27-6. Schematic diagram of an adult and an aging cholinergic ending showing various steps involved in the regulation of ACh synthesis. The intracellular ACh is shown as deriving from: (a) intrasynaptic Ch formed by AChE and transported into the ending; this uptake of Ch is impaired in the aging synapse; (b) neuronal phosphorylcholine and phosphatidylcholine. The diagram shows that in spite of an increased intracellular pool of Ch and a decreased release of ACh, the levels of ACh are decreased in the aging ending.

peak release of ^{3}H-ACh than did the 4-month tissue. The time needed for the 3-year tissue to reach its peak release was significantly longer than at 4-months and its rate of release was significantly slower in both types of experiments.

These neurochemical results demonstrate a decreased ability, dependent only on age and not experimental design (I or II), of the 3-year tissue to release ^{3}H-ACh when compared to 4-month tissue (figure 27-5). This correlates well with the morphological data, which demonstrate that two important features for neurotransmitter release (vesicular volume and length of the appositional membrane) were decreased in the 3-year (or 5-year) tissue. These results support the hypothesis that cholinergic transmission in the chicken iris is age dependent and declines with increasing age. This decline is related to modifications of presynaptic mechanisms of release and uptake of the neurotransmitter and its precursor (Giacobini, 1983).

A diagram of the changes in ACh synthesis and release in the aging cholinergic synapse is presented in figure 27-6. As shown in the diagram, the decrease in endogenous levels of ACh and Ch in the aged iris correlates with changes in kinetics of Ch uptake (Marchi et al., 1980b). Our present release experiments clearly demonstrate that only the younger tissue is able to replenish its ACh levels following periods of repeated depletion and release (figure 27-5). Not only is this ability progressively lost with age, but the release mechanism itself becomes less and less efficient with time (figure 27-5). The diagram in figure 27-6 suggests adaptive changes in choline mobilization.

These data indicate that cholinergic endings in older tissues handle Ch and ACh in a different manner than in the younger tissue (figure 27-6). Thus, it appears that the control mechanism that regulates and maintains ACh and Ch levels in cholinergic synapses is impaired in older animals. The results of our release experiments clearly show that the capability of the older junction to sustain prolonged depolarization is decreased and the amount of ACh released declines with age (figure 27-5). These changes are associated with severe alterations in the structure of the synapse, mainly a strongly reduced number of vesicles and a decreased appositional area. The characteristic sequence of age-dependent changes in the avian iris NMJ leading to the functional impairment of the cholinergic synapse is depicted in figure 27-7. Together with the demonstrated reduction in neurotransmitter level and in uptake of Ch (Marchi et al., 1980b), the reduction in release forms a characteristic triade (figure 27-6) of age-related defects that may have important implications for our understanding of aging processes. Age-dependent changes in peripheral organ innervation are not unique, but are seen for other neurotransmitters as well.

It appears that, in the peripheral nervous system, as in the central nervous system, noradrenergic terminals are more affected than the cell bodies during aging since total tyrosine hydroxylase activity in the iris declines, whereas it

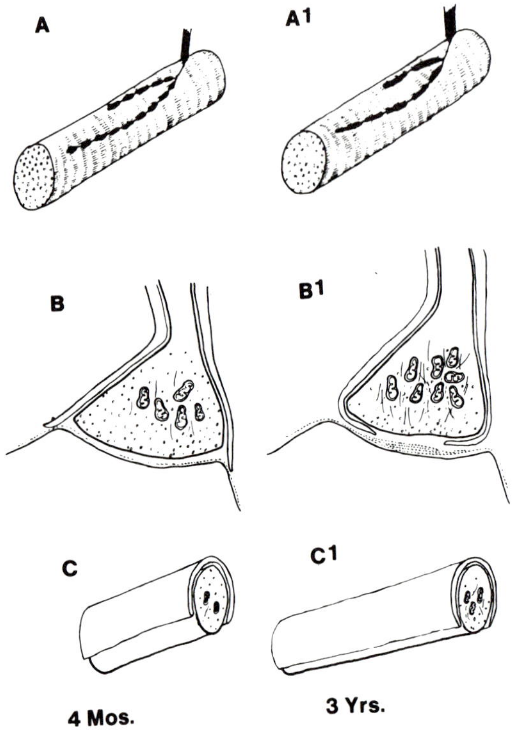

Figure 27-7. Comparison between NMYs at 4 months and 3 years of age in the iris of the chicken (**A** and **A1**) Nerve terminal arborization. A reduction in the number and an increase in the diameter of the synaptic boutons are seen in light microscopy in the 3-year-old NMY with zinc iodide–osmic acid staining. (**B** and **B1**) Decrease in number and polymorphism of synaptic vesicles. Accumulation of mitochondria and an increase in neurofilaments are seen in the 3-year-old NMY. The Schwann cell cytoplasmic processes envelop the terminal progressively reducing the appositional membrane area in the 3–year–old NMY (**C** and **C1**).

continually increases in the sympathetic ganglia (Yurkewicz et al., 1981). This may be explained by a decrease in enzyme synthesis or in transport to the terminals as was suggested for choline acetyltransferase in the ciliary ganglion and iris during aging (Marchi and Giacobini, 1981). However, when comparing cell bodies and terminals, the peripheral nervous system displays more marked differences in total tyrosine hydroxylase activity than does the central nervous system.

ACKNOWLEDGMENTS

The investigations carried on in E.G's laboratory were supported by PHS grants NS-11496, NS-11430, NS-15086, NSF GB-41475; grants from the University of Connecticut Research Foundation; National Institute of Aging grant AG05416-01A1; Air Force grant 83-0051; grants from S.I.U. School of

Medicine including Nowatski Eye Foundation, Central Research Committee; and funds from the S.I.U. Alzheimer Foundation and E.F. Pearson Foundation. This study was also supported by grants from the National Research Council of Italy (CNR) to the Unit for Muscle Biology and Physiopathology (I.M.). The authors gratefully acknowledge the help of Mrs. Diana Smith in compiling and editing the manuscript.

REFERENCES

Bjorklund, H., Hoffer, B., Olson, L., Palmer, M., and Seiger, A. (1984a) Enkephalin immunoreactivity in iris nerves: distribution in normal and graftd irides, persistence and enhanced fluorescence after denervations. Histochemistry 80:1–7.

Bjorklund, H., Dahl, D., Olson, L., and Seiger, A. (1984b) Glial fibrillary acidic protein-like immunoreactivity in the iris: development, distribution and reactive changes following transplantation. J. Neurosci. 4:978–988.

Campbell, M.J., McComas, A.J., and Petito, F. (1973) Physiological changes in ageing muscles. J. Neurol. Neurosurg, Psychiatr. 36:174–182.

Chiappinelli, V., Giacobini, E., Pilar, G., and Uchimura, H. (1976) Induction of cholinergic enzymes in chick ciliary ganglion and iris muscle cells during synapse formation. J. Physiol. 257:749–766.

Coers, C., Telerman-Toppet, M., Gerard, J.M. (1973) Terminal innervation ratio in neuromuscular disease. Arch. Neurol. 29:210–222.

Duncan, D. (1934) Determination of number of nerve fibers in eighth thoracic and largest lumbar ventral roots. J. Comp. Neurol. 59:47–60.

Ehinger, B. (1967) Adrenergic nerves in the avian eye and ciliary ganglion. Z. Zellforsch. 82:577–588.

Ehinger, B., and Falck, B. (1966) Concomitant adrenergic and parasympathetic fibres in the rat iris. Acta Physiol. Scand. 67:201–207.

Fahim, M.A., Holley, J.A., and Robbins, N. (1983) Scanning and light microscopic study of age changes at a neuromuscular junction in the mouse. J. Neurocytol. 12:13–15.

Frolkis, V.V., Bezrukof, V.V., Duplenko, Y.K., Shghegoleva, I.V., Shevtchuk, V.G., and Verkhratsky, N.S. (1973) Acetylcholine metabolism and cholinergic regulation of functions in aging. Gerontologia 19:45.

Frolkis, V.V., Martynenko, O.A., and Zamostyan, V.P. (1976) Aging of the neuromuscular apparatus. Gerontology 22:244–279.

Fujisawa, K. (1976) Some observations on the skeletal musculature of aged rats. Exp. Gerontol. 11:43.

Giacobini, E. (1982a) Cellular and molecular mechanisms of aging of the nervous system: toward a unified theory of neuronal aging. In: Giacobini, E., Filogamo, G., Giacobini, G., and Vernadakis, A. (eds.) The aging brain: cellular and molecular mechanisms of aging in the nervous system. New York: Raven, pp. 271–284.

Giacobini, E. (1982b) Aging of autonomic synapses. In: Fedoroff, S., and Hertz, L. (eds.) Advances in cellular neurobiology, vol. 3. New York: Academic Press, pp. 173–214.

Giacobini, E. (1983) A new hypothesis of aging of the cholinergic synapse. In: Samuels, D., Algeri, S., Gershon, S., Grimm, V., and Toffano, G. (eds.) Aging of the brain, New York: Raven, pp 197–210.

Giacobini, E., Mussini, I., and Mattio, T. (1984) Aging of cholinergic synapses in the avian iris. In: Caciagli, F., Giacobini, E., and Paoletti, R. (eds.) Developmental neuroscience: physiological, pharmacological and clinical aspects, pp. 89–93.

Giacobini, E., Mussini, I., and Mattio, T. (1986) Aging of cholinergic synapses: fiction or reality? In: Hanin, I. (ed.) Dynamics of cholinergic function. New York: Plenum, pp. 177–190.

Gutmann, E., and Hanzlikova, V. (1965) Age changes of motor end plates in muscle fibers of the rat. Gerontology 11:12–24.

Gutmann, E., and Hanzlikova, V. (1966) The motor unit in old age. Nature 209:921.

Kirby, M.L., Diab, I.M., and Mattio, T.G. (1978) Development of adrenergic innervation of the iris and fluorescent ganglion cells in the choroid of the chick eye. Anat. Rec. 191:311–319.

Lucchi, M.L., Bortolami, R., and Callegari, E. (1974) Fine structure of intrinsic eye muscles of birds: development and postnatal changes. J. Submicrosc. Cytol. 6:205–218.

Malmfors, T. (1965) Studies on adrenergic nerves: the use of rat and mouse iris for direct observations on their physiology and pharmacology at cellular and subcellular levels. Acta Physiol. Scand. 64:248.

Marchi, M., and Giacobini, E. (1980) Development and aging of cholinergic synapses. II. Continuous growth of acetylcholine and choline levels in autonomic ganglia and iris of the chick. Dev. Neurosci. 3:38–48.

Marchi, M., and Giacobini, E. (1981) Aging of cholinergic synapses in the peripheral autonomic nervous system. In: Pepeu, G., and Ladinsky, H. (eds.) Cholinergic mechanisms. New York: Plenum, pp. 25–45.

Marchi, M., Giacobini, E., and Hruschak, K. (1979) Development and aging of cholinergic synapses. I. Endogenous levels of acetylcholine and choline in developing autonomic ganglia and iris of the chick. Dev. Neurosci. 2:201–212.

Marchi, M., Hoffman, D.W., and Giacobini, E. (1980a) Aging of the peripheral nervous system: age dependent changes in cholinergic synapses. In: Neural regulatory mechanisms during aging. New York: Alan R. Liss, pp. 215–219.

Marchi, M., Hoffman, D.W., Giacobini, E., and Fredrickson, T. (1980b) Age dependent changes in choline uptake of the chick iris. Brain Res. 195:423–431.

Marchi, M., Hoffman, D.W., and Giacobini, E. (1980c) The peripheral nervous system: a model system of aging. In: Amaducci, L., et al. (eds.) Aging of the brain and dementia, vol. 13, pp. 159–166.

Marchi, M., Hoffman, D.W., Giacobini, E., and Fredrickson, T. (1980d) Development and aging of cholinergic synapses. IV. Acetylcholinesterase and cholineacetyltransferase activities in autonomic ganglia and iris of the chick. Dev. Neurosci. 3:235–247.

Marchi, M., Hoffman, D.W., Giacobini, E., and Volle, R. (1981a) Development and aging of cholinergic synapses. VI. Mechanisms of acetylcholine biosynthesis in the chick iris. Dev. Neurosci. 4:442–450.

Marchi, M., Yurkewicz, L., Giacobini, E., and Fredrickson, T., (1981b) Development and aging of cholinergic synapses. V. Changes in nicotinic cholinergic receptor binding in ciliary ganglia and irises of the chicken. Dev. Neurosci. 4:258–266.

Moyer, E.K., and Kaliszewski, B.F. (1958) The number of nerve fibers in motor spinal nerve roots of young, mature and aged rats. Anat. Rec. 131:681–699.

Mussini, I., and Giacobini, E. (1985) Changes with age in neuromuscular junction. In: Carraro, U., and Angelini, C. (eds.) Proceedings of cell biology and clinical management in functional electro stimulation of neurones and muscles. Padova: Cleup, pp, 1–4.

Rebeiz, J.J., Moore, M.J., Holden, E.M., and Adams, R.D. (1972) Variations in muscle status with age and systemic diseases. Acta Neuropathol. 22:127–144.

Schultzberg, M., Hokfelt, T., Olson, L., Alund, M., Nilsson, G., Terenius, L., Elde, R., Goldstein, M., and Said, S. (1980) Substance P, VIP, enkephalin and somatostatin immunoreactive neurons in intestinal tissue transplanted to the anterior eye chamber. J. Auton. Nerv. Syst. 1:293–303.

Seiger, A., Selin, U.-B., Kessler, J., Black, I., and Ayer-LeLievre, C. (1985) Substance P-containing sensory nerves in the rat iris: normal distribution, ontogeny and innervation of intraocular iris grafts. Neuroscience 15:519–528.

Tuffery, A.R. (1971) Growth and regeneration of motor end plates in normal cat and hind limb muscles. J. Anat. 110:221.

Yurkewicz, L., Marchi, M., Lauder, J.M., and Giacobini, E. (1981) Development and aging of noradrenergic cell bodies and axon terminals in the chicken. J. Neurosci. Res. 6:621–641.

Zenker, W., and Krammer, E. (1967) Untersuchungen uber Feinstruktur and Innervation der inneren Augenmuskulatur des Huhnes. Z. Zellforsch. Mikrosk. Anat. 83:147–168.

28. IN VITRO MODEL OF NEURONAL AGING AND DEVELOPMENT IN THE NERVOUS SYSTEM

REGINO PEREZ-POLO AND KARIN WERRBACH-PEREZ

1. INTRODUCTION

There is ample support for the hypothesis that free radical molecules play a major role in aging (Harman, 1981; Pryor, 1984; Packer, 1984; Cutler, 1985). This is also true for the pathological changes associated with aging brain (Agranoff, 1984; Cutler, 1985) although the bulk of studies on free radical processes in the nervous system have focused on acute neural tissue injury (Demopoulos et al., 1979; Naftchi and Gennaro, 1985; Willmore et al., 1983; Chan et al., 1984; Anderson and Means, 1985; Anderson et al., 1985). Although the relative contributions of the different components of auto-oxidation events are not fully resolved at the chemical level [see Aust et al. (1985) and Halliwell and Gutteridge (1985) for two views], there is little doubt that lipid peroxidation events, irrespective of the sources of catalysis, and endogenous levels of "scavenging enzymes" are predictive of pathological consequences. Given the cellular heterogeneity and level of intercellular communication peculiar to the nervous system, there are advantages to the development and use of an in vitro system where pertinent variables can be controlled.

2. HUMAN NEUROBLASTOMA MODEL

The human neuroblastoma clonal line SK-N-SH-SY-5Y, SY5Y, has been a useful model system for the study of neurite elongation in vitro (Perez-Polo et al., 1979; Recio-Pinto and Ishii, 1984), regulation of neurotransmitter

Vernadakis, A. (ed.), *Model Systems of Development and Aging of the Nervous System*.

expression (Biedler et al., 1978; Perez-Polo et al., 1982), expression of receptors to the nerve growth factor protein, NGF (Sonnenfeld and Ishii, 1982; Lyons et al., 1983) and other NGF-induced effects (West et al., 1976; Kuramoto et al., 1977 and 1981; Schulze and Perez-Polo, 1982). The properties of NGF have been amply reviewed elsewhere (Levi-Montalcini, 1982; Yankner and Shooter, 1982; Perez-Polo and Haber, 1983; Calissano et al., 1984; Perez-Polo, 1985; Varon et al., 1983/1984; Bradshaw et al., 1985; Thoenen and Edgar, 1985). NGF is a neuronotrophic factor required for the survival and differentiation of sensory and sympathetic neurons. In brain, NGF promotes expression of choline acetyltransferase (Hefti et al., 1985) and modulates behavioral responses to lesions (Stein and Will, 1983). In spinal cord, ambient levels of NGF modulate axonal sprouting by sensory neurons following spinal lesions (Hulsebosch et al., 1984). Also, NGF plays a role in the development of adrenal chromaffin cell precusors (Unsicker et al., 1985). For sympathetic neurons and their transformed counterparts, SY5Y cells, regeneration following axotomy depends on extracellular concentrations of NGF (Hendry, 1975; Levi-Montalcini, 1983; Perez-Polo et al., 1983).

3. 6-OH DOPAMINE AND NERVE GROWTH FACTOR

In vitro and in vivo, axotomy can be accomplished using the catecholamine derivative 6-hydroxydopamine, 6OHDA (Thoenen and Tranzer, 1968; Saner and Thoenen, 1970; Angeletti, and Levi-Montalcini, 1970; Levi-Montalcini et al., 1975; Tiffany-Castiglioni and Perez-Polo, 1979–1982). Catecholamines and their derivatives, such as 6OHDA, generate superoxide radicals and hydrogen peroxide by transferring electrons to oxygen (Sachs and Jonsson, 1973). The toxic effects of 6OHDA on SY5Y are fairly specific to this cell type (figure 28-1) using an assay system that is responsive at several concentrations of 6OHDA (figure 28-2), and gives comparable results when compared to other assay systems (figure 28-3). As was the case for the in vivo situation (Levi-Montacini et al., 1975), NGF offers protection in vitro at similar concentrations to those that elicit neurite outgrowth (figure 28-4) by a reversible mechanism. This protection is different from neurite induction by NGF in that, when SY5Y cells are exposed to dibutyryl cyclic AMP (cAMP), cells will extrude neurites, but will not be protected (figure 28-4). Pretreatment of SY5Y with NGF prior to 6OHDA treatment is effective in protecting cells in the presence of colcemid at concentrations that disrupt neurotubules but are not toxic in themselves, suggesting that protection is not a transport-mediated process. Further evidence for this is the fact that blocking catecholamine uptake with cocaine has only a small protective effect on 6OHDA toxicity (figure 28-5).

4. ROLE OF FREE RADICALS

A more direct chemical interpretation is based on the fact that 6-OHDA spontaneously reacts with molecular oxygen to form a 1,4-paraquinone and

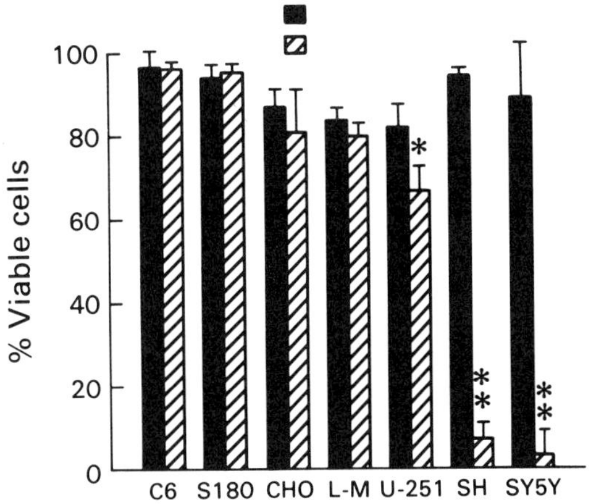

Figure 28-1. Cultures were seeded at 2×10^5 cells per 35-mm dish and exposed 24 hours later to 6-OHDA for 1 hour. Viability was determined by the Trypan blue dye exclusion test. Significance levels were determined by analysis of variance with a priori comparisons not significant ($^{**}p < 0.001$). C6 is rat glioma; S180 is mouse sarcoma; CHO is hamster fibroblast; L-M is mouse fibroblast; U257 is human glioma; and SH is the SY5Y parent line. *Solid bars* are controls and *hatched bars* are 6OHDA treated cells.

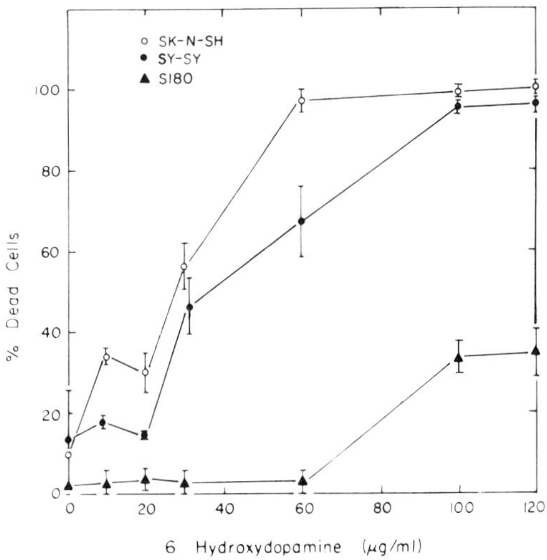

Figure 28-2. Dose–response effects of 6-OHDA. Cultures were seeded at 2×10^5 cells per 35-mm dish and exposed 24 hours later to 6-OHDA for 1 hour.

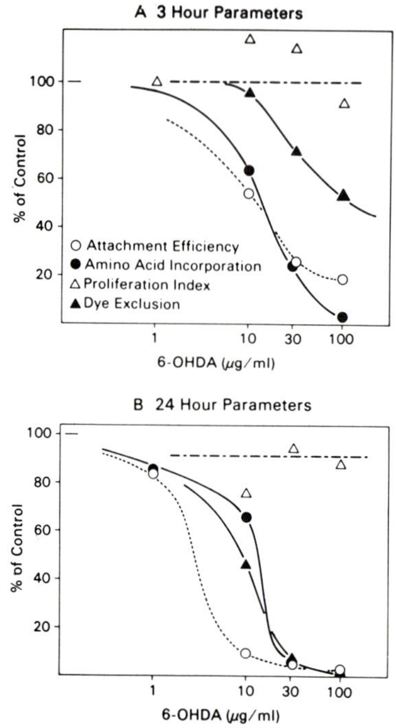

Figure 28-3. Effects of 6-OHDA at 3 and 24 hours. Cultures were seeded at 2×10^5 cells per 35-mm dish and exposed 24 hours later to 6-OHDA for 1 hour.

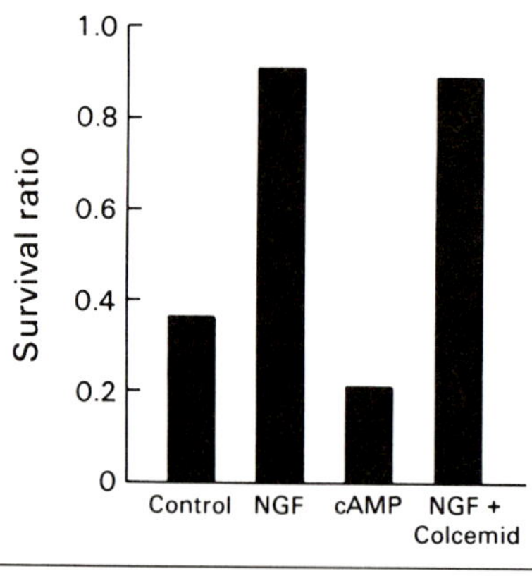

Figure 28-4. Ratio of percent viable cells treated with agent and 6-OHDA to percent viable cells treated with 6-OHDA alone. SY5Y cells were treated with NGF (1.9 μg/ml), dibutyryl cyclic AMP (cAMP) (1 m*M*), colcemid (0.1 μg/ml), or not treated for 6 days prior to testing effects of 0.12 m*M* 6/OHDA. Percentage of viable cells were determined by Trypan blue dye exclusion. Significance levels were determined by analysis of variance.

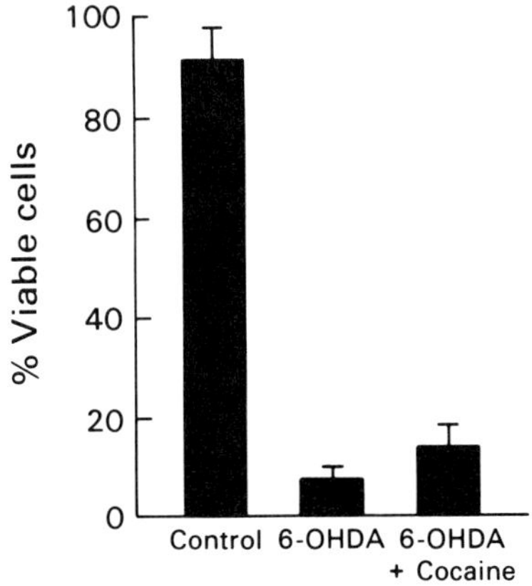

Figure 28-5. Viability of cells treated for 1 hour with 6-OHDA (0.12 m*M*) in the presence or absence of cocaine (1 m*M*). Percentage of viable cells were determined by Trypan blue dye exclusion. Significance levels were determined by analysis of variance. Results for cocaine did not differ significantly from those where it was absent.

its products (Saner and Thoenen, 1970) as well as hydrogen peroxide (H_2O_2), the superoxide radical ($O_2^{\dot{-}}$), and the hydroxyl radical ($\cdot OH$). The generation of these species is proposed to occur by the following reaction steps (Heikkila and Cohen, 1971; Cohen and Heikkila, 1974):

$$6-\text{OHDA} + O_2 \rightarrow SQ^{\cdot} + O_2^{\dot{-}} + H^+ \quad \text{(I)}$$

$$H^+ + \text{6-OHDA} + O_2^{\dot{-}} \rightarrow SQ^{\cdot} + H_2O_2 \quad \text{(II)}$$

$$SQ^{\cdot} + O_2 \rightarrow Q + O_2^{\dot{-}} + H^+ \quad \text{(III)}$$

$$SQ^{\cdot} + O_2^{\dot{-}} + H^+ \rightarrow Q + H_2O_2 \quad \text{(IV)}$$

where $SQ^{\cdot}$ is a semiquinone free radical of 6-OHDA and Q is a quinone. The formation of hydroxyl radicals can be catalyzed from the breakdown of H_2O_2 by trace metals in a Fenton-type reaction (Haber and Weiss, 1934)

$$H_2O_2 + Fe^{2+} \rightarrow \cdot OH + OH^- + Fe^{3+} \quad \text{(V)}$$

or by the Haber-Weiss reaction (1934),

$$H_2O_2 + O_2^{\dot{-}} \rightarrow \cdot OH + OH^- + O_2 \quad \text{(VI)}$$

Hydrogen peroxide may be formed additionally by the reduction of the superoxide radical:

$$2H^+ + 2O_2^{\dot{-}} \rightarrow H_2O_2 + O_2 \quad \text{(VII)}$$

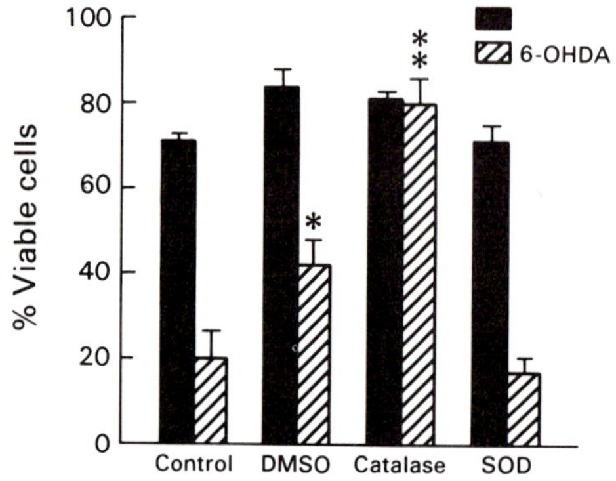

Figure 28-6. SY5Y cells were treated concomitantly with dimethylsulfoxide (DMSO; $1.2 \times 10^{-4} M$) and 6-OHDA at an equal molar ratio for 1 hour. The percentage viable cells were determined after 24 hours by the Trypan blue dye exclusion test. Significance levels were determined by analysis of variance ($^{**}p < 0.001$; $^{*}p < 0.01$ when compared to no treatment).

Table 28-1. Toxic effects of hydrogen peroxide.[a]

	SY5Y		CHO	
H_2O ($\times 10^{-4} M$)	Amino acid incorporation (3 hours)	Dye exclusion (24 hours)	Amino acid incorporation (3 hours)	Dye exclusion (24 hours)
0 (control)	100	100	100	100
1.2	89	64		
4.0	43	41	44	94
8.0	9	17	17	83

[a]Viability of cells treated for 1 hour with H_2O_2. Values are percentages of control values. Differs from control, $p < 0.001$, from Tiffany-Castiglioni et al. (1982).

Relative contributions of the various species can in principle be deduced from the different effects of superoxide dismutase (SOD), which catalyzes the conversion of O_2 to H_2O_2 (Fridovich, 1972); catalase, which catalyzes the conversion of H_2O_2 to H_2O in the presence of Fe^{3+} (Heikkila and Cohen, 1971; Cohen and Heikkila, 1974; O'Neil, 1972); or DMSO, which sequesters ·OH, preventing cascading of lipid peroxidation events. The levels and pattern of protection offered by these agents (figure 28-6) and the fact that neither SOD nor catalase had permanent effects on quinone generation (Tiffany-Castiglione et al., 1982) would suggest that H_2O_2-induced peroxidation is a key aspect of neuronal damage in this system. In agreement with this hypothesis is the finding that H_2O_2 is also toxic to SY5Y (table 28-1) and that these cells have low endogenous levels of catalase and glutathione transferase (table 28-2). It is significant that NGF, which protects from H_2O_2 effects

Table 28-2. Catalase, GSH S-transferase, and GSH peroxidase activities.[a]

Substrates		GSH S-TR H_2O_2	GSH S-TR L-Dopa	GSH-Px tBHP	GSH-Px H_2O_2
Cell line	Catalase	U/mg protein		U/mg protein	
SY5Y	1.58	0	0.17	0.32	0.13
CHO	3.64	0	0.37	0.44	0.055

[a] Enzyme activities for untreated cells. Two substrates were used for GSH-Px assays: tertiary butyl hydroperoxide (tBHP) and H_2O_2. The standard deviations were <1%. Differs significantly from SY5Y value, $p < 0.001$, from Tiffany-Castiglioni et al. (1982).

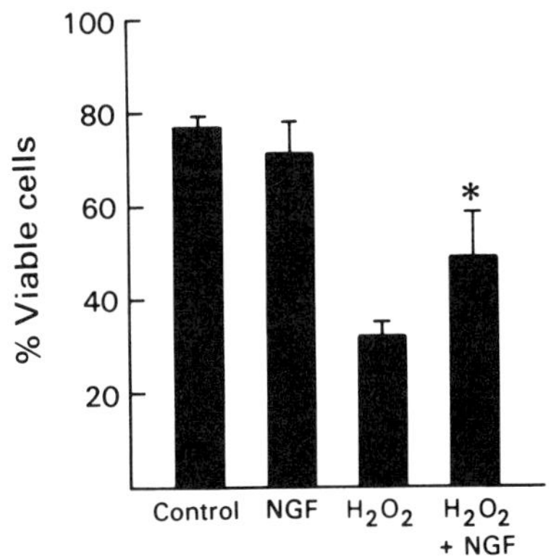

Figure 28-7. Ratio of percent viable cells treated with agent and H_2O_2 to percent viable cells treated with H_2O_2 alone. SY5Y cells were treated with NGF (1.9 μg/ml) for 6 days prior to testing effects of $4 \times 10^{-4}\,M$ H_2O_2. Percentage of viable cells were determined by Trypan blue dye exclusion. Significance levels were determined by analysis of variance.

(figure 28-7), has no effects on SOD (data not shown), but does stimulate both catalase and glutathione S-transferase activity in these cells (table 28-3).

5. CONCLUSIONS

In recent experiments, Varon and coworkers have presented evidence that catalase activity has trophic properties on cultured neurons of central nervous system origin (1985). In vivo, SOD application to cerebral cortex diminishes trauma-associated localized lesions in dogs (Sharp et al., unpublished results), suggesting that outcome of injury to neurons of both the peripheral and central nervous system may be modified directly by manipulation of local availability of catalase, SOD, and/or organic hydroxyperoxidases, or indirectly by the effects of trophic factors, such as NGF, on the endogenous levels of these enzymes. Given the dramatic effects that enzyme activities

Table 28-3. Effects of NGF on levels of catalase and GSH S-transferase in SY5Y.[a]

Treatment	Catalase (U × 10^3/dish)	GSH S-Tr (U × 10^3/dish)
Control	17	5.6
NGF	68*	20*

[a] Enzyme activities were determined in SY5Y cells or SY5Y cells treated with NGF for 11 days. *$p < 0.001$ when compared to controls.

such as catalase can have on cell survival and the known abrogation of the neuronal cell death program by neuronotrophic factors such as NGF during the development of the nervous system (Hamburger and Oppenheim, 1982; Perez-Polo, 1985) one might speculate that regulation of enzymes like catalase by neuronotrophic factors might be relevant not just to aging and injured nervous system, but also in development.

ACKNOWLEDGMENTS

Work carried out in our laboratories was funded in part by a grant from the National Institutes of Health, NS18708, and Robert A. Welch Foundation grant H698. It is a pleasure to acknowledge the collaboration of M. Albin, T. Sharp, and E. Tiffany-Castiglioni. Thanks to D. Masters for manuscript preparation.

REFERENCES

Agranoff, B.W. (1984) Lipid peroxidation and membrane aging. Neurobiol. Aging 5:337–338.

Anderson, D.K., and Means, E.D. (1985) Iron-induced lipid peroxidation in spinal cord: protection with mannitol and methyl predmisolone, Free Radicals Biol. Med. 1:59–64.

Anderson, R.E., Maude, M.B., and Nielsen, J.C. (1985) Effect of lipid peroxidation on rhodopsin regeneration. Curr. Eye Res. 4:65–71.

Angeletti, P.U., and Levi-Montalcini, R. (1970) Cytolytic effect of 6-hydroxydopamine on neuroblastoma cells. Cancer Res. 30:2863–2869.

Aust, S.D., Morehouse, L.A., and Thomas, C.E. (1985) Role of metals in oxygen radical reactions. J. Free Radicals Biol. Med. 1:3–25.

Biedler, J.L., Roffler-Tarlow, S., Schachner M., and Freedman, L.S. (1978) Multiple neurotransmitter synthesis by human neuroblastoma cell lines and clones. Cancer Res. 38:3751–3757.

Bradshaw, R.A., Dunbar, J.C., Isackson, P.J., Kouchalakos, R.N., and Morgan, C.J. (1985) Nerve growth factor: mechanism of action. In: Ford, R.J., and Maizel, A.L. (eds.) New York: Raven.

Calissano, P., Cattaneo, A., Biocca, S., Aloe, L., Mercanti, D., and Levi-Montalcini, R. (1984) The nerve growth factor established findings and controversial aspects. Exp. Cell Res. 154:1–9.

Chan, P.H., Schmidley, J.W., Fishman, R.A., and Longar, S.M. (1984) Brain injury, edema, and vascular permeability changes induced by oxygen-derived free radicals. Neurology 34:315–320.

Cohen, G., and Heikkila, R.E. (1974) The generation of hydrogen peroxide, superoxide radical, and hydroxyl radical by 6-hydroxydopamine, dialuric acid, and related cytotoxic agents. J. Biol. Chem. 249:2447–2452.

Cutler, R.G. (1985) Peroxide-producing potential of tissues: inverse correlation with longevity of mammalian species. Proc. Natl. Acad. Sci. USA 82:4798–4802.

Demopoulos, H.B., Flamm, E.S., Seligman, M.L., Mitamura, J.A., and Ransohoff, J. (1979) In: Popp, A.J., et al. (eds.) Neural rauma. New York: Raven, pp. 63–78.
Fridovich, I. (1972) Superoxide radical and superoxide dismutase. Accounts Chem. Res. 5:321–326.
Haber, F., and Weiss, J. (1934) The catalytic decomposition of hydrogen peroxide by ion salts. Proc. R. Soc. Lond. [A] 147:332.
Halliwell, B., and Gutteridge, M.C. (1985) Oxygen radicals and the nervous system. TINS 8:22–26.
Hamburger, V., and Oppenheim, R.W. (1982) Naturally occurring neuronal death in vertebrates. Neurosci. Comment. 1:39–55.
Harman, D. (1981) The aging process. Proc. Natl. Acad. Sci. USA 78:7124–7128.
Hefti, F., Hartikka, J., Eckenstein, F., Gnahn, H., Heumann, R., and Schwab, M. (1985) Nerve growth factor increases choline acetyl-transferase but not survival or fiber outgrowth of cultured fetal septal cholinergic neurons. Neruoscience 14:55–68.
Heikkila, R., and Cohen, G. (1971) Inhibition of biogenic amine uptake by hydrogen peroxide: a mechanism for toxic effects of 6-hydroxydopamine. Science 172:1257–1258.
Hendry, I.A. (1975) The response of adrenergic neurones to axotomy and nerve growth factor. Brain Res. 94:87–97.
Hulsebosch, C.E., Coggeshall, R.E., and Perez-Polo, J.R. (1984) Effects of nerve growth factor and its antibodies on sprouting of sensory axons following spinal cord hemisection. Brain Res. 232:1–10.
Kuramoto, T., Perez-Polo, J.R., and Haber, B. (1977) Membrane properties of a human neuroblastoma. Neurosci. Lett. 4:151–159.
Kuramoto, T., Werrbach-Perez, K., Perez-Polo, J.R., and Haber, B. (1981) Membrane properties of a human neuroblastoma II. Effects of differentiation. J. Neurosci. Res. 6:441–449.
Levi-Montalcini, R. (1982) Developmental neurobiology and the natural history of nerve growth factor. Annu. Rev. Neurosci. 5:341–362.
Levi-Montalcini, R. (1983) The nerve growth factor–target cell interaction: a model system for the study of directed axonal growth and regeneration.
Levi-Montalcini, R., Aloe, L., Muganini, E., Oesch, F., and Thoenen, H. (1975) Nerve growth factor induces volume increase and enhances tyrosine hydroxylase synthesis in chemically axotomized sympathetic ganglia of newborn rats. Proc. Natl. Acad. Sci. USA 72:595–599.
Lyons, C.R., Stach, R.W., and Perez-Polo, J.R. (1983) Binding constants of isolated NGF-receptors from different species. Biochem. Biophys. Res. Commun. 115:368–374.
Naftchi, N.E., and Gennaro, J.F. (1985) Prevention of damage in acute spinal cord injury by peptides and pharmacologic agents. Peptides 3:235–247.
O'Neil, S.P. (1972) Inactivation of immobiolized catalase by hydrogen peroxide in continuous reactors. Biotechnol. Bioeng. 14:201–205.
Packer, L. (1984) Oxygen radicals in biological systems Methods Enzymol. 105.
Perez-Polo, J.R. (1985) Neuronotrophic factors. In: Bottenstein, J.E., and Sato, G. (eds.) Cell culture in the neurosciences. New York: Plenum.
Perez-Polo, J.R., and Haber, B. (1983) Neuronotorphic interactions. Clin. Neurosci. 5:37–51.
Perez-Polo, J.R., Werrbach-Perez, K., and Tiffany-Castiglioni, E. (1979) A human clonal line model of differentiating neurons. Dev. Biol. 71:341–355.
Perez-Polo, J.R., Tiffany-Castinglioni, E., Ziegler, M., and Werrbach-Perez, K. (1982) Effect of neurve growth factor on catecholamine metabolism in a human neuroblastoma clone (SYSY). Dev. Neurosci. 5:418–423.
Perez-Polo, J.R., Tiffany-Castiglioni, E., and Werrbach-Perez, K. (1983) Model clonal system for study of neuronal cell injury. Birth Defects 19:201–220.
Pryor, W.A. (1984) Free radicals in biology. In: Pryor, W.A. (ed.). New York: Academic Press, 5.
Recio-Pinto, E., and Ishii, D.N. (1984) Effects of insulin-like growth factor-II and nerve growth factor on neurite outgrowth in cultured human neuroblastoma cells. Brain Res. 320:323–334.
Sachs, C., and Jonsson, G. (1973) Changes in central noradrenaline neurons after systemic 6-hydroxydopamine administration. J. Neurochem. 21:1517–1524.
Saner, A., and Thoenen, H. (1970) Model experiments on the molecular mechanism of action of 6-hydroxydopamine. Mol. Pharmacol. 7:147–154.
Schulze, I., and Perez-Polo, J.R. (1982) Nerve growth factor and cyclic AMP: opposite effects on neuroblastoma-substrate adhesion. J. Neurosci. Res. 8:393–411.

Slater, T.F. (1984) Free-radical mechanisms in tissue injury. Biochem. J. 222:1–15.

Sonnenfeld, K.H., and Ishii, D.N. (1982) Nerve growth factor effects and receptors in cultured human neuroblastoma cell lines. J. Neurosci. Res. 8:375–391.

Stein, D.G., and Will, B.E. (1983) Nerve growth factor produces a temporary facilitation of recovery from entorhinal cortex lesions. Brain Res. 261:127–131.

Thoenen, H., and Edgar, D. (1985) Neurotrophic factors. Science 229:238–242.

Thoenen, H., and Tranzer, J.P. (1968) Chemical sympathectomy by selective destruction of adrenergic nerve endings with 6-hydroxydopamine. Arch. Pharmakol. Exp. Pathol. 261:271–288.

Tiffany-Castiglioni, E., and Perez-Polo, J.R. (1979) The role of nerve growth factor in vitro in cell resistance to 6-hydroxydropamine toxicity. Exp. Cell Res. 121:179–189.

Tiffany-Castiglioni, E., and Perez-Polo, J.R. (1980) Evaluation of methods for determining 6-hydroxydopamine cytotoxicity. In Vitro 16:591–599.

Tiffany-Castiglioni, E., and Perez-Polo, J.R. (1981) Stimulation of resistance to 6-hydroxydopamine in a human neuroblastoma cell line by nerve growth factor. Neurosci. Lett. 26: 157–161.

Tiffany-Castiglioni, E., Saneto, R.P., Proctor, P.H., and Perez-Polo, J.R. (1982) Participation of active oxygen species in 6-hydroxydopamine toxicity to a human neuroblastoma cell line. Biochem. Pharmacol. 31:181–188.

Unsicker, K., Millar, T.J., Muller, T.H., and Hofmann, H.D. (1985) Embryonic rat adrenal glands in organ culture: effects of dexasmethasone, nerve growth factor and its antibodies on pheochromoblast differentiation. Cell Tissue Res. 241:207–217.

Varon, S., Manthorpe, M., and Williams, L.R. (1983/84) Neuronotrophic and neurite-promoting factors and their clinical potentials. Dev. Neurosci. 6:73–100.

West, G.J., Uki, J., Herschman, H.R., and Seeger, R.C. (1976) Adrenergic, cholinergic, and inactive human neuroblastoma cell lines with the action-potential Na ionophore. Cancer Res. 37:1372–1376.

Willmore, L.J., Hiramatsu, M., Kochi, H., and Mori, A. (1983) Formation of superoxide radicals after $FeCl_3$ injection into rat isocortex. Brain Res. 277:393–396.

Yankner, B.A., and Shooter, E.M. (1982) The biology and mechanism of action of nerve growth factor. Annv. Rev. Biochem 51:845–868.

29. GANGLIOSIDES AS A TOOL FOR STUDYING THE REPAIR OF ADULT MAMMALIAN BRAIN AFTER INJURY

ADRIANA CONSOLAZIONE, CATERINA ALDINIO,
LUCIO CAVICCHIOLI, ALBERTA LEON, AND GINO TOFFANO

1. INTRODUCTION

Cajal's concept that mature central neurons, unlike developing ones, are not able to regenerate after axotomy is presently under revision. Evidence has emerged indicating that several types of adult central neurons also possess an intrinsic capability for axon sprouting, regeneration, and formation of new connections after injury (Björklund and Stenevi, 1979; Cotman et al., 1981). This anatomic remodeling, involving mostly sprouting from surrounding undamaged axons, is accompanied by behavioral recovery (Sabel et al., 1985). However, the recovery process seems to depend on, among other factors, on the extent and localization of the injury (Nygren and Olson, 1977; Reis et al., 1978) and to be determined by appropriate signals, such as neuronotrophic factors, derived from the humoral microenvironment, the denervated target areas, and other cell–cell contacts. In fact, the concentration of neuronotrophic factors (NTFs) has been reported to increase in the adult central nervous system (CNS) following injury (Nieto-Sampedro et al., 1982), indicating the involvement of such factors not only for appropriate neuronal development, but also in the adult for neuronal survival and adequate repair processes (Whittemore et al., 1985; Toffano et al., 1986). In addition, deficiency or inhibition of NTFs in the adult CNS has been proposed as a possible cause for neurodegenerative diseases (Appel, 1981). Consequently, the identification of these factors controlling neuronal survival and

Vernadakis, A. (ed.), *Model Systems of Development and Aging of the Nervous System*.

neurite outgrowth both during development and in the adult state has become a major goal in current neurobiological studies.

2. NEURONOTROPHIC FACTORS IN THE ADULT CNS: ROLE OF PLASMA MEMBRANE COMPONENTS

To date, the only purified and well-characterized NTF is the nerve growth factor (NGF). Until recently, NGF was thought to be essential for survival, development, and regeneration only of sensory and sympathetic neuronal cells (Levi-Montalcini, 1966; Thoenen and Barde, 1980; Ebendal et al., 1980). At present, NGF seems, however, to be also essential for development and repair of CNS cholinergic neurons (Schwab et al., 1979; Hefti et al., 1985; Hefti, 1985). Nevertheless, tissue culture experiments suggest that NGF is only one of a number of macromolecules able to function as trophic factors on specific CNS neurons in vitro and perhaps also in vivo (Crutcher and Collins, 1982, Barde et al., 1982; Varon et al., 1983; Müller et al., 1984; Lindsay and Peters, 1984; Kligman and Marshak, 1985; Dal Toso et al., 1986). Independently of the nature of these molecules, the responsiveness of neurons to NTF must depend on recognition and transduction of the NTF signal at the level of the plasma membrane. Consequently, the neuronal plasma membrane is thought to play a key role in cell survival and anatomic remodeling after injury and many efforts are today devoted in assessing the functional role of intrinsic membrane components in regulating CNS repair processes. Among these components, gangliosides have recently proved to be of relevant interest.

3. INVOLVEMENT OF GANGLIOSIDES IN REGULATION OF NEURONOTROPHIC INTERACTIONS IN VITRO

Gangliosides, sialic acid containing glycolipids, are normal components of the mammalian plasma membranes, especially abundant in the nervous system (Ledeen, 1978). Due to their chemical structure and asymmetrical localization in the outer leaflet of the plasma membrane, gangliosides have been suggested to play a prominent role in the transduction of membrane-mediated information (Brady and Fishman, 1978). Gangliosides interact with various factors such as bacteria toxins, glycoprotein hormones, serotonin, and other biologically active agents (Yamakawa and Nagai, 1978). Furthermore, gangliosides, in particular the monosialoganglioside GM_1 (nomenclature according to Svennerholm, 1963), presumably play a role in the regulation of neuronal development, including neurite outgrowth and cell–cell recognition (Hakomori, 1981). In fact, 1) GM_1 content and pattern undergo pronounced modifications during CNS development (Dreyfus et al., 1980); 2) expression of GM_1 parallels neurite outgrowth in the developing cerebellum (Willinger and Schachner, 1980); and 3) administration of anti-GM_1 antibodies to developing animals causes behavioral dysfunction with morphological abnormalities (Karpiak et al., 1982) and inhibition of neurite outgrowth from regenerating axons in vitro (Spirman et al., 1982) and in vivo (Sparrow et al., 1984).

A role of GM_1 in regulation of neuronal development and repair is also supported by experiments involving addition of exogenous gangliosides to tissue cultures in vitro. In fact, under appropriate conditions, GM_1 stimulates biochemical and morphological differentiation of both clonal (Leon et al., 1982; Ferrari et al., 1983; Byrne et al., 1983; Spoerri, 1983) and primary neuronal cell cultures (Roisen et al., 1981; Dal Toso et al., 1984; Doherty et al., 1985; Skaper and Varon, 1985). The GM_1 effects follow a stable insertion of the ganglioside into the plasma membrane (Facci et al., 1984), and have been interpreted as due to enhancement of neuronal cell responsiveness to neuronotrophic factors (Ferrari et al., 1983; Katoh-Semba et al., 1984; Doherty et al., 1985). In pheochromocytoma (PC_{12}) cells, the ganglioside effect on neurite outgrowth and regeneration requires the presence of NGF. Likely, the exogenous GM_1 modifies the supramolecular organization and the functional properties of the membrane in defined microdomains. In this context the insertion of GM_1 into membranes changes cell membrane characteristics, including enzyme activity and receptor function (Partington and Daly, 1979; Leon et al., 1981; Goldenring et al., 1985; Berry-Kravis and Dawson, 1985). However, the action of GM_1 on PC_{12} cells does not involve a direct interaction with the NGF receptor, suggesting a GM_1 effect in the transduction of the NGF signals necessary for the acquisition and/or maintenance of the differentiated state (Rybak et al., 1983).

Interestingly, exogenous GM_1 in vitro modulates cell responsiveness to trophic factors other than NGF. In particular, GM_1 affects platelet-derived growth factor (PDGF)-stimulated cell proliferation via effects on PDGF receptor and phosphorylation (Bremer et al., 1984). Moreover GM_1 enhances neuronal cell responsiveness to a non-NGF neuronotrophic factor extracted from the adult mammalian CNS (Leon et al., submitted for publication.)

4. EFFECT OF GM_1 ON CNS REPAIR

Based on the above observations, GM_1 ganglioside administration has been tested for its ability to affect CNS neuronal repair processes in adult mammals. Chronic systemic administration of GM_1 facilitates biochemical and functional recovery of specific brain areas following different experimental lesions (Agnati et al., 1983; Toffano et al., 1983; Jonsson et al., 1984; Karpiak, 1984; Oderfeld-Nowak et al., 1984; Pedata et al., 1984; Reivich et al., 1984; Sabel et al., 1984; Casamenti et al., 1985). The GM_1 effect occurs independently of the neuronal system lesioned (i.e., dopaminergic, cholinergic, or noradrenergic) and may partly be due to enhanced cell survival. In order to gain information on the cellular and molecular mechanisms underlyuing such effects in vivo, we have assessed, in different conditions, the action of GM_1 after injury of the nigrostriatal pathway.

TABLE 29-1. Dose-dependent effect of GM_1 treatment on striatal TH activity after partial unilateral hemitransection of the nigrostriatal system.

Treatment	V_{max} as % of control — Days after hemitransection: 8	14
Saline	45 ± 3.5	50 ± 3.0
GM_1 (5 mg/kg)	50 ± 3.0	65 ± 5.0*
GM_1 (30 mg/kg)	48 ± 3.0	85 ± 4.8**

The intraperitoneal treatment, either with saline or GM_1, started on the second day after surgery and ended 24 hours before sacrifice. Data represent V_{max} mean values of tyrosine hydroxylase (TH) expressed as % of the respective unlesioned side. All means are from 6 to 8 determinations. The V_{max} of TH determined in the unlesioned side of the saline-treated group (19.7 ± 1.01 nmol $CO_2 \cdot h^{-1} \cdot mg^{-1}$ protein) was not significantly different from that of the GM_1-treated group at any time. *$p < 0.05$ and **$p < 0.01$ versus the respective group treated with saline (Student's *t*-test).

In previous studies, we have investigated the effect of GM_1 administration on the recovery of the nigrostriatal dopaminergic parameters in adult rats after partial unilateral hemitransection (Toffano et al., 1983; Savoni et al., 1985). In the lesioned side of the control saline-treated animals, striatal tyrosine hydroxylase (TH) activity, striatal TH-immunopositive nerve terminals and homovanillic acid (HVA) content underwent a marked decrease with respect to the unlesioned contralateral side. Treatment with GM_1 induced in the lesioned striatum a significant increase of TH activity in a dose-dependent manner (table 29-1) already apparent after 2 weeks. This effect was associated with: 1) an increase in HVA content, 2) an increased number of TH-immunofluorescent terminals in the partially denervated striatum (Toffano et al., 1983), and 3) an increased number of TH-positive cells in the substantia nigra ipsilateral to the lesion. The wheat-germ agglutinin conjugated to horseradish peroxidase (WGA-HRP) technique indicated that all the TH-immunoreactive nigral neurons possessed an intact connectivity to the striatum (figure 29-1). The sensitivity to apomorphine-induced rotational behavior, as an index of striatal dopaminergic denervation, was also assessed (Ungerstedt, 1968). In the GM_1-treated rats, the number of apomorphine-induced turns was significantly reduced with respect to the saline-treated group (Toffano et al., 1984a). The above findings suggest that GM_1 favors the functional recovery of nigrostriatal transmission after partial deafferentation. Whether this is due to a GM_1 effect on the sprouting phenomenon is currently under investigation. Preliminary experiments suggest that this is a possibility (Sabel et al., 1985).

Interestingly, the above GM_1 effect seems to be dependent on the extent of the lesion, suggesting that perhaps an essential prerequisite for triggering the repair process is the survival of an adequate part of the ipsilateral neurons. The integrity of the contralateral nigrostriatal system is also a fundamental

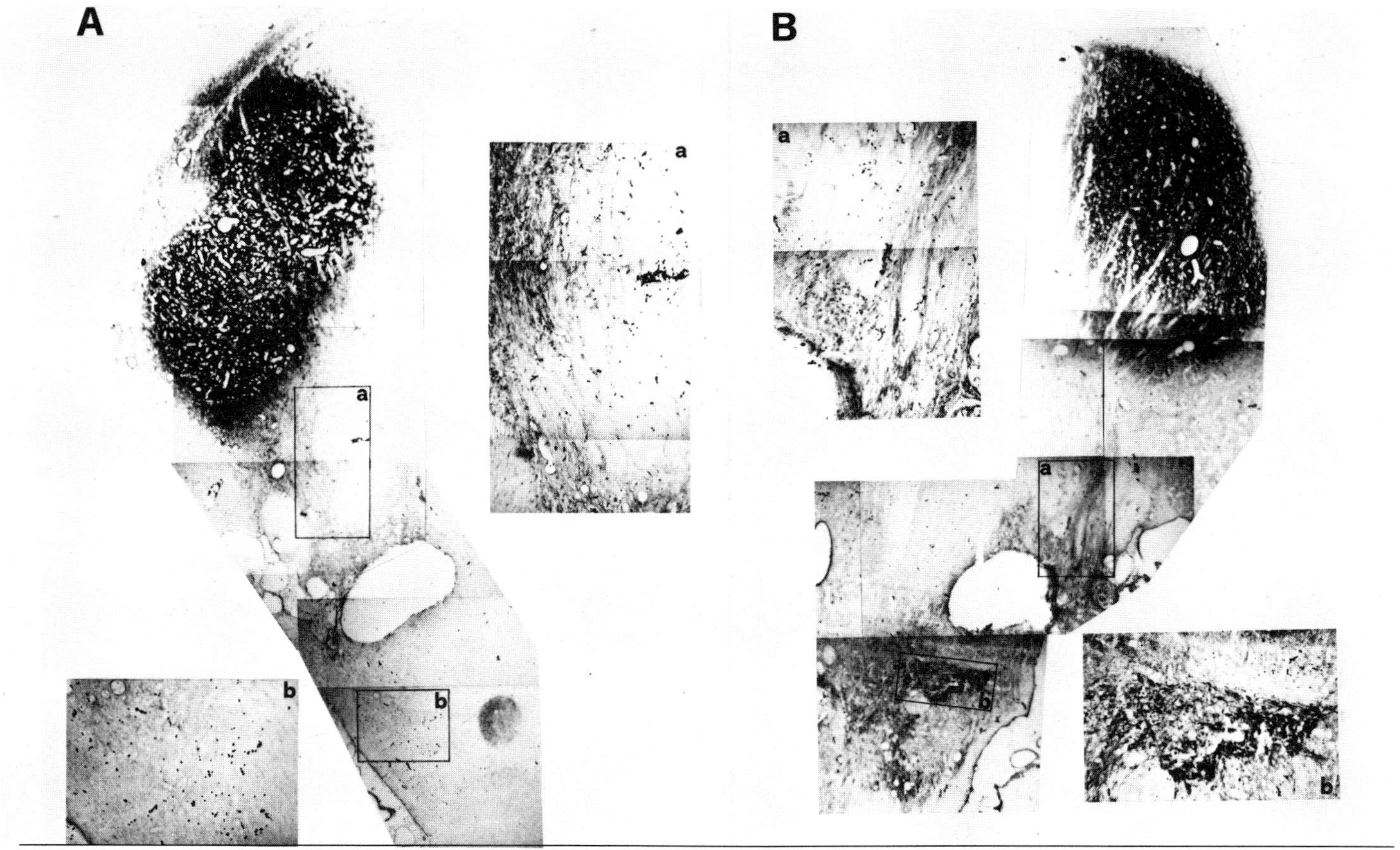

Figure 29-1. Horizontal sections of rat brain through striatum and substantia nigra processed for WGA-HRP histochemistry and tyrosine hydroxylase immunocytochemistry. The rats were hemitransected and started on saline or GM_1 treatment (30 mg/kg i.p.) 48 hours after surgery; 5% WGA-HRP (0.1 μl) was injected intrastriatally after 28 days, when the treatment was ended. Rats were perfused the day after with 4% paraformaldehyde containing 0.1% glutaraldehyde. (**A**) Lesioned side of a saline-treated animal. Note the few fibers and cell bodies in **a** and **b**, respectively. (**B**) Lesioned side of a GM_1-treated animal. Note the increased number of fibers (**a**) and nigral cell bodies (**b**) as compared to respective *insets* in A.

prerequisite for the action of the ganglioside (Toffano et al., 1984b). Furthermore the GM_1 effect occurs only in animals undergoing normal light–dark cycle (12 hours each), indicating that adequate sensory perception plays a relevant role in brain repair (Toffano et al., 1984a). GM_1 administration was ineffective when the nigrostriatal pathway was chemically damaged with 6-hydroxydopamine (6-OHDA) indicating that the effect of GM_1 depends also on the nature of the neuronal injury (Toffano et al., 1984b).

Noteworthy, after a mechanical injury of the nigrostriatal system, NTF induction was always observed not only in the lesioned but also in the contralateral unlesioned striatum and substantia nigra of untreated animals. In contrast, no induction of neuronotrophic activity occurred in the substantia nigra of the 6-OHDA-lesioned animals (Consolazione et al., 1985). This suggests that adequate NTF production in appropriate brain areas is needed for functional repair. In accord with this view are also the data on the polyamine levels, such as putrescine, in partial unilateral hemistransected rats (Agnati et al., 1985). Induction of polyamine biosynthesis has been suggested to be essential for the survival of lesioned neurons (Gilad and Gilad, 1983) presumably due to a permissive role of polyamines on the action of neuronotrophic factors. The intraventricular administration of NGF induces a marked increase of the activity of polyamine biosynthetic enzyme, ornithine decarboxylase (ODC) (Lewis et al., 1978). GM_1 administration enhances the putrescine content in rats following partial unilateral hemitransection while administration of an ODC inhibitor prevents the GM_1 effect (Agnati et al., 1985).

5. CONCLUSION AND FUTURE PERSPECTIVES

Although the cellular and molecular mechanisms underlying the repair processes of the adult mammalian brain are still largely obscure, it is nonetheless evident that CNS repair is dependent on a considerable number of variables. Among them, the induction of NTFs seems to be particularly relevant. However, although NTF activity is indeed induced following certain types of CNS injury in adult mammals, this induction does not per se seem to be sufficient in sustaining the brain repair process. This may be due either to a still inadequate titer of NTFs or to "environmental" inadequacy for neuronal cell reception, translation, and/or execution of the NTF signal. The results reported in this study concerning chronically administered GM_1 effects suggest that it is possible to modulate the activity of the lesion-induced NTFs, thereby enhancing brain repair. As such, a more detailed understanding of the mechanisms of action of GM_1, as well as of the NTFs, on lesion–induced neuronal plasticity may yield pertinent information with regard to the possibility of optimizing brain repair processes in adult mammals.

However, it must be recalled that, independently of the NTFs, another important variable is the entity of the secondary damage, e.g., edema, following trauma or stroke to the brain. Very recently, it has been reported that GM_1 administration decreases brain edema following trauma and anoxic conditions (Borzeix, 1984; Karpiak and Mahadik, 1984). In addition, GM_1 normalizes electrolytic unbalances and membrane-associated enzyme activities altered by the injury (Borzeix, 1984; Karpiak et al., 1986). Hence GM_1 facilitation of CNS recovery may occur not only via modulation of neuronotrophic interactions, but perhaps also through reduction of secondary phenomena that together worsen the primary neuronal damage. Which of these two possibilities, i.e., amplification of lesion-induced NTF activity and reduction of secondary neuronal damage, plays a major role in the GM_1 enhancement of neuronal repair processes in vivo is currently under investigation.

REFERENCES

Agnati, L.F., Fuxe, K., Calzà, L., Benfenati, F., Cavicchioli, L., Toffano, G., and Goldstein, M. (1983) Gangliosides increase the survival of lesioned nigral dopamine neurons and favour the recovery of dopaminergic synaptic function in striatum of rats by collateral sprouting. Acta Physiol. Scand. 119:347–363.

Agnati, L.F., Fuxe, K., Zini, I., Davalli, P., Corti, A., Calzà, L., Toffano, G., Zoli, M., Piccinini, G., and Goldstein, M. (1985) Effects of lesions and ganglioside GM_1 treatment on striatal polyamine levels and nigral DA neurons: a role of putrescine in the neurotrophic activity of granliosides. Acta Physiol. Scand. 124:499–506.

Appel, S.H. (1981) A unifying hypothesis for the cause of amyotrophic lateral sclerosis, parkinsonism and Alzheimer's disease. Ann. Neurol. 10:499–505.

Barde, Y.A., Edgar D., and Thoenen, H. (1982) Purification of a new neuronotrophic factor from mammalian brain. EMBO J. 1:549–553.

Berry-Kravis, E., and Dawson, G. (1985) Possible role of gangliosides in regulating an adenylate cyclase-linked 5-hydroxytryptamine (5-HT_1) receptor. J. Neurochem. 45:1739–1747.

Björklund, A., and Stenevi, U. (1979) Regeneration of monoaminergic and cholinergic neurons in the mammalian central nervous system. Physiol. Rev. 59:62–100.

Borzeix, M.G. (1984) Drugs effect in an experimental stroke resulting from a transient cerebral oligaemia. Drugs Dis. 1:44–53.

Brady, R.O., and Fishman, P.H. (1979) Gangliosides as biotransducers of membrane-mediated information: cholera toxin as a model. In: Silverstein, S.C. (ed.) Transport in macromolecules in cellular systems, vol. 2. Berlin: Dahlem Konferenzen, pp. 69–84.

Bremer, E.G., Hakomori, S., Bowen-Pope, D.F., Raines, E., and Ross, R. (1984) Ganglioside-mediated modulation of cell growth, growth factor binding, and receptor phosphorylation. J. Biol. Chem. 259:6818–6825.

Byrne, M.C., Ledeen, R.W., Roisen, F.J., Yorke, G., and Sclafani, J.R. (1983) Ganglioside-induced neuritogenesis: verification that gangliosides are the active agents and comparison of molecular species. J. Neurochem. 41:1214–1222.

Casamenti, F., Bracco, L., Bartolini, L., and Pepeu, G. (1985) Effects of ganglioside treatment in rats with a lesion of the cholinergic forebrain nuclei. Brain Res. 338:45–52.

Consolazione, A., Aldinio, C., Presti, D., Dal Toso, R., Leon, A., and Toffano, G. (1985) Correlation between long-term GM_1 ganglioside effects on CNS repair and neurotrophic activity. Soc. Neurosci. Abstr. 279:13.

Cotman, C.W., Nieto-Sampedro, M., and Harris, E.W. (1981) Synapse replacement in the nervous system of adult vertebrates. Physiol. Rev. 61:684–784.

Crutcher, K.A., and Collins, F. (1982) In vitro evidence for two distinct hippocampal growth factors: basis of neuronal plasticity? Science 217:67–68.

Dal Toso, R., Leon, A., Presti, D., Benvegnù, D., Facci, L., Giorgi, O., Nowak, L., and Toffano, G. (1984) Effect of monosialoganglioside on expression of dopaminergic characteristics of embryonic mesencephalic cells in cultures. Inserm 126:297–308.

Dal Toso, R., Giorgi, O., Presti, D., Benvegnù, D., Favaron, M., Soranzo, C., Leon, A., and Toffano, G. (1986) Neuronotrophic activity for CNS neurons extracted from bovine caudate nucleus: partial purification and characterization. In: Molecular aspects of neurobiology. Giovanni Lorenzini Foundation (in press).

Doherty, P., Dickson, J.G., Flanigan, T.P., and Walsh, F.S. (1985) Ganglioside GM_1 does not initiate but enhances neurite regeneration of nerve growth factor-dependent sensory neurones. J. Neurochem. 44:1259–1265.

Dreyfus, H., Luis, J.C., Harth, S., and Mandel, P. (1980) Gangliosides in cultured neurons. Neuroscience 5:1647–1655.

Ebendal, T., Olson, L., Sieger, A., and Hedlund, K.O. (1980) Nerve growth factors in the rat iris. Nature 286:25–28.

Facci, L., Leon, A., Toffano, G., Sonnino, S., Ghidoni, R., and Tettamenti, G. (1984) Promotion of neuritogenesis in mouse neuroblastoma cells by exogenous gangliosides: relationship between the effect and the cell association of ganglioside GM_1. J. Neurochem. 42:299–305.

Ferrari, G., Fabris, M., and Gorio, A. (1983) Gangliosides enhance neurite outgrowth in PC12 cells. Dev. Brain Res. 8:215–221.

Gilad, G.M., and Gilad, V.H. (1983) Polyamine biosynthesis is required for survival of sympathetic neurons after axonal injury. Brain Res. 273:191–194.

Goldenring, J.R., Otis, L.C., Yu, R.K., and De Lorenzo, R.J. (1985) Calcium/ganglioside-dependent protein kinase activity in rat brain membrane. J. Neurochem. 44:1229–1234.

Hakomori, S. (1981) Glycosphingolipids in cellular interaction differentiation and oncogenesis. Annu. Rev. Biochem. 50:733–764.

Hefti, F. (1985) Nerve growth factor (NGF) promotes survival of septal cholinergic neurons after injury. Soc. Neurosci. Abstr. 197:7.

Hefti, F., Hartikka, J., Eckenstein, F., Gnahm, H., Heumann, R., and Schwab, M. (1985) Nerve growth factor (NGF) increases choline acetyltransferase but not survival of fiber outgrowth of cultured septal cholinergic neurons. Neuroscience 14:55–68.

Jonsson, G., Gorio, A., Hallman, H., Janigro, D., Kojima, H., Luthman, J., and Zanoni, R. (1984) Effects of GM_1 ganglioside on developing and mature serotonin and noradrenaline neurons lesioned by selective neurotoxins. J. Neurosci. Res. 12:459–475.

Karpiak, S.E. (1984) Exogenous gangliosides enhance recovery from CNS injury. In: Ledeen, R., Rapport, M., Yu, R., and Suzuki, K. (eds.) Ganglioside structure, function and biomedical potential. New York: Plenum, pp. 489–497.

Karpiak, S.E., and Mahadik, S.P. (1984) Reduction of cerebral edema with GM_1 ganglioside. J. Neurosci. Res. 12:485–492.

Karpiak, S.E., Huang, Y.L., and Rapport, M.M. (1982) Immunological model of epilepsy. J. Neuroimmunol. 3:15–21.

Karpiak, S.E., Li, Y.S., Aceto, P., and Mahadik, S.P. (1986) Acute effects of gangliosides on CNS injury. In: Tettamanti, G., et al., (eds) Neuronal plasticity and ganglioside. Liviana (in press).

Katoh-Semba, R., Skaper, S.D., and Varon, S. (1984) Interaction of GM_1 ganglioside with PC12 pheochromocytoma cells: serum and NGF-dependent effects on neuritic growth (and proliferation). J. Neurosci. Res. 12:299–310.

Kligman, D., and Marshak, D.R. (1985) Purification and characterization of a neurite extension factor from bovine brain. Proc. Natl. Acad. Sci. USA 82:7136–7139.

Ledeen, R.W. (1978) Ganglioside structure and distribution: are they localized at the nerve endings: J. Supramol. Struct. 8:1–17.

Leon, A., Facci, L., Toffano, G., Sonnino, S., and Tettamanti, G. (1981) Activation of (Na^+, K^+) ATPase by nanomolar concentrations of GM_1 ganglioside. J. Neurochem. 37:350–357.

Leon, A., Facci, L., Benvegnù, D., and Toffano, G. (1982) Morphological and biochemical effects of gangliosides in neuroblastoma cells. Dev. Neurosci. 5:108–114.

Levi-Montalcini, R. (1966) The nerve growth factor: its mode of action on sensory and sympathetic nerve cells. Harvey Lect. 60:217–259.

Lewis, M.E., Lakshmanan, J., Nagaiah, K., McDonnell, P.C., and Guroff, G. (1978) Nerve growth factor increases activity of ornithine decarboxylase in rat brain. Proc. Natl. Acad. Sci. USA 75:1021–1023.

Lindsay, R.M., and Peters, C. (1984) Spinal cord contains neurotrophic activity for spinal nerve sensory neurons: late developmental appearance of a survival factor distinct from nerve growth factor. Neuroscience 12:45–51.

Müller, H.W., Beckh, S., and Seifert, W. (1984) Neuronotrophic factor for central neurons. Proc. Natl. Acad. Sci USA 81:1248–1252.

Nieto-Sampedro, M., Lewis, E.R., Cotman, C.W., Manthorpe, M., Skaper, S.D., Barbin, G., Longo, F.M., and Varon, S. (1982) Brain injury causes a time dependent increase in neuronotrophic activity at the lesion site. Science 217:860–861.

Nygren, L.G., and Olson, L. (1977) Intracisternal neurotoxins and monoamine neurons innervating the spinal cord: acute and chronic effects on cells and axon counts and nerve terminal densities. Histochemistry 52:281–306.

Oderfeld-Nowak, B., Skup, M., Ulas, J., Jezierska, M., Gradkowska, M., and Zaremba, M., (1984) Effect of GM_1 ganglioside treatment on postlesion responses of cholinergic enzymes in rat hippocampus after various partial deafferentations. J. Neurosci. Res. 12:409–420.

Partington, C.R., and Daly, J.W. (1979) Effect of gangliosides on adenylate cyclase activity in rat cerebral cortical membranes. Mol. Pharmacol. 15:484–491.

Pedata, F., Giovannelli, L., and Pepeu, G. (1984) GM_1 ganglioside facilitates the recovery of high affinity choline uptake in the cerebral cortex of rats with a lesion of the nucleus basalis magnocellularis. J. Neurosci. Res. 12:421–427.

Reis, D.J., Gilad, G., Pickel, V.M., and Joh, T.H. (1978) Reversible changes in the activities and amounts of tyrosine hydroxylase in dopamine neurons of the substantia nigra in response to axonal injury as studied by immunochemical methods. Brain Res. 144:325–342.

Reivich, M., Dora, E., Tanaka, K., Urbanics, R., and Greenberg, J. (1984) The action of GM_1 ganglioside on the hemodynamic and metabolic effects of ischemia. Clin. Neuropharmacol. [Suppl. 1] 7:597.

Roisen, F.J., Bartfeld, H., Magelle, R., and Yorke, G. (1981) Ganglioside stimulation of axonal sprouting in vitro. Science 214:577–578.

Rybak, S., Ginzburg, I., and Yavin, E. (1983) Gangliosides stimulate neurite outgrowth and induce tubulin mRNA accumulation in neural cells. Biochem. Biophys. Res. Commun. 116:980–994.

Sabel, B.A., Slavin, M.D., and Stein, D.G. (1984) GM_1 ganglioside treatment facilitates behavioral recovery from bilateral brain damage. Science 225:340–342.

Sabel, B.A., Dunbar, G.L., Fass, B., and Stein, D.G. (1985) Gangliosides, neuroplasticity and behavioral recovery after brain damage. In: Will, et al. (eds.) Brain plasticity, learning and memory. New York: Plenum, pp. 481–493.

Savoini, G. Fuxe, K., Agnati, L.F., Calza, L., Moroni, F., Lombardi, G., Goldstein, M., and Toffano, G. (1985) Effect of GM_1 ganglioside on the recovery of dopaminergic nigrostriatal neurons after lesion. In: Bignami, A., Bloom, F.E., Bolis, C.L., and Adeloye, A. (eds.) Central nervous system platicity and repair. New York: Raven, pp. 75–84.

Schwab, M.E., Otten, U., Agid, Y., and Thoenen, H. (1979) Nerve growth factor (NGF) in the rat CNS: absence of specific retrograde axonal transport and tyrosine hydroxylase induction in locus coeruleus and substantia nigra. Brain Res. 168:473–483.

Skaper, S.D., and Varon, S. (1985) Ganglioside GM_1 overcomes serum inhibition of neuritic outgrowth. Int. J. Dev. Neurosci. 3:187–198.

Sparrow, J.R., McGuinnes, C., Schwartz, M., and Grafstein, B. (1984) Antibodies to gangliosides inhibit goldfish optic nerve regeneration in vivo. J. Neurosci. Res. 12:233–243.

Spirman, N., Sela, B.A., and Schwartz, M. (1982) Antiganglioside antibodies inhibit neuritic outgrowth from regenerating goldfish retinal explants. J. Neurochem. 39:874–877.

Spoerri, P.E. (1983) Effects of gangliosides on the in vitro development of neuroblastoma cells: an ultrastructural study. Int. J. Dev. Neurosci. 1:383–391.

Svennerholm, L. (1963) Chromatographic separation of human brain gangliosides. J. Neurochem. 10:613–623.

Thoenen, H., and Barde, Y.A. (1980) Physiology of nerve growth factor. Physiol. Rev. 60:1284–1335.

Toffano, G., Savoini, G., Moroni, F., Lombardi, G., Calza, L., and Agnati, L.F. (1983) GM_1 ganglioside stimulates the regeneration of dopaminergic neurons in the central nervous system. Brain Res. 261:163–166.

Toffano, G., Savoini, G., Aporti, F., Calzolari, S., Consolazione, A., Maura, G., Marchi, M.,

Raiteri, M., and Agnati, L.F. (1984a) The functional recovery of damaged brain: the effect of GM_1 monosialoganglioside. J. Neurosci. Res. 12:397–408.

Toffano, G., Agnati, L.F., Fuxe, K., Aldinio, C., Consolazione, A., Valenti, G., and Savoni, G. (1984b) Effect of GM_1 ganglioside treatment on the recovery of dopaminergic nigro-striatal neurones after different types of lesion. Acta Physiol. Scand. 122:313–321.

Toffano, G., Agnati, L.F., and Fuxe, K. (1986) On the effect of ganglioside GM_1 on neuronal plasticity. Int. J. Develop. Neurosci. (in press).

Ungerstedt, U. (1968) 6-Hydroxydopamine-induced degeneration of central monoamine neurons. Eur. J. Pharmacol. 5:107–110.

Varon, S., Adler, R., Manthorpe, M., and Skaper, S. (1983) Culture strategies for trophic and other factors directed to nerve cells. In: Pfeiffer, S.E. (ed.) Neuroscience approached through cell culture, vol. 2 Boca Raton, FL: CRC Press, pp. 53–67.

Whittemore, S.R., Nieto-Sampedro, M., Needels, D.L., and Cotman, C.W. (1985) Neuronotrophic factors for mammalian brain neurons: injury induction in neonatal, adult and aged rat brain. Dev. Brain Res. 20:169–178.

Willinger, M., and Schachner, M. (1980) GM_1 ganglioside as a marker for neuronal differentiation in mouse cerebellum. Dev. Biol. 74:101–107.

Yamakawa, Y., and Nagai, Y. (1978) Glycolipids at the cell surface and their biological functions. Trends Biochem. Sci. 3:128–131.

30. AGING-RELATED PATHOLOGY IN HUMAN NEUROBLASTOMA AND TERATOCARCINOMA CELL LINES

GREGORY M. COLE AND PAOLA S. TIMIRAS

1. INTRODUCTION: IN VITRO MODEL SYSTEMS FOR AGING AND AGE-RELATED PATHOLOGY

Humans and other primates take decades to age and even the 2–3 years required in a short-lived mammal like the mouse or rat is an impractically long period for real experimentation and innovation. Thus, one major obstacle to a better understanding of the complicated and lengthy processes governing mammalian aging and age-related pathology is the paucity of useful model systems wherein events can be accelerated and hopefully simplified. For such analysis to be possible, markers reflecting adequate indices of some segment of the aging or pathological process must be found. For example, cancer is a major age-related pathology and the process of carcinogenesis may be studied both in vivo and in vitro because transformed cells exhibit a set of known properties that readily allow them to be identified, assayed, and serendipitously selected for. Thus, established in vitro, models for carcinogenesis (Kennedy et al., 1980) and the related but more general process of mutagenesis (Ames et al., 1973) have been instrumental in the explosive growth in our understanding of oncogenes, their activation, and the resultant cancer.

Perhaps, because it is less amenable to such in vitro analysis, expansion of our knowledge about atherosclerosis and atherosclerotic lesions has made important, but less spectacular, gains at the molecular level. Nevertheless, despite the complexity of cellular interactions, model systems have demon-

Vernadakis, A. (ed.), *Model Systems of Development and Aging of the Nervous System*.

strated that chronic abnormal cholesterol metabolism is central to the development of atherosclerotic pathology (Bierman and Ross, 1977).

Aging is a more general phenomenon than the pathologies and we lack sufficiently specific markers to characterize the induction of aging processes in a model system; therefore the only really convincing work in mammalian experimental gerontology comes from life extension experiments which, if successful in rodents, may be of several years duration (Segall and Timiras, 1976). Because aging is complex and different tissues may age or develop age-related pathology from multiple causes, induction of real aging in a whole mammal may actually prove to be impossible. The limited "segmental" progeroid pathology observed in all of the known human syndromes manifesting some signs of accelerated aging (Martin, 1978) reinforces this point. Hence, the induction of a limited set of markers for aging or an age-related pathology has become our more attainable goal. Whether the in vitro "senescence" model of fibroblast replicative limit is a valid represention of essential aspects of mammalian aging (Mehlhorn and Cole, 1985), it will probably leave open the problem of intrinsic senescence in postmitotic cells, in particular in neurons that have reached their replicative limit at birth. It should, however, be possible to model certain aspects of neuronal senescence in vitro, provided adequate markers can be found. If a process of neuronal aging involves reactions to specific types of damage, e.g., due to free radicals (Mehlhorn and Cole, 1985) or hormones (Sapolsky et al., 1985) or other toxins (Langston et al., 1983; Moroni et al., 1984), segmental neuronal aging should be possible to induce in vitro.

At present, the only index of general neuronal aging that we have is lipofuscin accumulation, inductable in cultured neurons and neuroblastoma (Nandy et al., 1978), but lipofuscin accumulation is, unfortunately, a poor correlate of functional neuronal aging and a nonspecific change and, thus, clearly insufficient to identify a cell as aged. We have, therefore, focused on neuronal pathologies where more specific markers are currently available, but we believe basic aging will likewise be amenable to similar analysis when suitable markers have been demonstrated in vivo.

2. AGE-RELATED NEURONAL PATHOLOGIES

Currently the best markers for significant age-related neuronal pathologies are the Lewy bodies of idiopathic Parkinson's disease (Goldman et al., 1983) and the amyloid lesions of senile dementia of the Alzheimer type (SDAT) (Terry and Katzman, 1983), consisting of neuritic plaque core amyloid (Masters et al., 1985), vascular amyloid (Glenner and Wong, 1984), and neurofibrillary tangles of paired helical filaments (PHF) (Mann, 1985). The incidence of all of these lesions increases during normal aging, but large numbers of plaques and tangles in the cortex are found only in SDAT and large numbers of Lewy bodies in the substantia nigra and other monoaminergic nuclei (Ohama and Ikuta, 1976) are typical of idiopathic Parkinson's disease.

Huntington's disease is an age-related neuronal pathology where significant genetic information is available, but it lacks really good markers and appears to affect such restricted populations of neurons that it may be more difficult to approach in vivo.

Because there are no animal models where SDAT plaques and tangles (Ball et al., 1983) or Lewy bodies are found, we decided to use human neuronal cells despite anticipation of considerable experimental difficulties. We solved the problem by using human tumor cell lines that we have shown can be induced with retinoic acid to differentiate and produce reasonable facsimiles of normal neurons (Cole et al., 1985a). By reasonable facsimiles, we mean only that they are probably sufficient for our purpose: they contain the proteins we are interested in, distributed approximately normally.

Some comments on the use of tumor cells may help assuage the skepticism of critics. The notion that tumor-derived postmitotic cells retain the properties of continuously dividing, "immortal" cancer cells is misguided. The only immortal thing about a cancer cell is its continued replicative capacity and our differentiated neuronal cells do not divide (Ino et al., 1986). Cell transformation and the properties of cancer cells are caused by the activation of specific oncogenes, and retinoic acid-induced differentiation is known to shut down this oncogene expression (Thiele et al., 1985). Thus, if the tumor cells are close to genetically normal and oncogene expression is shut down, then, to begin with, approximately normal cells could be produced by this means. (In the case of teratocarcinoma cells the possibility of total normalization has been repeatedly proven in chimeric mice where much or all of the mouse can be produced from teratocarcinoma cell lines.) However, we do not require or expect our differentiated tumor cells to become completely normal, only that they be capable of expressing markers for age-related pathology. In fact, tumor-derived neurons have been found to generate lipofuscin (Nandy et al., 1978) and SDAT-like Hirano bodies and Pick's bodies (Powers et al., 1976; Bender and Ghatak, 1978). Tumor-derived neurons may be abnormal, but we know of no examples of any abnormalities or mutations that might have arisen by chance in any of our cell lines and that can prevent neuronal aging or the development of age-related pathology. On the contrary, we know trisomy 21 results in the accelerated development of more age-related pathologies than any other syndrome (Martin, 1978) and, in particular, accelerated development of Alzheimer's pathology (Burger and Vogel, 1973). Hence, there is a genetic defect, trisomy 21, that we can manufacture in a cell line, and that leads to accelerated development of SDAT markers in vivo. Should other genetic defects affecting age-related pathology be found, they could also be introduced into a human neurogenic cell line. We are confident that the power and usefulness of this ability for genetic manipulation far outweigh any speculative disadvantages concerning the possibility that abnormalities in most or all tumor-derived neurons might somehow prevent the induction of age-related pathology.

3. CAN AGE-RELATED NEURONAL PATHOLOGY BE INDUCED?

If aging is the result of general cumulative damage of many types, and more extensive specific damage generates specific pathologies as is known for atherosclerosis and cancer, then it should be possible to induce age-related neuronal pathologies. Such damage might be produced by endogeneously generated toxins in normal aging and by environmental agents and/or abnormal gene products in pathologies. Here we summarize some of the evidence that suggests neurofibrillary changes. Lewy bodies, and Pick's bodies are reactions to damage.

3.1 Neurofibrillary changes in reaction to injury

Alzheimer's neurofibrillary changes are not unique to SDAT, but have been reported in over 20 different conditions (Wisniewski et al., 1979) including individuals as young as 15 years of age and healthy elderly individuals (Mann and Yates, 1983). Paired helical filament (PHF) formation has been directly associated with repeated trauma in prizefighters (Wisniewski et al., 1976), in single incidents of head injury (Mortimer et al., 1985, Rudelli et al., 1982), adjacent a sclerosing angioma (Liss et al., 1979), and in the ipsilateral substantia nigra after basal ganglia infarct (Forno, 1983). Significant prior evidence of head injury was markedly elevated in a retrospective study of SDAT patients (Heyman et al., 1984). Collectively, these findings are consistent with the view that PHF are formed in response to damage. They are also in agreement with the evidence that they can be formed in culture in human fetal neurons exposed to excitotoxic amino acids (De Boni and McLachlan, 1985).

3.2 Lewy body formation in response to damage

Lewy bodies, once thought to be specific for idiopathic Parkinson's disease, also occur in normal aging (Forno, 1982) and in unrelated conditions such as Shy-Drager syndrome (Sung et al., 1979), ataxia telangiectasia (x-ray sensitivity) (Agamanolis and Greenstein, 1979), and Hallervorden-Spatz syndrome (Park et al., 1975). Although little is known about the pathogenesis of these lesions, there is a growing body of evidence linking oxidative free radical damage to normal aging (Mehlhorn and Cole, 1985), Parkinson's disease (Mann and Yates, 1983; Perry et al., 1982), Hallervorden-Spatz syndrome (Park et al., 1975), and ataxia telangiectasia (Bridges and Harnden, 1981). The typical distribution of Lewy bodies in monoaminergic neurons (Ohama and Ikuta, 1976), including sympathetic ganglia (Forno and Norville, 1976) in Parkinson's disease, makes it plausible that appropriately damaged, differentiated human catecholaminergic cells might be able to express this lesion. The induction of Parkinson-like syndrome with 1-methyl-4-phenylpyridine (MPTP) made this agent a major candidate for testing.

Pick's bodies are also considered to be a reactive response of neurons

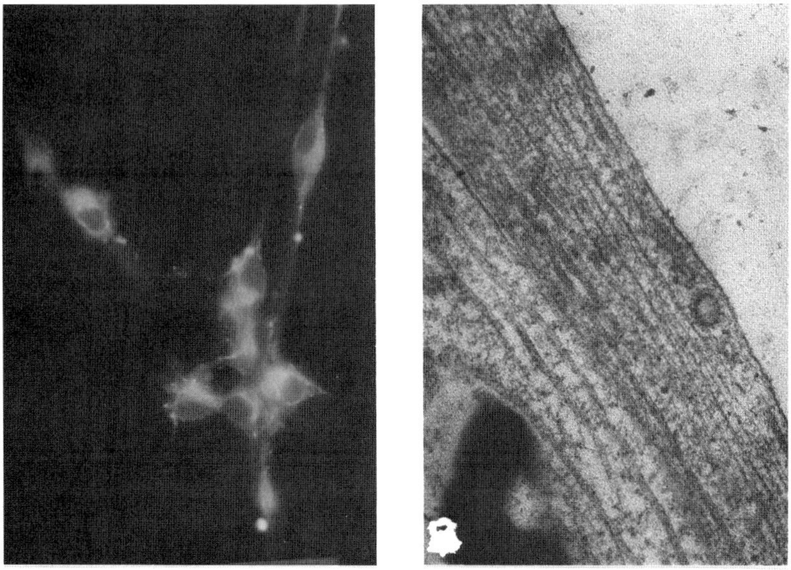

A B

Figure 30-1. (**A**) Indirect neurofilament fluorescence of a cluster of SY5Y human neuroblastoma cells differentiated in 10^{-5} *M* retinoic acid for 3 weeks. (**B**) Part of a neuritic process from similarly differentiated LAN-5 human neuroblastoma cells showing the presence of domains of 10-nm intermediate filaments and adjacent microtubules. Both differentiated cell lines stain with polyclonal antisera to the 68 k neurofilament protein obtained from Dr. Doris Dahl and with monoclonal antibodies to the 150 k and 200 k neurofilament peptides obtained from Dr. Brian Anderton.

secondary to damage caused by an unknown agent (Wisniewski et al., 1972) and Pick's bodies contain straight 15-nm filaments that cross-react with antisera to PHF (Rasool and Selkoe, 1984).

Thus, there is considerable evidence to suggest age-related neuropathology can be induced in vivo.

4. MODELING NEUROPATHOLOGY IN DIFFERENTIATED HUMAN CELL LINES

Our initial efforts to produce age-related human neuropathology focused on nerve growth factor (NGF) and retinoic acid-differentiated human neuroblastoma cells (Cole et al., 1985a), which contain all three neurofilament proteins in reasonably well-organized processes in differentiated cells (Cole et al., 1985b) (figure 30-1A and B). As expected, we found lipofuscin was readily induced (figure 30-2A and B) by different agents as were neurofibrillary tangles. Thus, colchicine-induced tangles (figure 2C) as did aluminum, while zinc (Cole et al., 1985b), oubain, and the calcium ionophore, A23187, caused the neurofilaments to collapse into a ball-like distribution (figure 2D).

However, despite repeated attempts with many different toxins, including

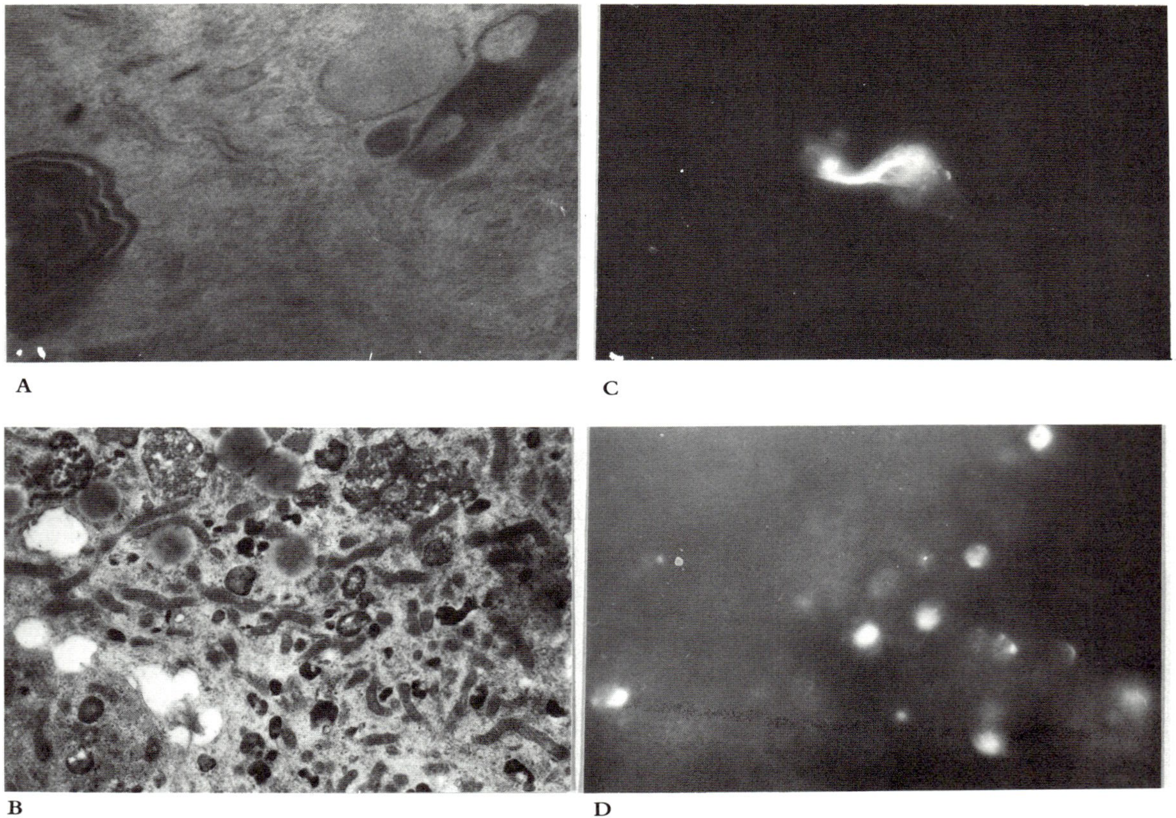

Figure 30-2. (**A**) Lamellar bodies and lipofuscin granules with classic morphology were found in SY5Y cells pulsed with 5 μg/ml mitomycin C for 30 minutes and maintained in 1 μg/ml 7S mouse nerve growth factor for 2 months. (**B**) The cytoplasm of retinoic acid-differentiated LAN-5 human neuroblastoma cells treated with 50 μg/ml of the cysteine protease inhibitor leupeptin for 1 week showed large numbers of dense bodies and lipofuscin-like structures (as well as increased autofluorescence). Similar protein shows tangle-like structures in retinoic acid-differentiated SY5Y neuroblastoma cells treated with 0.05 μg/ml colchicine for 1 week. More commonly, cells show the neurofilaments collapsed into a tight bundle around the nucleus. (**D**) Indirect neurofilament immunofluorescence depicting the collapsed cytoplasmic balls seen in retinoic acid-differentiated LAN-5 cells treated with 10^{-5} *M* A23187 calcium ionophore for 1 week. Typically, these cells or retinoic acid-differentiated SY5Y or

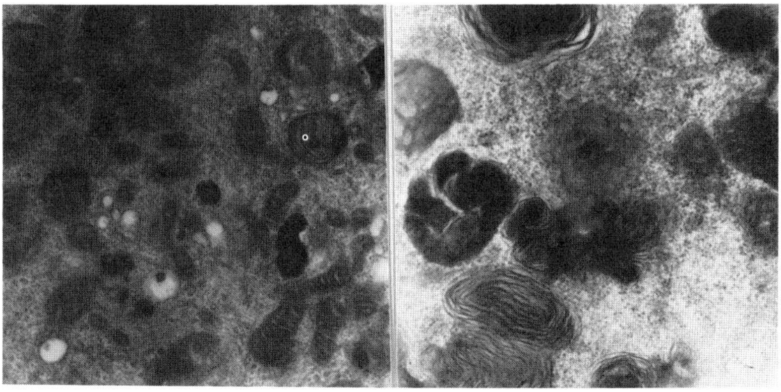

A B

Figure 30-3. Retinoic acid-differentiated SY5Y cells treated with 100 μg/ml N-MPTP showed abnormal degenerating mitochondria (**A**) consistent with the mitochondrial role in activating the drug and membranocytoplasmic and/or zebra bodies (**B**), but no Lewy-type inclusions were found.

colchicine, aluminum, leupeptin, quinolinic acid, calcium ionophore, iron, manganese, menadione, dicumarol, oubain, thallium, zinc, and doxorubicin, we found no staining with polyclonal antisera that labels SDAT tangles of PHF and plaque amyloid (generously provided by Dr. D.J. Selkoe) (Ihara et al., 1983; Selkoe and Abraham, 1985). While disappointing, these results were not surprising because PHF and Alzheimer's amyloid have never been demonstrated in the peripheral nervous system (PNS) and available human neuroblastoma cell lines are all PNS derived.

Our attempts to produce Lewy bodies in differentiated neuroblastoma with MPTP resulted in laminar and zebra bodies at high doses (figure 30-3), but no Lewy bodies and, in fact, no low-dose-specific toxicity was observed. We now suspect this may be due to the required activation of N-MPTP to 1-methyl-4-phenylpyridine (MPP+) (Bradbury et al., 1986) by glial cell monoamine oxidase B because neuroblastomas contain only monoamine oxidase A (Ino et al., 1986). However, no Lewy bodies have been found in humans developing Parkinson's syndromes after N-MPTP or in primate models (Dr. L. Forno, personal communication) so Lewy body induction may not be possible with MPP+.

5. TERATOCARCINOMA-DERIVED HUMAN NEURONS

To obtain a human cell line capable of generating human central nervous system (CNS) neurons to use as targets, we employed a human teratocarcinoma cell line, NT2/D1 (the gift of Dr. P. Andrews), which differentiates to make neurons in response to retinoic acid (Andrews, 1984). We confirmed the NT2/D1-derived neurons express all three neurofilament proteins (figure

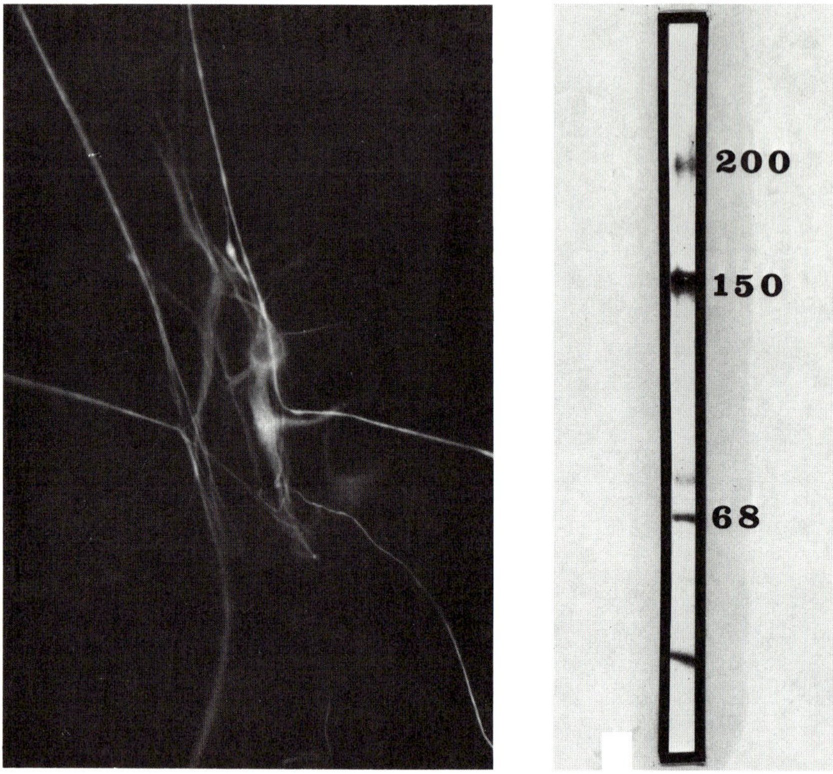

A B

Figure 30-4. (**A**) Retinoic acid-differentiated NT2/D1 human teratocarcinoma cells labeled with Anderton's BF10 anti-155k neurofilament antibody by indirect immunofluorescence show intense process fluorescence. (**B**) Western blot of Triton-insoluble cytoskeletal material developed with Dahl's polyclonal antisera to neurofilament proteins shows bands for all three neurofilament proteins as well as several lower molecular weight bands that may represent degradation products.

30-4A and B) and generate cells that possess a surface marker, A4 (Miller et al., 1984), present only in the CNS (figure 30-5) (antisera to A4 were the gift of Dr. J. Cohen). The differentiated NT2/D1 cells had ultrastructurally well-ordered processes (figure 30-6) and some possessed cholinergic surface markers (Kushner et al., 1985) as well as choline acetyltransferase (CAT) activity and immunoreactive CAT (data not shown). In the presence of serum, our NT2/D1 cultures differentiated with retinoic acid did not make cells bearing the O4 oligodendrocyte marker (Sommer and Schachner, 1981) (detectable by an antiserum generously provided by Dr. M. Schachner) and, in basic agreement with Andrews' (1984) results, only very rarely made glial fibrillary acidic protein (GFAP)-positive cells (identified by anti-GFAP

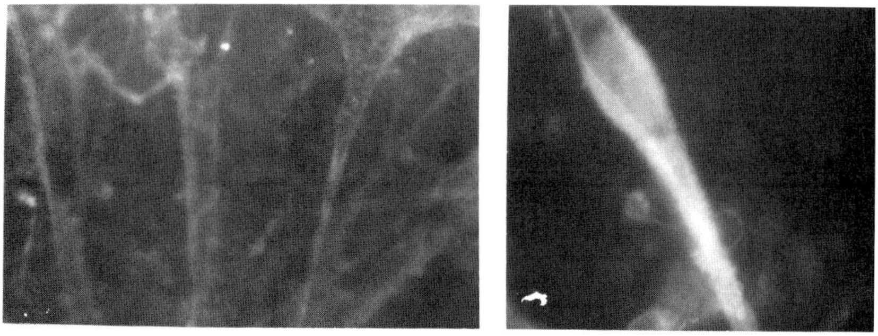

A B

Figure 30-5. (**A** and **B**) Apparent indirect immunofluorescence of retinoic acid-differentiated NT2/D1 human teratocarcinoma cells with 1:100 A4 monoclonal antibody, a CNS marker. The cells tend to lose the A4 marker in the continued presence of serum.

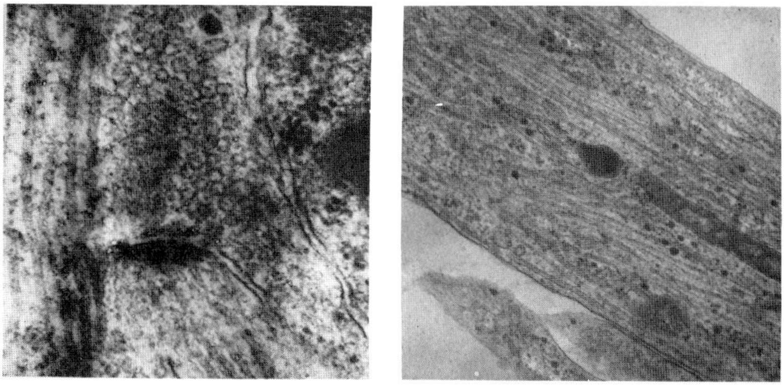

A B

Figure 30-6. (**A**) Differentiated NT2/D1 human teratocarcinoma cell processes with generally parallel arrays of microtubules and 10 nm filaments, probably neurofilaments. (**B**) Synapses are frequently encountered in the neuronal ganglia in the older cultures. Some have classic morphology.

serum, the kind gift of Dr. D. Dahl). These cultures make a variety of types of neurons identified by morphology and specific antisera.

Retinoic acid-differentiated neuronal NY2/D1 cultures were treated with a variety of toxins that were screened for their capacity to induce SDAT pathology with one of Selkoe's anti-PHF sera ("Shira"), which labels both plaque core amyloid and tangles. Most toxins and untreated cultures showed essentially no or very rare staining, but doxorubicin-treated cultures consistently produced stainable cytoskeletal fibers in processes and staining of rare tangles that also double labeled with antineurofilament monoclonal antibodies

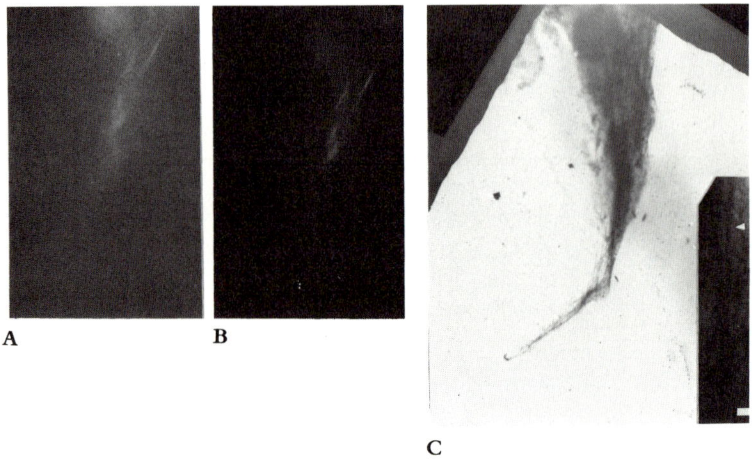

Figure 30-7. Differentiated NT2/D1 cells were replated onto polylysine–formvar-coated gold electron-microscopy grids and allowed to form processes for several weeks. They were then treated with doxorubicin for 2 weeks and allowed to recover for another 2 weeks. Cultures were then extracted with 0.5% Triton X-100 to make cytoskeletal whole cell mount preps and then double labeled with Selkoe's anti-paired helical filament (PHF) antisera and neurofilament monoclonal. (**A**) Tangle in cell identified with anti-PHF serum (1:200). (**B**) Double labeling with antineurofilament monoclonal antibodies. (**C**) Low power electron micrograph of the labeled neuronal whole cell mount in **A** and **B**. (**Inset**) Although the unsectioned material is difficult for the beam to penetrate, twisted filaments can be discerned in the anti-PHF-staining region and these appear to have dimensions close to paired helical filaments. Bar = 1000 Å.

(Dahl, 1983) (the generous gift of Dr. D. Dahl) (figure 30-7A and B). Some tangles also labeled with Anderton's BF10 monoclonal antibody which recognizes phosphorylated epitopes on the 155 k neurofilament protein and tangles in Alzheimer's disease, but nothing in the perikaryon of normal cells (Haugh et al., 1985) (figure 30-8A and B). We are presently investigating the ultrastructure of these neurofibrillary tangles in doxorubicin-treated cells using immunogold labeling. Their reactivity with Selkoe's anti-PHF serum strongly suggests a relationship with SDAT tangles. This hypothesis is supported by the observation of twisted filaments in high-voltage electron micrographs of anti-PHF labeled cells grown on gold grids and extracted with Triton x-100. (figure 30-7).

To better determine the nature of the common fibrillar staining observed with Selkoe's antiserum in the processes of doxorubicin-treated cells, we employed two polyclonal antisera against a synthetic peptide raised to residues 11–23 of the known sequence of Alzheimer's vascular and plaque amyloid protein (beta protein) (Masters et al., 1985) (graciously supplied by Dr. C. Masters), and an antiserum raised to residues 1–28 of the Glenner sequence for Alzheimer's amyloid (the kind gift of Dr. J. Kates), and demonstrated that these antisera stain similar fibrils in doxorubicin-treated cells and

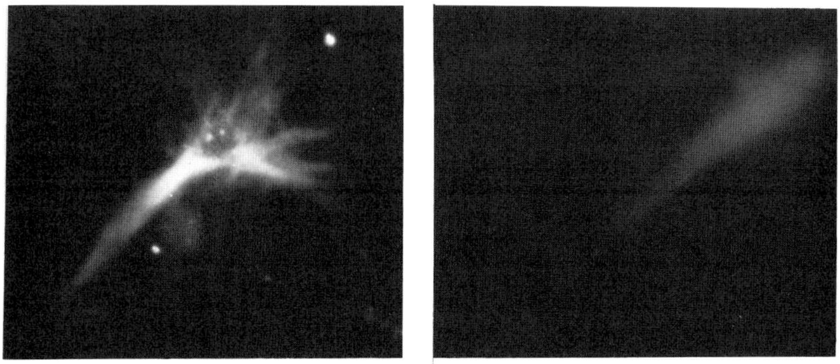

A B

Figure 30-8. (**A** and **B**) Doxorubicin-treated NT2/D1 cells incubated with antineurofilament monoclonal antibody sometimes show masses of neurofilament fluorescence in cell perikaryon. This is not seen in normal brain with BF10 monoclonal antibody, which labels phosphorylated epitopes. Note also the neurofilament fluorescence in nucleoli in **A**. This is typical of this type of cell. The cell in **B** was a large pyramid-shaped structure and the "tangle" was so bright that it was shot through two neutral density filters. It did not react with anti-PHF serum.

only the low molecular weight A4 or beta protein (Wong et al., 1985; Masters et al., 1985) and its dimer in immunogels (Cole et al., 1986). These antiamyloid peptide sera did not stain tangles or other material in the perikaryon in agreement with the findings in SDAT (Kates, personal communication; Masters et al., 1985; Wong et al., 1985), but the fibrils double labeled with BF10 antineurofilament monoclonal. In contrast, doxorubicin treatment of retinoic acid-differentiated PA-6 cells, a human teratocarcinoma cell clone that produces glia but not neurons (Cole et al., 1986), did not result in fibril staining with antiamyloid peptide sera. Elimination of dividing cells with arabinosylcytosine left ganglia intact with no beta-protein fibril induction.

Thus, we have demonstrated the induction of Alzheimer's amyloid protein (APC, A4, beta protein) in our doxorubicin-treated cells cultured with or without serum and found it to be in neuritic processes of neurofilament-positive cells. This strongly supports the hypothesis that plaque amyloid in Alzheimer's disease brain derives from degenerating neuritic processes, rather than from microglia, endothelial cells, exogenous prions, or a serum precursor. Other evidence supports the hypothesis of extracellular amyloid arising from degenerating neurites because neuritic plaques precede amyloid plaques (Burger and Vogel, 1973; Gibson, 1985). Vascular amyoid originating from degenerating neurites from the locus coeruleus and central cholinergic tracts innervating blood vessels in SDAT may lead to a compromised blood–brain barrier because amyloid-encrusted vessels are apparently more subject to damage (Kalyan-Raman and Kalyan-Raman, 1984). Alternatively, vascular amyloid may be generated independently, but from the same precursor and by a similar process in endothelial or other vessel related cells.

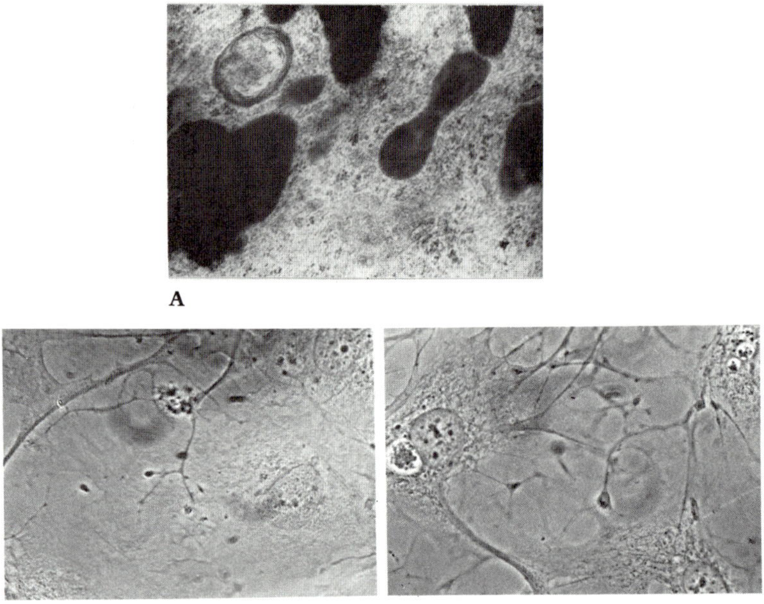

Figure 30-9. Doxorubicin-treated differentiated NT2/D1 cells. (**A**) Cell bodies often show lamellar bodies, dense bodies, and lipofuscin-like structures (and increased autofluorescence.) (**B**) Degenerating neurites are common in doxorubicin damaged cultures. Here a "plaque-like" cluster is visible (phase, orig. mag. × 630) (**B** and **C**) Cell nuclei often show either focal clearing or a phase-dense clumping of chromatin consistent with increased heterochromatization. In general, the nucleoli show fragmentation or separation and cells may develop vacuolar or even granulovacuolar degeneration (phase, orig. mag. × 630) process fluorescence.

6. OTHER APSECTS OF ALZHEIMER'S PATHOLOGY

In addition to neurofibrillary tangle and plaque protein production, our doxorubicin-treated cultures show several other aspects of SDAT pathology including lipofuscin and Iamellar bodies (Mann, 1985) (figure 39-9A), distended and degenerating neurites (Probst et al., 1983) (figure 30-9B), cell loss, heterochromatization of nuclei (McLachlan et al., 1984) (figure 30-9C), and reduced protein synthesis measured by ^{35}S-methionine incorporation into tricarboxylic acid (TCA)-precipitable total protein (data not shown). It is plausible that additional SDAT-related phenomena will be found in this system. Doxorubicin was initially tested because it is known to produce a marked axonopathy in rodents with neurofilamentous tangles and heterochromatization of the nuclei (Parhad et al., 1984; Yamamoto et al., 1984; Zagoren et al., 1984) and decreased transcription (Sajdel-Sulkowska and Marotta, 1984) while Alzheimer's disease also shows tangles, nuclear heterochromatization and reduced transcription. Doxorubicin is known to generate free radicals (Smith et al., 1985), which may account for lipofuscin produc-

Table 30-1. Comparison of some selected doxorubicin-induced effects and (SDAT).

Effect	Doxorubicin	SDAT
Protein synthesis	− (2)	− (8, 14)
Nissl substance	− (16)	− (7)
Transcription	− (2)	− (12)
Heterochromatin	− (15, 16)	+ (7, 8)
Dense bodies	+ (15)	+ (7, 14)
Lamellar bodies	+ (1)	+ (7)
Lipofuscin	+ (3)	+ (7)
Beta amyloid protein	+ (3)	+ (5, 9)
Neurofibrillary tangles	+ (10, 15)	+ (7, 14)
Neuron number	− (10, 15, 16)	− (7, 14)
Reactive gliosis	+ (15)	+ (14)
Neuritic pathology	+ (1)	+ (11, 14)
Disrupted actin	+ (4)	+ (6)
Paracrystalline nuclear inclusions of Seite type (actin)	?	+ (7)
Nucleolar fragmentation	+ (1, 16)	+ (17)
Slowed axonal neurofilament transport	+ (13, 16)	+ (Gajdusek, 1985, postulated)
Vacuolation	+ (15, 16)	− (7, 14)

Increase is represented by (+) and decrease by (−). Numbers in parentheses represent references listed below:

1. Bigotte and Olsson (1983, 1984)
2. Cho et al. (1980)
3. Cole (1986)
4. Colombo et al. (1984)
5. Glenner and Wong (1984)
6. Goldman (1983)
7. Mann (1985)
8. McLachlan et al. (1984)
9. Masters et al. (1985)
10. Parhad et al. (1984)
11. Probst et al. (1983)
12. Sajdel-Sulkowska and Marotta (1984)
13. Sidhenius (1982)
14. Terry and Katzman (1983)
15. Yamamoto et al. (1984)
16. Zagoren et al. (1984)
17. Mann (1978)

tion as well as other effects. The well-studied cardiotoxic effect of doxorubicin produces a myofibrillar pathology that appears to depend on free radical production, possibly involving calcium release from mitochondria and sarcoplasmic reticulum (Doroshow et al., 1983; Harris and Doroshow, 1985). We are currently testing the hypothesis that doxorubicin-induced free radical generation or reduced transcription is the critical factor in amyloid plaque core (APC) protein production. There are, of course, other interesting possibilities, including the direct nuclear effects and inhibition of RNA polymerase as well as direct and indirect effects on other cytoskeletal elements (Colombo et al., 1984) and on calcium, but our preliminary evidence favors a role for free radicals.

7. ADVANTAGES OF AN IN VITRO MODEL

A summary of some of the similarities between doxorubicin-induced damage and SDAT pathology is given in table 30-1.

The development of an in vitro model capable of expressing tangles and plaque protein provides a mechanism for exploring the molecular basis of Alzheimer's amyloid lesion production using radiolabeling, pulse chase, protein synthesis and protease inhibitors, antioxidants, and a battery of other pharmacological and molecular biological tools. For example, we are currently screening for protease inhibitors that reduce the production of APC protein. Oligonucleotide and cDNA probes for the amyloid proteins are now being synthesized that will greatly facilitate elucidation of the mechanism of APC production on this system.

A second potential use of this in vitro model may be as a test system for agents capable of inducing Alzheimer's amyloid lesions, which we will call *alzogens* for convenience. Potential alzogens may be environmental toxins, cloned genes, chromosome segments, or infectious agents. For example, because cultured trisomy-21 neurons are known to be sensitive to free radical-generating agents (Brooksbank and Balazs, 1984) and individuals with trisomy 21 show the pathology of precocious senile dementia (Burger and Vogel, 1973), we have cloned a karyotypically stable, essentially euploid neurogenic human teratocarcinoma cell line and introduced a selectable marker, pSV2 neo, into human chromosome 21 to enable the manufacture of trisomy 21 neurogenic cell lines that may be employed in this system. Similarly, should a gene segment containing a marker associated with dominant familial Alzheimer's be isolated, it could be introduced into these or NT2/D1 cells to search for an effect on the production of SDAT-related proteins.

8. DAMAGE AND THE CELLULAR RESPONSE TO STRESS

For theoretical purposes, one must place our hypothesis of neuronal aging and age-related pathology as a response to specific types of damage against the background of the general response of cells to stress. At the systemic physiological level, many investigators, beginning with Selye, have noted an apparent relationship between stress and aging, but the cellular response has been sadly neglected by most gerontologists and pathologists.

The generalized cellular response to stress (analagous to Selye's general adaptation syndrome) has been initially explored primarily in terms of heat shock in *Drosophila*, but also obtains for most living systems from bacteria to man (Nover, 1984). It involves the induction of a small set of major heat-shock proteins and a larger set of minor heat-shock proteins (*hsps*), an inhibition of normal protein synthesis, and assorted reprogramming of cellular metabolism including a major rearrangement of the cytoskeleton and nucleolar changes (Welch and Suhan, 1985). All or part of the classic heat-shock response is induced by a number of different agents, including transition metals, chelating agents, sulfhydryl reagents, inhibitors of gene expression, ionophores, many viruses, and oxidative or free radical stresses (viz., hydrogen peroxide and menadione). Doxorubicin and menadione both auto-

Table 30-2. Comparison of some similarities between the cellular stress response and SDAT.

Effect	Stress	SDAT
New nuclear DNA-binding proteins	+ (7)	+ (8)
rRNA	− (4)	− (2)
Transcription	− (4)	− (5)
Rough endoplasmic reticulum	− (4)	− (2)
Paracrytalline inclusions of Seite	+ (7)	+ (2)
Intermediate filament cytoskeletal collapse	+ (7)	+ (6)
Altered histones	+ (4)	+ (3)
Cytoplasmic ribonuclease	+ (1)	+ (5)
Induction of heat-shock proteins	+ (4)	??
Nucleolar fragmentation	+ (7)	+ (9)

Increase is represented by (+) and decrease by (−). Numbers in parentheses represents references listed below:
1. Gross and Schulz-Harder (1983)
2. Mann (1985)
3. McLachlan et al. (1984)
4. Nover (1984)
5. Sajdel-Sulkowska and Marotta (1984)
6. Terry and Katzman (1983)
7. Welch and Suhan (1985)
8. Wetmur et al. (1984)
9. Mann (1978)

Note: Stress response here is always acute rather than chronic; elevated ribonuclease has only been measured in yeast. We do not know anything about the effects of a chronic cellular stress response in the CNS and nothing about the human CNS.

xidize through semiquinone intermediates to produce a similar free radical stress (Smith et al., 1985). The basic response to free radical-induced stress in bacteria has been shown to overlap and include many heat-shock proteins (Christman et al., 1985). Thus, there is substantial evidence to suggest that doxorubicin will also induce the heat-shock response.

The heat-shock or cellular stress response occurs in neurons from a variety of species (Greenberg and Lasek, 1985) and one may anticipate many neuronal stresses both experimentally, in vitro, and in neuropathologies, in vivo, might include features unique to the target cells and the agents involved against the background of the cellular stress response. Of course, one must bear in mind that a pathology may represent the sum of chronic damage while the general features of the stress response have all been described under conditions of acute treatment. And, although they are nonspecific to a particular pathology, heat-shock or stress proteins may interact with more specific components of a pathological response and play an important role in the genesis of lesions.

9. SDAT AND THE CELLULAR STRESS RESPONSE

Some areas of apparent overlap between SDAT and the general stress response are summarized in table 30-2. The most salient common phenomena are the reduced transcription and protein synthesis, nuclear alteration, and cytoskeletal disruption. We speculate that part of the heat-shock response is chronically activated in SDAT and plays a significant role in the disease process. In support of this new idea we note the 68 k microtubule, which is homologous, if not identical, to the principal intermediate filament-associated stress protein *hsps* 68/70 (Lim et al., 1984), is a cytoskeletal component of many cells, and is thought to play a regulatory role (Napolitano et al., 1985a and b) in the cytoskeletal collapse in the stress response. Evidence for cytoskeletal collapse in SDAT comes from intermediate filaments and microtubule-associated proteins in the neurofibrillary tangle (Kosik et al., 1984; Perry et al., 1985; Selkoe et al., 1985), microtubules, and in granulovacular degeneration (Price et al., 1985), and actin in Hirano bodies (Goldman, 1983). The mRNA for the 68/70 *hsps* associated with intermediate filaments and their regulation is elevated in trisomy-21 brain in a manner consistent with a gene dosage effect, but the protein itself is not (Whatley et al., 1984), indicating that abnormal metabolism might be involved in SDAT pathology. The 68/70 *hsps* tightly binds mRNA and polyribosomes (Schonfelder et al., 1985), which could account for the strong association between PHF and polyribosomes observed by several investigators (see Perry et al., 1985).

The low molecular weight *hsps* are also cytoskeletal-associated proteins that migrate to the nucleus during stress-induced cytoskeletal collapse (Leucht et al., 1986) and these proteins are homologous to alpha crystalline, including the strongly hydrophobic domains involved in aggregation (Ingolia and Craig, 1982). Probes for alpha-crystalline A2 chain have recently been shown to map to chromosome 21 (Quax-Jeuken et al., 1985). If the low molecular weight cytoskeletal-associated *hsps* are also chromosome-21 encoded, as we suspect, we speculate these proteins or special neuronal homologues will play an important role in the pathogenesis of SDAT, including the cytoskeletal rearrangement and the inhibition of protein synthesis and subsequent generalized deficits in luxury protein synthesis that have been observed (Dekosky and Bass, 1985).

10. CONCLUSIONS

In vitro models where human neuronal aging or age-related pathology may be accelerated should facilitate progress in our understanding of aging and age-related pathology. We present an example of this type of model and show several markers for SDAT, including that the beta plaque core amyloid protein may be induced in response to a doxorubicin stress. This system should therefore prove useful in elucidating Alzheimer's amyloid beta-protein metabolism. We argue that production of Alzheimer's lesions in response to selected types of damage probably also occurs in vivo. This damage or stress response may be conceptualized in the context of the general cellular stress response. The general cellular stress response to damage has therefore been applied to put SDAT pathology in a useful heuristic framework, suggesting further experimentation on SDAT material. The agent or agents that produce the specific SDAT pathology in vivo are unknown, but might be detected with this or an analogous test system for alzogen activity.

ACKNOWLEDGMENTS

We would like to thank Drs. Brian Anderton, Jim Cohen, Doris Dahl, Joe Kates, Pinky Kushner, Colin Masters, Melitta Schachner, and Dennis Selkoe for their generous gifts of the antibodies that made this work possible. We are also grafteful to the G. and L. Pfeiffer Research Foundation for their support.

REFERENCES

Agamanolis, D.P., and Greenstein, J.I., (1979) Ataxia-telangiectasia: report of a case with Lewy bodies and vascular abnormalities within cerebral tissue. J. Neuropathol. Exp. Neurol. 38:475–498.

Ames, B.N., Durston, W.E., Yamasaki, E., and Lee, F.D., (1973) Carcinogens are mutagens: a simple test system combining liver homogenates for activation and bacteria for detection. Proc. Natl. Acad. Sci. USA 70: 2281–2285.

Andrews, P.W. (1984) Retinoic acid induces neuronal differentiation of a cloned human embryonal carcinoma cell line in vitro. Dev. Biol. 103:285–293.

Ball, M.J., MacGregor, J., Fyfe, I.M., Rapoport, S.I., and London, E.D., (1983) Paucity of morphological changes in the brains of ageing beagle dogs: further evidence that Alzheimer lesions are unique for human central nervous system. Neurobiol. Aging 4:127–132.

Bender, B.L., and Ghatak, N.R. (1978) Light and electron microscopic observations on a ganglioneuroma. Acta Neuropathol. 42:7–10.

Bierman, E.L., and Ross, R. (1977) Aging and atherosclerosis. In: Paoletti, R., and Gotto, A.M., Jr. (eds.) Atherosclerosis reviews, vol. 2. New York: Raven, pp. 77–111.

Bigotte, L., and Olsson, Y. (1983) Toxic effects of adriamycin on the central nervous system. Acta Neuropathol. 61:291–299.

Bigotte, L., and Olsson, Y. (1984) Cytotoxic effects of adriamycin on the central nervous system of the mouse: cytofluorescence and electron microscopic observations after various modes of administration. Acta Neurol. Scand. [Suppl. 100] 70:55–67.

Bradbury, A.J., Costall, B., Domeney, A.M., Jenner, P., Kelly, M.E., Marsden, C.D., and Naylor, R.J. (1986) 1-Methyl-4-phenylpyridine is neurotoxic to the nigrostriatal dopamine pathway. Nature 319:56–57.

Bridges, B.A., and Harnden, D.G. (1981) Untangling ataxia-telangiectasia. Nature 289:222.

Brooksbank, B.W.L., and Balazs, R. (1984) Superoxide dismutase, glutathione peroxidase and lipoperoxidation in Down's syndrome fetal brain. Dev. Brain Res. 16:37–40.

Burger, P.C., and Vogel, S. (1973) The development of the pathologic changes of Alzheimer's disease and senile dementia in patients with Down's syndrome. Am. J. Pathol. 73:457–476.

Cho, E.S., Spencer, P.S., and Jortner, B.S. (1980) Doxorubicin. In: Spencer, P.S., and Schaumberg, H.H. (eds.) Experimental and clinical neurotoxicology. Baltimore: Wilkins and Wilkins, pp. 430–439.

Christman, M.F., Morgan, R.W., Jacobson, F.S., and Ames, B.N. (1985) Positive control of a regulon for defenses against oxidative stress and some heat-shock proteins in *Salmonella typhimurium*. Cell 41:753–62.

Cole, G.M. (1986) An in vitro model for Alzheimer's disease pathology. Ph. D. dissertation. University of California, Berkeley.

Cole, G.M., Wu, K., and Timiras, P.S. (1985a) A culture model for age-related human neurofibrillary pathology. Int. J. Dev. Neurosci. 3:23–32.

Cole, G.M., Dahl, D., and Timiras, P.S. (1985b) Neurofilaments in human tumor-derived neuronal cells. J. Neurochem. [Suppl.]44:S137.

Cole, G.M., Dahl, D., and Timiras, P.S. (1986) Neuro- and gliogenesis in retinoic acid differentiated human teratocarcinoma cells [abstr.]. Fed. Proc. 45:888.

Colombo, R., Necco, A., Vailati, G., Saracco, B., Milzani, A., and Seari, G. (1984) Doxorubicin affects actin assembly in vitro. Cell Biol. Int. Rep. 8:127–134 (1984)

Dahl, D. (1983) Immunohistochemical differences between neurofilaments in perikarya, dendrites and axons. Exp. Cell Res. 149:397–408.

De Boni, U., and McLachlan, D.R.C. (1985) Controlled induction of paired helical filaments of the Alzheimer type in cultured human neurons by glutamate and aspartate. J. Neurol. Sci. 68:105–118.

Dekosky, S.T., and Bass, N.H. (1985) Biochemistry of senile dementia. In: Lajtha, A. (ed.) Handbook of neurochemistry, vol. 10, 2nd edn. New York: Plenum, pp. 617–650.

Doroshow, J.H., Davies, K.J.A., and Hochstein, P. (1983) Adriamycin cardiac toxicity: the role of mitochondrial reactive oxygen metabolism., In: Greenwald, R.A., and Cohen, G. (eds.) Oxyradicals and their scavenger systems, vol. 2: Cellular and medical aspects. Amsterdam: Elsevier, pp. 113–118.

Forno, L.S. (1982) Pathology of Parkinson's disease. In: Movement disorders. Neurology 2. London: Butterworth. pp. 25–40.

Forno, L.S. (1983) Reaction of the substantia nigra to massive basal ganglia infarction. Acta Neuropathol. 62:96–102.

Forno, L.S., and Norville, R.L. (1976) Ultrastructure of Lewy bodies in the stellate ganglion. Acta Neuropathol. 34:183–197.

Gajdusek, D.C. (1985) Hypothesis: interference with axonal transport of neurofilament as a common pathogenetic mechanism in certain diseases of the central nervous system. N. Engl. J. Med. 714–719.

Gibson, P.H. (1985) Relationship between numbers of cortical argentophilic and congophilic senile plaques in the brain of elderly people with and without senile dementia of the Alzheimer type. Gerontology 31:321–324.

Glenner, G.G., and Wong, C.W. (1984) Alzheimer's disease and Down's syndrome: sharing of a unique cerebrovascular amyloid fibril protein. Biochem. Biophys. Res. Commun. 122:1131–1135.

Goldman, J.E. (1983) The association of actin with Hirano bodies. J. Neuropathol. Exp. Neurol. 42:146–152.

Goldman, J.E., Yen, S.H., Chiu, F.C., and Peress, N.S. (1983) Lewy bodies of Parkinson's disease contain neurofilament antigens. Science 221:1082–1084.

Greenberg, S.G., and Lasek, R.J. (1985) Comparison of labeled heat shock proteins in neuronal and non-neuronal cells of *Aplysia californica*. J. Neurosci. 5:1239–1245.

Gross, T., and Schulz-Harder, B. (1983) Heat-shock induction of ribonuclease and protease in yeast. Hoppe Seylers Z. Physiol. Chem. 364:1132–1133.

Harris, R.N., and Doroshow, J.H. (1985) Effect of doxorubicin-enhanced hydrogen peroxide and hydroxyl radical formation on calcium sequestration by cardiac sarcoplasmic reticulum. Biophys. Biochem. Res. Commun. 130:739–745.

Haugh, M.C., Ulrich, J., Probst, A., Khan, J., and Anderton, B. (1985) Immunostaining of Alzheimer neurofibrillary tangles with monoclonal antibodies that recognize normal neurofilaments. Biochem. Soc. Trans. 13:951–952.

Heyman, A., Wilkinson, W.E., Stafford, J.A., Helms, M.J., Sigman, A.H., and Weinberg, T. (1984) Alzheimer's disease: a study of epidemiological aspects. Ann. Neurol. 15:335–341.

Ihara, Y., Abraham, C., and Selkoe, D.J. (1983) Antibodies to paired helical filaments in Alzheimer's disease do not recognize normal brain proteins. Nature 304:727–730.

Ingolia, T.D., and Craig, E.E. (1982) Four small *Drosophila* heat shock proteins are related to each other and to mammalian alpha crytallin. Proc. Natl. Acad. Sci. USA 79:2360–2364.

Ino, M., Cole, G.M., and Timiras, P.S. (1986) Tyrosine hydroxylases and monoamine oxidase-A activity increases in differentiating human neuroblastoma after elimination of dividing cells. Dev. Brain Res. 30:120–123.

Ivy, G.O., Schottler, F., Wenzel, J., Baudry, M., and Lynch, G. (1984) Inhibitors of lysosomal enzymes: accumulation of lipofuscin-like dense bodies in the brain. Science 220:985–987.

Kalyan-Raman, U.P., and Kalyan-Raman, K. (1984) Cerebral amyloid angiopathy causing intracranial hemorrhage. Ann. Neurol. 16:321–329.

Kennedy, A.R., Fox, M., Murphy, G., and Little, J.B. (1980) Relationship between x-ray exposure and malignant transformation in C3H10T1/2 cells. Proc. Natl. Acad. Sci. USA 77:7262–7266.

Kosik, K.S., Duffy, L.K., Dowling, M.M., Abraham, C., McCloskey, A., and Selkoe, D.J. (1984) Microtubule-associated protein 2: monoclonal antibodies demonstrate the selective incorporation of certain epitopes with Alzheimer neurofibrillary tangles. Proc. Natl. Acad. Sci. USA 81:7941–7945.

Kushner, P.D., Sternberg, H., and Stephenson, D. (1985) A rare cell surface antigen found on motor neurons. [abstr. 331.9]. Neurosci Abstr. 11:1137.

Langston, J.W., Ballard, P., Tetrud, J.W., and Irwin, I. (1983) Chronic Parkinson in humans due to a product of meperidine-analog synthesis. Science 219:979–981.

Leucht, B., Biessmann, H., Palber, K., and Bonner, J.J. (1986) Small heat shock proteins of *Drosophila* associate with the cytoskeleton. Proc. Natl. Acad. Sci. USA 83:90–94.

Lim, L., Hall, C., Leung, T., and Whatley, S. (1984) The relationship of the rat brain 68 kDa microtubule-associated protein with synaptosomal plasma membranes and with the *Drosophila* 70kDa heat-shock protein. Biochem. J. 224:677–680.

Liss, L., Ebner, K., and Couri, D. (1979) Neurofibrillary tangles induced by a sclerosing angioma. Hum. Pathol. 10:104–108.

Mann, D.M.A. (1978) Granulovaculor degeneration in pyramidal cells of the hippocampus. Acta Neuropathol. 42:149–151.

Mann, D.M.A. (1985) The neuropathology of Alzheimer's disease: a review with pathogenetic, aetiological, and therapeutic considerations. Mech. Ageing Dev. 31:213–255.

Mann, D.M.A., and Yates P.O. (1983) Possible role of neuromelanin in the pathogenesis of Parkinson's disease. Mech. Ageing Dev. 21:193–203.

Mann, D.M.A., Yates, P.O., and Marcynuik, B. (1984) Age and Alzheimer's disease. Lancet 1:281–282.

Masters, C.L., Multhaup, G, Simms, G., Pottgieser, J., Martins, R.N. and Beyreuther, K. (1985) Neuronal origin of a cerebral amyloid: neurofibrillary tangles of Alzheimer's disease contain the same protein as the amyloid of plaque cores and blood vessels. EMBO J. 4:2757:2763.

Martin, G.M. (1978) Genetic syndromes in man with potential relevance to the pathobiology of aging. Birth Defects. 14:5–39.

McLachlan, D.R.C., Lewis, P.N., Lukiw, W.J., Simm, A., Bergeron, C., and De Boni, U. (1984) Chromatin structure in dementia. Ann. Neurol. 15:329–334.

Mehlorn, R.J., and Cole, G. (1985) The free radical theory of aging: a critical review. Adv. Free Radical Biol. Med. 1:165–223.

Miller, R.H., Williams, B.P., Cohen, J., and Raff, M. (1984) A4: an antigenic marker of neural tube derived cells. J. Neurocytol. 13:329–338.

Moroni, F., Lombardi, G., Monetti, G., and Aldinio, C. (1984) The excitotoxin quinolinic acid is present in the brain of several mammals and its cortical content increases during the aging process. Neurosci. Lett. 47:51–55.

Mortimer, J.A., French, L.R., Hutton J.T., and Schuman, L.M., (1985) Head injury as a risk factor in Alzheimer's disease. Neurology 35:264–267.

Nandy, K., Baste, C., and Schneider, F.H., (1978) Further studies of centrophenoxine on lipofuscin pigment in neuroblastoma cells in culture: an electron microscopic study. Exp. Gerontol. 13:311–322.

Napolitano, E.W., Pachter, J.S., Chin, S.S.M., and Liem, R.K.H., (1985a) Beta-internexin, a ubiquitous intermediate filament assoicated protein. J. Cell Biol. 101:1323–1331.

Napolitano, E.W., Hutchison, S.B., and Liem, R.K.H. (1985b) Association of an endogenous heat shock protein (HSP70) with neuronal and non-neuronal cytoskeletal elements. J. Cell Biol. 101:30a.

Nover, L. (1984) *Heat Shock Response*, Springer-Verlag, Stuttgart, pp. 1–82.

Ohama, E., and Ikuta, F. (1976) Parkinson's disease: distribution of Lewy bodies and monoamine neuron system. Acta Neuropathol. 34:311–319.

Parhad, I.M., Griffin, J.W., Clark, A.W., and Koves, J.F. (1984) Doxorubicin intoxication: neurofilamentous axonal changes with subacute neuronal death. J. Neuropathol. Exp. Neurol. 43:188–200.

Park, B.E., Netsky, M.G., and Bestill, W.L. (1975) Pathogenesis of pigment formation in Hallervorden-Spatz syndrome and related disorders. Neurology 25:1172–1178.

Perry, G.N., Rizzuto, N., Autilio-Gambetti, L., and Gambetti, P. (1985) Paired helical filaments from Alzheimer disease patients contain cytoskeletal components. Proc. Natl. Acad. Sci. USA 82:3916–3920.

Perry, T.L., Godin, D.V., and Hanson, S. (1982) Parkinson's disease: a disorder due to nigral glutathione deficiency? Neurosci. Lett. 33: 305–310.

Powers, J.M., Balentine, J.D., Wisniewski, and Terry, R.D. (1976) Retroperitoneal ganglioneuroblastoma: a kaleidoscope of neuronal degeneration. J. Neuropathol. Exp. Neurol. 35:14–25.

Price, D.L., Struble, R.G., Altschuler, R.J., Casanova, M.F., Cork, L.C., and Murphy, D.B. (1985) Aggregation of tubulin in neurons of Alzheimer's disease [abstr. 178]. J. Neuropathol. Exp. Neurol. 44:366.

Probst, A., Basker, U., Bron, R., and Ulrich, J. (1983) Neuritic plaques in senile dementia of Alzheimer type: a Golgi analysis in the hippocampal region. Brain Res. 268:249–254.

Quax-Jeuken, Y.E.F.M., Quax, W., Van Rens, G., Khen, P.M., and Bloomendal, H. (1985) Cytogenet. Cell Genet. 40:727–728.

Qaux-Jeuken, Y.E.F.M., Quax, W., Van Rens, G., Khen, P.M., and Bloomendale, H. (1985) Assignment of the human alphaA-crystalline gene (CRYAL) to chromosome 21. Cytogenet. Cell Genet. 40:727–728.

Rasool, C.G., and Selkoe, D.J. (1984) Sharing of specific antigens by degenerating neurons in Pick's disease and Alzheimer's disease. N. Engl. J. Med. 312:700–705.

Rasool, C.G., Abranham, C., Anderton, B.H., Haugh, M., Kahn, J., and Selkoe, D.J., (1984) Alzheimer's disease: immunoreactivity of neurofibrillary tangles with anti-neurofilament and anti-paired helical filament antibodies. Brain Res. 310:249–260.

Rudelli, R.J., Strom, O. Welch, P.T., and Ambler, M.W. (1982) Posttraumatic premature Alzheimer's disease: neuropathologic findings and pathogenetic considerations. Arch. Neurol. 39:570–575.

Sajdel-Sulkowska, E.M., and Marotta, C.A. (1984) Alzheimer's disease brain: alterations in RNA levels and in a ribonuclease–inhibitor complex. Science 225:947–949.

Sapolsky, R.M., Krey, L.C., and McEwen, B.S., (1985) Prolonged glucocorticoid exposure reduces hippocampal neuron number: implications for aging. J. Neurosci. 5:1222–1227.

Schonfelder, M., Horsch, A., and Schmid, H.P. (1985) Heat shock increases the synthesis of the poly(A)-binding protein in HeLa cells. Proc. Natl. Acad. Sci. USA 82:6884–6888.

Segall, P.E., and Timiras, P.S. (1976) Pathophysiologic findings after chronic tryptophan deficiency in rats: a model for delayed growth and aging. Mech. Aging Dev. 5:109–124.

Selkoe, D.J., and Abraham, C. (1985) Plaque amyloid in alzheimer's disease (AD): purification by flow cytometry and protein characterization [abstr.] Neurology [Suppl.1] 35:217.

Selkoe, D.J., Abraham, C., Rasool, C.G., McCluskey, A., Kosik, K.S., and Duffy, L.K. (1985) Paried helical filaments in human neurons: relationship to neurofilaments. Ann. N.Y. Acad. Sci. 455:583–596.

Selkoe, D.J., Abraham, C.R., Podlisny, M.B., and Duffy, L.K. (1986) J. Neurochemistry (in press).

Sidenius, P. (1982) Effect of doxorubicin on slow axonal transport [abstr. 36.5]. In: Fifth international congress of neuromuscular diseases [abstracts], Marseilles, France, 12–18 September 1982.

Smith, M.T., Evans, C.G., Thor, H., and Orrenius, S. (1985) Quinone-induced oxidative injury to cells and tissues, In: Sies, H. (ed.) Oxidative stress, New York: Academic Press, 91–113.

Sommer, I., and Schachner, M. (1981) Monoclonal antibodies (01 to 04) to oligodendrocyte cell surfaces: an immunocytological study in the central nervous system. Dev. Biol. 83:311–327.

Sung, J.H., Mastri, A., and Segal, E. (1979) Pathology of Shy-Drager syndrome. J. Neurophathol. Exp. Neurol 38:353–368.

Terry, R., and Katzman, R. (1983) Senile dementia of the Alzheimer type: defining a disease. In Katzman, R., and Terry, R. (eds.) The neurology of aging. Philadelphia: F.A. Davis, pp. 51–84.

Thiele, C.J., Reynold, C.P., and Israel, M.A. (1985) Decreased expression of N-myc precedes retinoic acid-induced morphological differentiation of human neuroblastoma. Nature, 313:404–406.

Welch, W.J., and Suhan, J.P. (1985) Morphological study of the mammalian stress response: characterization of changes in cytoplasmic organelles, cytoskeleton, and nucleoli, and appearance of intranuclear actin filaments in rat fibroblasts after heat-shock treatment. J. Cell Biol. 101:1198–1211.

Wetmur, J.G., Casals, J., and Elizan, T.S. (1984) DNA binding protein profiles in Alzheimer's disease. J. Neurol. Sci. 66:201–208.

Whatley, S.A., Hall, C., Davison, A.N., and Lim, L. (1984) Alterations in the relative amounts of specific mRNA species in the devleoping human brain in Down's syndrome. Biochem. J. 220:179–187.

Wisniewski, H.M., Coblentz, J.M., and Terry, R.D. (1972) Pick's disease: clinical and ultrastructural study. Arch. Neurol. 26:97–108.

Wisniewski, H.M., Narang, H.K., Corsellis, J., and Terry, R.D. (1976) Ultrastructural studies of the neuropil and neurofibrillary tangles in Alzheimer's disease and post-traumatic dementia. J. Neuropathol. Exp. Neurol. 35:367.

Wisniewski, K., Jervis, G.A., Moretz, R.C., and Wisniewski, H.M. (1979) Alzheimer neurofibrillary tangles in diseases other than senile and presenile dementia. Ann. Neurol. 5:288–294.

Wong, C.W., Quaranta, U., and Glenner, G.G. (1985) Neuritic plaques and cerebrovascular amyoid in Alzheimer disease are antigenically related. Proc. Natl. Acad. Sci. USA 82:8729–8732.

Yamamoto, T., Iwasaki, Y., and Konno, H. (1984) Retrograde axoplasmic transport of adriamycin: an experimental form of motor neuron disease. Neurology 34:1299–1304.

Zagoren, J.C., Seeling, M., Bornstein, M.B., and Spencer, P.S., (1984) The evolution of cellular degeneration in dorsal root ganglia exposed to doxorubicin in tissue culture. J. Neuropathol. Exp. Neurol. 43:384–394.

ADDENDUM

Following completion of this manuscript, several groups have reported on molecular cloning of the amyloid plaque core protein precursor gene from brain. APC precursor is reported to be located on chromasome 21 and closely linked to restriction fragment polymorphic markers for familial Alzheimer's disease (Tanzi et al., 1987, *Science* 235:880–884). The complete sequence for the precursor codes for approximately 80 Kd protein with a membrane spanning region which includes most of the fragment found in amyloid plaque fibrils (Kang et al., 1987, *Nature* 325:733–736). Further, in sporadic Alzheimer's disease, three cases have been reported to contain a duplication of a small region of chromasome 21 including the APC precursor gene (Delabar et al., 1987, *Science* 235:1390–1393). An apparently central role for the APC precursor gene in S.D.A.T. emphasizes the critical need for a culture model to examine the expression and processing of this protein.

31. COMPARISONS OF GLIAL CELLS FROM NEWBORN AND AGED MOUSE IN CULTURE

ANTONIA VERNADAKIS, NIKOS SAKELLARIDIS,
DIMITRA MANGOURA, AND DAVID DAVIES

1. INTRODUCTION

Only in the last decade has consideration been given to glial cells in the aging brain, and the role of glial cells in the neuronal aging process is far from being understood. Morphological changes in number and volume of astroglia cells have been reported in various brain areas in the aging rat (Geinisman et al., 1978; Landfield et al., 1977; Lindsey et al., 1979). The phenomenon of "gliosis" with aging has been interpreted to reflect a compensatory response of astrocytes to their microenvironment and specifically to neuronal loss. On the other hand, the decline in synaptic function with aging (see review by Vernadakis, 1985) may partially be attributed to the lack of responsiveness of glial cells to synaptic plasticity including "reactive synaptogenesis," a phenomenon observed in neuronal regeneration (Cotman and Nadler, 1978; Cotman and Scheff, 1979). Studies using the hippocampus as a model system to compare reactive synaptogenesis in the mature and aged rat brain have shown that reactive synaptogenesis is slower in the aged animals. Reactive synaptogenesis observed after partial denervation in brain-lesion studies appears to depend upon trophic signals provided by hormones and other growth-regulating factors, and the regulatory role of glial cells, i.e., the ability of glial cells to prepare the neuropil for the growth of new synapses. It is likely that, in the aging brain, glial cells may not supply the required

Vernadakis, A. (ed.), *Model Systems of Development and Aging of the Nervous System.*

growth factors for new synapse formation. Thus the proposal has been put forth that glial cells change with aging and may ultimately contribute to neuronal aging through changes in the microenvironment.

Cultured human fibroblasts have been used as a model to study cellular aging (Swim and Parker, 1957; Hayflick and Moorhead, 1961; Strehler, 1977). Various in vitro alterations of the cell membrane assoicated with age have been reported such as cell size (Simons and Swim, 1967; Guillouzo et al., 1972), sensitivity to Con A-mediated agglutination (Yamamoto and Yamamoto, 1977), surface topography (Bowman and Daniel, 1975), functional specialization (Johnson et al., 1974), carbohydrate content (Courtois and Hughes, 1976; Azencott et al., 1975; Spataro et al., 1979), cell adhesion (Azencott et al., 1975), and hormone receptors (Rosenbloom and Goldstein, 1976). In early studies, we used the glioma cell line (C-6, 2B clone courtesy of Dr. Jean de Vellis) at early passage (20–25) and late passage (50–90) as a model to study changes in glial cells with aging in culture (Parker et al., 1980; Vernadakis et al., 1982). We have found differences in doubling time (12–16 hours for late passage versus over 24 hours for early passage), cell size and protein content (late passage greater than early), responsiveness to hormones, and nucleotide decreases with cell passage. In subsequent studies we have been comparing changes in glial cells in primary cultures derived from newborn and aged mouse brain with days in culture and with cell passage. We have compared the glial enzyme activities of glutamine synthetase (GS) (Martinez-Hernandez et al., 1977; Norenberg and Martinez-Hernandez, 1979) and cyclic nucleotide phosphohydrolase (CNP) (Poduslo, 1975; Poduslo and Norton, 1972), astrocytic and oligodendrocytic markers, respectively, and also have used the astrocytic marker glial fibrillary acidic protein (GFA) (Bignami et al., 1972) and oligodendrocytic marker glycerolphosphate dehydrogenase (GPDH) (de Vellis, 1973; McGinnis and de Vellis, 1974), to identify astrocytes and oligodendrocytes morphologically.

2. MORPHOLOGICAL CHARACTERISTICS

Glial-enriched cultures were prepared from newborn and aged (18-month-old) mouse cerebral hemispheres (Vernadakis et al., 1984). After 3–4 weeks in culture, when cells were at confluency, cultures were trypsinized, 0.1%, and the dispersed cells were replated at a density of 2–3 million. At cell passage 5, cells were frozen in liquid nitrogen and maintained for over two years before subsequent cell passage (Vernadakis et al., 1986). As also reported by several other investigators, in cultures derived from newborn mouse brain, the predominant cells are glial with 5%–10% connective tissue cells, i.e., fibroblasts. The dispersed cells adhere to the dish within 3–4 days, and the various glial populations become apparent within 10–15 days in culture (figure 31-1a). However, in the cultures derived from aged mouse brain, the dispersed cells did not adhere to the dish for at least 9–10 days. Then small patches, or islands, of large flat cells began to appear sporadically

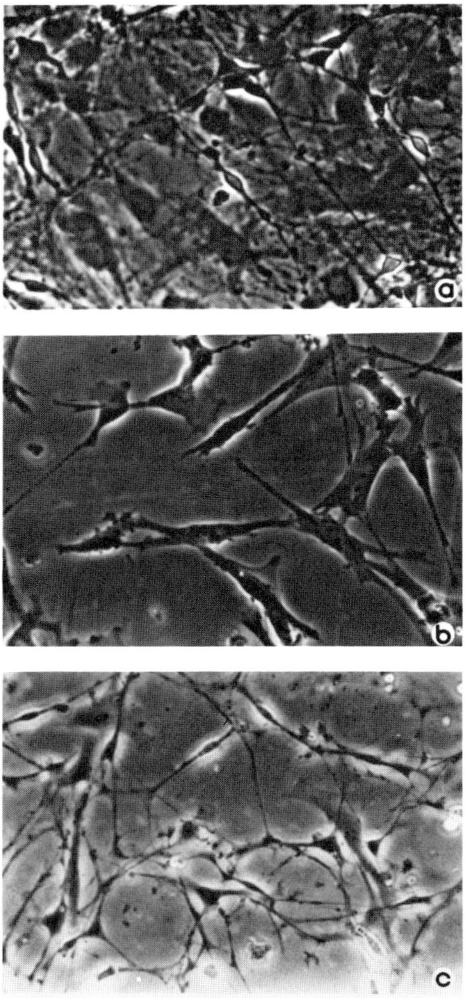

Figure 31-1. (**a**) Glial cell cultures derived from dissociated newborn mouse cerebral hemisphere. A 15-day primary culture (P_0) under phase contrast. Dark process-bearing cells (oligodendrocytes) are seen on top of the flat background cells (astrocytes and connective tissue cells). × 470. (**b**) Flat cell cultures derived from 18-month-old mouse cerebral hemispheres. A 9-day culture under phase contrast. × 470. (**c**) Glial cell cultures derived from 18-month-old mouse cerebral hemispheres. A 14-day primary culture (P_0) under phase contrast. Primarily process-bearing cells and flat background cells are seen. × 470. From Vernadakis et al. (1984).

on the dish (figure 31-1b). By 2–3 weeks in culture, an almost confluent monolayer of flat cells formed, on top of which of smaller number of round dark process-bearing cells, presumptively oligodendrocytes, could be seen (figure 31-1c).

GFA-positive and GPDH-rhodamine-positive cells were present in both types of cell cultures, newborn and aged mouse, up to passage 4 (figure 31-2). However, with progressive cell passage, the number of GFA-positive cells declined significantly as did the intensity and pattern of immunoreactivity (figure 31-3). An attempt to stain these cultures with GPDH antiserum was not successful. Whether this lack of GPDH-positive immunoreactivity reflects a decline in oligodendrocytes (as shown biochemically; see figures 31-9 and 31-10) or a cellular change enzyme configuration cannot be detected immunocytochemically, and thus remains to be further investigated. Recently, Lindsay et al. (1982) have also reported that only a very small percentage (less than 1%) of cells stained with antigalacto-cerebroside antisera, indicating oligodendroglia.

Phase microscopy revealed different patterns of growth in cells cultured under various substratum conditions: plastic stratum, glass coverslips, or sandwiched between plastic and glass (figures 31-4 through 31-8). The cell population on glass coverslips at passage 10 or 11, 24 days in culture (figure 31-4) consisted of carpets of flat epithelioid cells with fusiform, bipolar cells lying on top (figure 31-4, arrows). In the special microenvironment provided in the space between the glass coverslip and the plastic petri floor (figures 31-5 and 31-6), cells of three types were found: at early passages (P5–7), large veil-like epithelioid cells (heavy arrows); at late passages (P10 or 11), large epithelioid, phase-dark cells with processes resembling mature astrocytes (light arrows), and cells with small somata similar to oligodendrocytes (arrowheads). The cell populations grown on plastic (figures 31-7 and 31-8) consisted predominantly of spindle-shaped cells and a smaller number of polyhedral cells. An increasing number of the polyhedral cells were multinucleated at the late passages especially in the cultures derived from the aged mouse (figures 31-7 and 31-8); however, similar cells were also observed in the cultures of late passage initially prepared from newborn (figure 31-6). Somatic polyploidy occurs widely among both plants and animals, but its function is unclear. The phenomenon has been assoicated and/or correlated with such diverse factors as age, hormonal action, cell enlargement, diet, and resistance to mutagenesis (D'Amato, 1977; Epstein, 1967; Evans, 1976; Gerzeli and Barni, 1976; Mysliwski et al., 1977; Nagl, 1978). Extensive evidence from studies of fibroblasts during in vitro aging supports the view that polyploidy increases during in vitro aging (Kaji and Matsuo, 1981; Matsuo et al., 1982).

These findings indicate that the morphological expression of glial cells in culture is dependent upon the culture microenvironment. It appears that each of the three types of microenviornment favored a specific cell population:

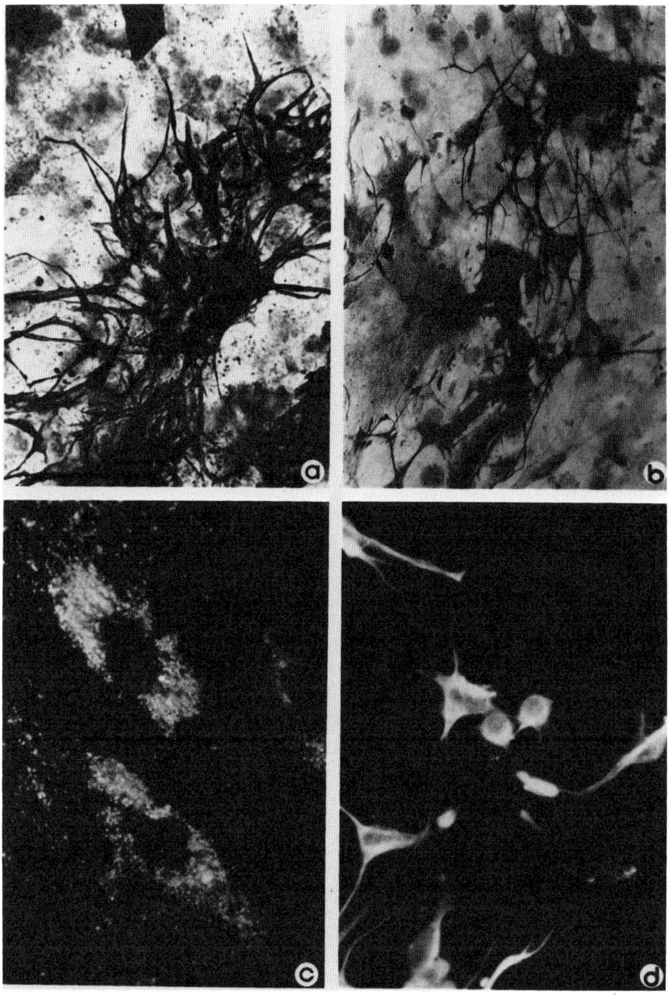

Figure 31-2. (**a**) GFA-positive cells from dissociated newborn mouse cerebral hemispheres, passage 3, 22 days in culture. × 1860. (**b**) GFA-positive cells from dissociated aged mouse (18 months old) cerebral hemispheres, passage, 2, 11 days in culture. × 1900. (**c**) GPDH-rhodamine-positive cells from dissociated newborn mouse cerebral hemispheres, passage 4, 8 days in culture. × 2,700. (**d**) GPDH-rhodamine-positive cells from dissociated aged (18 months old) mouse cerebral hemispheres, passage 5, 25, days in culture. × 1.900. From Vernadakis et al. (1984).

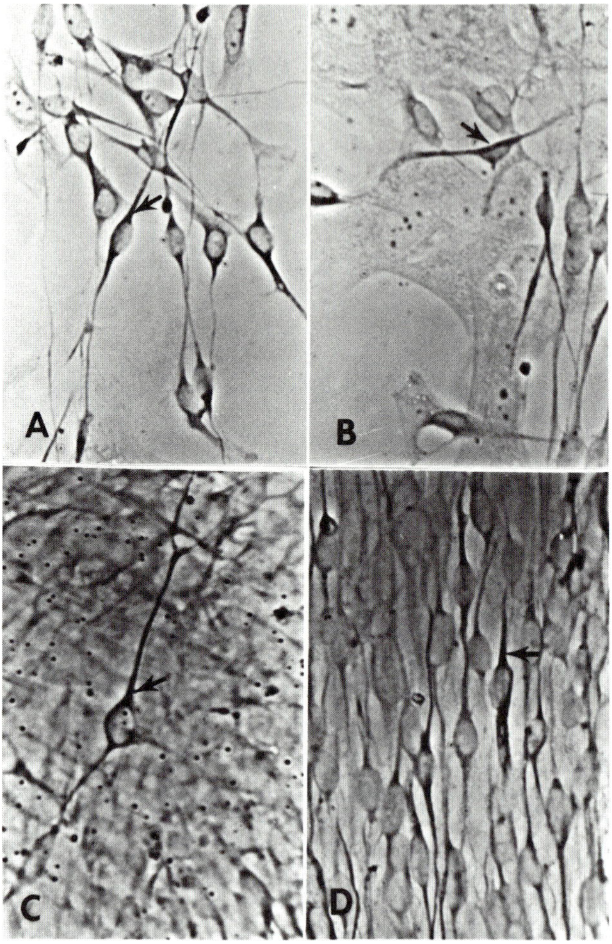

Figure 31-3. Phase-contrast photomicrographs of cultures prepared from mouse cerebral hemispheres. (**a** and **b**) Cultures from newborn mouse, cell passage 10, 24 days in culture. (**c** and **d**) Cultures from aged mouse (18 months), cell passage 11, 24 days in culture. Arrows demonstrate GFA-positive immunoreactivity. From Vernadakis et al. (1986).

glass stratum favored proliferation of flat cells, plastic-favored cells that became multinulceated with passage, and many of the cells that were sandwiched under the glass coverslips resembled mature astrocytes. The influence of cell–stratum interactions on cell morphology remains to be further investigated. However, such interactions have been recently considered in axonal growth and guidance (Letourneau, 1983) and by implication can also be involved in glial differentiation in culture.

The decline in GFA-positive cells with cell passage (figure 31-3) is of

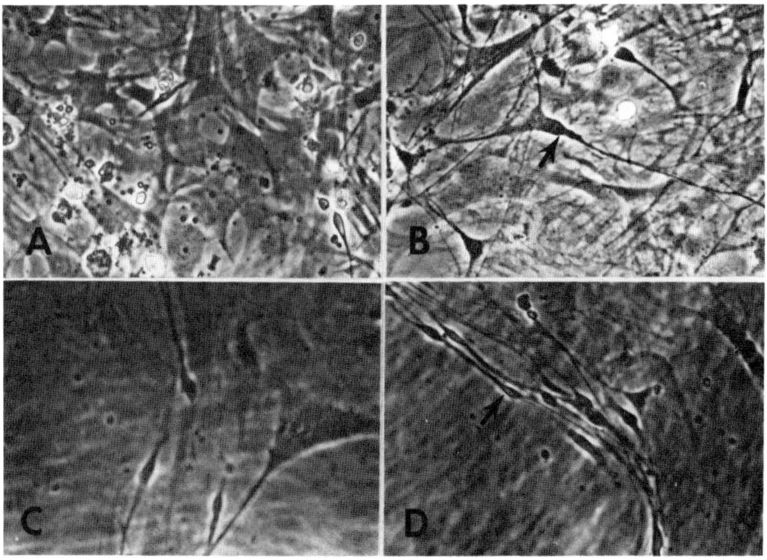

Figure 31-4. Phase-contrast photomicrographs of cultures prepared from mouse cerebral hemispheres. Cells were from late passage (P10 or 11) and were grown on glass coverslips. (**a**) Cultures from newborn mouse, cell passage 10, 12 days in culture, and (**b**) 24 days in culture. (**c**) Cultures from aged mouse (18 months), cell passage 11, 12 days in culture; and (**d**) 24 days in culture. *Arrows* indicate fusiform bipolar cells resting on top of a carpet of epithelioid cells. From Vernadakis et al. (1986).

interest and warrants further exploration. The question that should be considered is whether there is a decline in GFA astrocytes or whether there is a cellular change in the GFA molecule so that the antibody no longer recognizes it and thus the decline in GFA-positive immunoreactivity. In a recent study, Lindsay et al. (1982) characterized astrocyte cultures derived from corpus callosum of adult rat after surgical lesion and a period of postoperative "priming" in vivo. They describe two distinct morphologies in early cultures: process-bearing cells with small rounded perikarya, and very large flattened cells with a readily discernible nuclei. By 3 weeks in culture, 90% of the GFAP-stained cells adopted a flattened morphology and, after 4 or more weeks in culture, process-bearing cells were virtually absent. In our cultures, flat epithelioid cells as described by Lindsay et al. (1982) were observed on the plastic floor of the culture dish (figures 31-7 and 31-8). Meller and Waelsch (1984) have recently reported cyclic morphological changes of glial cells in long-term cultures of rat brain. After four subcultures and 8–10 weeks of cultivation, the following cell types could be distinguished: 1) flat epitheliod cells proposed to be precursors of 2) astroglial cells and GFA-positive and 3) oligodendroglia cells. After 6 weeks, the most prominent cells are the flat epithelioid cells. Our findings also show that with cell passage the

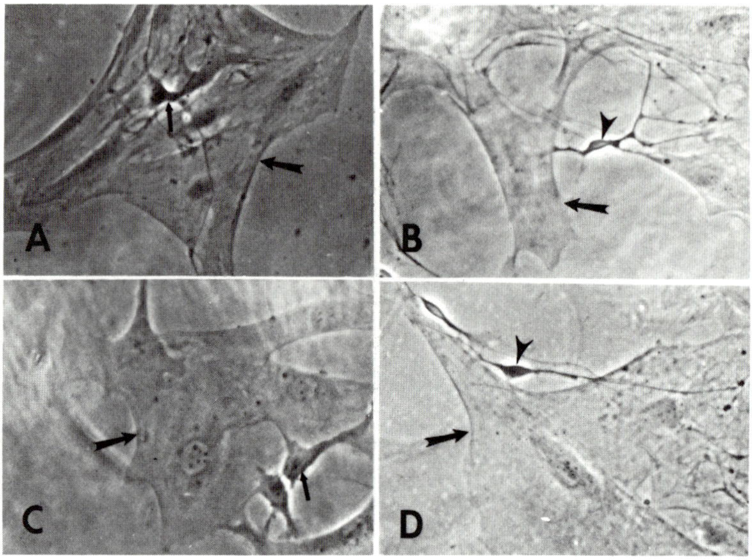

Figure 31-5 Phase-contrast photomicrographs of cultures prepared from mouse cerebral hemispheres. Cells were from early passage (P5–P7) and were grown between glass coverslip and culture dish floor. (**a**) Cultures from newborn mouse, cell passage 6, 12 days in culture; and (**b**) Cell passage 5, 24 days in culture. (**c**) Cultures from aged mouse (18 months), cell passage 7, 12 days in culture; and (**d**) cell passage 6, 24 days in culture. *Heavy arrows* indicate epithelioid cells; *light arrows* show presumptive astrocytes; and *arrowheads* point to presumptive oligodendrocytes. From Vernadakis et al. (1986).

appearance of flat epithelioid cells increases. Based on these findings, we proposed that these changes with cell passage represent cellular sensecence. Changes in cell membrane that ultimately would be reflected in the size and shape of the cell have been reported to occur in various somatic cells with senescence in vitro and inlcude: changes in cell size (Simons and Swim, 1967; Guillouzo et al., 1972), sensitivity to Con A-mediated agglutination (Yamamoto et al., 1977), surface topography (Bowman and Daniel, 1975), functional specialization (Johnson et al., 1974), carbohydrate content (Courtois and Hughes, 1976; Azencott et al., 1975; Spataro et al., 1979), cell adhesion (Azencott et al., 1975), and lectin RCAI-binding sites (Aubery et al., 1980).

3. BIOCHEMICAL CHARACTERISTICS

The biochemical characterization of cultures throughout cell passage 11 is presented in figures 31-9 and 31-10 for CNP activity in figures 31-11 and 31-12 for GS activity. In cultures derived form newborn mouse cerebral hemispheres, CNP activity increased and remained rather constant with cell passage. In contrast, in cultures derived from aged mouse cerebral hemispheres, CNP increased at P2, decreased at P4, and remained at that level at

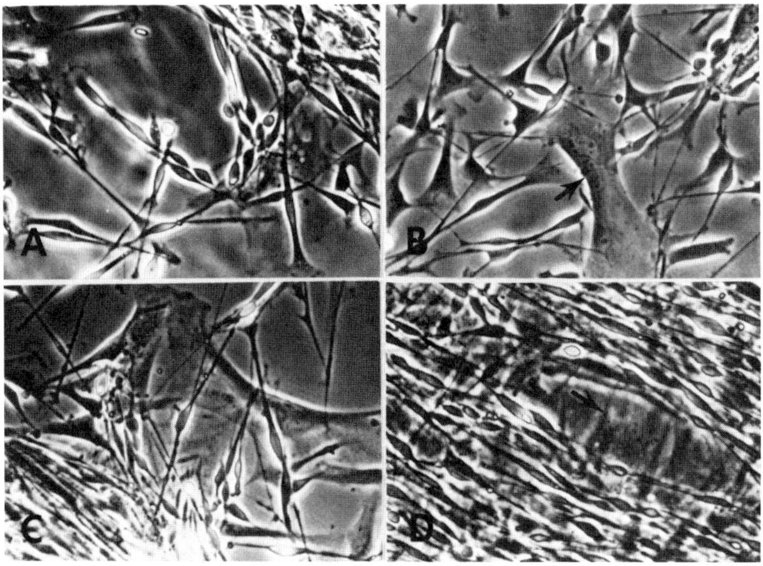

Figure 31-6. Phase-contrast photomicrographs of cultures prepared from mouse cerebral hemispheres. Cells were from late passage P10 or P11, and were grown between glass coverslip and culture dish floor. (**a**) Cultures from newborn mouse, cell passage 10, 12 days in culture; and (**b**) 24 days in culture. (**c**) Cultures from aged mouse (18 months old), cell passage 11, 12 days in culture; and (**d**) 24 days in culture. *Arrow* indicates multinucleated cell.

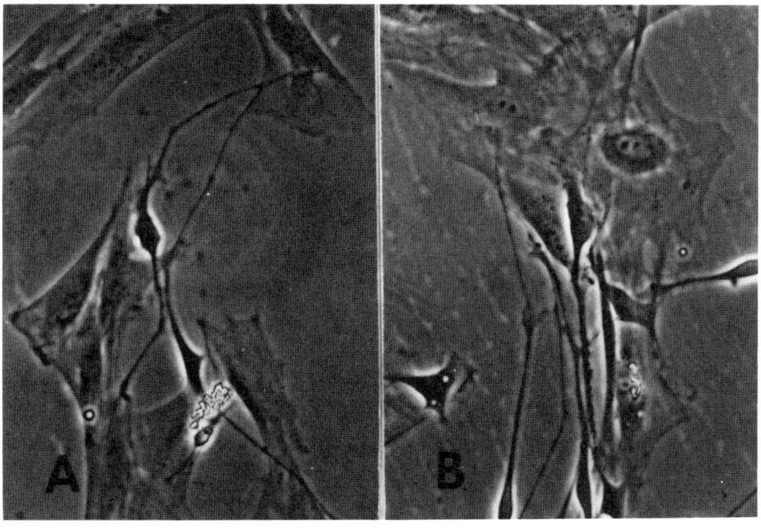

Figure 31-7. Phase-contrast photomicrographs of cultures prepared from mouse cerebral hemispheres. Cells were from early passage (P6 or P7) and were grown on the floor of plastic culture petri dish. (**a**) Cultures from newborn mouse, cell passage 6, 1 day in culture. (**b**) Cultures from agd mouse (18 months), cell passage 7, 1 day in culture. From Vernadakis et al. (1986).

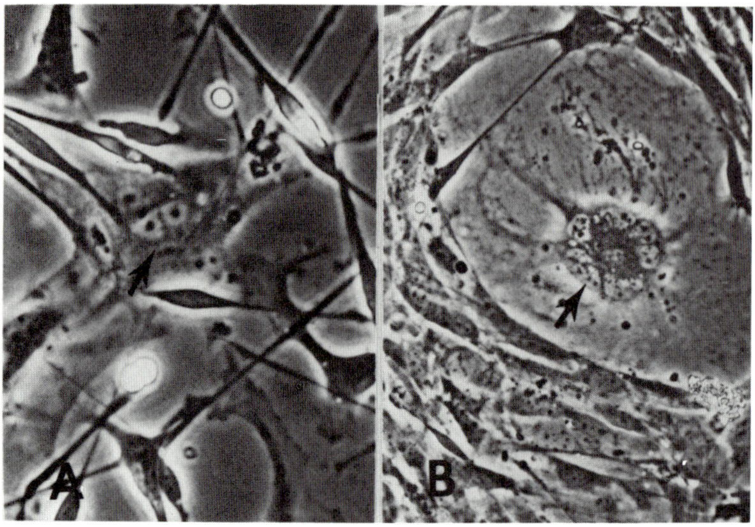

Figure 31-8. Phase-contrast photomicrographs of cultures prepared from mouse cerebral hemispheres. Cells were from late passage (P10 or P11) and were grown on the floor of plastic culture petri dish. (**a**) Cultures from newborn mouse, cell passage 10, 24 days in culture. (**b**) Cultures from aged mouse (18 months), cell passage 11, 24 days in culture. *Arrow* indicates multinucleated cell.

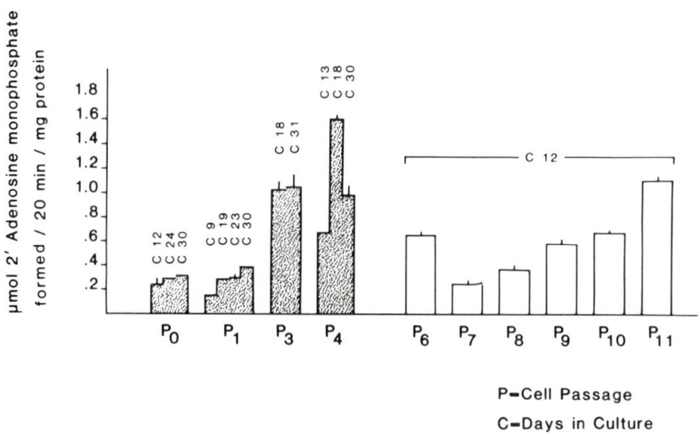

Figure 31-9. Cyclic nucleotide phosphohydrolase activity in glial cell cultures. Cells were dissociated from newborn mouse cerebral hemispheres and used either as primary cultures passage 0, (P0) or passaged cultures (passages 1–11). Data for passaged P0 to P4–5 were previously published (J.N.R. 11: 253–262, 1984). They were presented here for comparison to passages 6–11. *Points with lines* represent means ± SE for 3–5 cultures. From Vernadakis et al. (1986).

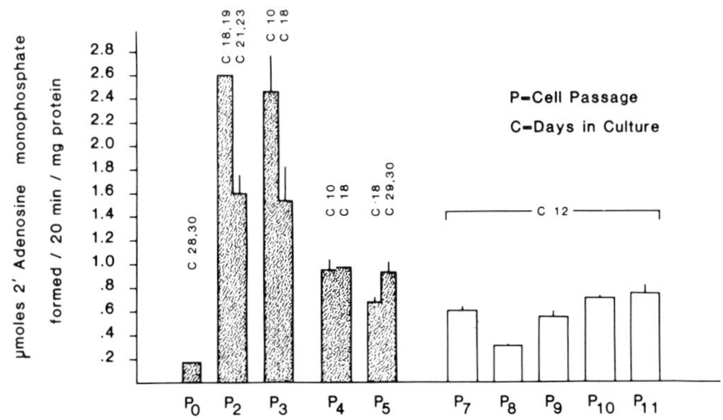

Figure 31-10. As in figure 31-9, except that cells were dissociated from aged mouse (18 months) cerebral hemispheres. From Vernadakis et al. (1986).

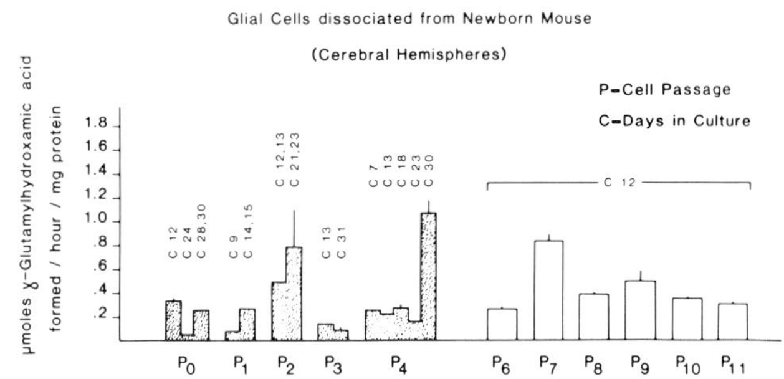

Figure 31-11. Glutamine synthetase activity in glial cell culture. Details as in figure 31-9. From Vernadakis et al. (1986).

all subsequent cell passages. Thus, the responsiveness of CNP-containing glia to cell passage, presumptively oligodendrocytes, to cell passage remained more or less constant. In contrast, the changes in GS activity with cell passage were dramatic (figures 31-10 and 31-11). Whereas GS activity did not change with cell passage in cultures derived from newborn mouse, GS markedly increased between P5 and P8 and again between P8 and P11, reaching a 20-fold value in cultures derived from aged mouse. Whether the increase in GS activity represents increase in cell activity induced by the experimental conditions or increase in number of cells containing GS cannot be ascertained from this study. This marked GS activity coincides temporally with the appearance of large polyhedral multinucleated cells (figures 31-6 and 31-8). Lapham (1962) has reported that, in reactive protoplasmic astrocytosis, there occurs an increased number of nuclei without increased proliferation, suggesting that nuclear division in reactive protoplasmic astrocytes may have

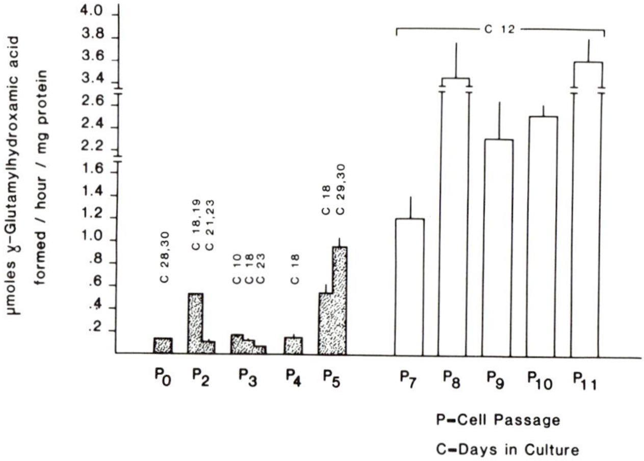

Figure 31-12. As in figure 31-11, except that cells were dissociated from aged mouse (18 months) cerebral hemispheres. From Vernadakis et al. (1986).

a special adaptive significance with greater functional capacity. The possibility that in our study multinucleated cells are GS-containing glial cells remains to be evaluated. The possibility that in culture there is a shift in the astrocyte population with a decline in GFA-positive cells and an increase in GS-containing cells remains to be further explored.

Recently, Raff and associates (1983) have reported that a glial progenitor cell in neonatal rat optic nerve can be induced to develop into a fibrous astrocyte or an oligodendrocyte, depending on the culture medium. Moreover, they have found that some cells in the optic nerve cultures showed a mixed astrocyte–oligodendrocyte expression. Recently, Hallermayer and Hamprecht (1984) have reported colocalization of GS and GFA in primary cultures of dissociated cells from newborn mice. In cultures grown in serum-free medium containing dibutyryl-cAMP, GS was found in about half of the GFA-containing astroblasts. After addition of dexamethasone to the cultures. GS appeared also in another cell type, in which no GFA, but fibronectin, was detectable. Oligodendrocytes revealed by staining for galactocerebroside never contained GS. These reported findings demonstrate not only the heterogeneity of primary glial cells, but also their plasticity in vitro. Extrapolating from these observations to our findings reported here, it is conceivable that the decline in GFA-positive glial cells and the increase in GS activity represent a shift toward GS-containing glial and other cells, reflecting a phenomenon of cellular aging.

Although the increase of astrocytes with cell passage in culture cannot be equated with the astrogliosis occurring in vivo in the aging brain, it may be the equivalent of an in vitro astrogliosis as a response to a stressful environment. One of the most striking characteristics of astrocytes is the reactive hyperplasia they exhibit in response to tissue injury. This reactive response has been interpreted to express the involvement of these cells in the adaptive mechanisms leading to the restoration of tissue homeostasis (Lapham, 1962). From our in vitro findings, it appears that this function of the astrocyte is also expressed in vitro. Based on these parallel in vivo and in vitro phenomena, we are tempted to suggest that the astrocyte is a cell of great potential and adaptability and may be a key cell in the survival of the neuron during the aging process.

The growth patterns of oligodendrocytes with cell passage were striking. In the cultures from newborn mouse, the oligodendrocyte population increased with passage (figure 31-4, passages 3 and 4). In contrast, in the cultures derived from the aged mouse, there was a great "burst" of oligodendrocytes in passages 2 and 3 (figure 31-4); the oligodendrocytes then began to decline. Again, Lindsay et al. (1982) reported that, after two or three passages, anti-GC-positive reactive cells were rare in adult astrocyte cultures. The decline in oligodendrocytes has been a general phenomenon observed in culture. One possibility advocated by us and others is that the decline reflects the dependence of oligodendrocytes on neuronal factors. In another study using glia-enriched cultures derived from embryonic chick brain, we found that exposure of these cultures to neuron-conditioned medium enhanced the activity of CNP and is interpreted to reflect proliferation of oligodendrocytes (Sakellaridis et al., 1986). Also in glia-enriched cultures derived from chick embryo, CNP activity decreases with days in culture, again perhaps reflecting the absence of neurons in the cultures (Sakellaridis et al., 1983). On the other hand, Abney et al. (1983) reported that precursor cells in 17- to 18-day embryonic rat optic nerve can develop into oligodendrocytes in vitro in the absence of living neurons. Thus, the possibility of other factors, possibly present in the medium, that may play a role in oligodendrocyte survival warrants consideration. De Vellis and associates (Sanetto and de Vellis, 1985; Morrison and de Vellis, 1981) have recently reported oligodendrocyte and astrocyte growth and proliferation in chemically defined media. Of interest is also their finding that 30% of isolated oligodendrocytes cultured in serum-supplemented medium express GFA immunoreactivity. Taken into a broader perspective of cell growth phenomena, these findings appear to support the view that glial cells show tremendous plasticity in their phenotypic expression.

4. GENERAL CONCLUSIONS

In vitro model systems for studying cellular aging have provided significant clues about in vivo aging processes. The cell model presented in this chapter demonstrates that astrocytes and oligodendrocytes undergo changes with cell passage in culture. The changes observed can be correlated to changes observed with aging in vivo and thus further justify the validity of this cell system as an aging model.

As also discussed in other chapters of this book, several roles have been attributed to glial cells in CNS function including myelination, modulation of extraneuronal microenvironment by secreting factors, neurotransmitter concentration in the synaptic cleft, and ionic and amino acid regulation. Changes therefore in glial cells with aging will influence the function of the neuron and may contribute to its malfunction and ultimate degeneration. Thus, although neuronal degeneration is not a primary finding in CNS aging, the decreased response of the neuron to its microenvironment with age may be a function of the glial cell, a major regulator of the neuronal microenvironment.

ACKNOWLEDGMENTS

The original work in this chapter was supported by USPHS training grant T32 HD 07072, research grant USPHS HD 18894 from the NICHD, and the Developmental Psychobiology Research Group Endowment Fund.

REFERENCES

Abney, E.R., Williams, B.P., and Raff, M. (1983) Tracing the development of oligodendrocytes from precursor cells using monoclonal antibodies, fluorescence-activated cell sorting, and cell culture. Dev. Biol. 100:166–171.

Aubery, M., Guillouzo, A., Bernard, B., and Front, J. (1980) Changes in *ricinus communis* lectin binding to the cell surface of human liver cells in culture. The relationship between in vitro cellular aging and the age of the donor. Exp. Cell Res. 129:273–280.

Azencott, R., Hughes, K.C., and Courtois, P. (1975) Cell surface alterations and "in vitro" aging of animal cells. Adv. Exp. Med. Biol. 53:147–153.

Bignami, A., Eng. L.F. Dahl. D., and Uyeda, C.T. (1972) Localization of the glial fibrillary acidic protein in astrocytes by immunofluoresence. Brain Res. 43:429–435.

Dowman, P.D., and Daniel, C.W. (1975) Aging of human fibroblasts in vitro: surface features and behavior of aging W138 cells. Mech. Ageing Dev. 4:147–158.

Cotman, C.W., and Nadler, J.V. (1978) Reactive synaptogenesis in the hippocampus. In: Cotman, C.W. (ed.) Neuronal plasticity. New York: Raven, pp. 227–271.

Cotman, C.W., and Scheff, S.W. (1979) Compensatory synapse growth in aged animals after neuronal death. Mech. Ageing Dev. 9:103–117.

Courtois, Y., and Hughes, R.C. (1976) Surface labelling of senescent chick fibroblasts by lactoperoxidase-catalysed iodination. Geronotology 22:371–379.

D'Amato, F. (1977) Nuclear cytology in relation to development. London: Cambridge University Press.

de Vellis, J. (1973) Mechanisms of enzymic differentiation in the brain and in cultured cells. In: Rockstein, M., and Sussman, M.L. (eds.) Development and aging of the nervous system. New York: Academic Press, pp. 171–198.

Epstein, C.J. (1967) Cell size, nuclear content, and the development of polyploidy in the mammalian liver. Proc. Natl. Acad. Sci. USA 57:327–334.

Evans, I.H. (1976) Polyploidization in rat liver: the role of binucleate cells. Cytobios 16:115–124.

Geinisman, Y., Bondareff, W., and Dodge, J.T. (1978) Hypertrophy of astroglial processes in the dentate gyrus of senescent rat. Am. J. Anat. 153:537–544.

Gerzeli, G., and Barni, S. (1976) Changes in liver cell ploidy of young rats following isoprenaline treatment. Cell Tissue Kinet. 9:267–272.

Guillouzo, A., Guguen, C., Lenoir, P., and Bourel, M. (1972) Modifications morphologiques et

ultrastructurales liecs à la senescence dans les ligness de foci human adulte. C.R. Acad. Sci. Paris 274:484–487.
Hallermayer, K., and Hamprecht, B. (1984) Cellular heterogeneity in primary cultures of brain cells revealed by immunocytochemical localization of glutamine synthetase. Brain Res. 295:1–11.
Hayflick, L., and Moorhead, P.S. (1961) The serial cultivation of human diploid cell strains. Exp. Cell Res. 25:585–621.
Johnson, R., Hammer, M., Sheridan, J., and Revel, J.P. (1974) Gap junction formation between reaggregated Novickoff hepatoma cells. Proc. Natl. Acad. Sci. USA 71:4536–4540.
Kaji, K., and Matsuo, M. (1981) Aging of chick embryo fibroblasts in vitro. V. Time course studies on polyploid nucleus accumulation. Exp. Cell Res. 131:410–412.
Landfield, P.W., Sandles, L., Wohlstadter, T.C., and Lynch, G. (1977) Patterns of astroglial hypertrophy and neuronal degeneration in the hippocampus of aged, memory-deficient rats. J. Gerontol. 32:3–12.
Lapham, L.W. (1962) Cytologic and cytochemical studies of neuroglia. I. A study of the problem of amitosis in reactive protoplasmic astrocytes. Am. J. Pathol. 21:1–21.
Letourneau, P.C. (1983) Axonal growth and guidance. Trends Neurosci. 6:451–455.
Lindsay, R.M., Barber, P.C., Sherwood, M.R.C., Zimmer, J., and Raisman, G. (1982) Astrocyte cultures from adult rat brain: derivation, characterization and neurotrophic properties of pure astroglial cells from corpus callosum. Brain Res. 243:329–343.
Lindsey, J.D., Landfield, P.W., and Lynch, G. (1979) Early onset and topographical distribution of hypertrophied astrocytes in hippocampus of aging rats: a quantitative study. J. Gerontol. 34:661–671.
Martinez-Hernandez, A., Bell, K.P., and Norenberg, M.D. (1977) Glutamine synthetase: glial localization. Science 195:1356–1358.
Matsuo, M., Kaji, K., Utakoji, T., and Hosoda, K. (1982) Ploidy of human embryonic fibroblasts during in vitro aging. J. Gerontol. 37:33–37.
McGinnis, J.F., and de Vellis, J. (1974) Purification and characterization of rat brain glycerolphosphate dehydrogenase. Biochim. Biophys. Acta 360:17–27.
Meller, K., and Waelsch, M. (1984) Cyclic morphological changes of glial cells in long-term cultures of rat brain. J. Neurocytol. 13:29–47.
Morrison, R.S., and de Vellis, J. (1981) Growth of purified astrocytes in a chemically defined medium. Proc. Natl. Acad. Sci. USA 78:7205–7209.
Mysliwski, A. Mysliwski, J., and Kmiec, Z. (1977) Age-related changes in chromatin of liver cell nuclei of different ploidity. Histochemistry 52:91–96.
Nagl, W. (1978) Endopolyploidy and polyteny in differentiation. Amsterdam: North Holland.
Novenberg, M.D., and Martinez-Hernandez, A. (1979) Fine structural localization of glutamine synthetase in astrocytes of the rat brain. Brain Res. 161:303–310.
Parker, K., Norenberg, M., and Vernadakis, A. (1980) "Transdifferentiation" of C-6 glial cells in culture. Science 208:179–181.
Poduslo, S.E. (1975) The isolation and characterization of a plasma membrane and a myelin fraction derived from oligodendroglia of calf brain. J. Neurochem. 24:647–654.
Poduslo, S.E., and Norton, W.T. (1972) Isolation and some chemical properties of oligodendroglia from calf brain. J. Neurochem. 19:727–736.
Raff, M., Miller, R.H., and Noble, M. (1983) A glial progenitor cell that develops in vitro into an astrocyte or an oligodendrocyte depending on culture medium. Nature 303:390–396.
Rosenbloom, A.L., Goldstein, S., and Yip, C.C. (1976) Insulin binding in cultured human fibroblasts increases with normal and precocious aging. Science 193:412–415.
Sakellaridis, N.S., Bau, D., Mangoura, D., and Vernadakis, A. (1983) Developmental profiles of glial enzymes in the chick embryo: in vivo and in culture. Neurochem. Int. 5:685–689.
Sakellaridis, N.S., Mangoura, D., and Vernadakis, A. (1986) Effects of neuron-conditioned medium and fetal calf serum content on glial growth in dissociated cultures. Dev. Brain Res. 27:31–41.
Saneto, R.P., and de Vellis, J. (1985) Characterization of cultured rat oligodendrocytes proliferating in a serum-free, chemically defined medium. Proc. Natl. Acad. Sci. USA 82:3059–3513.
Simons, J.W.I.M., and Swim, H.E. (1967) The use of frequency distribution of cell diameters to characterize cell populations in tissue culture. Exp. Cell Res. 45:336–350.
Spataro, A.C., Bosmann, H.B., and Myers-Robfogel, M.W. (1979) Sialyltransferase activities of aging diploid fibroblasts. Biochim. Biophys. Acta 553:378–387.
Strehler, B.L. (1977) Time cells and aging. New York: Academic Press.

Swim, H.E., and Parker, K.F. (1957) Culture characteristics of human fibroblasts propagated serially. Am. J. Hyg. 66:235–243.

Vernadakis, A. (1985) The aging brain. Clin. Geriatri. Med. 1:61–94.

Vernadakis, A., Parker, K., Arnold, E.B., and Norenberg, M. (1982) Role of glial cells in CNS aging in the nervous system. New York: Raven, pp. 57–68.

Vernadakis, A., Manguora, D., Sakellaridis, N., and Linderholm, S. (1984) Glial cells dissociated from newborn and aged mouse brain. J. Neurosci. Res. 11:253–262.

Vernadakis, A., Davies, D., Sakellaridis, N., and Mangoura, D. (1986) Growth patterns of glial cells dissociated from newborn and aged mouse brain with cell passage. J. Neurosci. Res. 15:79–85.

Yamamoto, K., Yamamoto, M., and Ooka, M. (1977) Cell surface changes associated with aging of chick embryo fibroblasts in culture. Exp. Cell Res. 108:87–93.

32. TISSUE CULTURE IN STUDIES OF AGING IN THE NERVOUS SYSTEM: TOXICITY OF MPTP TO DOPAMINE NEURONS

CATHERINE MYTILINEOU

1. INTRODUCTION

During the very early stages of the development of the nervous system, the dividing neuronal precursors abruptly stop the synthesis of DNA and lose the ability of cell division. From this point, the young neurons will undergo extensive changes in their morphology and biochemistry during the process of differentiation. During the lifespan of an organism, the neurons undergo changes that could be associated with the process of aging. These include morphological changes, like the accumulation of pigments such as lipofuscin, amyloid and senile plaques (Porta and Hartroft, 1969; Brizzee et al., 1975), and changes in the dendritic tree and ultrastructure of synapses (Scheibel and Scheibel, 1975; Feldman, 1976). Alterations in the levels of neurotransmitters and the enzymes associated with their synthesis and catabolism have also been attributed to the process of aging (Finch, 1973; Algeri et al., 1983). The number of neurons in the brain decreases as the age of an organism advances (Brody, 1955). Different neuronal groups show different sensitivity to neuronal loss in the process of advancing age.

Since the discovery that embryonic neurons can be kept in vitro by Harrison at the beginning of the century (Harrison, 1907), neuronal tissue cultures have been used extensively, mainly for studies of the process and control of neuronal development. Neurons from embryonic or early-postnatal brain tissue can be maintained in vitro for periods varying from a few weeks to several months, depending on the method of culture. In general, dissociated cells survive shorter periods in culture than explants; supplementing the feeding medium with serum also appears to promote prolonged in vitro survival.

One way of using tissue cultures in studying the process of neuronal aging is to culture neurons from adult nervous system (Silberberg and Kim, 1979; Kim, 1983). In these studies, dorsal root and sympathetic ganglia from adult animals and human postmortem tissue have been successfully kept in tissue culture. The advantage of this approach is that the cultured cells have already undergone the changes that result from the advancing age. One limitation of this method of study is that only peripheral nervous tissue, possessing the ability to regenerate, can be used. A different approach is to use embryonic cells that have reached a plateau of the growth process and study their response to experimental stress. Such studies could provide insights in the ways that neurons could respond to the natural environmental stress occurring during the process of aging.

We have used the latter experimental approach in our studies of the dopamine (DA)-containing neurons of the substantia nigra and ventral tegmental area grown in explant cultures. The main lesion in Parkinson's disease is destruction of the substantia nigra DA neurons, but other neuronal systems can be affected as well. Parkinsonism usually affects individuals of advanced age. Although the etiology of Parkinson's disease remains unknown, it is possible that during the process of aging the neurons of the substantia nigra zona compacta lose some of their defense mechanisms and become more susceptible to the toxic by-products of cellular metabolism or environmental toxins. The recent discovery that 1-methyl-4-phenyl-1,2,3,6-tetrahydropyridine (MPTP) can produce damage, specifically to the DA neurons of the substantia nigra zona compacta, provides a tool for the investigation of some of these possibilities.

In our recent studies we have examined the effect of MPTP on explant cultures from embryonic rat midbrain (Mytilineou and Cohen, 1984; Mytilineou et al., 1985; Mytilineou and Cohen, 1986). We have also attempted to modify the toxic effect of MPTP on the DA neurons by experimentally altering the cellular microenvironment. Such experiments could provide information on the mechanisms involved in neuronal toxicity and the defense mechanisms utilized by the neurons to survive environmental stress.

2. IN VIVO MPTP TOXICITY

The toxic effect of MPTP on DA neurons was discovered when it was accidentally self-administered by illicit drug users. Several of these people developed severe parkinsonian symptoms (Davis et al., 1979; Langston et al., 1983). In one case that was autopsied later, there was an extensive loss of pigmented neurons in the substantia nigra zona compacta (Davis et al., 1979). Thus the neurological symptoms produced after injection of MPTP appeared to be due to a lesion similar to that observed in patients with Parkinson's disease.

Reproducing the toxic effect of MPTP on DA neurons in an experimental animal model would obviously be a valuable tool in studies related to parkinsonism as well as neuronal toxicity in general. For this reason, several laboratories undertook to develop such a model.

The animal model that best reproduces the toxic effects of MPTP observed in humans is the monkey. After administration of relatively low doses of MPTP, the monkey shows both biochemical lesions (depletion of DA from the striatum and degeneration of substantia nigra pars compacta neurons) and behavioral effects that mimic parkinsonism (Burns et al., 1983; Langston et al., 1984a). Rodents are much less sensitive to MPTP toxicity (Chiueh et al., 1985). However, mice develop biochemical lesions of the nigrostriatal DA system after relatively high doses of the toxin (Heikkila et al., 1984a) and have been used extensively in studies of many aspects of MPTP toxicity.

Our current understanding of the toxicity produced by MPTP can be summarized briefly as follows. After systemic injection, MPTP permeates the brain where it is rapidly metabolized to 1-methyl-4-phenylpyridine (MPP^+; Langston et al., 1984b; Markey et al., 1984). This biotransformation is mediated by monoamine oxidase B (MAO B; Chiba et al., 1984). It is beleived the MPP^+ is formed by MAO B extraneuronally, possibly within glial cells (Uhl et al., 1985) and then it becomes available to accumulate within the DA neurons. MPP^+, but not MPTP, uses the high-affinity uptake pump of the catecholamine neurons (Javitch and Snyder, 1985; Javitch et al., 1985). Thus it can accumulate in very high concentrations within the DA terminals in the striatum and produce its toxic effects.

3. MPTP TOXICITY IN CULTURES

Several years ago our laboratory developed a tissue culture model of the nigrostriatal system (Whetsell et al., 1981; Mytilineou et al., 1983) for use in studies related to the pathogenesis of parkinsonism. The discovery of MPTP prompted us to examine the possibility of using the tissue culture model and studying the effects of the toxin when directly applied to the DA neurons.

Organotypic cultures from embryonic rat midbrain proved a useful in vitro model for studying the toxicity of MPTP (Mytilineou and Cohen, 1984; Cohen and Mytilineou, 1985). Although toxicity in the rodent occurs only after high doses of MPTP, in tissue culture DA neurons from both rats and mice are destroyed by micromolar concentrations of MPTP (Mytilineou and Cohen, 1986). The reason for the increased sensitivity of the cultured DA neurons to MPTP is not presently understood. However, some of the differences between the in vivo and an in vitro model could account for this effect. For example, an increase in the sensitivity could result from the continuous presence of relatively stable concentrations of the toxin in the tissue culture medium. In several animal species studied (rat, mouse, guinea pig, monkey), the toxicity of MPTP parallels the length of time that ^{3}H-MPTP is present in the brain after peripheral injection (Johannessen et al., 1985). Alternatively, the susceptibility of the cultures to the toxin could be due to the increased ratio of glial cells over neurons. If, indeed, MPTP is metabolized to MPP$^+$ within glial cells, the increased glia-to-neuron ratio could contribute to an increase in the availability of the toxic metabolite. Another possibility is that the cultured neurons, because of the deficient tissue culture environment, either never develop adequate defense mechanisms or undergo accelerated aging and cannot overcome the stress of the neurotoxin. It has been reported that aged mice are more sensitive to MPTP toxicity than are younger ones (Langston, 1985).

In our initial studies we characterized the toxicity of MPTP on the rat midbrain explants (Mytilineou and Cohen, 1984). MPTP, when applied to the cultures on day 8 in vitro, produces a severe depletion of several indices of dopaminergic neuronal intergrity. In order to visualize the DA neurons present in the cultures, we utilized their ability to accumulate exogenous catecholamines. We incubated the cultures with alpha-methylnorepinephrine at a 10-μM concentration for 30 minutes at 37°C. This treatment fills both neuronal somata and processes with the amine. Subsequent treatment for catecholamine histofluorescence renders visible all the neurons that still possess an amine uptake pump. In order to quantitate the degree of damage produced by the toxin, we measured the ^{3}H-DA uptake by the cultures. Since the high-affinity uptake pump is mainly localized on the axonal membrane, this measurement mostly represents the surviving DA axons. Finally, biochemical analysis of the cultures permitted us to determine whether application of the toxins resulted in changes in the levels of DA and its metabolites.

The effect of exposure to 10-μM MPTP on midbrain DA neurons in explant cultures is summarized in table 32-1. Exposure to 10-μM MPTP resulted in significant reduction of uptake sites on the DA neurons. The toxic effect was increased upon prolongation of exposure to the same concentration of the toxin. This toxic effect of MPTP was not reversed after 5 days of washing with drug-free medium, an indication that the depletion in ^{3}H-DA uptake was probably due to degeneration of the dopaminergic axons. Catecholamine histofluorescence, after incubation of the cultures with alpha-methylnorephrine, also indicated that MPTP treatment for 7 days resulted in depletion of dopaminergic neurons in the explants (Mytilineou and Cohen, 1984; Mytilineou et al., 1985).

In vitro, biochemical experiments have shown that MPTP is metabolized by brain monoamine oxidase to a pyridinium species, MPP$^+$ (Chiba et al., 1984). MPP$^+$ cannot cross the blood–brain barrier and therefore cannot be administered peripherally to experimental animals. In order to determine whether the metabolic product of MPTP is responsible for the toxicity, we applied MPP$^+$ to midbrain explants in a protocol similar to that used for

Table 32-1. Uptake of ^{3}H-DA by cultures of embryonic rat midbrain after treatment with MPTP.

Drug exposure (days)	No drug wash (days)	^{3}H-DA uptake		
		Control	MPTP	% Control
4	1	0.193 ± 0.29	0.118 ± 0.013	61
4	5	0.340 ± 0.088	0.119 ± 0.016	35
7	1	0.214 ± 0.019	0.055 ± 0.010	18
7	5	0.411 ± 0.070	0.072 ± 0.025	15

Cultures were treated with 10-μ*M* MPTP on day 8 in vitro. The cultures were processed for ^{3}H-DA uptake as described in the section on methodology. The values represent picomoles of ^{3}H-DA per culture ± SEM. The uptake of MPTP-treated cultures differs from the corresponding controls by at least $p < 0.05$ (two-tailed Student's *t*-test). The results presented in this table are from Mytilineou and Cohen (1984).

Table 32-2. Effect of different concentrations of MPP$^+$ and MPTP on the uptake of ^{3}H-DA by cultures of embryonic rat midbrain.

Drug	Concentration (μ*M*)	% Control	*n*
MPP$^+$	0.2	100 ± 15	6
	0.6	110 ± 21	5
	2.5	35 ± 4*	6
	10.0	5 ± 0.4*	5
MPTP	2.5	100 ± 14	6
	10.0	44 ± 9*	6
	25.0	42 ± 4*	7

Cultures were exposed to the drugs for 7 days and the uptake was measured 24 hours after removal of the drugs. Uptake for control cultures for MPP$^+$ was 0.52 ± 0.03 and for MPTP 0.24 ± 0.03 pmol ^{3}H-DA/culture. *Differs from control $p < 0.005$; Student's *t*-test. From Mytilineou et al. (1985).

MPTP (Mytilineou et al., 1985). These experiments (shown in table 32-2) demonstrated that MPP$^+$ is indeed toxic to the DA neurons in culture. The concentration of MPP$^+$ necessary for producing toxic changes in the DA neurons is lower than that of MPTP. The observation that MPP$^+$ is a more potent toxin than MPTP is consistent with the suggestion that it might be responsible for MPTP toxicity. MPP$^+$ is also a very potent neurotoxin with specificity for striatal DA neurons when injected intracysternally into mice (Bradbury et al., 1986).

4. IS DOPAMINE INVOLVED IN MPTP TOXICITY?

It has been proposed that DA, if not properly stored in the synaptic vesicles, could produce damage through the generation of hydrogen peroxide and hydroxyl radicals during its metabolic oxidation by MAO (Cohen, 1983). In our original experiments with MPTP, we wanted to investigate whether its toxicity was related to the formation of hydrogen peroxide and oxyradicals through the metabolism of DA by MAO. To examine that possibility, we

Table 32-3. Effect of exposure to L-DOPA on the levels of DA, DOPAC, and HVA in midbrain explants on day 8 in vitro.

L-DOPA (hours)	DA (ng/culture)	DOPAC (ng/culture)	HVA (ng/culture)	DOPAC/DA
Control (no DOPA)	3.5 ± 0.2	0.58 ± 0.05	1.7 ± 0.5	0.17
1	8.4 ± 1.0	4.8 ± 0.2	1.1 ± 0.3	0.57
2	9.1 ± 0.7	6.3 ± 0.5	1.3 ± 0.1	0.69
4	7.7 ± 0.5	6.8 ± 0.5	1.2 ± 0.1	0.88
24	5.3 ± 0.5	1.4 ± 0.2	0.9 ± 0.05	0.27

The cultures were established from embryos on gestational day 16 ($n = 3$–5 cultures per group). On day 8 in vitro, they were exposed to 100-μ*M* L-DOPA in feeding medium to which 0.2 mg/ml ascorbic acid were added. At the times indicated, the cultures were washed twice with balanced salt solution and processed for high-performance liquid chromatography as described in methodology.

treated cultures with an MAO inhibitor (parygyline or deprenyl) 24 hours prior and during exposure to MPTP. Treatment with MAO inhibitors significantly protected the DA neurons from the toxic effect of MPTP (Mytilineou and Cohen, 1984; Cohen and Mytilineou, 1985). Although these observations appeared to confirm our hypothesis of an involvement of DA in MPTP toxicity, at about the same time other laboratories demonstrated that MAO participated in the metabolism of MPTP (Chiba et al., 1984), a process necessary for the expression of its toxicity (Markey et al., 1984). Therefore the protective effect of MAO inhibitors on the toxicity of MPTP that we observed in the cultures could be the result of the inhibition of its metabolism to an active toxin. Administration of MAO inhibitors was also protective to monkeys and mice against the toxicity of MPTP (Cohen et al., 1984; Langston et al., 1984b; Heikkila et al., 1984b).

Since the experiments with MAO inhibitors could not provide direct evidence for the participation of DA in the toxicity of MPTP, we attempted to increase the intraneuronal content and metabolism of DA by incubating cultures with L-DOPA. Table 32-3 shows the effect of addition of L-DOPA (100 μ *M*) to the feeding medium on the DA, dihydroxyphenylacetic acid (DOPAC), and homovanillic acid (HVA) content of explant cultures during a 24-hour exposure. The addition of L-DOPA resulted in increased levels of DA, which peaked at 2 hours, began to decrease by 4 hours, and returned toward normal levels by 24 hours. DOPAC levels also increased, remained at very high levels from 2 to 4 hours, and returned toward normal levels by 24 hours. After 2 hours, DOPAC levels were more than ten times that of control cultures, while DA levels increased during this time only by 2.6 times. The DOPAC–DA ratio indicates that exposure to L-DOPA results in an increased rate of DA metabolism in the dopaminergic neurons. The metabolic rate of DA increases threefold by 1 hour and continues to rise for the next 3 hours. As with DA, by 24 hours DOPAC levels also return

Table 32-4. Effect of exposure to L-DOPA on the toxicity of MPTP to DA neurons in culture.

Treatment	^{3}H-DA uptake (cpm/culture ± SEM)
Control	5431 ± 459
MPTP	3699 ± 290
L-LOPA	5286 ± 650
L-DOPA + MPTP	3596 ± 433

Cultures were treated with 10-μM MPTP in the presence or absence of 100-μM L-DOPA. The drug treatment was repeated twice for 24-hour intervals. All drugs were removed 24 hours before the analysis for ^{3}H-DA uptake. Eight cultures were used for each drug treatment, except for controls (seven cultures.)

toward control values. The levels of HVA do not increase in the explants at any time after exposure to L-DOPA. However, as expected, the HVA levels in the feeding medium are greatly increased during this period (unpublished results).

Using the above experimental protocol, we incubated cultures with 10-μM MPTP in the presence or absence of L-DOPA. The cultures were treated for 48 hours and the culture medium was changed at 24 hours. Thus the DA neurons exposed to L-DOPA would undergo two cycles of very active DA synthesis and metabolism. As shown in table 32-4, exposure of the cultures to 10-μM MPTP for 48 hours caused only a small (32%) reduction in the ^{3}H-DA accumulation. The presence of L-DOPA during MPTP exposure did not in any way modify the toxicity of MPTP. These results suggest that probably the levels of DA and its rate of metabolism are not related to the toxic action of MPTP. A similar conclusion has been reached in experiments with mice, where depletion of DA by treatment with the tyrosine hydroxylase inhibitor alpha-methyltyrosine did not affect the toxicity of MPTP (Schmidt et al., 1985; Fuller and Hermick-Luecke, 1985).

In another set of experiments, we attempted to evaluate the involvement of hydroxyl radicals in the toxicity of MPTP. We used ethanol as a scavenger of hydroxyl radicals (Allis and Cohen, 1977). In these experiments, ethanol failed to protect the cultures from the toxicity of MPTP at all the concentrations that were not producing damage to the cultures. However, these results should be reevaluated, in view of recent studies that indicate that antioxidants partially protect mice from the toxicity of MPTP (Perry et al., 1985).

Since the proposal by Harman (1956) that one factor responsible for the process of aging could be free radical attacks on the cell constituents, many studies have been designed to examine the relationship of free radical generation and cell damage or death (Tappel, 1968; Del Maestro et al., 1980; Cohen, 1983; Vistnes et al., 1983). The studies described in this chapter demonstrate the possibility of using an in vitro model of central nervous system neurons in experiments that could provide insights on the generation of free radicals, their effect on neuronal integrity, and the possibility of increasing the effectiveness of the neuronal defense mechanisms against such insults.

5. CONCLUSIONS

This chapter discusses the possibility of using tissue culture preparations from embryonic nervous system in studies of neuronal aging. One approach to such studies is to examine the response of cultured neurons to experimentally induced environmental stress. Experiments using the neurotoxin 1-methyl-4-phenyl-1,2,3,6-tetrahydropyridine (MPTP), which, when administered in vivo, can destroy specifically the dopaminergic neurons of the substantia nigra, are presented. MPTP toxicity can be reproduced in cultures of embryonic rat midbrain. The dopaminergic neurons in these cultures are destroyed by exposure to micromolar concentrations of MPTP. Bioactivation of MPTP by monoamine oxidase is necessary for the toxicity. The product of oxidation of MPTP, 1-methyl-4-phenylpyridinium (MPP^+), is more toxic to the cultured dopamine neurons than is MPTP. The toxicity of MPTP is not modified by an increase in the intraneuronal concentration and metabolism of dopamine. The use of such experimental studies to obtain insights into the intraneuronal self-defense mechanisms is evaluated.

AKNOWLEDGMENTS

This work was supported by NIH grants NS-11631 and NS-18979, a Javits Award, and by the American Parkinson Disease Association and American Federation for Aging Research.

REFERENCES

Algeri, S., Calderini, G., Toffano, G., and Ponzio, F. (1983) Neurotransmitter alterations in aging rats. In: Samuel, D., Gershon, S., Algeri, S., and Grim, V.E. (eds.) Aging of the brain. New York: Raven, pp. 227–243.

Allis, B., and Cohen, G. (1977) The neurotoxicity of 5,7-dihydroxytryptamine: protection by 1-phenyl-3-(2-thiazolyl)-2-thiourea and by ethanol. Eur. J. Pharmacol. 43:269–272.

Brizzee, K.R., Harkin, J.C., Ordy, J.M., and Kaack, B. (1975) Accumulation and distribution of lipofuscin, amyloid and senile plaques in the aging nervous system. In: Brody, H., and Harman, D. (eds.) Clinical, morphological and neurochemical aspects in the aging central nervous system. New York: Raven, pp. 34–78.

Brody, H. (1955) Organization of the cerebral cortex. III. A study of aging in the human cerebral cortex. J. Comp. Neurol. 102:511–556.

Burns, R.S., Chiueh, C.C., Markey, S.P., Ebert, M.H., Jacobowitz, D.M., and Kopin, I.J. (1983) A primate model of parkinsonism: selective destruction of dopaminergic neurons in the pars compacta of the substantia nigra by *N*-methyl-4-phenyl-1,2,3,6-tetrahydropyridine. Proc. Natl. Acad. Sci. USA 80:4546–4550.

Chiba, K., Trevor, A., and Castagnoli, N., Jr. (1984) Metabolism of the neurotoxic tertiary amine, MPTP, by brain monoamine oxidase. Biochem. Biophy. Res. Commun 120:574–578.

Chiueh, C.C., Burns, R.S., Markey, S.P., Jakobowitz, D.M., and Kopin, I.J. (1985) Primate model of parkinsonism: selective lesion of nigrostriatal neurons by 1-methyl-4-phenyl-1,2,3,6-tetrahydropyridine produces an extrapyramidal syndrome in rhesus monkeys. Life Sci. 36:213–218.

Cohen, G. (1983) The pathobiology of Parkinson's disease: biochemical aspects of dopamine neuron senescence. J. Neural Transm. [Suppl.] 19:89–103.

Cohen, G., and Mytilineou, C. (1985) Studies on the mechanism of action of 1-methyl-4-phenly-1,2,3,6-tetrahydropyridine (MPTP). Life Sci. 36:237–242.

Cohen, G., Pasik, P., Cohen, B., Leist, A., Mytilineou, C., and Yahr, M.D. (1984) Pargyline and deprenyl prevent the neurotoxicity of 1-methyl-4-phenyl-1-2,3,6-tetrahydropyridine (MPTP) in monkeys. Eur. J. Pharmacol. 106:209–210.

Davis, G.C., Williams, A.C., Markey, S.P., Ebert, M.H., Caine, E.D., Reichert, C.M., and Kopin, I.J. (1979) Chronic parkinsonism secondary to intravenous injection of meperidine analogues. Psychiat. Res. 1:249–254.

Del Maestro, R.F., Thaw, H.H., Björk, J., Planker, M., and Arfors, K.-E. (1980) Free radicals as mediators of tissue injury. Acta Physiol. Scand. [Suppl.] 482:43–57.

Feldman, M.L. (1976) Aging changes in the morphology of cortical dendrites. In: Terry, R.D., and Gershon, S. (eds.) Neurobiology of aging. New York: Raven, pp. 211–227.

Finch, C.E. (1973) Catecholamine metabolism in the brains of ageing male mice. Brain Res. 52:261–276.

Fuller, R.W., and Hemrick-Luecke, S.K. (1985) Mechanisms of MPTP (1-methyl-4-phenyl-1,2,3,6-tetrahydropyridine) neurotoxicity to striatal dopamine neurons in mice. Prog. Neuro-Psychopharmacol. Biol. Psychiat. 9:687–690.

Harman, D. (1956) Aging: a theory based on free radical and radiation chemistry. J. Gerontol. 11:298–299.

Harrison, R.G. (1907) Observations on the living, developing nerve fiber. Proc. Soc. Exp. Med. Biol. 4:140–143.

Heikkila, R.E., Hess, A., and Duvoisin, R. (1984a) Dopaminergic neurotoxicity of 1-methyl-4-phenyl-1,2,3,6-tetrahydropyridine in mice. Science 224:1451–1453.

Heikkila, R.E., Manzino. L., Cabbat, F.S., and Duvoisin, R.C. (1984b) Protection against the dopaminergic neurotoxicity of 1-methyl-4phenyl-1,2,3,6-tetrahydropyridine by monoamine oxidase inhibitors. Nature 311:467–469.

Javitch, J.A., and Snyder, S.H. (1985) Uptake of MPP^+ by dopamine neurons explains selectivity of Parkinsonism-inducing neurotoxin, MPTP. Eur. J. Pharmacol. 10:455–456.

Javitch, J.A., D'Amato, R.J., Strittmatter, S.M., and Snyder, S.H. (1985) Parkinsonism-inducing neurotoxin, *N*-methyl-4-phenyl-1,2,3,6-tetrahydropyridine: uptake of the metabolite *N*-methyl-4-phenylpridine by dopamine neurons explains selective toxicity. Proc. Natl. Acad. Sci. USA 82:2173–2177.

Johannessen, J.N., Chiueh, C.C., Burns, R.S., and Markey, S.P. (1985) Differences in the metabolism of MPTP in the rodent and primate: parallel differences in sensitivity to its neurotoxic effects. Life Sci. 36:219–224.

Kim, S.U. (1983) Neuronal aging in tissue and cell cultures: a review. In Vitro 19:73–82.

Langston, W.J. (1985) Mechanism of MPTP toxicity: more answers, more questions. TIPS 6:375–378.

Langston, J.W., Ballard, P., Tetrud, J.W., and Irwin, I. (1983) Chronic parkinsonism in humans due to a product of meperidine-analogue synthesis. Science 219:979–980.

Langston, J.W., Forno, L.S., Rebert, C.S., and Irwin, I. (1984a) Selective nigral toxicity after systemic administration of 1-methyl-4-phenyl-1,2,3,6-tetrahydropyridine (MPTP) in the squirrel monkey. Brain Res. 292:390–394.

Langston, J.W. Irwin, I., Langston, E.B., and Forno, L.S. (1984b) Pargyline prevents MPTP-induced parkinsonism in primates. Science 225:1480–1482.

Markey, S.P., Johannessen, J.N., Chiueh, C.C., Burns, R.S., and Herkenham, M.A. (1984) Intraneuronal generation of a pyridinium metabolite may cause drug induced parkinsonism. Nature 311:464–467.

Mytilineou, C., and Cohen, G. (1984) 1-Methyl-4-phenyl-1,2,3,6-tetrahydropyridine destroys dopamine neurons in explants of rat embryo mesencephalon. Science 225:529–531.

Mytilineou, C., and Cohen, G. (1986) A tissue culture model for studying MPTP toxicity to dopamine neurons. Adv. Neurol. 45:145–148.

Mytilineou, C., Dembiec-Cohen, D., Van Woert, M., and Hwang, E. (1983) Explant cultures of dog substantia nigra and striatum: a model for the study of nigro-striatal dopamine neurons. J. Neural Transm. [Suppl.] 19:37–51.

Mytilineou, C., Cohen, G., and Heikkila, R.E. (1985) 1-Methyl-4-phenylpyridine (MPP^+) is toxic to mesencephalic dopamine neurons in culture. Neurosci. Lett. 57:19–24.

Perry, T.L., Yong, V.W., Clavier, R.M., Jones, K., Wright, J.M., Foulks, J.G., and Wall, R. (1985) Partial protection from the dopaminergic neurotoxin 1-methyl-4-phenyl-1,2,3,6-

tetrahydropyridine by four different antioxidants in the mouse. Neurosci. Lett. 60:109–114.

Porta, E.A., and Hartford, W.S. (1969) Lipid pigments in relation to aging and dietary factors (lipofuscins). In: Wolman, M. (ed.) Pigments in pathology. New York: Academic Press, pp. 191–235.

Scheibel, M.E., and Scheibel, A.B. (1975) Structural changes in the aging brain. In: Brody, H., and Harman, D. (eds.) Clinical, morphological and neurochemical aspects in the aging central nervous system. New York: Raven, pp. 11–37.

Schmidt, C.J., Bruchwick, E., and Lovenberg, W. (1985) Lack of evidence supporting a role for dopamine in 1-methyl-4-phenyl-1,2,3,6-tetrahydropyridine neurotoxicity. Eur. J. Pharmacol. 113:149–150.

Silberberg, D.H., and Kim, S.U. (1979) Studies of aging in cultured nervous system tissue. Int. Rev. Cytol. 10:117–130.

Tappel, A.L. (1968) Will antioxidant nutrients slow aging process? Geriatrics 97–105.

Uhl, G.R., Javitch, J.A., and Snyder, S.H. (1985) Normal MPTP binding in parkinsonian substantia nigra: evidence for extra-neuronal toxin conversion in human brain. Lancet 956–957.

Vistnes, A.I., Henriksen, T., Nicolaissen, B., and Armstrong, D. (1983) Free radicals and aging: electron spin resonance studies on neuronal lipopigments and cells grown in vitro. Mech. Ageing Dev. 22:335–345.

Whetsell, W.O., Jr., Mytilineou, C., Shen, J., and Yahr, M.D. (1981) The development of the dog nigrostriatal system in organotypic cultures. J. Neural Transm. 52:149–161.

INDEX